AF606780

SAP® Build Process Automation

SAP PRESS is a joint initiative of SAP and Rheinwerk Publishing. The know-how offered by SAP specialists combined with the expertise of Rheinwerk Publishing offers the reader expert books in the field. SAP PRESS features first-hand information and expert advice, and provides useful skills for professional decision-making.

SAP PRESS offers a variety of books on technical and business-related topics for the SAP user. For further information, please visit our website: *www.sap-press.com*.

Glavanovits, Haider, Koch, Krancz
SAP Build: No-Code Development, Centralized Access, and Process Automation
2024, 801 pages, hardcover and e-book
www.sap-press.com/5772

Strasser, Sokollek, Sänger, Spierling, Schönwalder
SAP Signavio: Business Process Transformation
2024, 424 pages, hardcover and e-book
www.sap-press.com/5855

Dutta, Ghosh, Goon, Jana, Mukherjee, Rao, Rao, Sane, Veshala, Viswanathan
Workflow for SAP S/4HANA
2024, 686 pages, hardcover and e-book
www.sap-press.com/5697

Banda, Chandra, Gooi
SAP Business Technology Platform (2nd Edition)
2024, 729 pages, hardcover and e-book
www.sap-press.com/5919

Acharya, Bajaj, Dhar, Ghosh, Lahiri
Application Development with SAP Business Technology Platform
2023, 574 pages, hardcover and e-book
www.sap-press.com/5504

Steven Jacobson, Chaitanya Puvvada, Seshadri Ramanarayanan,
Stephan Schluchter, Archana Shukla

SAP® Build Process Automation

The Comprehensive Guide

Editor Rachel Gibson
Acquisitions Editor Hareem Shafi
Copyeditors Julie McNamee
Cover Design Graham Geary
Photo Credit iStockphoto: 1482448571/© bernie_photo
Layout Design Vera Brauner
Production Eric Wyche
Typesetting SatzPro, Germany
Printed and bound in the United States of America, on paper from sustainable sources

ISBN 978-1-4932-2629-0
1st edition 2025

© 2025 by:
Rheinwerk Publishing, Inc.
2 Heritage Drive, Suite 305
Quincy, MA 02171
USA
info@rheinwerk-publishing.com
+1.781.228.5070

Represented in the E.U. by:
Rheinwerk Verlag GmbH
Rheinwerkallee 4
53227 Bonn
Germany
service@rheinwerk-verlag.de
+49 (0) 228 42150-0

Library of Congress Cataloging-in-Publication Control Number: 2024057835

Contents at a Glance

Contents

Preface

In the fast-paced, ever-evolving landscape of modern business, organizations are always looking to improve efficiency in a robust and flexible way. Those looking to become intelligent, sustainable enterprises are achieving these goals by embracing automation. SAP Build Process Automation has emerged as a leading tool to streamline business processes and accelerate digital transformation.

SAP Build Process Automation allows organizations to automate repetitive tasks, integrate systems, and build end-to-end workflows that are scalable, adaptable, and robust. This book is designed to serve as a comprehensive guide to understanding and implementing automation using SAP Build Process Automation. We'll delve into the technical and functional concepts needed to use SAP Build Process Automation in this book. However, we encourage you to frequently brainstorm how these concepts can be specifically applied to your business needs. While we'll discuss real-world examples to help illustrate these concepts at times, we invite you to frequently inspect how these examples can be molded to fit your organizational needs.

The automation journey is one all organizations can benefit from. While some organizations may choose to weave it through the very fabric of their enterprise, others may look to implement it in smaller ways. Some may already be far along their journey, while others may just be beginning theirs. This book is designed to help anyone along that journey, guiding them to understand the power of SAP Build Process Automation and how to apply it within their organization.

Objective of This Book

This book aims to teach you everything—really, everything—about SAP Build Process Automation. Whether you're new to the product or have already gained some experience, this book has been written for you. It will teach you how to use the product to build, extend, and automate business processes. To achieve the same, you'll learn about a combination of theory and practical approaches, spiced up with examples based on the experience and customer engagements of the authors.

On one hand, you'll get guidance, sometimes step-by-step, on how to use the product. You'll benefit from some practical tips and tricks, get helpful advice, and learn more about the purpose of the different capabilities and how to gain the most benefit and value when using them for your scenario. This book is much more than an instruction manual, as on the other hand, you'll learn a lot about the concepts and ideas behind the product, its applications, and its artifacts. Before getting into the details of how to perform automation in your business, we first discuss what SAP Build Process Automation is, when and where to use it, what's the benefit, and even why to go on this journey.

The goal is to incrementally extend your knowledge on SAP Build Process Automation by first starting with a general introduction, which already provides a good foundation for knowledge of the product. Then, it's up to you to determine which topics you want to deepen your knowledge in. Although we recommend reading the complete book, the chapters are structured in a way that you can start your deep-dive journey wherever you want. You'll find referrals within the book to the different chapters to provide you a better context and truly enable you to learn everything (you want) about SAP Build Process Automation and understand the bigger picture.

This book will be your starting point to becoming an expert on SAP Build Process Automation. You'll soon follow this up by transforming and adopting the know-how you've gained with this book into practical experience when it comes to building, extending, and automating business processes with SAP Build Process Automation.

Target Audience

The book is targeted for both technical and nontechnical professionals who are involved with SAP systems and want to automate business processes and ensure that these processes are secure, governable, and well-monitored. In particular, individuals in the following roles within an organization will find this book of interest:

- **Developers, technical consultants and business process analysts**
  The book provides step-by-step instructions for creating business process projects. It's tailored for business process developers, consultants, and business process analysts who are interested in building and customizing business processes efficiently using SAP Build Process Automation.
- **IT administrators**
  IT administrators will learn about the architecture and lifecycle of SAP Build Process Automation and how to monitor and manage them. The sections on governance and security are particularly relevant to administrators responsible for overseeing system integrity and user roles.
- **Enterprise and technical architects**
  The book describes in detail how SAP Build Process Automation can work with external systems and applications. By understanding how SAP Build Process Automation integrates into the broader system landscape, architects can design effective, scalable, and sustainable automation solutions.
- **Project managers and team leads**
  Project managers and team leads often oversee the implementation of new technologies. Understanding the basics of how SAP Build Process Automation works will help them understand how to manage timelines, resources, and deliverables associated with the implementation.

- **Business leaders and decision-makers**
  The leaders and decision-makers who oversee digital transformation initiatives will understand how SAP Build Process Automation adds value in terms of operational efficiency toward their strategic objectives.
- **SAP BTP users and enthusiasts**
  Those familiar with SAP Business Technology Platform (SAP BTP) who want to explore new tools and methodologies to enhance productivity in their environment will also find this book beneficial.

Structure of the Book

This book is structured in 17 chapters, starting from a more general introduction followed by a detailed description of the different areas of SAP Build Process Automation accompanied by practical guidance and real-life examples:

- In **Chapter 1, Introduction to Enterprise Automation**, you'll see the bigger picture. We'll discuss what enterprise automation is all about, how it combines different products in a smart way, and which role SAP Build Process Automation is playing there. This chapter describes how SAP is accelerating enterprise automation.
- In **Chapter 2, SAP Business Technology Platform**, we'll take a look under the hood of the technical foundation. While gaining an overview of the different capabilities and services, you'll learn what SAP Build is all about. Additionally, you'll see how these services help you keep the core of your business application, in particular SAP S/4HANA, clean, along with what "clean" means. Fusion development will be introduced as well, along with how no-code, low-code, and pro-code environments support this.
- Now that the groundwork has been done, we'll come to **Chapter 3, Overview of SAP Build Process Automation.** This chapter lays the foundation for the remaining parts of this book. By understanding the purpose of SAP Build Process Automation, we'll also take a guided tour through all of its capabilities and the architecture. You'll get to know the different applications and artifacts that will be revisited and discussed in more depth in the following chapters.
- **Chapter 4, Creating a Business Process Project** explains how to design, test, and deploy the business process project. It's the starting point to automate your business processes and a wrapper for all automation artifacts in SAP Build Process Automation.
- **Chapter 5, The Process and Its Artifacts,** teaches you about the different artifacts and activities of SAP Build Process Automation and dives deeper into creating and configuring processes using the low-code process designer. Different sections of the chapter introduce best practices for an effective process design.

- **Chapter 6, Action Artifacts**, introduces actions, which form the core building block in SAP Build to connect to external systems. This chapter explains in detail how to create and configure actions. Various sections of this chapter explain several connectivity options for actions and how easily you can build extensions to tailor your business needs.
- **Chapter 7, Automation Artifacts**, introduces the automation artifact, which helps users automate repetitive, rule-based tasks. This chapter dives into the technical and functional details related to automating, such as tasks using screen-capturing technology, activities provided to us through software development kits (SDKs), flow controls, and data handling capabilities. We'll also discuss other concepts that help users deploy and execute automations.
- In **Chapter 8, Decision Artifacts**, we introduce the decisions in SAP Build Process Automation, explaining the purpose of decision artifacts in maintaining the business rules. Decision artifacts are the central repository for all business rules in SAP Build Process Automation and help manage and streamline business processes. The chapter explains the importance of decision management in optimizing business operations and takes you through the journey of selecting the right decision solutions for various SAP scenarios. It also covers the types of decision deployments and the types of decision models while also providing hands-on experience in creating and managing business rules. By the end of the chapter, you'll have a good understanding of how to effectively leverage decision artifacts for the most impactful decision management and automation with SAP Build Process Automation.
- **Chapter 9, Business Process Lifecycle Management**, describes how to manage the lifecycle of a business process project and introduces the different lifecycle phases. Then, it goes into key phases that are usually important in the creation and configuration of business process project, the release and deployment of projects into runtime environments, and the execution and processing of tasks. After the lifecycle phases, the chapter covers technical monitoring, explaining how to track technical aspects and performance metrics that are crucial for nondisruptive running of processes. Finally, it introduces business monitoring, emphasizing that there should be monitoring of business outcomes and achievements of strategic objectives. This chapter provides you with key strategies and methods to manage and monitor the entire lifecycle of business process projects effectively.
- **Chapter 10, Executing Processes, Automations, and Decisions**, provides you with a comprehensive guide to starting and managing processes, automations, and decisions in SAP Build Process Automation from their applications. The chapter begins with an introduction to various execution strategies. It then delves into understanding the runtime environments that support these operations, giving you a basic knowledge of the infrastructure involved. Then, the chapter explores the application programming interfaces (APIs) of SAP Build Process Automation, outlining how these interfaces enable integration and execution of processes, automations, and

decisions from any SAP or non-SAP applications. You're then guided through practical steps on running these artifacts, providing hands-on knowledge to effectively implement them.

- **Chapter 11, Monitoring**, introduces business and technical monitoring where you'll learn how to monitor runtime instances of processes and automation. The section dives deeper into using the process visibility feature to track end-to-end business processes and detect bottlenecks for process optimization.
- **Chapter 12, Governance**, explains how to enable the governance framework during process automation. You'll learn to use various approaches such as access management, collaboration, using environments during deployment, transport management, and other governance options to ensure that all the processes are correctly executed and aligned with organizational strategy.
- **Chapter 13, Authentication, Authorization, and Auditing Logging**, introduces authentication mechanisms and authorizations that should be assigned to different personas such as business users and developers of SAP Build Process Automation. This chapter also teaches audit logging capabilities in the platform that are required for security, compliance, and troubleshooting.
- **Chapter 14, Working with Store Content**, focuses on SAP Build Store and prebuilt content packages. Enterprise-ready, industry-based prebuilt content is a key feature of SAP Build Process Automation. In this chapter, you'll learn about the need for content stores to accelerate overall process development. Various sections of this chapter provide details on different types of stores and content, including how to use stores to discover and import content packages, extend the content based on your unique requirements, and avail the commercial benefits of publishing the content in the store.
- With all the knowledge of SAP Build Process Automation you'll have gained at this point, **Chapter 15, Extending Standard Processes** visits one of the main use cases of the book. Starting with an introduction to standard business processes and how they are covered in business applications, we'll conclude that sometimes extensions are necessary to cross the boundaries of customizing. The chapter covers their risks and benefits and clarifies what is meant by on-stack extensions and side-by-side extensions and how they are covered with SAP products via an example. We'll go into more detail regarding fusion development, learn what process variants templates are good for, get to know the extensibility wizard for SAP S/4HANA, and make this topic more tangible with some example scenarios.
- **Chapter 16, Scaling Process Automation**, covers different aspects of how to involve different personas. Starting with the end user or business user, you'll get introduced to SAP Task Center, its architecture, and how to enable it. Finally, we dive into the topic of fusion development and discuss how professional and citizen developers, key users, or business process experts could and should work together and how they benefit from this approach.

- In **Chapter 17, Conclusion**, we'll revisit the topics of every chapter, highlight the learnings, and take a look back on the discussed topics.

Acknowledgements

I would like to thank my family and friends for all their support over the years. From the Buffoon Bar to X40, you know who you are. I would also like to express my gratitude to my co-authors Archana, Sesh, Priya, and Stephan. Without them, this wouldn't have been possible. Thank you.

—**Steven Jacobson**

I would like to express my heartfelt gratitude to SAP, where my journey began as a fresher in 2007. It was during my early days, while immersing myself in ABAP books, that the idea of writing books took root. From that moment, I envisioned one day seeing my name on the cover of a book. This dream evolved into a deeper desire to document my knowledge and experiences in SAP Build Process Automation, a field that has led me to be recognized as an APJ Champion, a Women in Robotic Process Automation honoree, and an SAP Community Eminent Member. This book is dedicated to my colleagues who have always believed in me.

First and foremost, I would like to express my heartfelt appreciation to my parents who instilled in me the values that have been foundational to my success and have shaped the person I am today. I'm also deeply thankful to my spouse, whose unwavering support has been instrumental in making all my achievements possible. This book wouldn't have come to fruition without his encouragement, patience, and belief in me. I also want to extend my appreciation to my children for their understanding, especially as I spent countless weekends away from them to complete this project. Their support and love have been invaluable. A special mention must go to my sister, whose constant encouragement lifted my spirits during the toughest moments of my journey. She has been my rock, helping me to overcome every obstacle along the way.

I am also immensely grateful to my coauthors Archana, Sesh, Steve, and Stephan for their collaboration and invaluable contributions throughout this project. Their dedication and partnership were key to turning this project into a success.

I owe a deep debt of gratitude to my mentors, who have guided me with their wisdom and support, and I'm forever thankful for their contributions in my professional journey.

—**Chaitanya Priya Puvvada**

First and foremost, I want to convey my heartfelt thanks to my wife, Gayathri, and our son, Adhrit. Your persistent support and motivation, especially during the long hours I dedicated to writing this book, fueled my energy and kept my spirits high. I am thankful to my parents and sister for their unwavering belief in me and their constant encouragement throughout my life. This journey would have been impossible without all your love, support, and encouragement.

My sincere thanks to my friends and my mentors in guiding me and shaping who I am today. I would like to express my deep appreciation and acknowledge the hard work of my coauthors Archana, Priya, Stephan, and Steve who contributed to writing the book.

A huge shout-out goes to the amazing members of the SAP Build Process Automation team and my colleagues across Bulgaria, France, Germany, and India. This book is a testament to the incredible work you've done for the past two years.

Finally, I wish to express my gratitude to the *almighty* for providing me with all that I need.

—**Seshadri Sreenivas Ramanarayanan**

Dear Ines—my wife, love, true friend, and partner in crime—a heartfelt thank you goes to you for all your support and encouragement during this project and everything else. More is yet to come. To our daughters, Zoé and Yael, thanks for your understanding when I was in my office in the "not to be disturbed mode." You are great, giving us fun, love, and inspiration in our life. Thanks also go to my family and friends, who are my backbone and safe harbor.

It all began earlier last year with a phone call from my colleague and friend, Archana Shukla, who described it as a "once-in-a-lifetime opportunity." A huge thank you, Archana, for trusting me to join as a coauthor alongside our awesome colleagues: Puvvada Chaitanya Priya, Seshadri Sreenivas, and Steven Jacobson. You guys rock! I am grateful to be part of this journey together with you.

Sharing our product knowledge and helping our customers, partners, and colleagues is in our DNA. We've done this for many years in various formats, so it's great to see this now reflected in our book.

A big thank you goes out to all the incredible colleagues supporting me and us during so many years to reach the point to write this book. Teamwork makes the dream work—it's true.

And a very special thanks goes to my manager, André Borchert, for his tremendous support and being a great mentor and role model. To Thomas Volmering, who has encouraged me from day one at SAP, I'm just grateful.

And finally, a big thank you to Rachel Gibson from Rheinwerk Publishing, who guided, motivated, and pushed (when necessary) us through this whole journey.

—**Stephan Schluchter**

Who shall I start with, there are so many friends and colleagues who have been instrumental in shaping my career and standing with me at various stages of my professional and personal journey. I would dedicate this book to everyone who has supported me, encouraged me, and believed in me.

I must obviously start with my parents first who have done everything to ensure that we get a good education and prosper in life, backed with a solid value system. I wouldn't be where I am today without their unconditional love and support. Next is my partner and dear friend who has been my backbone so that I can manage to achieve whatever big or small I can in my life. This book is the result of his unwavering support and encouragement. I can't go without thanking my brother for his sheer pursuance and thoughtful guidance in all important phases of my life. Many decisions in my life are owed to him. Through this book, I would like him to know that I love him and thank him for always being beside me from our growing years.

Sincere thanks go to my coauthors Sesh, Stephan, Steve, and Priya for putting the entire book together. It wasn't an easy journey, but we found ways to live our dream of writing our first book. I have known them for several years, and it was a relishing experience to work together on the book. Through their commitment and focus, we were able to complete the book.

—**Archana Shukla**

Chapter 1
Introduction to Enterprise Automation

In this first chapter, we introduce you to enterprise automation and discuss what's behind this term, how it evolved, and what's the relation to SAP Build Process Automation. You'll also learn how SAP supports enterprise automation.

You might have heard about *enterprise automation* already. It's sometimes called hyperautomation, more generally. Gartner describes hyperautomation as follows (see *http://s-prs.co/v592805*):

> *Hyperautomation is a business-driven, disciplined approach that organizations use to rapidly identify, vet, and automate as many business and IT processes as possible. Hyperautomation involves the orchestrated use of multiple technologies, tools, or platforms, including artificial intelligence (AI), machine learning, event-driven software architecture, robotic process automation (RPA), business process management (BPM) and intelligent business process management suites (iBPMS), integration platform as a service (iPaaS), low-code/no-code tools, packaged software, and other types of decision, process, and task automation tools.*

In short, hyperautomation is a combination of different technologies to automate as many processes as possible within an organization.

While hyperautomation marks the overall definition, the concrete offering from SAP with products, prebuilt packages, and a guided approach is called *enterprise automation*. We'll use this term going forward in this book.

But why should you even automate anything? Well, there are good reasons to do so. As you're reading this book, you probably already have your own motivation, but let's take a more objective view.

We all try to avoid mistakes, but we're human and, therefore, full of mistakes or at least prone to make them from time to time. One aspect of automation and digitalization overall is to reduce these errors. So, instead of having someone maintain data manually in a system, we have an automated activity—a bot that extracts the data from a document, for example, an invoice, and then posts it into a business system.

We all want to be as fast as possible with our work, which is also sometimes the source of mistakes. Coming back to our example of maintaining some data manually, it can be time-consuming. Others are waiting for the stuff to get it done, processes are delayed, customers are annoyed, business is going down, people are fired, and companies are closed. Okay, it doesn't need to be that dramatic, but you got the point. We want to *speed up the processes.*

We all want to have work that challenges and encourages us. If the system is doing bigger parts of the work for you, you're free to concentrate and focus on more value-adding work, beneficial tasks, and activities that are more satisfying for you. You can free up your time and be more productive.

These are some of the main promises of automation. In a perfect world, we could end here, close the chapter and the book, and concentrate on other things. But life isn't perfect, as we all know. Following, we'll see whether the promises are fulfilled or something is missing, as well as whether enterprise automation is helping us out—sneak preview, it is! First, we define what enterprise automation stands for, followed by clarifying the need for it and how it evolved, and finally showing how it's realized with SAP accompanied with some use-case examples.

1.1 What Is Enterprise Automation?

Why does it make sense to combine different technologies for a broader scope and goal, as well as greater business benefits at the end? And is it a must to do so? Let's get some more clarity on this. We'll derive a more concrete definition of what enterprise automation is all about here. Going back to the earlier statement by Gartner, we've seen that there are several technologies involved. If we combine them, we might come up with three major areas, as shown in Figure 1.1:

- Integration with event-driven software architecture
- Process analysis in intelligent business process management suites (iBPMS)
- Process and task automation tools, decisions in iBPMS and business process management (BPM) software

All of these are enriched with AI functionality, prepackaged software, and possibility for low-code/no-code development. You can use all of them standalone, and you're probably already doing so. But sometimes, a smart combination of these makes more sense or enables your organization to become more efficient. Why?

Let's compare this with an analogy from the world of music. Have you ever heard a wind band or concert band? These bands are based on three major registers: woodwind, brass, and percussion. All of these can exist and make some beautiful music on their own, but the full power and pleasure for the audience comes when they are combined. Nevertheless, the piece of music you want to play provides the guidance on how they are used and played together.

The same holds true for the world of business processes: you can combine integration, process analysis, and process automation; use them in different combinations; focus on individual capabilities; and leave some of them out, depending on your composition, business scenario, audience and business people you want to reach, and the outcome you're aiming for.

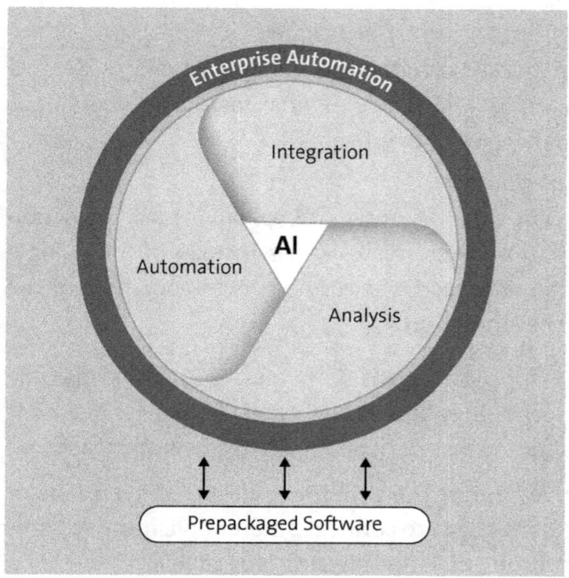

Figure 1.1 Enterprise Automation with Three Major Areas

The tricky thing is now to make the right choice. But there's no need to panic because in enterprise automation, the approach is flexible enough to change the band and its focus as you go. A wind band can have different members: very talented, well-educated amateurs who have great musical knowledge and experience but do something else in their regular job, or professional musicians, with an education who do this for their living. The same is true for enterprise automation and other areas of process or application development, which attracts and includes both *professional developers* and *citizen developers*. The latter often have a great understanding of the processes and some technical talent or enthusiasm but lack coding skills. When both come together, we call this *fusion development*.

To summarize, with enterprise automation, you bring together capabilities for integration, process analysis, and process automation, followed by enhancement with AI functionalities and enrichment with prebuilt content packages.

1.2 Why Enterprise Automation Is Needed and How It Has Evolved

Let's revisit the promises of automation: speed up processes, reduce errors, and free up time. Hopefully, you've already experienced them. If not, you might be facing these issues and challenges:

- **Fragmented and incomplete automation**
  Processes have been digitalized and automated but are only focused on individual tasks, some efforts have been duplicated, and an end-to-end view or end-to-end automations overviews are missing.
- **Bot proliferation**
  Bots have been developed and are focused on solving individual problems. Several bots are located here and there with no knowledge of what tasks other bots have already done, causing increased management overhead.
- **Brittle and expensive automation approach**
  Disconnected toolsets, costly point-to-point integrations, and no connection between processes and the integration layer creates siloes instead of following a comprehensive process automation approach. There's no concrete analysis of the business processes before and after automation.

Clearly, this needs to improve. The approach needs to be stable and also flexible enough to incorporate the required technology based on the actual business needs. At this time, the idea of enterprise automation was born and formed a new approach.

The product offering by SAP also evolved during this time. *Integration* has been a major SAP cornerstone for decades to address the need to connect different business systems in a meaningful way. With SAP Integration Suite, this offering is available.

Starting with workflow, business rules, and process visibility services, the next milestone was achieved—when a robotic process automation (RPA) product became available. Already in 2019, the notion of BPM and RPA being the two sides of the same process automation coin started with the offering of SAP Workflow Management and SAP Intelligent Robotic Process Automation. The next step was to release SAP Process Automation (launched in 2021 and generally available in February 2022), which is a no-code/low-code offering that lowered the entry barrier for citizen developers while also being open and useful for professional developers. This evolved into SAP Build Process Automation, as one service within SAP Build, which was released in 2022 and featured AI capabilities and prebuilt content packages. SAP Build also includes SAP Build Apps to create enterprise-grade apps with drag-and-drop simplicity and SAP Build Work Zone to design business sites using a visual editor and prebuilt templates. We'll discuss this further later in the chapter.

At the same time, during 2022, SAP started to bring SAP Signavio solutions—its process analysis, intelligence, and process mining and best practices offering—closer to the other two products. This led to the approach of enterprise automation unveiled at SAP Sapphire in 2023.

1.3 How SAP Accelerates Enterprise Automation

Enterprise automation with SAP isn't a new product. It's a smart and flexible combination of three existing products to help customers become intelligent and automated enterprises (see Figure 1.2). This approach will help organizations integrate applications, discover inefficiencies, and automate business processes. You can use it across diverse industries, lines of business (LoBs), and technologies.

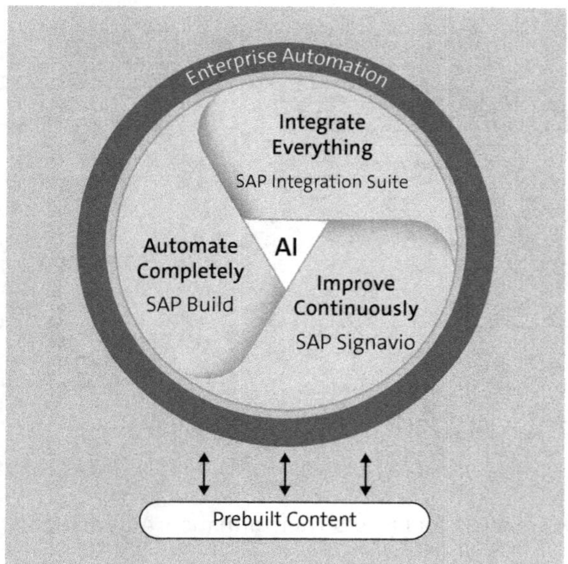

Figure 1.2 Enterprise Automation with SAP

With SAP Integration Suite, you're able to integrate everything, SAP and non-SAP applications across complex multicloud or on-premise environments. SAP Signavio supports you in analyzing your business processes using benchmark processes against industry best practices to uncover any inefficiencies and take the right action. For the action part, you can include SAP Build by creating automated processes or extending applications using a low-code/no-code environment and prebuilt content packages.

Although it's completely up to you where you start, depending on the business scenario, you probably want to get your integrations straight first. Afterward, analyze these integrated business processes, and derive the first trigger points for process improvements. These improvements could be to use a capability within SAP S/4HANA or to use SAP Build Process Automation to increase the automation rate of the process or create an extension.

You might also need to extend a standard business process within your ERP system, SAP S/4HANA, for example, journal entry bookings. To ensure every new entry in the business journal gets checked up front, and if necessary, approved, you can use SAP

Build Process Automation. Once successfully up and running, you'll use SAP Signavio to check the performance and do some process data analysis. Or, if you want to include other business applications in this process or make use of other means to send out notifications, you can use SAP Integration Suite.

For a better understanding of the possible use cases, let's get into the details of the three pillars of the enterprise automation framework and especially how they work together, as shown in Figure 1.3. We'll discuss these more in the following sections.

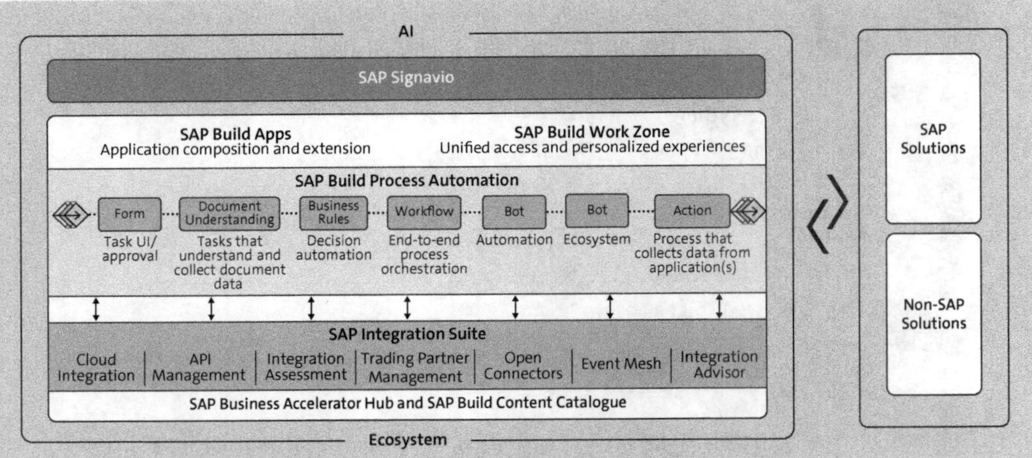

Figure 1.3 The Enterprise Automation Framework for Process Transformation, Automation, and Integration

1.3.1 SAP Integration Suite

It's true that automation often starts with integration. SAP Integration Suite aims to integrate everything: SAP and non-SAP applications, on-premise or in the cloud, connecting hybrid landscapes, and so on. It's one single solution for all integration needs containing the required prebuilt integrations, APIs, and connectors. Technically, SAP Integration Suite is part of SAP Business Technology Platform (SAP BTP), which we'll discuss in Chapter 2. You might want to leverage this service for the integration of your business processes across different systems to pull or write data or to trigger transactions directly in a system of records, that is, the business application. This could be any system, for example, SAP S/4HANA or SAP ERP, to be integrated in a purchase order scenario (see Figure 1.4).

Besides the integration capabilities for enterprise automation, the advanced event mesh features are also relevant. When a business application exposes an event, such as a purchase order getting created, this is an event. SAP Event Mesh is listening to these events and provides the events to SAP Build Process Automation, where they can trigger an extension process for some additional approvals.

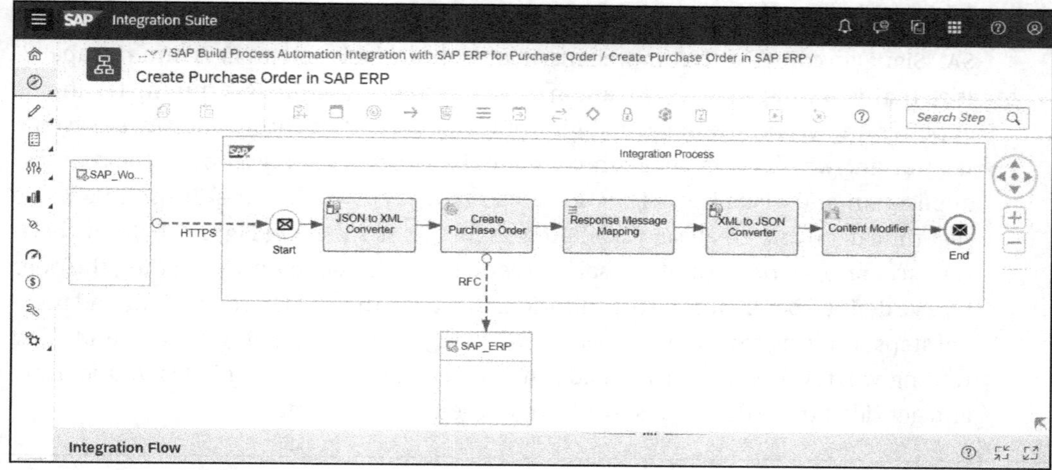

Figure 1.4 Integration Flow to Create a Purchase Order in SAP

Finally, you can make use of the APIs, packaged integration content, and adapters that are available in SAP Business Accelerator Hub (*https://www.api.sap.com/integrations*), as shown in Figure 1.5.

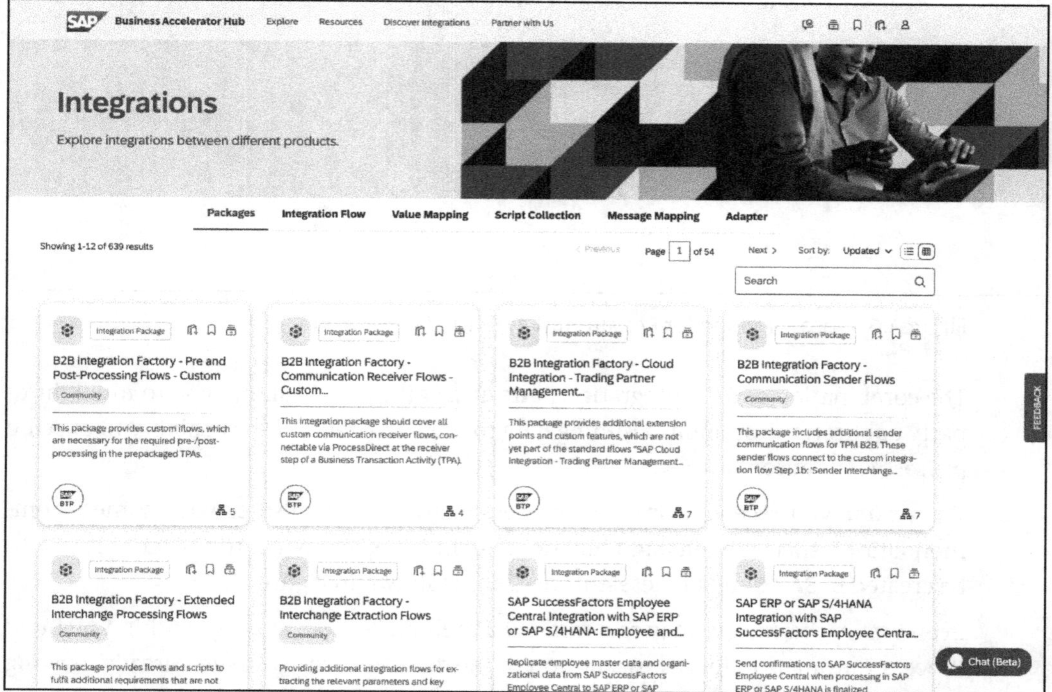

Figure 1.5 SAP Business Accelerator Hub with Access to Prebuilt Integration Content

1.3.2 SAP Signavio

SAP Signavio comes with comprehensive process analysis and process mining capabilities that you can use with SAP and non-SAP systems. The process mining technique explores and visualizes the actual flow of certain activities within a business system. It's tracking which route, for example, a purchase order is going through the business application and visualizes all the instances (already executed or still running using real-time data) and creates a visual process model, as shown in Figure 1.6. In the best-case scenario, all or most of the sales orders follow the same flow, including the ones you've defined before for the organization. In reality, there will be deviations and manual steps resulting from employees not following the process, doing workarounds, and finding whatever way works outside the process. Process mining helps you identify and get rid of these deviations to get your process back on track.

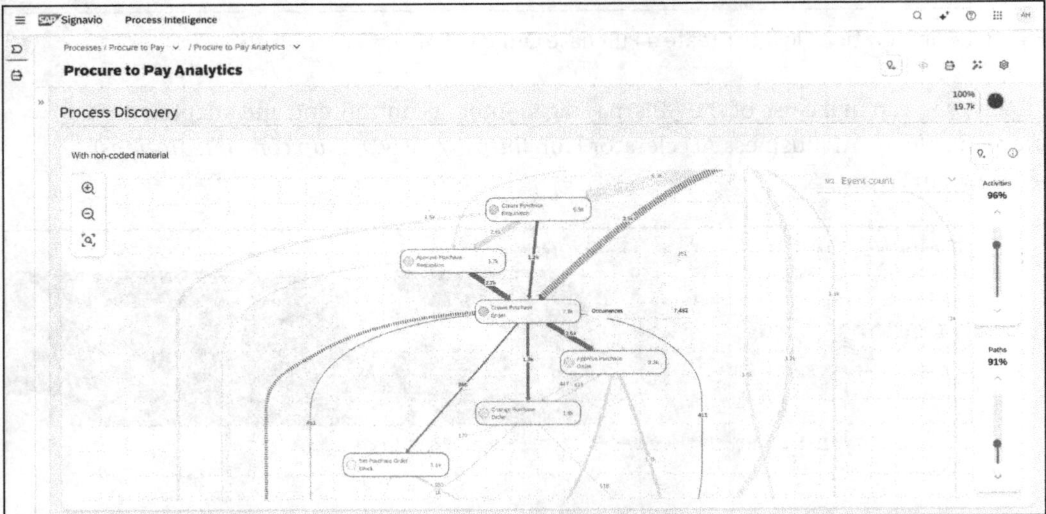

Figure 1.6 Process Mining for Procure-to-Pay with SAP Signavio

The combination of SAP Integration Suite and SAP Signavio allows you to mine third-party applications or publish mining data to the SAP Event Mesh capability to execute process steps automatically.

The results you receive from your process analysis could also provide some recommendations, such as to enable a functionality in SAP S/4HANA or to trigger a process to be created in SAP Build Process Automation, as shown in Figure 1.7.

Even better, the system may recommend a prebuilt content package within SAP Build Process Automation for fast resolution of the detected process issue. Already existing processes can be triggered in the same way. For this, you define conditions in SAP Signavio to automatically trigger a process in SAP Build Process Automation.

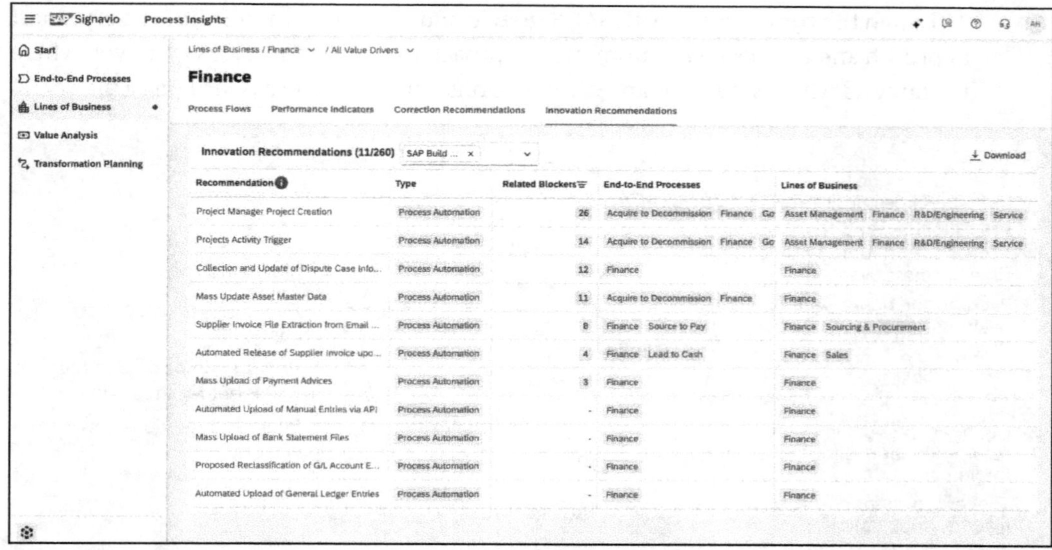

Figure 1.7 Innovation Recommendation in SAP Signavio to Dissolve Process Blockers with SAP Build Process Automation

1.3.3 SAP Build Process Automation

SAP Build Process Automation is your service within SAP Build to extend standard business processes, automate processes, or orchestrate activities across different systems and organizational units. See Figure 1.8 for a first glimpse of the design-time environment, which uses prebuilt content packages and makes use of certain AI functionalities (e.g., for document information extraction). As this book is about SAP Build Process Automation, you'll learn much more in the following chapters, of course.

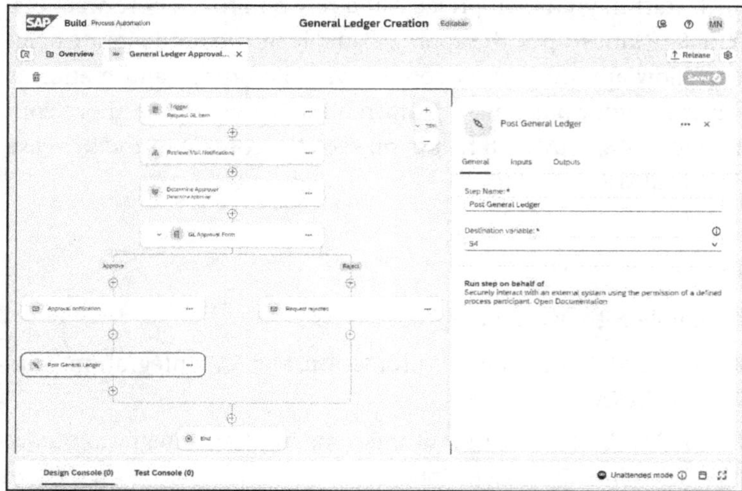

Figure 1.8 General Ledger Approval Process in SAP Build Process Automation

Although the combination with SAP Signavio and SAP Integration Suite is a very smart approach and pushes your automation approach to the next level, it's up to you where to start and which service or any combinations you want to use (see Figure 1.9).

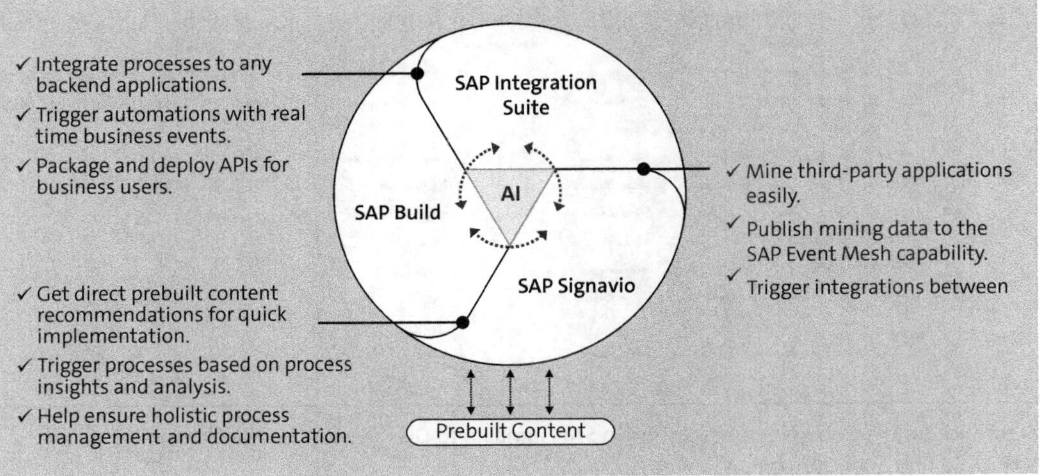

Figure 1.9 Touchpoints between SAP Integration Suite, SAP Signavio, and SAP Build for Fostering Enterprise Automation

Enterprise automation with SAP fosters the smart combination of process automation (SAP Build Process Automation), process analysis (SAP Signavio), and integration (SAP Integration Suite). It supports a holistic hyperautomation strategy within any organization.

Fusion team development becomes a reality and is accelerated with business AI capabilities. The extensive industry and LoB prebuilt content and templates for these three pillars allow you a quick-start in your own projects and rely on proven best practices. Although best-suited for SAP landscapes, the approach and included products are open and flexible enough to allow any third-party connectivity. Enterprise automation is also an excellent enabler for your business transformation activities and more concrete for your transition to SAP S/4HANA. It helps you keep the core clean and increase the level of automation in all areas.

1.3.4 Use Case Examples

Let's make this more tangible with two concrete use cases:

- Combining SAP Signavio, SAP Build Process Automation, and SAP Integration Suite with postings in SAP S/4HANA
- Using SAP Build Apps, SAP Build Process Automation, and SAP Integration Suite with raising a service ticket in SAP Service Cloud

We'll discuss these use cases in the following sections.

Approval-Based Journal Entry Postings

Financial processes, such as the posting of new journal entries, often require several manual steps and reiterations. Before something is posted in the business journal, it should be checked and sometimes even approved based on certain parameters because it's extra effort to delete entries made in the journal entry, which is recording all the business transactions in the accounting of a business.

With enterprise automation, you can manage these postings. In our example, you use SAP Signavio to identify any bottlenecks, inefficiencies, and manual activities with the goal to increase the level of automation. SAP Build Process Automation is used to check and approve these entries and trigger their posting. You might want to even include SAP Integration Suite for secure communication and use a preconfigure iFlow, which triggers the actual posting in the SAP S/4HANA system. The complete product flow and related activities are outlined in Figure 1.10.

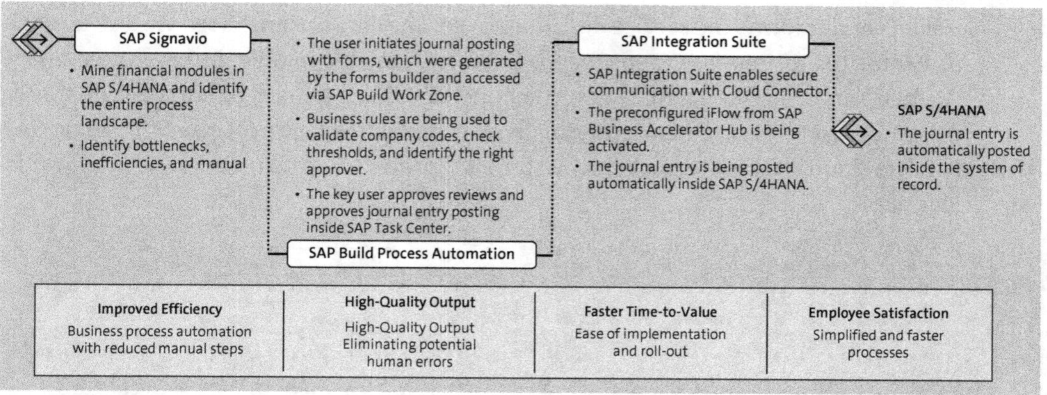

Figure 1.10 Approval-Based Journal Entry Postings

You start with SAP Signavio to evaluate what's going on or, to be more precise, what's going wrong in the journal entry posting process. Here, you recognize that there too many manual and repetitive tasks slowing down the overall process. The recommended solution is to create a business process in SAP Build Process Automation to automate the same. Once the data is either uploaded or posted manually in SAP S/4HANA, the approval process starts. First, there will be checks on whether all the needed data is provided and to identify the right approver. This is done completely automatically behind the scenes with bots and business rules. If data is missing or approvals are necessary, the end user will be involved. SAP Integration Suite then handles the secure communication with the cloud connector and uses a preconfigured iFlow stored in SAP Business Accelerator Hub. Finally, the journal entry is posted in SAP S/4HANA.

With this, you improve efficiency as manual steps are reduced, increase the quality of the output as potential human errors are eliminated, realize faster time to value, and increase employee satisfaction with a simplified and faster business process.

Smart Waste Management

Garbage containers collected by sanitation workers are unfortunately not always at the expected place. Sometimes, they are also damaged or contain the wrong type of garbage. So far, the sanitation workers were asked to provide a manual report on these issues. With enterprise automation, this sounds like a problem of the past.

Using SAP Build Apps, you can create an easy-to-use frontend with a comprehensive application logic, leveraging different technologies to localize the garbage container or maintain any damages, issues, and so on. Here, we're using, for example, the Google Maps API connectivity to mark the container on the map. The app triggers a process instance in SAP Build Process Automation to send mail notifications to different stakeholders and informs the service desk employee. The data is transferred via SAP Integration Suite to SAP Service Cloud where a ticket is created to handle the case and manage the ticket by a service agent. Under the hood, API management handles all the traffic, including to the Google Maps API. Finally, the incident will be processed further in SAP Service Cloud, and we're back in the standard process. A rough overview is shown in Figure 1.11.

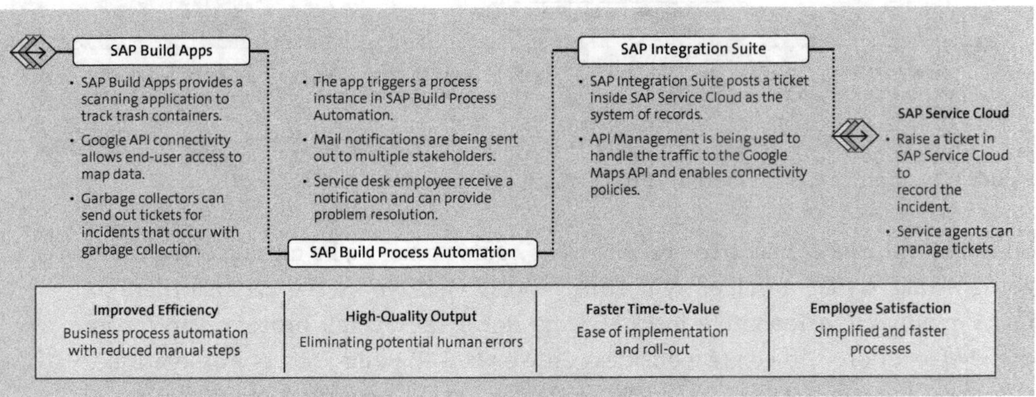

Figure 1.11 Smart Waste Management

Again, the automation level has been increased, leading to improved efficiency, increased quality, happy employees, and a digitalized process that is flexible enough to be adapted for new business requirements.

The differentiators in this use case include the holistic hyperautomation strategy, fusion team development, integrated business AI capabilities, fault toleration user experience (UX) automation, extensive industry and LoB solutions, and third-party connectivity.

1.4 Summary

In this chapter, we've discussed the what, why, and how of enterprise automation, including why it's needed and how it has evolved so far. In detail, we've discussed how SAP delivers enterprise automation with the flexible combination of process analysis, process automation, and integration, as shown in the use cases at the end.

In the following chapters, we'll concentrate on one of these products: SAP Build Process Automation. Additionally, several technologies, offerings, and approaches briefly mentioned will be explained in more detail.

Chapter 2
SAP Business Technology Platform

SAP Business Technology Platform is the home of SAP Build Process Automation, as well as other services that are provided there. We'll discuss the SAP BTP architecture and visit SAP Build and its related services.

Innovations are key for every successful enterprise, but what sounds like a truism isn't that easy in real life. Reacting to changing market requirements, new competitors, and diverse customer expectations makes organizing smooth end-to-end business processes challenging already. But, in addition, you should also be innovative, as well as able to handle some of the mentioned challenges and find the right and successful way for your organization to outperform competitors. For all of this, a stable foundation helps.

Think about your own career, which is most likely built on the solid foundation of the education you've experienced and the skills you've trained on while being open and embracing changes and new opportunities. The same is true when you think of building a house. With a solid foundation, you ensure stability, but you're also open to other plans and dreams you want to realize for your home.

The solid technology foundation for your business to innovate, integrate, and transform your operations is SAP Business Technology Platform (SAP BTP). Using SAP BTP, you'll be equipped with the latest technology while remaining flexible enough to participate in changing environments and have the foundation to innovate and outperform competitors.

Let's take a closer look at the SAP BTP capabilities and services to understand its architecture and discuss the available set of services—SAP Build—which is a solution to extend applications and processes.

2.1 Introduction

The solutions offered by SAP cover enterprise resource management, human capital management, spend management, supply chain management, customer relationship management, and much more. These are centered around SAP's cloud ERP solutions, SAP S/4HANA, and other SAP solutions and third-party applications, enriched with industry-specific offerings (see Figure 2.1).

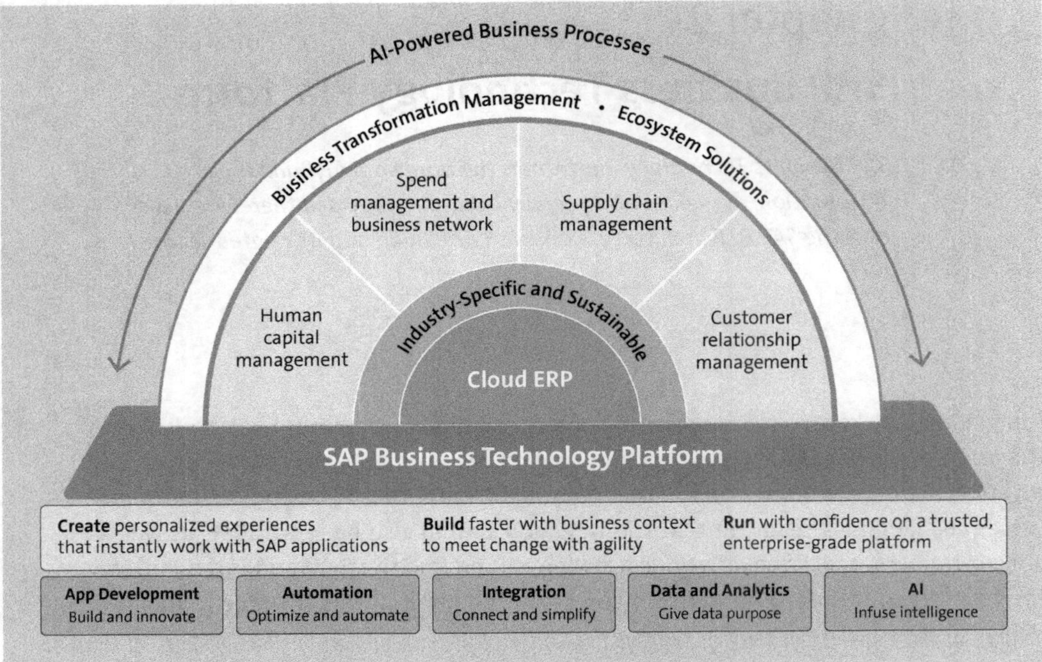

Figure 2.1 SAP BTP Supporting Digital Innovations

This includes all business processes belonging to recruit to retire, source to pay, design to operate, or lead to cash.

The SAP BTP open platform combines several capabilities and connects, extends, and infuses artificial intelligence (AI) with these solutions. Although optimized for SAP applications, SAP BTP is also open for non-SAP ones.

With a set of services to create personalized solutions, there are five foundational pillars covered by SAP BTP:

- **Application development**
  An environment for low-code/no-code development, as well as pro-code tooling, enhanced with capabilities to create delightful user interfaces (UIs).
- **Automation**
  Automate, extend, or orchestrate business processes across SAP or non-SAP landscapes to improve the same.
- **Integration**
  Connect SAP, non-SAP, on-premise, cloud, and hybrid solutions for seamless communication and data exchange.
- **Data and analytics**
  Derive insights from business data for informed decision-making.

- **Artificial intelligence (AI)**
  A comprehensive set of services and capabilities to infuse AI in applications and extensions.

Establishing the solid technology foundation supports organizations in their agile business and in being prepared for change and innovation. With SAP BTP, you won't just get a cloud platform with a bunch of technology, but instead you'll get technology that truly supports your business. That's the purpose.

> **SAP BTP and Clean Core**
>
> Not only is SAP BTP *the* technology platform for innovation but it's also *the* side-by-side platform for your clean core strategy (we'll cover extensions in detail in Chapter 15). Clean core relates to a concept and strategy to extend SAP S/4HANA Cloud systems and upgrade them in a safe, nondisruptive way. This approach allows you to always have access to the latest innovations. Being on the latest SAP version provides faster update cycles and drives innovations via the platform. It will simplify upgrades by following a standard upgrade path. A lower total cost of ownership (TCO) can be achieved with less manual effort, reduced costs for maintenance, and the actual operation of the extensions. Improved performance, enhanced security, and simplified data are due to streamlined data structures, reduced redundancies, and improved quality of data.

2.2 Architecture

If you're now looking for a technical architecture model (TAM) diagram, you might be disappointed. If you're looking for an illustration to better understand the capabilities of SAP BTP and the related services, then you're at the right place in this chapter.

SAP BTP is based on these five pillars: application development, automation, integration, data and analytics, and AI, as shown in Figure 2.2. It goes without saying that these five pillars and the services underneath them can be used in any combination and are complementary to each other. We've already seen a short glimpse of such approaches in Chapter 1, where we've combined several services to realize enterprise automation.

To reiterate, SAP BTP aims to support your business. That's why the services are best-suited to become an intelligent enterprise that is not only innovative but also resilient enough to also go through difficult and challenging times. This was an important topic during and after the pandemic, and it's also true now and will remain so the future.

In the following sections, we'll take a closer look at each pillar of SAP BTP.

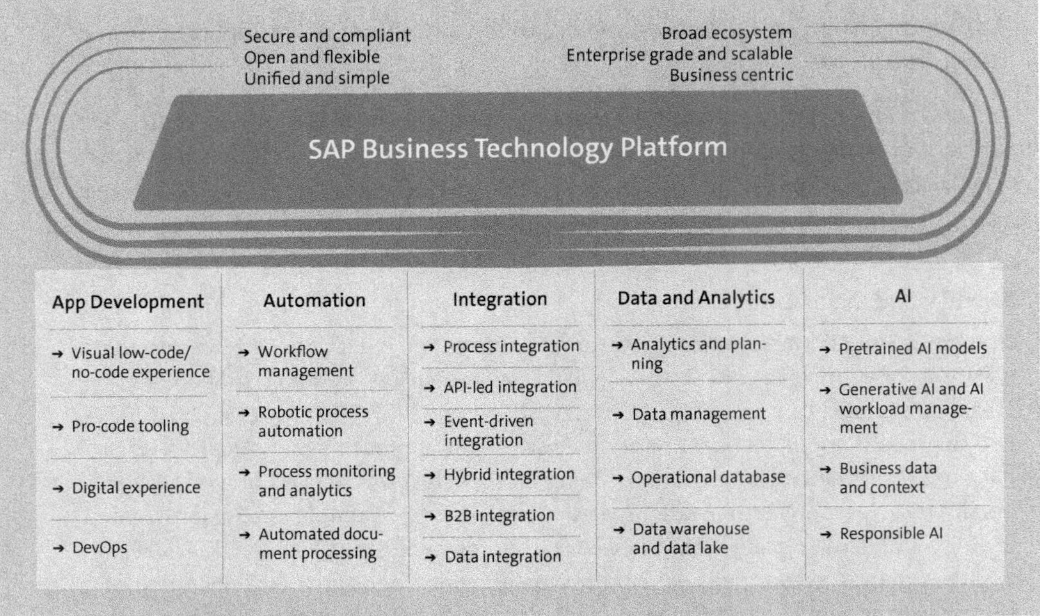

Figure 2.2 Under the Hood of SAP BTP

2.2.1 Application Development

Application development (app dev) provides the needed development environments to create applications and UI, supporting different levels of professionality for developers or different skills levels. This means you could either start with a low-code or even no-code development approach. This might be attractive for those who have a concrete understanding of the business and its needs, but they only like to use technology without going too deep into coding and are also not experts in a programming language. These people are sometimes called *key users* or *citizen developers*.

On the other hand, others will be in favor of a pro-code development approach if they like to code using application frameworks (e.g., SAP Cloud Application Programming Model or ABAP RESTful application programming model) or development kits, or if they're familiar with a programming language (e.g., Java or ABAP). In the end, it's up to you because SAP BTP and SAP Build in particular (discussed in the next chapter) cover all worlds (no-code, low code, and pro-code), as shown in Figure 2.3. Even better, you're able to intertwine them, which is called *fusion development*.

Increasing the developer efficiency is paramount, supported by productivity tools and the infusion of AI capabilities, for example, Joule, to generate code based on input prompts in SAP Build Code. Thinking in another direction, you can also make use of prebuilt content, packaged applications, application programming interfaces (APIs), and more available in SAP Business Accelerator Hub for different industries and lines of business (LoBs).

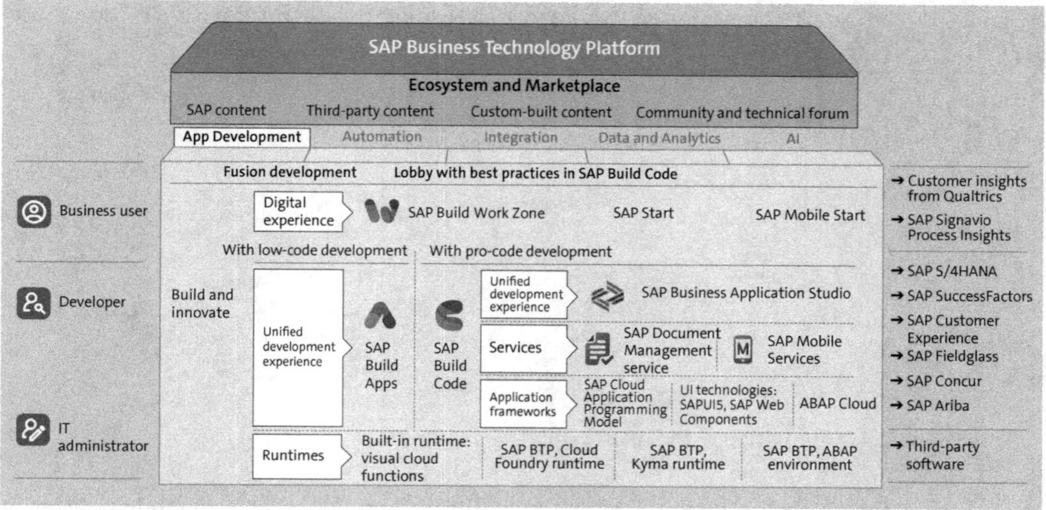

Figure 2.3 Application Development with SAP BTP

In addition, the business user is considered here, as SAP BTP provides a seamless user experience (UX) within the developed applications or on top with a smooth digital experience by creating business sites, portal pages, digital workplaces, or mobile access. The UX provides a unified, personalized, and central entry point to all applications. In addition, the IT administrator will be thrilled by the lifecycle management capabilities, transport management, and reporting dashboard to track the projects to master the runtime.

Whatever you want to develop here, it can focus on SAP applications, their extensions, and net-new apps but also span across them or even include third-party applications. Remember the picture of the solid foundation: everything is there for a great basis and flexible and open enough to go beyond.

2.2.2 Automation

With the automation capabilities, everyone, especially you, can automate business processes and tasks without writing code in a very intuitive environment. However, if you insist on writing code, this is also possible in some places. We see here again the option for fusion development, as outlined in Chapter 1. In general, you'll get all the capabilities to extend standard business processes, automate certain tasks, or orchestrate across different systems, as shown in Figure 2.4. Here again, this could either be SAP or non-SAP systems.

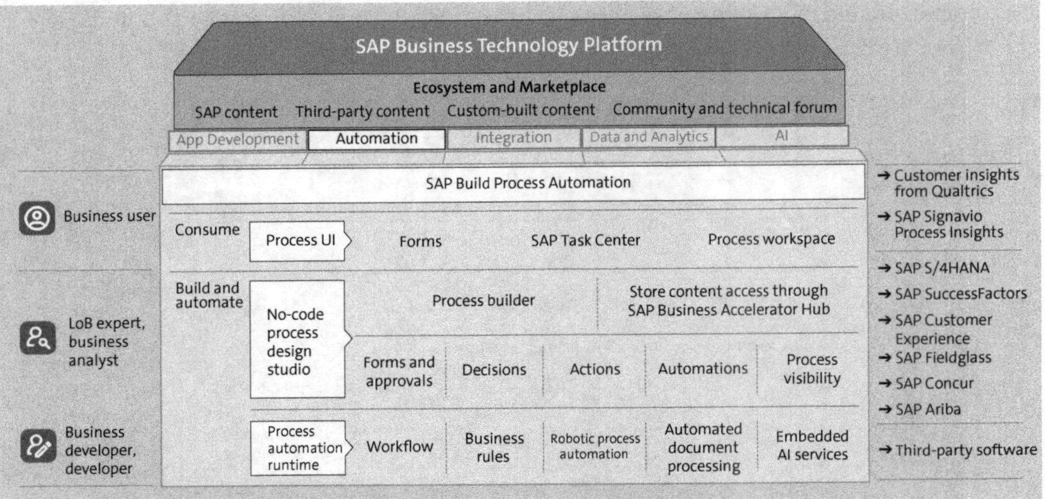

Figure 2.4 Automation with SAP BTP

These processes can be enriched with conditions and decisions to include policies in the process execution. You can make use of forms to reach out to the end user. If no end user is required or no one should be involved by design, create bots to do the work (for you).

Infusing AI capabilities or making use of prebuilt content packages also boosts your productivity and invites complete newbies to automate their business processes. We won't go into detail about this here, as the following chapters in this book will drill down into every aspect of SAP Build Process Automation, which is SAP's tool for automation.

2.2.3 Integration

A connected ecosystem is key to driving business success. Just think about your own IT landscape and how many SAP and non-SAP applications are handled there. To accelerate your innovation efforts, the integration between these applications and beyond your organizational boundaries must be robust and flexible enough to cater to the ever-changing business landscape. We're again in our solid foundation illustration from the very start.

The good news is that integration in SAP BTP provides what's needed to integrate everything, as shown in Figure 2.5. This comprehensive, versatile, hybrid, and enterprise-grade capability is also known as SAP Integration Suite.

Application-to-application (A2A), business-to-business (B2B), or business-to-government (B2G) integrations are front and center here, supported by all relevant open standards as well as SAP proprietary connectivity technologies. This includes process-, data-, API-, or event-based scenarios. The rich set of 170+ out-of-the-box connectors for non-SAP applications and the ability to create new connectors on your own provides great flexibility.

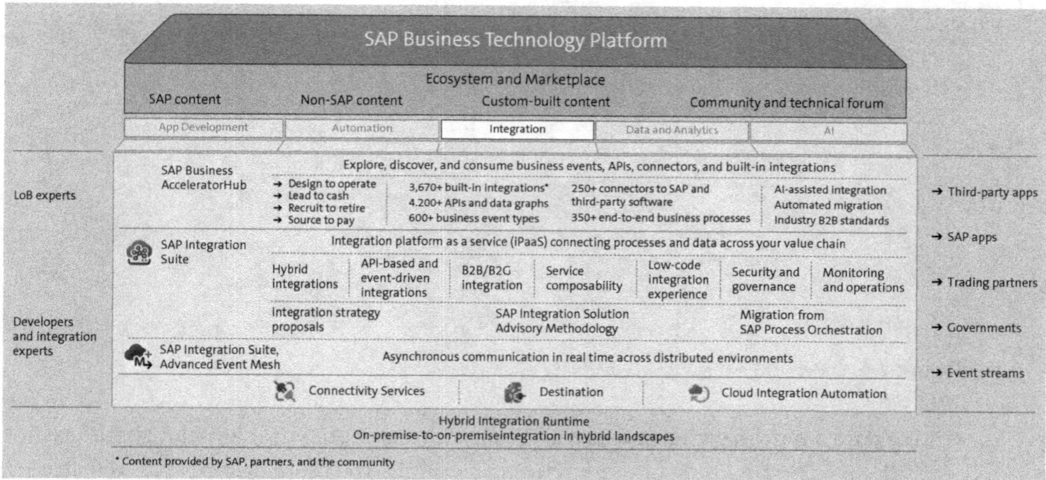

Figure 2.5 Integration with SAP BTP

In addition, a systematic approach is offered that includes best practices. To accelerate the effort, thousands of proven integrations, built and maintained by SAP and partners, can be used. You'll find them in the SAP Business Accelerator Hub, which is, by the way, also the home for prebuilt process content packages (covered in Chapter 3). API management features to securely expose, manage, and monitor APIs are also included. Complemented by SAP Integration Suite, advanced event mesh, as a message and event broker to support eventing and messaging technologies for asynchronous and event-based scenarios.

The scalable architecture and high-performance capabilities of SAP Integration Suite ensures everyone can access information, data, and systems across organizational boundaries and systems when they need it—in real time. If you think about the combination of this powerful integration platform with your business processes, it's clearly a perfect match.

2.2.4 Data and Analytics

Getting a consolidated view across all data assets and deriving the right insights to give this data a purpose is challenging for many organizations. This might not come as a surprise when you consider all the different areas where data is produced. Data and analytics within SAP BTP take care of this, as shown in Figure 2.6.

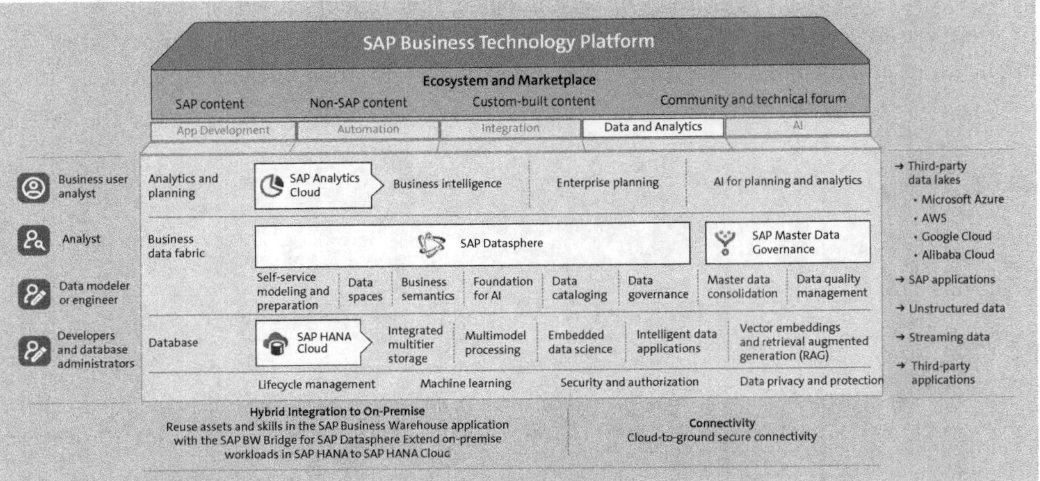

Figure 2.6 Data and Analytics in SAP BTP

Giving data a purpose means deriving the maximum value of that data. Therefore, it's essential to combine data and analytics capabilities to create dashboards, tell stories, and even get deeper data insights by using data warehousing and analytics solutions.

Within data and analytics, SAP BTP uses SAP HANA Cloud as the strategic database of choice, which is a fully managed in-memory multitier cloud database as a service (DBaaS). It integrates data across the organization and provides live data for proven and fast decision-making.

The business data fabric, covered by SAP Datasphere, enriched the data warehouse cloud with new, sophisticated semantic modeling capabilities, including the analytical model, a new data catalog, and application and data integration. These key features are completed with data warehousing and federation, business content, and a data marketplace.

Finally, the prepared and managed data can be consumed in the SAP Analytics Cloud analytics software. Gaining insights from your enterprise will help improve your planning and result in better decision-making. The integrated planning and augmentation capabilities are part of this cloud-based, self-service business intelligence solution. Although it's a cloud solution, you can access on-premise analytics content, just to complete the picture. Information makes the difference, when the right people have it handy at the right when they need it. The collaboration features ensure this, making it easy for different stakeholders and interested parties to have intuitive and easy access to the analyzed data and to discuss that data.

2.2.5 Artificial Intelligence

AI is clearly the next game changer in the software industry, but how can you make it transformational for your business? In the best case, you should be able to leverage business-ready, trusted, and easy-to-use AI capabilities. With SAP BTP, this is exactly what you get (see Figure 2.7) in two forms: (1) using AI services that are running on top of SAP BTP, for example, document information extraction, covered as SAP AI Services, or (2) using AI capabilities as an integral part of the services offered in SAP BTP, for example, Joule within SAP Build Code to generate application code.

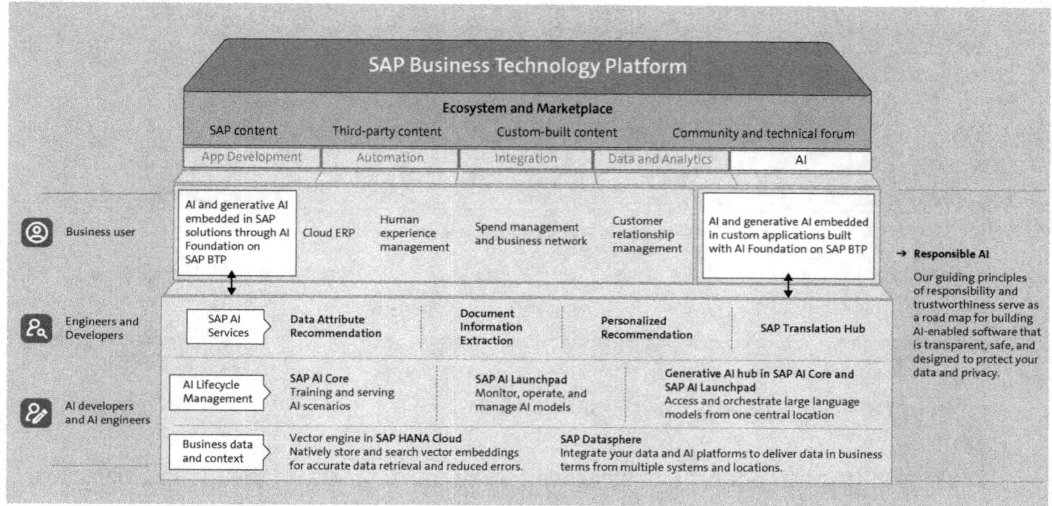

Figure 2.7 AI with SAP BTP

There are ongoing investments in AI capabilities in all SAP BTP pillars. So, AI not only serves as a pillar on its own but also contributes to and infuses into application development, automation, integration, and data and analytics. This provides another advantage, as SAP BTP is focused on your business and access to related business applications, SAP or non-SAP. The increased and boosted developer productivity is obvious. The power of generative AI capabilities in all areas of SAP BTP is also a game changer for your business.

Regarding AI, SAP follows these three principles:

- **Relevant**
  AI is directly supporting your business, across the organization.
- **Reliable**
  Every decision based on AI is covered by the business data.
- **Responsible**
  The highest ethical, security, and privacy standards are included.

In addition, Joule, as shown in Figure 2.8, is planned to become more prominent across SAP's product portfolio. It's your new copilot that works for all solutions to help you increase your immediate productivity with a consistent UX. Joule is accessible directly in applications and the tools to create applications, automate processes, cover integration aspects, and much more.

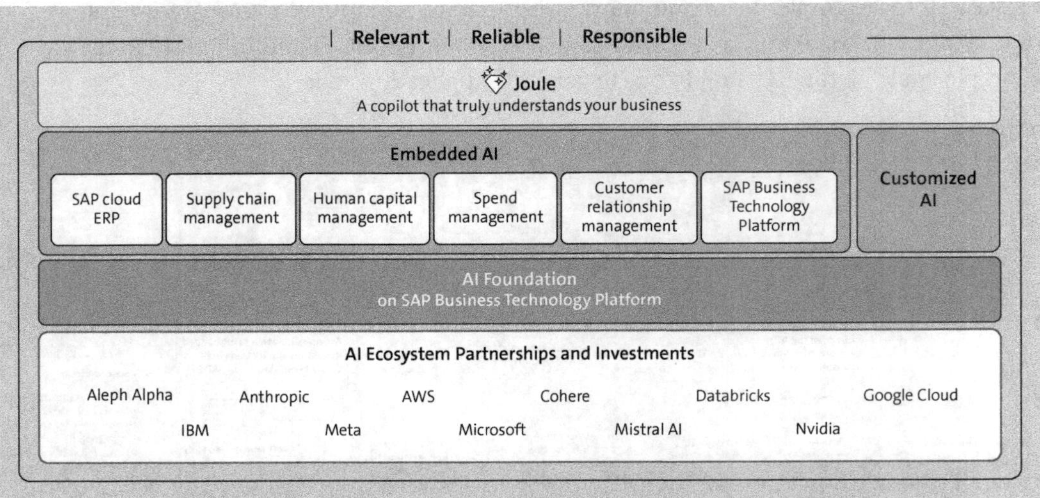

Figure 2.8 AI Embedded across the SAP Portfolio

While writing this book, there have been some cool features on the AI road map, particularly also for SAP Build Process Automation. On the next layer, the AI capabilities are embedded within different LoB solutions, such as cloud ERP or human capital management, which make them easy to be consumed directly without too much configuration effort. All of this relies on a strong ecosystem with different partners.

Tip

You'll find all SAP BTP services, related missions, links to learning journeys, information on licensing, applicable data centers, and so on in SAP Discovery Center at *https://discovery-center.cloud.sap* (see Figure 2.9).

Figure 2.9 SAP Discovery Center for SAP BTP

2.3 SAP Build

We've already discussed the application development and automation pillars and you know that behind automation, you'll find SAP Build Process Automation, the topic of this book. But what is actually covered with SAP Build?

You know that it's key for any organization to be innovative. One aspect of being innovative is to figure out and foster what differentiates you, your enterprise, or your organization. This might be your service to your customers, your product, the art of collaboration, or whatever makes you different and better than your competitor. This difference could also be a unique process or new approach to solving problems. These strengths must also be covered within your IT landscape, including the used (business) applications and the executed processes. This results in customizing the standard business applications to your requirements or even extending them to support you in the best possible way. These customizations could range from simple productivity apps or extensions up to business-critical apps, where you extend your core ERP system, for example. For this, you need a flexible framework or platform, which you'll find in SAP BTP. To be more precise, SAP Build (as part of SAP BTP) is *the* solution for extensions. Let's take a closer look at SAP Build in the following sections.

2.3.1 Overview

SAP Build is your choice for smooth extension development. You can create applications, automate business processes, or design digital workspaces, as shown in Figure 2.10.

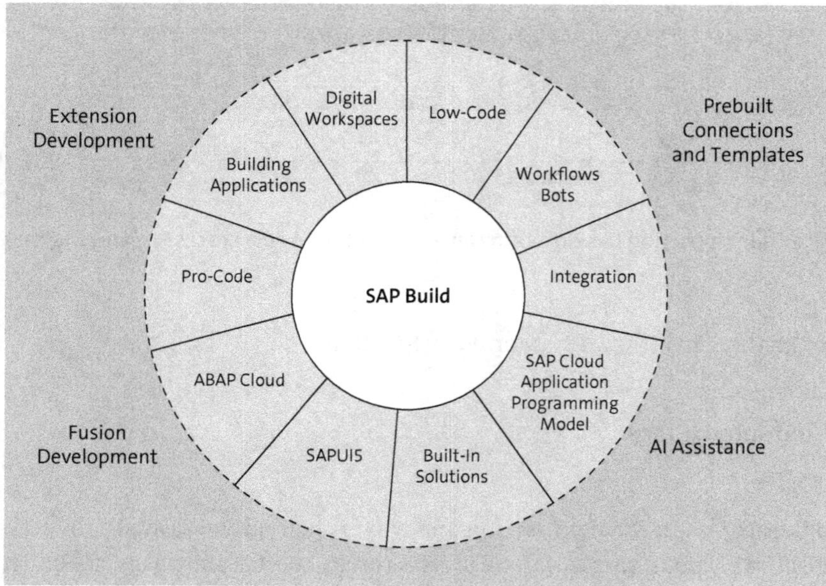

Figure 2.10 SAP Build for Smooth Extension Development

Therefore, SAP Build offers low-code, pro-code, and even generative AI tools, which are also working together (fusion development). You're supported by a huge amount of prebuilt content packages to accelerate your projects and drive faster time to value.

A more detailed view of SAP Build is shown in Figure 2.11. SAP Build is deeply integrated with the business applications, which is key to extending them. SAP Build also leverages the extensive library of connectors, iFlows, business events, and so on available in SAP Business Accelerator Hub for SAP and non-SAP applications. The prebuilt content you'll also find there caters to finance, human capital management, supply chain management, procurement, and many others. The intelligent copilot Joule is also included here, and more is planned to come. The idea is to support you in the creation of applications and business processes with generative AI.

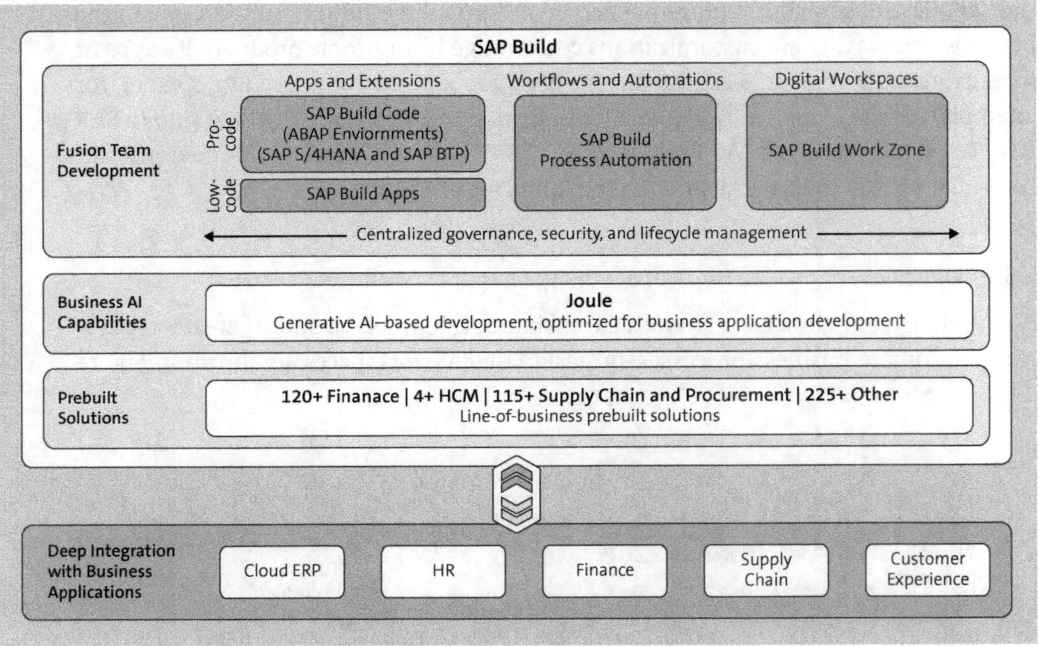

Figure 2.11 SAP Build Supporting Fusion Team Development for Apps, Processes, and Digital Workspaces

Following are the three main areas supported by SAP Build:

- Apps and extensions
- Workflow and automations
- Digital workspaces

We'll explain these areas in the next section and what is behind the related products. The access and starting point for all of these tools is the SAP Build **Lobby** page, as shown in Figure 2.12. From the lobby, you can create, manage, and work collaboratively on your projects.

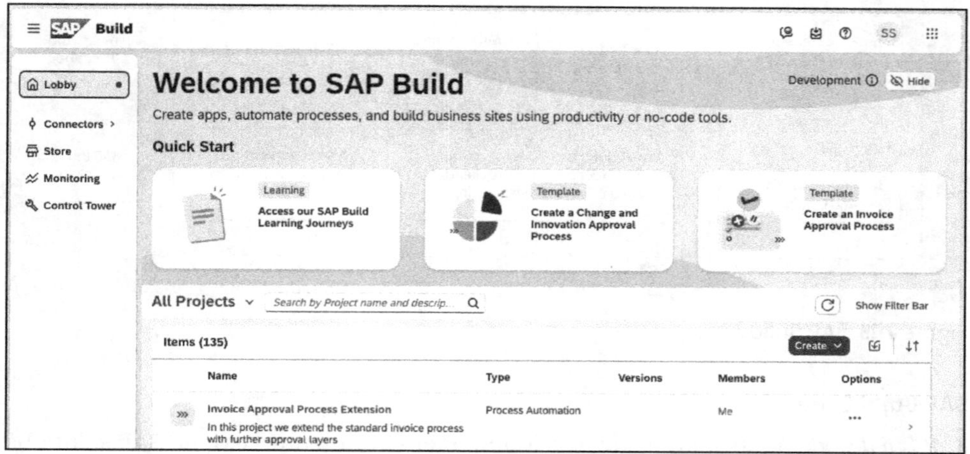

Figure 2.12 SAP Build Lobby, the Starting Point for Extension Development

2.3.2 Apps and Extensions

Whether you're familiar with programming languages and know how to code, for example, ABAP or Java, or whether you rather prefer a low-code, no-code approach, SAP Build gives you the choice. This is a great advantage because you don't need to search for the right tool for different developer personas or different skill levels.

In the following sections, we'll walk you through SAP Build Apps, SAP Build Code, and ABAP environments, as well as highlight when to use what.

SAP Build Apps

If you're in favor of the low-code, no-code approach or probably at least just want to start there, then *SAP Build Apps* is your starting point, as shown in Figure 2.13. With an intuitive visual drag-and-drop experience, you craft standalone apps and app extensions. The integration to SAP or third-party systems is be done through REST APIs or OData. Built-in formulas and preconfigured flow functions support you in adding UI controls and business logic. Using design themes, also including an SAP Fiori theme, you're free to design your UI and make it highly customizable. If you want to create full-stack apps with data models and business logic, you also find the features to do so in SAP Build Apps.

It's up to you whether you then create a mobile or web version of the created application.

Integration with other SAP Build services is provided, and you can even move on with a pro-code approach in SAP Build Code or the ABAP environments.

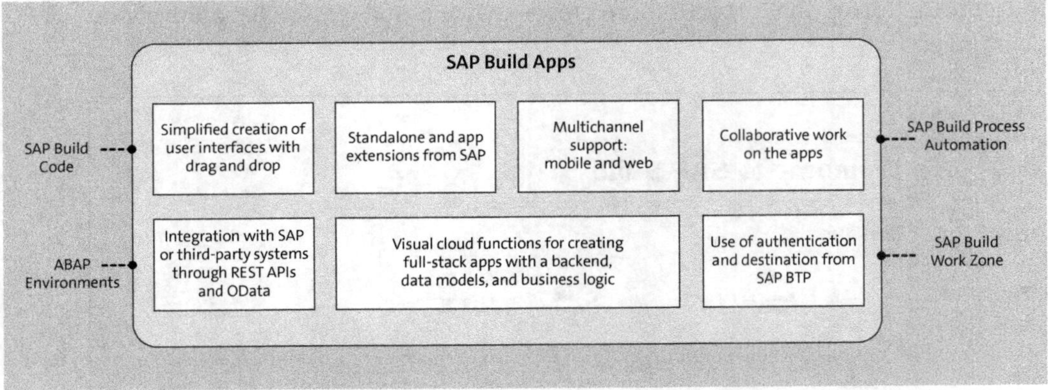

Figure 2.13 SAP Build Apps

SAP Build Code

SAP Build Code offers a pro-code development environment to extend SAP solutions (see Figure 2.14). You can build SAP Fiori apps while leveraging out-of-the-box templates based on SAP best practices, extend existing SAP Fiori apps via predefined extensions points, or even adapt the UIs of SAP Fiori elements applications. If you want to develop mobile apps or applications close to SAP HANA, SAP Build Code provides what you need. You can leverage Java and JavaScript in the integrated development environment. Again, you might want to use them to complement the other services that are part of SAP Build or SAP BTP in general. For SAP Build Code, SAP Build Apps, and ABAP environments, the fact that they are running on SAP BTP provides trusted security for authentication, authorization, and data protection.

SAP Build Code comes with a generative AI–based code development. Leveraging Joule, you can generate code and app logic aligned with SAP-centric programming models by just maintaining the prompt in natural language. In addition, you can generate data models or rapid unit tests.

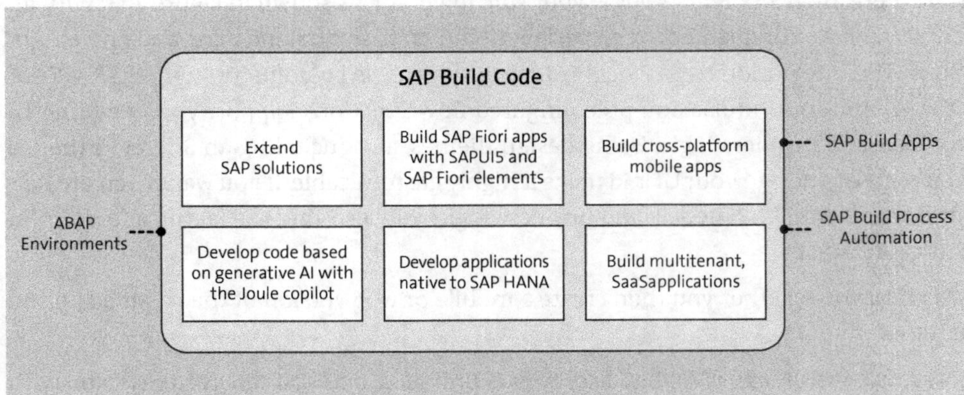

Figure 2.14 SAP Build Code

The addition of SAP Build Code in the SAP Build family of services marks the next evolutionary step to include pro-code capabilities.

ABAP Environments

The next milestone of SAP Build is the addition of *ABAP environments*, as shown in Figure 2.15. By using the ABAP Cloud development model, you can either develop on-stack ABAP extensions using the SAP S/4HANA Cloud ABAP environment or side-by-side extensions with the SAP BTP, ABAP environment. This accelerates the cloud transformation and aims to keep the core clean. With this addition, not only no-code/low-code Java developers but also the big community of ABAP developers leverage SAP Build, starting in the lobby.

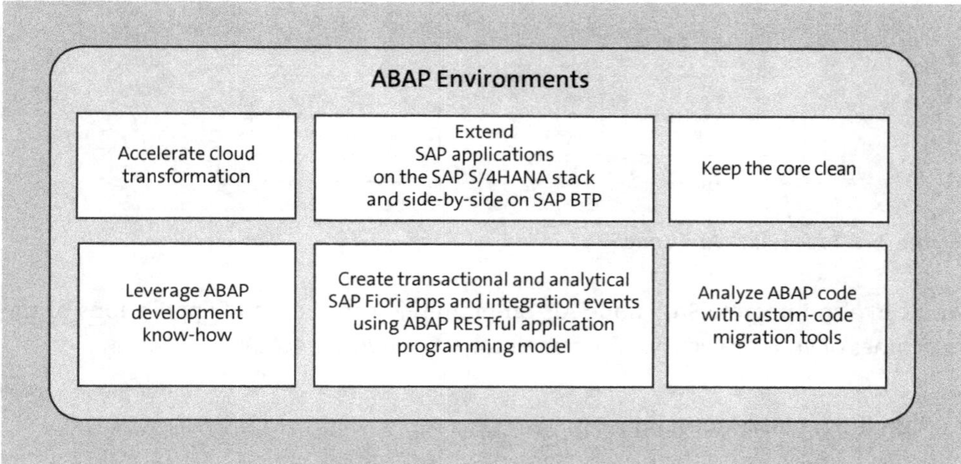

Figure 2.15 ABAP Environments as Part of SAP Build

Once the applications are developed with SAP Build Apps, SAP Build Code or the ABAP environment, the SAP Build Work Zone central access point becomes important.

2.3.3 Digital Workspaces

Having one UX across the enterprise will make it easy to use all of your applications, extensions, or business processes. Therefore, you choose *SAP Build Work Zone*, as shown in Figure 2.16. Digital workspaces, business sites, and portals are most often the first touch of your end users with the connected business applications or whatever you've developed with SAP Build. Make this first contact as easy and delightful as possible. Lower the entry barrier or have none at in the best case, which will also have a positive impact on productivity.

The created business sites provide tremendous flexibility, enriched with content widgets and guided experiences. You can have the entry and access point for your business users to share content and documents across interactive workspaces and spread the knowledge through your organization. With a personalized, role-based UX, you "talk" directly to your business user and acknowledge their needs and even include your own corporate identity design, themes, colors and so on. It doesn't matter whether this experience will be traditionally on a desktop or via a mobile device.

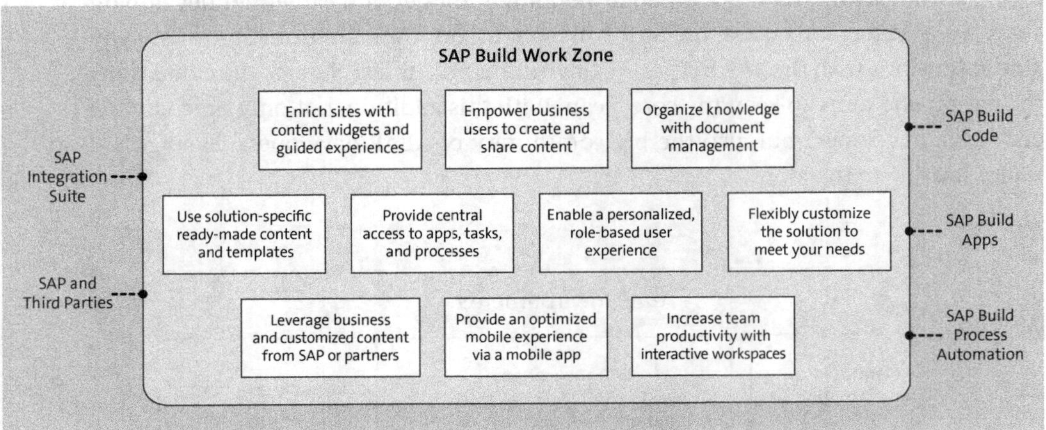

Figure 2.16 SAP Build Work Zone

You can also integrate SAP, non-SAP, on-premise, or in the cloud applications or the outcomes of your projects for the remaining SAP Build services.

Workflows and Automations

SAP Build Process Automation simplifies business process and task automation without the need to write code. We won't cover it here, as we'll explore it in detail next in Chapter 3.

2.4 Summary

In this chapter, we started with the technical foundation—SAP BTP—to innovate your business and become an intelligent enterprise. You know now that SAP BTP is not only a foundation but a rich platform with capabilities in the area of application development, process automation, integration, data and analytics, and AI. It enriches and also extends the solutions offered by SAP, starting with a cloud ERP and surrounded by industry-specific and sustainable solutions for HR, spend management, supply chain management, or customer relationship management. We've touched upon keeping the core of these solutions clean and how to drive extensions with SAP Build either in a low-code/no-code or pro-code approach or bringing these approaches together in fusion development. For providing digital workspaces, you'll choose SAP Build Work Zone. For application development, you'll choose either SAP Build Apps, SAP Build Code, ABAP environments, or their combination. For process automation, you'll go for SAP Build Process Automation. We'll start to explore what this means in the next chapter.

Chapter 3
Overview of SAP Build Process Automation

In this chapter, we'll answer the following questions: What is SAP Build Process Automation all about? Why did we write a book about it? Why you should consider using SAP Build Process Automation in your project, and where? But first and foremost, you'll be introduced to its components, applications, and artifacts.

Business processes are the lifeline of every organization—just think of all the different activities and tasks you're asked to work on each day. Most are related to business processes or derived from lead-to-cash, hire-to-retire, design-to-operate, or source-to-pay end-to-end scenarios. Luckily, most of these processes are standardized and covered within an ERP system, although from time to time, you'll need to extend these standard processes, for example, by adding several approval levels. Or, you might want to orchestrate several processes across different units and therefore also across different systems. You may want to automate dedicated tasks within a process, which are repetitive, mundane, and requiring robot-like actions. You're innovative, so you probably have a great idea for an efficient business process that isn't covered yet in any system. These are the cases, on a high level, where SAP Build Process Automation will support you and make a difference.

Let's now take a closer look at this service, which is part of SAP Build, and get an understanding of its purpose and architecture, as well as get an overview of the included applications, the different artifacts and their characteristics, and the lifecycle phases.

3.1 Introduction

Whether it's extending standard business processes, creating net-new ones, or automating dedicated tasks, SAP Build Process Automation provides the necessary capabilities. In fact, it helps you simplify these activities. However, this doesn't mean the processes you're improving with the service are simplistic; in fact, they might be comprehensive and complex because they are reflecting your day-to-day business.

The nice thing about SAP Build Process Automation is that you don't need to know or determine at the beginning what characteristics your process may have or which artifacts you might need. Of course, you should know what you want to do, what the result should look like, and which business requirements are in play, but you have the flexibility to outline your process as you go.

Whether you want to create a workflow, add some forms, speed up with a bot automation, include business rules to reflect certain policies, or even leverage some AI features, everything is there. Even better, you can do this without writing a single line of code—just drag and drop the required artifacts.

Note

If you want to code, you can do so here and there, but it's not needed. This also lowers the entry barrier for the those who want to work with the service. You'll experience this in the upcoming chapters. Remember fusion development? As stated during our introduction to SAP Build, this is also supported with SAP Build Process Automation, meaning the combination of low-code/no-code with pro-code development either via the other SAP Build services—*SAP Build Code, SAP Build Apps, SAP Build Work Zone*—or other development environments.

When you're using SAP Build Process Automation, you're doing this with a purpose, for example, extending standard processes. You also need to connect with applications, processes, and data across SAP and non-SAP systems when you want to orchestrate processes. Therefore, you'll leverage several bots and connectors, make use of application programming interfaces (APIs), or include *SAP Integration Suite*, as described in Chapter 1. The important point here is that you'll connect your process to the related business data and make it business-relevant, as everything will be part of the process context. We'll discuss the details later in Chapter 5.

If you're using *SAP Signavio solutions* to analyze your business process, you might get a recommendation to create a business process in SAP Build Process Automation or trigger an already created one. In addition, you might just come to the conclusion that the best way to fix the current issues is to define a net-new process, probably as an extension.

But how does this actually happen? Let's take a closer look at the capabilities as outlined in Figure 3.1.

No surprise, with SAP Build Process Automation, you can automate *processes and workflows*, as we said earlier. It's completely up to you for which scenarios you use it, including in which industry, for which line of business (LoB), in which landscape, and at what level of complexity. The tool has the flexibility to give you the needed freedom to do what you want, especially for process extensions in which you might want to include *forms and multistep approvals*. This is where a business user will interact with the process. For the forms, you can either make use of the integrated forms editor or include an SAPUI5-based form (fusion development).

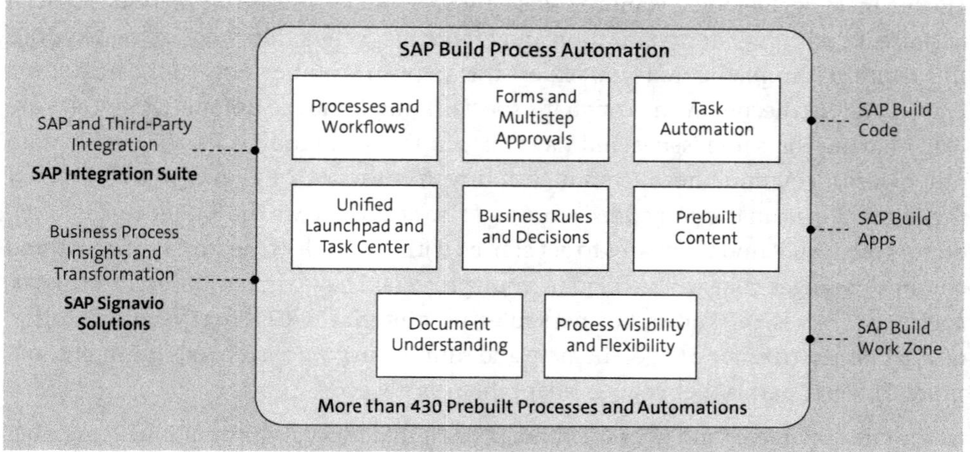

Figure 3.1 Capabilities of SAP Build Process Automation

There are processes that require no user intervention, repetitive and mundane tasks that no one really wants to do, and the need to collect, compare, check, and distribute or consolidate data from different sources. Think about collecting hundreds of individual sales orders in a spreadsheet and maintaining them manually in your ERP system (e.g., SAP S/HANA) one by one, every day. Sound boring? Instead, you can leverage *task automation* in SAP Build Process Automation to free up your employees from these kinds of tasks and give them more time for tasks that add more value and are more engaging, which will also increase their motivation.

Getting back to the tasks, once they are executed in a process instance, somehow you want to access them and be able to work on them. Examples include maintaining some data to start a process, reviewing the inputs of your employee's purchase order request before approving it, and getting the request processed in the related business solution. For this, you'll access a unified launchpad within SAP Build Process Automation or have central access via SAP Task Center (a *unified launchpad and task center*).

When discussing business processes and automation, we most often also talk about policies. *Policies* can be created within an organization, refer to legal regulations, or ensure a dedicated outcome when a certain event is happening. For example, all your premium customers located in Spain should receive a discount of 10% of their purchase price above 249 euros during your winter special starting November 21 and running until January 12. Customers located in other countries should also get discounts with other conditions. This policy is just a sentence, but others can be quite complex and should also be reflected in your business processes. Best practice calls for not touching the process definition in general. This is where *business rules and decisions* will help you. Whatever policies or regulations are in place, you can maintain them there, referring to and including the business-related data but also keeping it independent from the process flow itself.

The business problem you want to solve with SAP Build Process Automation is special because it's within your organization and tailored to well-defined requirements. Once the solution is implemented, you expect this to push your business to the next level. This might be true, but even though it's special, it might not be unique. Other folks are likely having the same issues and have chosen the same tool to fix the problem. In other words, you don't need to start each time from scratch. SAP has *prebuilt content* for you, available in the embedded store and also accessible via the SAP Business Accelerator Hub. You'll find templates to get started with, examples to learn the product, and content packages that will accelerate your projects. They contain different artifacts, such as processes/workflows, business rules, user interfaces (UIs), bots, documentation of what needs to be done to get them up and running in your system environment, and more. The best part is that you can adapt them as needed.

AI simplifies our lives and does the same in SAP Build Process Automation. Remember the sales order to be collected in a spreadsheet, which we now have automated? This data becomes challenging when the sales orders are in different formats. For this and similar scenarios, there is *document understanding*, which leverages the Document Information Extraction service to automatically extract the information, structured or unstructured data, from different documents and process the data moving forward.

Once the business process is up and running, you want to know what's going on. Depending on your role, there are different information needs. An administrator is more interested in smooth execution with no error messages or connectivity issues with the backend system. The business user also wants a smooth execution, but has information needs beyond pure technical monitoring. To get information about the process performance, measure some key performance indicators (KPIs), or get some status updates, you'll use *process visibility and flexibility*. Starting from a high-level overview dashboard, you can even drill down into the details of each running process instance to check the related business data.

3.2 Architecture

For a better technical understanding, let's open the hood of SAP Build Process Automation. The service itself is running on SAP Business Technology Platform (SAP BTP) and using the same security concepts: authorization mechanisms with a dedicated roles concept, user management, disaster recovery, audit logs, and so on. We are talking about a multitenant application, so every user within an organization is using and sharing the same application components, the same database, and the same network. The overall architecture is illustrated in Figure 3.2.

Regarding sharing, SAP Build Process Automation also shares some central components with the other services of SAP Build, for example, the lobby, embedded store, project repository, monitoring, and settings. We'll explain their purpose in Section 3.3, so stay tuned.

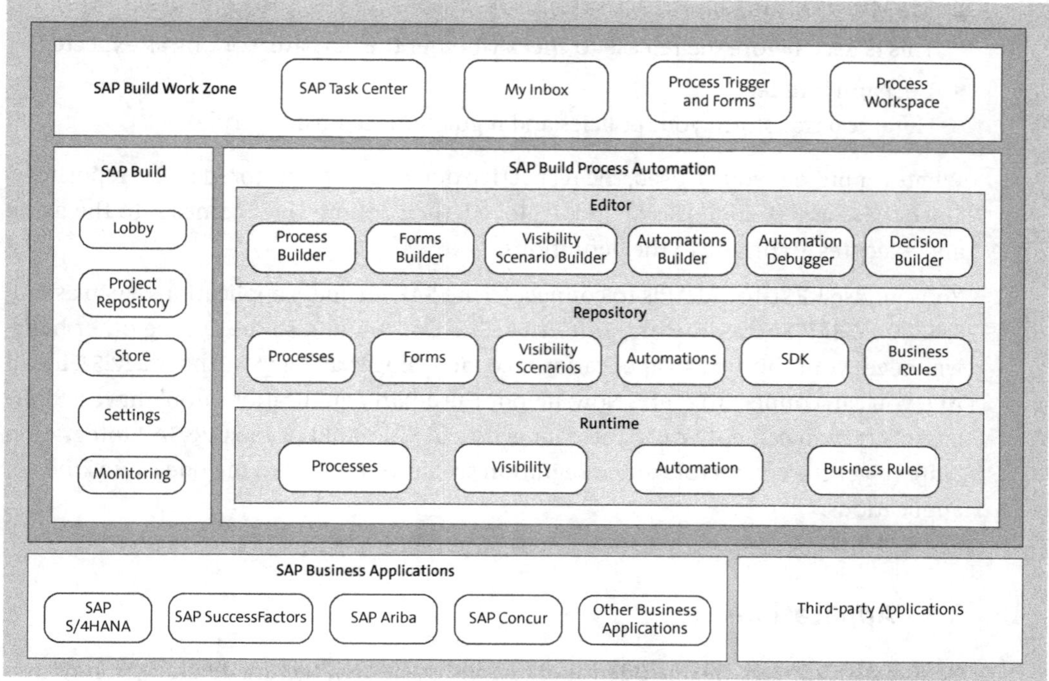

Figure 3.2 Architecture of SAP Build Process Automation

In Section 3.1, we already discussed how to access the tasks. If you're looking for a central, harmonized and state-of-the-art entry point, you should use SAP Build Work Zone. With this personal workspace setup, you'll ensure a delightful experience for your business user and a smooth access to every aspect, information need, or action related to the usage of SAP Build Process Automation. Within SAP Build Work Zone, you can include the My Inbox application to open the created forms, access SAP Task Center to federate tasks from different systems in one single inbox, publish tiles to start processes, or collaborate to solve questions around your process and other topics.

In SAP Build Process Automation, you'll get access to all the required editors:

- **Visibility scenario builder**
  This is used to create the visibility scenarios, based on the process context.
- **Process builder**
  This is *the* editor to define your process with all the needed artifacts, which are just one click away.
- **Form builder**
  This editor opens whenever you design a form for a user task.
- **Automation builder**
  Here, you define your bot automations.

- **Automation debugger**
  This is used before the release to check whether the bots are working as expected.
- **Decision builder**
  Here, you transform your policies and regulations to business rules.

Whatever builder you've used, the respective outcome is then stored in the repository. Once these assets are released and deployed, the runtime layer comes into the game and executes according to the definitions.

You can use REST-based APIs to connect to the SAP business applications, such as SAP S/4HANA, SAP SuccessFactors, SAP Ariba, SAP Concur, and so on. On the other hand, whenever your third-party applications (i.e., anything that isn't SAP) have a REST-based API, you can connect to them. Now, if your third-party applications don't have a REST-based API, you can either use the connectors in SAP Build or include SAP Integration Suite to get the connectors and integration scenarios needed to include these kinds of applications.

3.3 Applications

You've seen some of the included applications in the architecture diagram, so now let's discuss what you can do with them? We'll use the analogy of a hotel to introduce you to the applications of SAP Build Process Automation. Thanks for visiting us, we wish you a pleasant stay.

3.3.1 Lobby

Remember your last trip to a hotel? As soon as you go in the main door, you're in the lobby. The same is true for SAP Build Process Automation: you open it, and you're in the **Lobby** screen, as you can see in Figure 3.3. In fact, it's true for SAP Build in general, as the lobby is the starting or entry point for every service of SAP Build.

When you enter a nice, bright, tidy, welcoming lobby in a hotel, chances are high you'll experience the same in your hotel room. Look at this bright, tidy, welcoming, and nice lobby of SAP Build, where you want to start to create apps as well as automate processes and business sites using productivity or no-code/low-code tools immediately. The idea was to provide a single entry point that provides all necessary information and accesses to start immediately and intuitively. You'll continue with this experience once you're working on the different projects—in the hotel rooms, if you will.

You can even use the different options offered in the **Quick Start** area at the top, as shown in Figure 3.4. If you remember, we introduced the prebuilt content packages earlier, and some of the most prominent, most requested, and most used packages are available here as well. The displayed content packages change from time to time. Just click on one, and you'll be directed to the content in the **Store**.

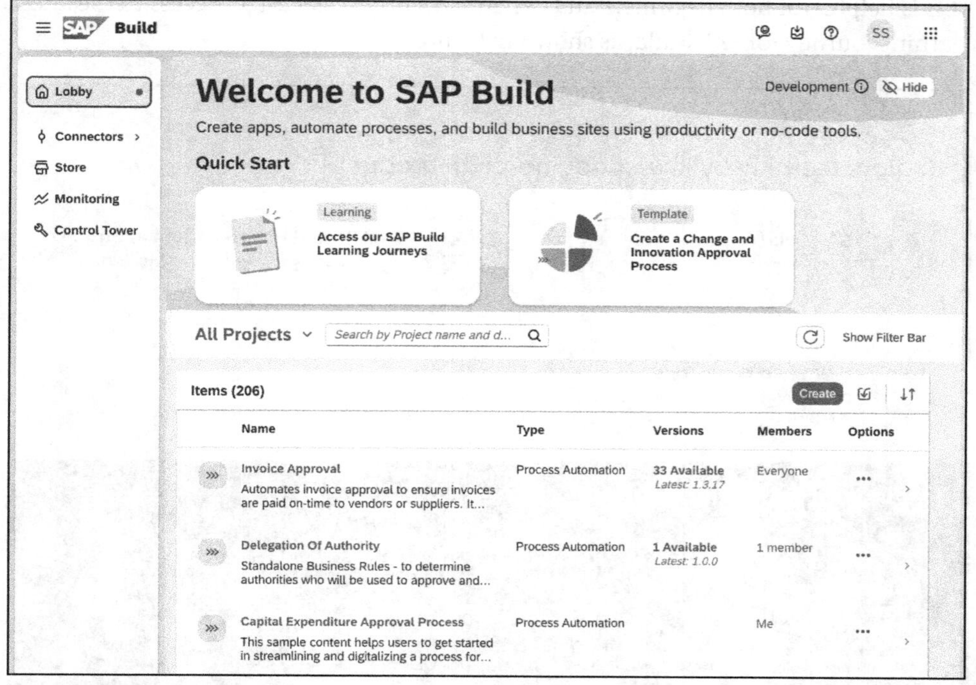

Figure 3.3 Lobby of SAP Build

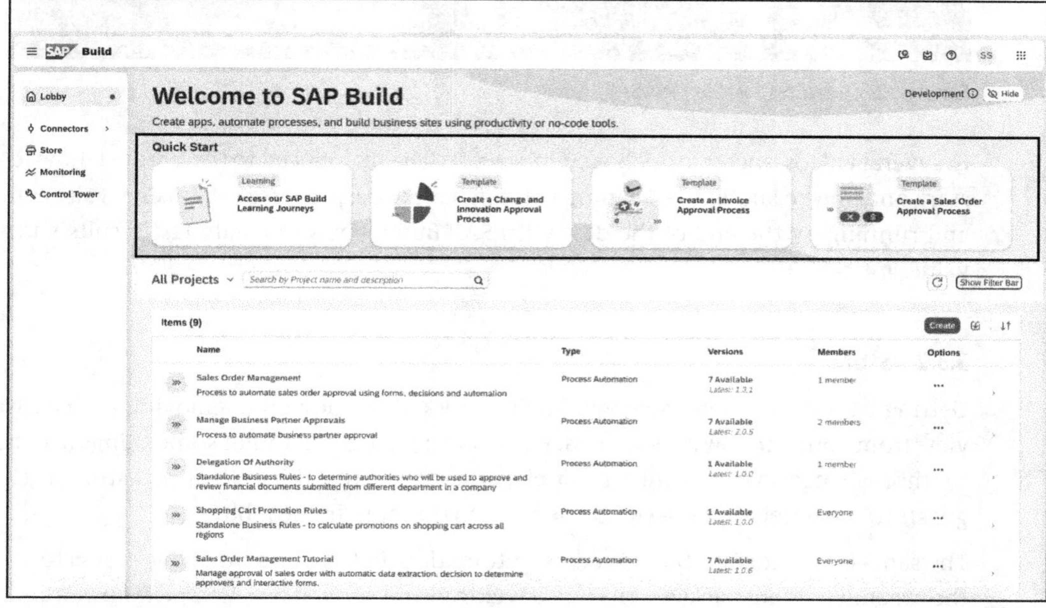

Figure 3.4 Quick Start Area

This is helpful for both beginners and experts. There, you get direct access to the latest learning journey for SAP Build, as shown in Figure 3.5.

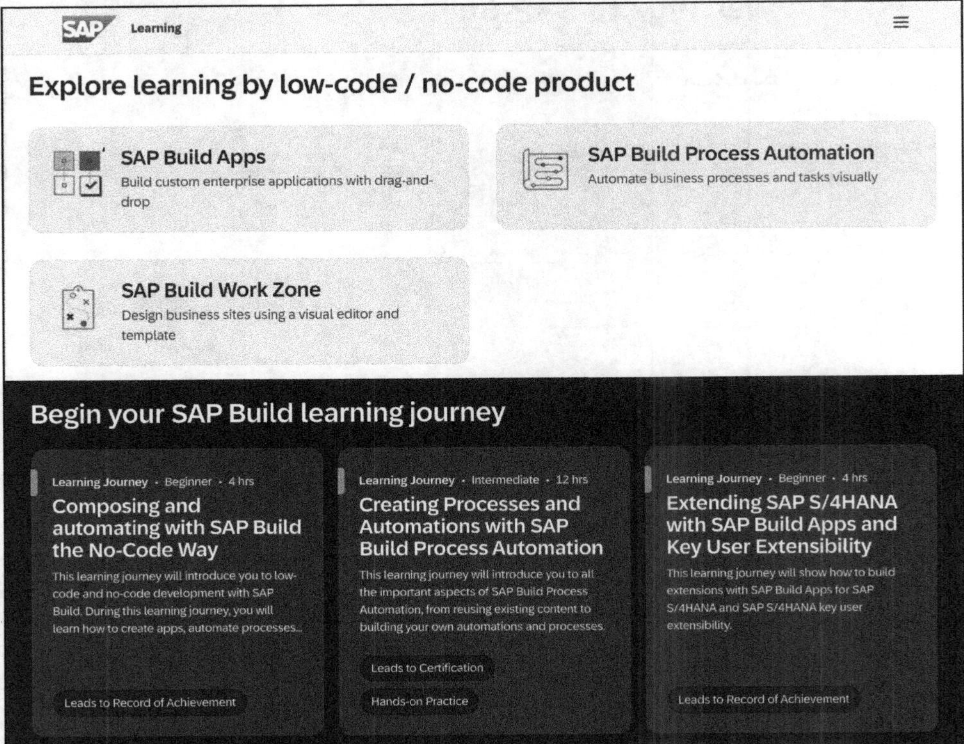

Figure 3.5 Learning Journey for SAP Build

In several videos, short tutorials, and more, you'll be introduced to the product, how to start, and how to succeed—meaning you can have apps, processes, or business sites up and running by the end of the day. With SAP Build, you can create fast results with value.

3.3.2 Store

Getting back to our hotel analogy: You've checked in and you're enjoying the beach view from your window as you unpack when you realize you forgot some things (e.g., a toothbrush, newspaper, drinks, razor blades, etc.). Luckily, this is the usual stuff other guests in the hotel might need, so you head to the store in the lobby.

The same is true for SAP Build Process Automation. In the **Store**, you have access to all the prebuilt content packages that are developed and curated by SAP and its authorized partners. These content packages are inspired and based on real customer examples that have occurred multiple times at different customers or have been requested by several customers. Most of the time, these are common use cases, such as invoice approval processes, creation of business partners, and so on. There are also more specific use cases for dedicated industries that are often created, offered, and published by SAP partners.

Besides the business processes that you also see filtered in Figure 3.6, there is much more to be discovered and explored.

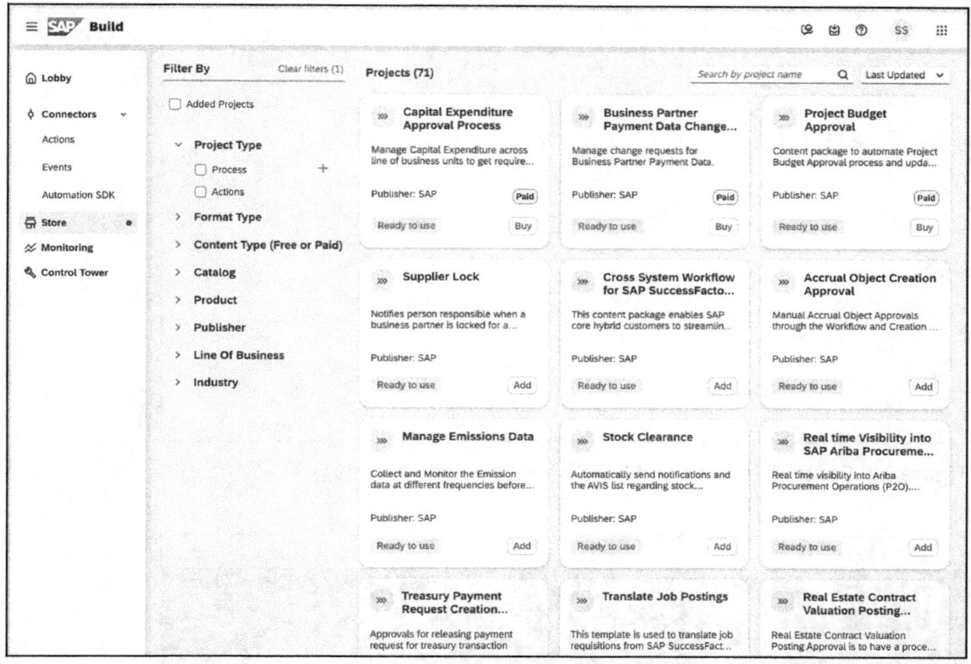

Figure 3.6 The Store (Filtered by Publisher: SAP, SAP S/4HANA)

Other available project types are *actions*, which are vehicles to encapsulate APIs and directly use them in the business process. In addition, the bots (i.e., task automation) contain only automation artifacts.

First, you can take a look at the content, but you can also directly import these content packages in SAP Build Process Automation and start a new project based on them. The content packages also include some documentation, for example, a configuration guide, and you'll also see all the related artifacts, such as automations, data types, decisions, forms or processes, and much more, as shown in Figure 3.7.

You might wonder whether you can also browse through this content without having a license for SAP Build Process Automation? Yes, this is possible. If you just want to go on a window-shopping expedition, you'll find the same content packages in SAP Business Accelerator Hub (*http://api.sap.com/build*), as shown in Figure 3.8.

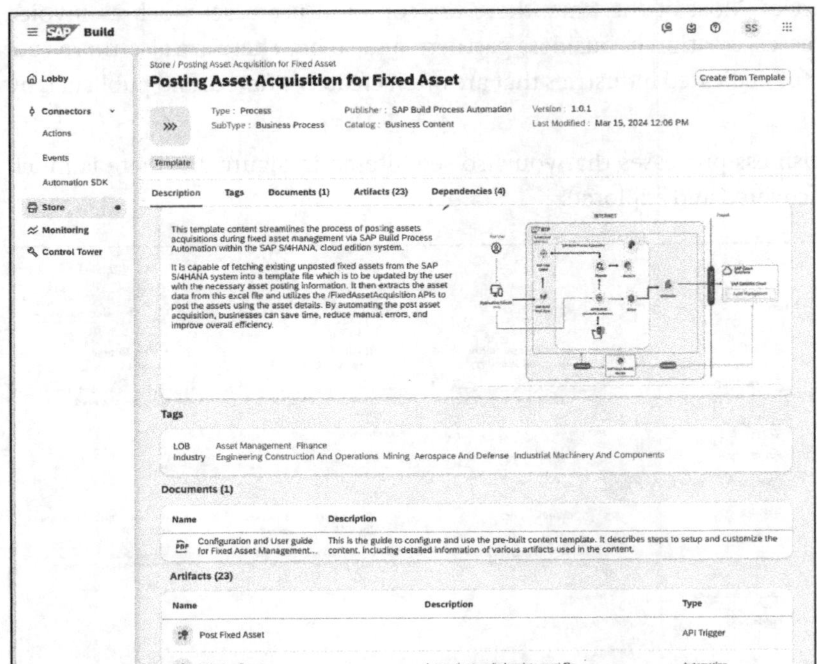

Figure 3.7 Content Package and Its Information and Artifacts

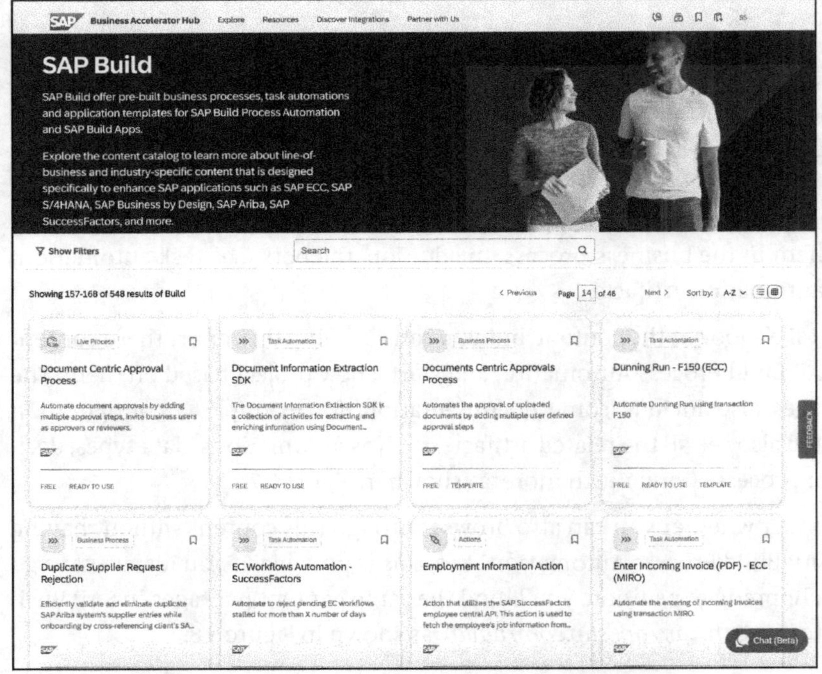

Figure 3.8 SAP Business Accelerator Hub with Prebuilt Content for SAP Build

So, the embedded store is your go-to area to accelerate your projects or start them out. Some of these content packages are also useful to get a better understanding of the product. We'll discuss the embedded store in more detail in Chapter 14.

3.3.3 Monitoring

Monitoring is essential to ensure control when executing your automated processes. In general, there are two types of monitoring: business monitoring and technical monitoring. In *business monitoring*, you're more interested in checking different KPIs to get information about the process performance (e.g., via throughput times), so it's rather tailored to the business user who can use the process visibility capability to carry it out. In *technical monitoring*, you're more interested in keeping systems and processes up and running, so the focus is on administrators in the IT department. Both types of monitoring have one commonality: they use real-time data, like a control room with monitors for the cameras in the hotel or observing the technical parameters of the heating, air conditioning, and lights. In this section, we'll introduce the out-of-the-box technical monitoring capabilities of SAP Build Process Automation, as shown in Figure 3.9.

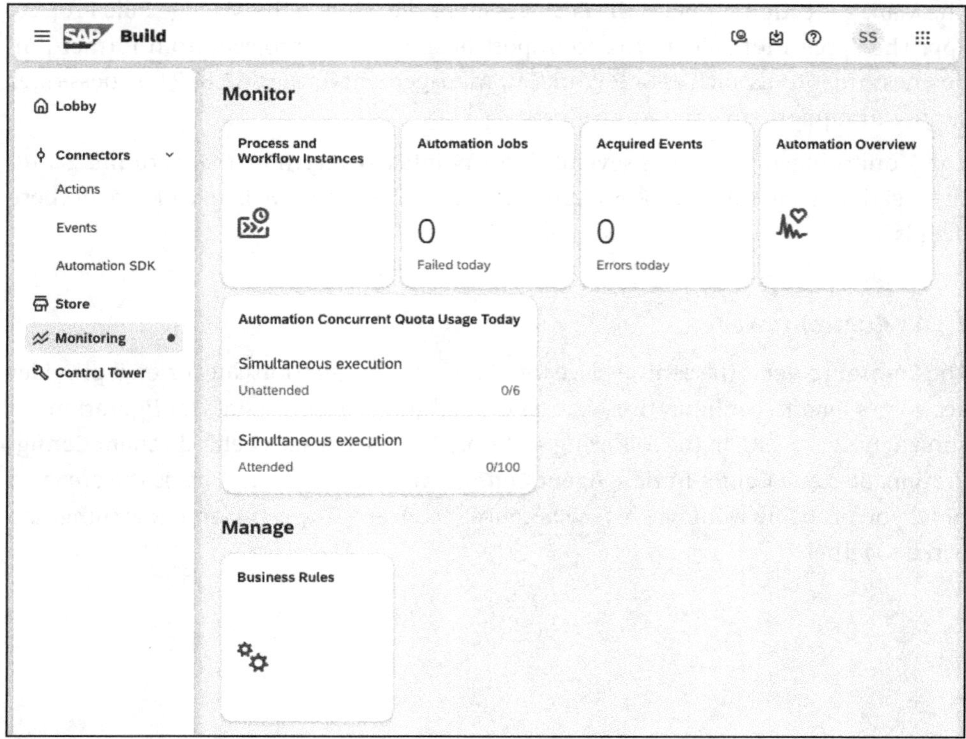

Figure 3.9 Technical Monitoring in SAP Build Process Automation

In technical monitoring, there are two areas catering to different needs:

- **Monitor**
  The technical real-time data can be consumed.
- **Manage**
  The administrator gets access to definitions.

In the **Monitor** area, you find four tiles. In the **Processes and Workflow Instances** tile, all the running process instances are listed, along with their current status and any errors. You can search for processes and filter by status, project, started by, started after, and so on. From this view, you can then drill down into the details of every running instance and even see the payload. Therefore, special user rights are also needed, that is, a dedicated role, to have access to this area. Besides the running instances, you can access the completed process instances here.

In the **Automation Jobs** tile, you get access to all the bots and the related scheduled jobs. You can also search for them, set filters, and more. The **Acquired Events** tile is needed when you're using the process visibility capability. The **Automation Overview** tile is related to the automation jobs and gives you a dashboard view of the overall executed jobs. We'll come back to these monitoring topics in detail in Chapter 11.

The **Manage** section provides the **Business Rules** tile to open the Manage Rule Projects app. The application allows you to import business rules projects from former SAP business rules tools such as SAP Workflow Management or the SAP BTP business rules service to support any migration efforts.

The Monitoring app provides several access points to everything that is running after the creation in SAP Build Process Automation. In Chapter 11, we'll take of care of these details.

3.3.4 Control Tower

The **Control Tower** is the central place for different configuration categories. It provides access to tenant configuration, backend configuration, or agent configuration, as shown in Figure 3.10. In the following sections, you'll see what's behind **Tenant Configuration, Backend Configuration**, **Agent Configuration**, and **Others**. This is the area in a hotel you probably would never see because it concerns the setup and how things are wired together.

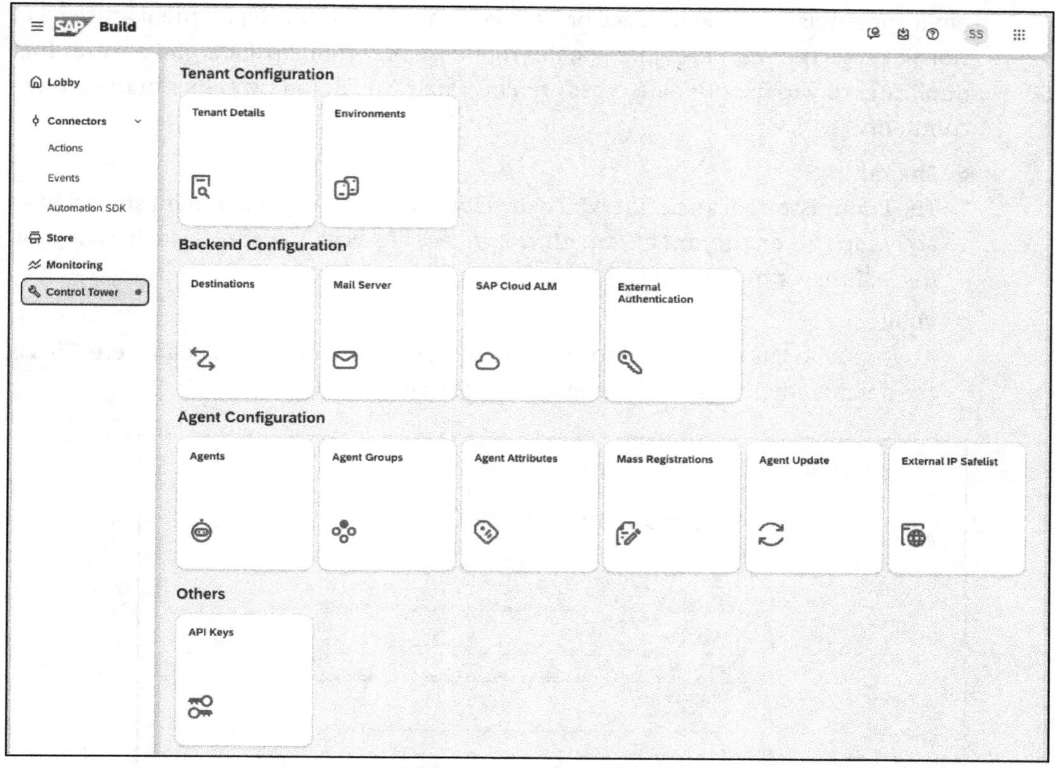

Figure 3.10 Control Tower with Different Configuration Possibilities

Tenant Configuration

In **Tenant Configuration**, you can provide some more information under **Tenant Details**. Most often, you'll work in a setup with three different tenants, one for development, one for testing, and one for production. Here you can define the type of tenant. The **Development** tenant type has no restrictions, and you can perform all operations; however, the **Test** and **Production** tenants come with some restrictions, as shown in Figure 3.11. For example, you're not allowed to create new projects in testing and production. This is quite logical, as a new project should be started in the development tenant and then be tested before it's available in the production tenant. Tenants will be explained further in Chapter 12.

Another mechanism to control access to different projects is the usage of environments. These environments can be based on departments, LoBs, industries, regions, and so on. In fact, this also provides some secure guardrails when working with other people or team members on the projects in SAP Build Process Automation. In short, the environment is a dedicated space or isolated container for the deployment and execution of projects. It also provides enables the reuse of certain artifacts and ensures that duplicates of variables or data types aren't created. You'll find two types of these environments:

- **Shared**
  The administrator can decide which developer on the particular tenant should have access to this environment to perform project lifecycle operations, such as deploying, editing, or monitoring.
- **Public**
  Anyone who has the needed roles can deploy, monitor, edit, or execute here. This is the default environment and doesn't need to be created.

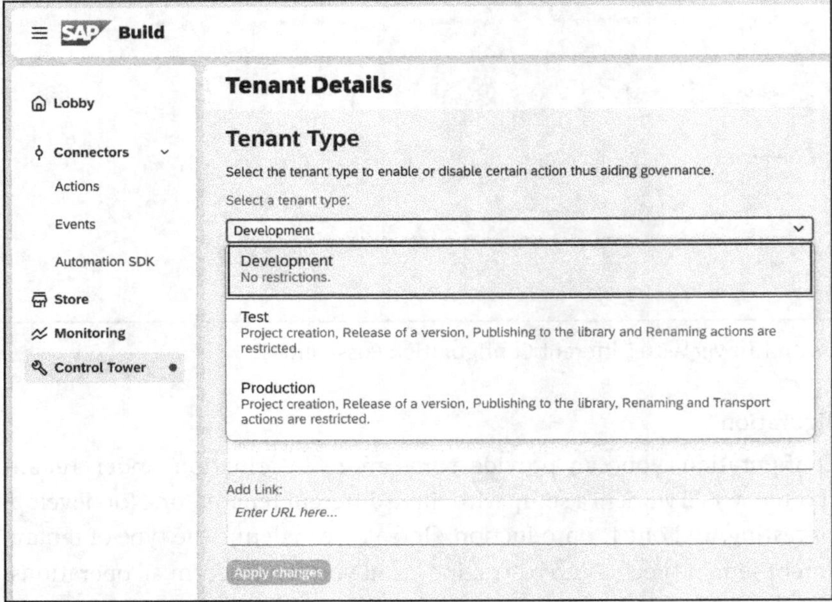

Figure 3.11 Tenant Details with Different Tenant Types

Both tenant type definition and environments of your tenant configuration are means to establish governance for your projects. We'll revisit this in Chapter 12.

Backend Configuration

Backend Configuration, as shown in Figure 3.12, provides four options: **Destinations**, **Mail Server**, **SAP Cloud ALM**, and **External Authentication**.

The destinations you can add or access here come from SAP BTP, where they have been defined. They cover information on how to connect to a system, for example, credentials, keys, and so on. In fact, these are the configured endpoints that you choose from in SAP Build Process Automation and add them to your tenant.

As you've learned, there are forms available to communicate via user tasks with the end user. But sometimes this isn't sufficient or not the right channel. For example, you want to send out an email notification to point to a specific task or just to provide an update of the current status of the process flow. With **Mail Server**, you can configure this Simple Mail Transfer Protocol (SMTP) server. Finally, you're creating another destination in SAP BTP, where you define the name and proxy type, provide a description, decide on the authentication method, and assign some further properties.

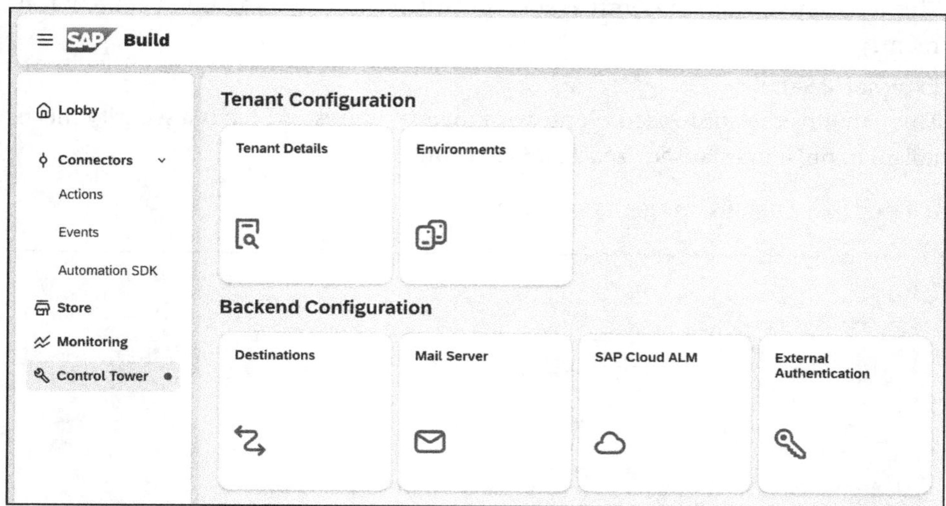

Figure 3.12 Backend Configuration

SAP Build Process Automation provides an integration with the SAP Cloud ALM platform (**SAP Cloud ALM** area) to monitor jobs.

SAP Build Process Automation not only provides different capabilities but also connects with a broad ecosystem of other products by different software partners (e.g., Google or Microsoft), especially in the area of bot automations. The authentication against these systems is maintained in **External Authentication**.

Agent Configuration

An agent, to be more precise, the desktop agent, is a component of SAP Build Process Automation that is installed locally on your desktop. It's needed to execute the automation projects that launch or run different kinds of applications, read information from screens, make data entries, click on the screen, or process data.

With **Agent Configuration**, as shown in Figure 3.13, you can manage the following:

- **Agents** and **Agent Groups**
  Here you register your local agent or your group of agents and create the connection.

- **Agent Attributes**
  This helps you better filter your agents and ensure that a specific agent executes the job.
- **Mass Registrations**
  Here, you register a bigger number of agents or agent groups and manage them later on.
- **Agent Update**
  This is used to ensure the latest agent regularly provided by SAP is in your environment.
- **External IP Safelist**
  This contains the authorized agents with their IP address, which is a security mechanism to prevent unauthorized agents to connect.

You'll find more details on agents in Chapter 7.

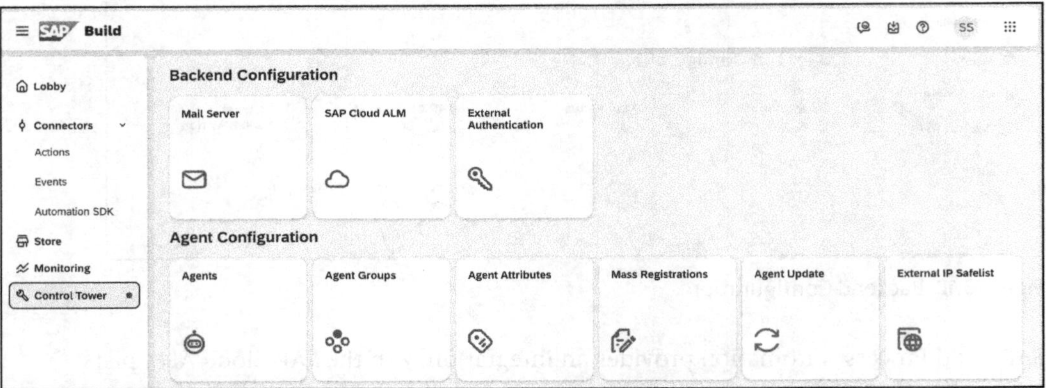

Figure 3.13 Agent Configuration

3.3.5 Connectors

As stated at the beginning of this chapter, connecting with other systems and their business processes is a very common use case for SAP Build Process Automation. The openness and ability to connect to these systems is a key capability that allows you to extend standard processes, orchestrate activities across different systems, and execute bot automations with a plethora of applications. This could either be applications for your personal productivity or other business applications. There are different types of connectors, either to perform certain actions, get information, or act on events occurring in another business system. We'll now walk you through **Actions**, **Events**, and **Automation SDKs**, listed under **Connectors**, as shown in Figure 3.14.

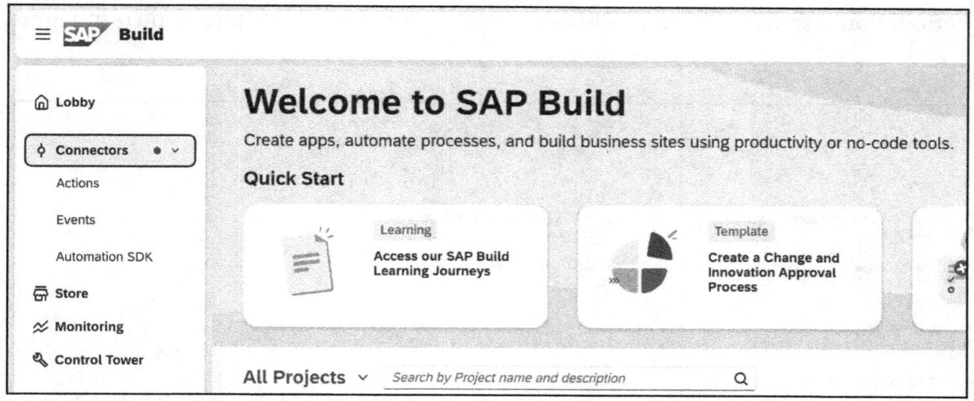

Figure 3.14 Connectors

Actions

Actions are used to connect to external systems. In Figure 3.15, you see the list of actions available in this tenant under **Items** and also the **Create** button to create new actions. The basis of every action is an API, which is the interface and the connection point to the other system. APIs come with calls, such as GET, PATCH, POST, and so on, as well as with a set of parameters that is often huge.

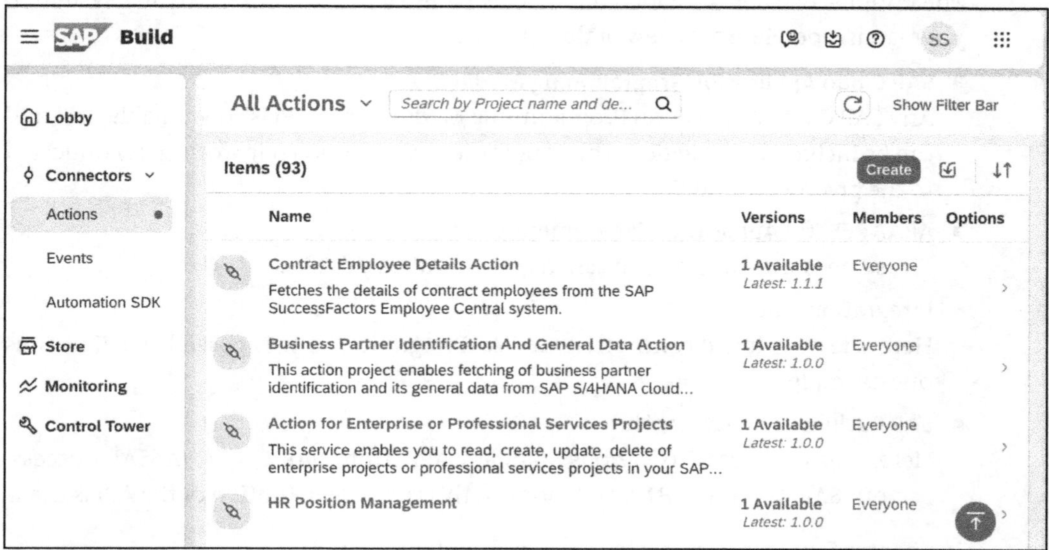

Figure 3.15 List of Actions

Within an action, you can reduce this huge set of parameters and bring it into a consumable format for the non-code/low-code development approach. The action can also be defined once and reused in several projects. When you click **Create** to start a new action, you also have several API source options to choose from, as shown in Figure 3.16.

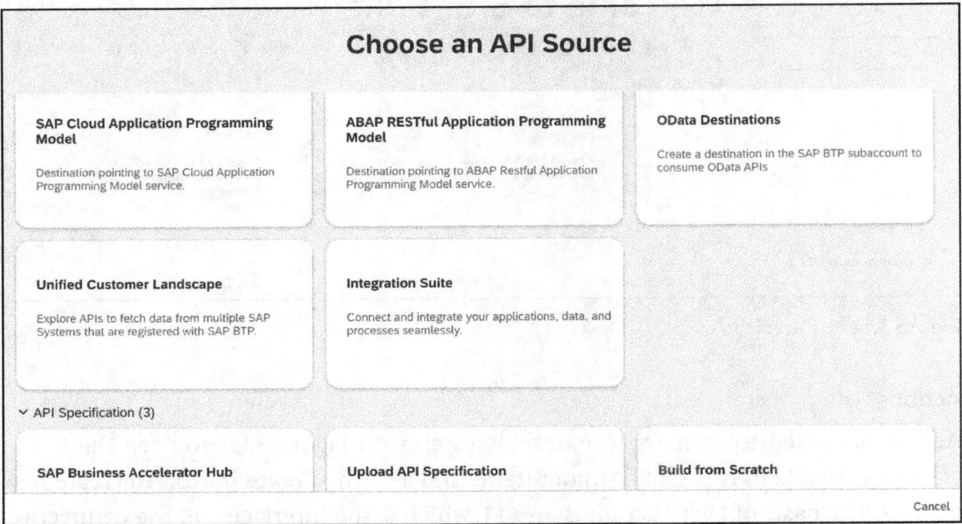

Figure 3.16 API Sources for Actions

The options available provide you the needed flexibility to connect to any system in your landscape. Here are a few of the options:

- **SAP Cloud Application Programming Model**
  Any SAP Cloud Application Programming Model–based API is consumable here and can be included in an action. This is how the low-code/no-code world gets connected to the pro-code world.
- **ABAP RESTful Application Programming Model**
  The same is true here, as you can include ABAP-based APIs.
- **Integration Suite**
  Here, the mentioned connectivity to SAP Integration Suite is possible via iFlows, as one example.
- **SAP Business Accelerator Hub**
  Here, you get access to various standard APIs from SAP S/4HANA, SAP SuccessFactors, SAP Ariba, SAP BTP, and more. In this case, the definition of the API is used.

In Chapter 6, we explain how to create and configure actions.

Events

Recall that the extension of standard processes is a key element here. Besides using actions, we also can leverage the events coming from SAP S/4HANA and others. Whenever an event is occurring, such as a status change of a business object, the information is sent to SAP Build Process Automation and can be used to trigger a process we've defined here. Due to the direct connection to SAP Business Accelerator Hub where these events are published, we can include them in the process definition. You can access the screen shown in Figure 3.17 by clicking **Create** in **Events**.

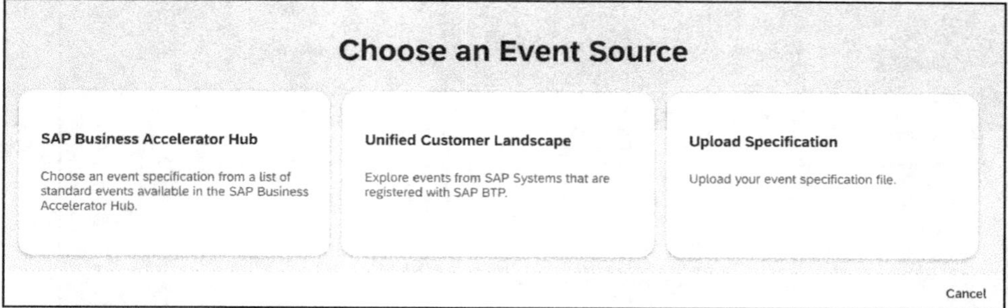

Figure 3.17 Creation of an Event

Next, you can filter based on systems—in our case, SAP S/4HANA—in the **Product** dropdown and get a list of available events that you can include in your process, as shown in Figure 3.18.

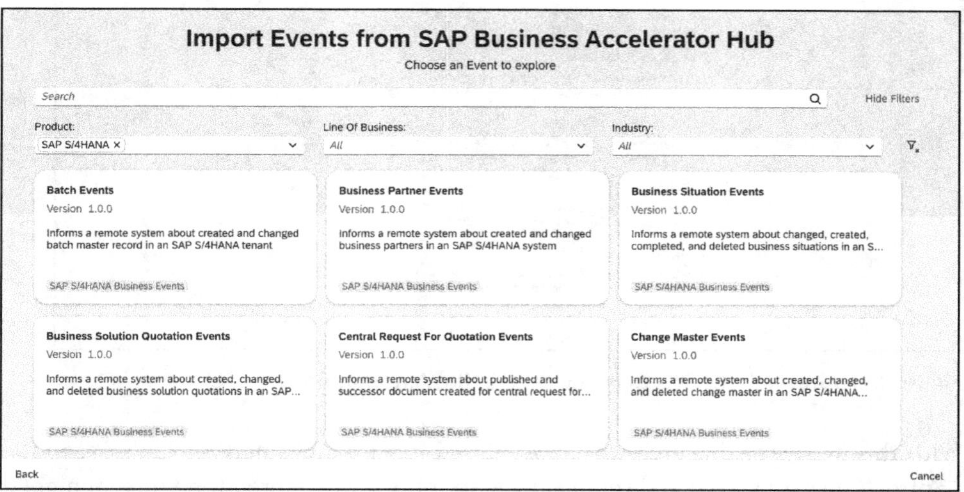

Figure 3.18 Import Events from SAP Business Accelerator Hub, Filtered for SAP S/4HANA

Automation Software Development Kits

You might have also seen software development kits in the content store, but you manage them under **Automation SDK**. These functions can be directly used in bot automations and let you perform certain activities in the related applications, such as Microsoft Excel, Google Workspace, SAP GUI, and many more. Figure 3.19 shows the list of SDKs already imported.

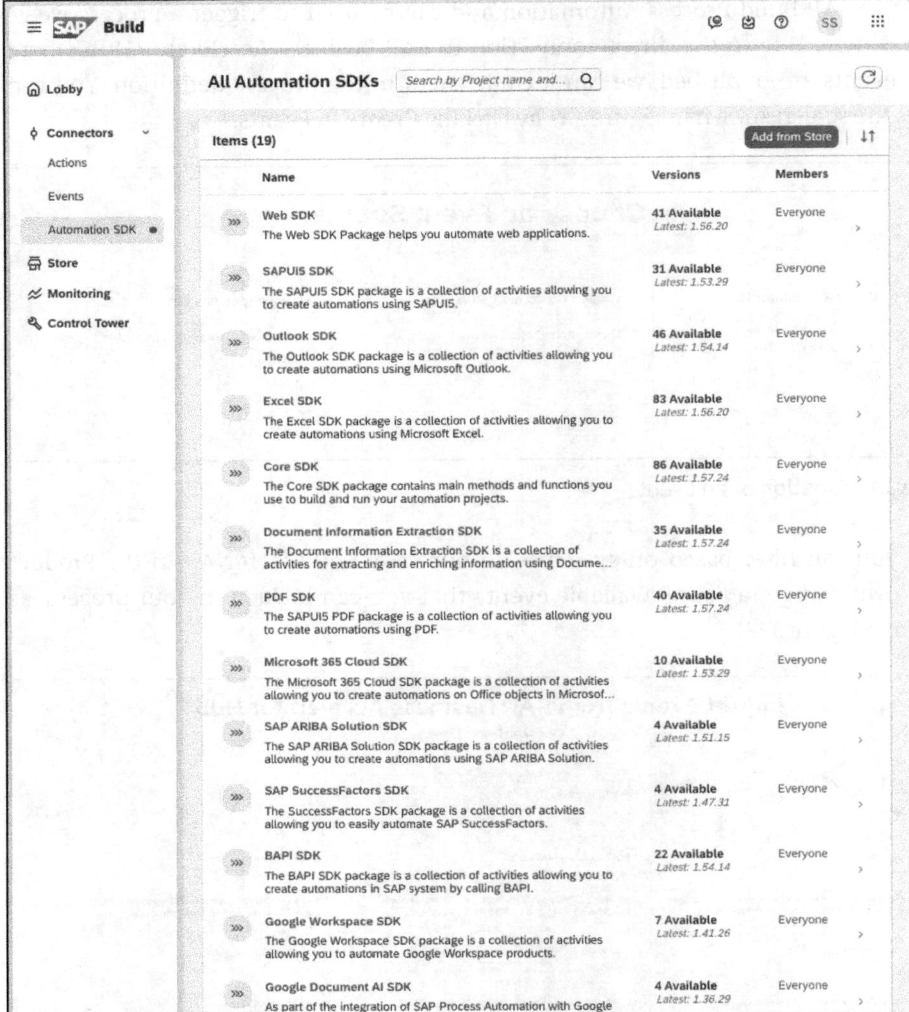

Figure 3.19 Automation SDKs to Be Included in Bot Automations

With this, we leave this area within SAP Build Process Automation. You've seen the **Control Tower**, we discussed the technical monitoring, explored the **Store**, and got a glimpse of the **Connectors**. These are mainly items meant for administrators or developers.

3.3.6 Inbox

With the inbox, also called **My Inbox**, we'll now include the end user into this introduction of SAP Build Process Automation lobby. The inbox appears in the top-right corner of the **Lobby** screen (Figure 3.20) and can be opened by clicking the icon. This gives you access to all the tasks assigned to your user (respectively, your user ID).

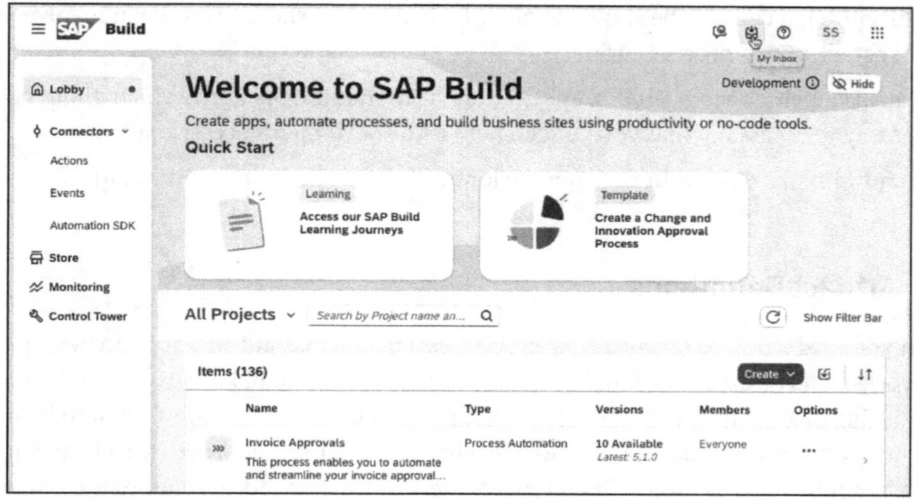

Figure 3.20 Opening the My Inbox Application

In the My Inbox application itself, shown in Figure 3.21, you'll see on the left-hand side all available tasks for you with some high-level information. You can also search for, filter based on certain attributes, or sort and group tasks here.

Figure 3.21 My Inbox Application

The right side shows the selected task, form, or UI related to it. You can also **Claim** a task. This means when a task has been sent to a group of recipients, it's normally displayed in everyone's inbox who is part of this group. By claiming the task, it will only be available for you and removed from the others inbox. Now it's up to you whether you want to **Approve** or **Reject** this task (the text of the buttons is editable), which is based on an approval. If it's just a task to inform you, you can only select **Confirm**.

In addition, a log can also be shown to get information on who else has been involved in other steps of this process so far.

You can also include the My Inbox application in SAP Build Work Zone, which might be a nicer way to access it. What you've seen here is the developer's access to the inbox.

We've now finalized our tour through the lobby of SAP Build Process Automation.

3.4 Artifact Definitions

When you're starting to create a new process extension, task automation, orchestrating process, and so on in SAP Build Process Automation, you'll make use of several artifacts during design time. In this overview chapter, we'll focus on a few of them to help you put them into context in the following chapters of this book. We'll introduce you to the business process project, data type, processes, decision, forms, automation and action.

And now: action! Well, not yet—we start with the business process project first.

3.4.1 Business Process Projects

The business process project is a kind of bracket or wrapper for all the artifacts you're creating and managing in SAP Build Process Automation, so everything is included here: processes, automations, decisions, data types, forms, actions, and even visibility scenarios. However, it's not a must to have every mentioned artifact in a project. In fact, sometimes you'll just create a process without automations or decisions. On the other hand, the project could also just include a decision or just have a task automation in the project to create the bot (see Figure 3.22).

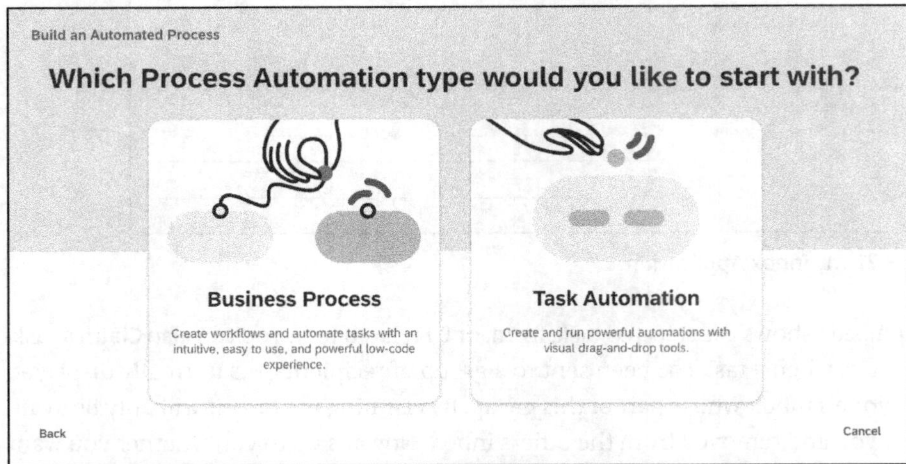

Figure 3.22 Choosing to Create a Business Process Project with All Possible Artifacts or Only a Task Automation

Once you've decided on your project and want to create it, you just need to provide a title and description if you want one (see Figure 3.23). Although not required, entering a description might be useful later to understand the purpose of this project or to help others who work with you on this project.

Figure 3.23 Creating a New Business Process Project

Once you click **Create**, the project is available, and you can start to define the required artifacts to fulfill your business requirements for this process, as shown in Figure 3.24.

Figure 3.24 Opened Business Process Project with Some Artifacts

The projects are then listed in the lobby, as shown in Figure 3.25. There, you can also perform further actions on projects:

- Publish projects to the library for download and reuse by others.
- Save projects as new with a different name.
- Export and import projects to your local system.
- Transport projects between two tenants, for example, from development to testing.

- Rename projects, or change their descriptions.
- Manage the **Members** to define who can work on the projects.
- Release projects with a version number, and make them ready for deployment.
- Delete projects if no longer needed.

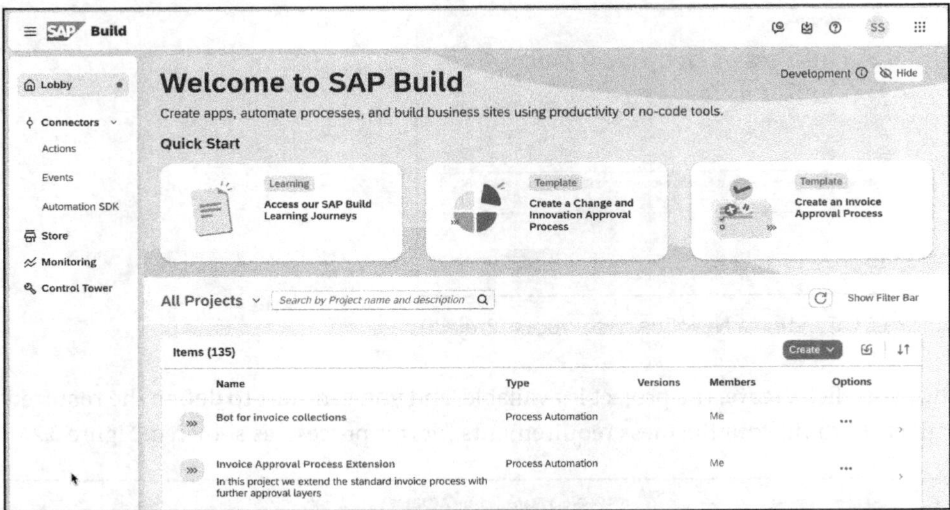

Figure 3.25 Projects Listed in the Lobby

> **Tip**
>
> Projects created with SAP Build Apps are also accessible in the lobby.

Now that you know artifacts are stored in a business process project (more details about this in Chapter 4), let's see which artifacts we're talking about. For the creation of these artifacts, no coding knowledge is required, as we are leveraging the advantages of a no-code/low-code environment.

3.4.2 Data Types

In a process, you not only need to consider the flow of steps (or, in automation, the activities of the bot) but also business data. This data can be complex. In SAP Build Process Automation, you can create your own data in data types as part of a business process project, as shown in Figure 3.26.

This approach allows you to better formalize and describe the data used and its manipulation or validation. The structure within a data type can contain different formats, for example, Boolean, number, string, date, time, and more. Sometimes, a data type can be quite complex, as shown in Figure 3.27. In this example, we see the details of a journal entry.

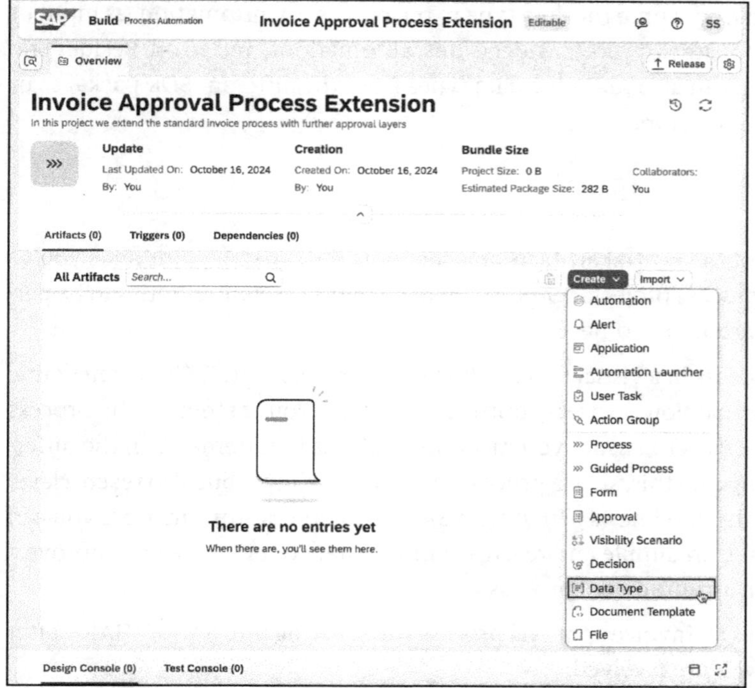

Figure 3.26 Creating a Data Type

Build Process Automation | Journal Entry Posting Approval | Editable | SS

Overview | *Journal Entry Details | Release

New Field | Other users won't be able to edit this artifact while you're editing it. | Import Excel File | Save

Name	Type	Sample	List	Required
headerDetails	Journal Entry Header Details		No	No
ApprovalStep	String		No	No
createdBy	String		No	No
postingDate	String		No	No
journalEntryDate	String		No	No
journalEntryType	String		No	No
fiscalPeriod	String		No	No
transactionCurrency	String		No	No
exchangeRate	String		No	No
refDocNum	String		No	No
currencyTranslationDate	String		No	No
documentHeaderText	String		No	No
ledgerGroup	String		No	No
approvalAmount	Number		No	No
companyCode	String		No	No
createdByEmail	String		No	No

New Child

Data Type Details

General Information

Name:*
Journal Entry Details

Identifier:*
journalEntryDetails

Description:
Data type for journal entry with header, items and attachment details

Data type is active

Strict

Information

Last Update: March 20, 2024
By:
Created:
By:

Design Console (1) | Test Console (0)

Figure 3.27 Data Type Details for a Journal Entry

Once created, you can consume the data type in the process or automation, as input or output parameters for steps, processes, decisions, automations, and so on. Besides creating data types, you can also reuse the data types from the different SDK packages or other packages from the **Store**.

3.4.3 Processes

The process is one of the central artifacts and serves as the starting point once you've created a business process project. A *process* itself is a defined set of activities to be performed to reach a certain outcome.

You work on the process in a visual no-code/low-code process editor. There you define the start triggers, actual flow, process controls, and more. You can enrich the process with all available artifacts to ensure you meet the business requirements in the different scenarios. Having said this, such a process can be quite simple but also reach a level of complexity. But the good news is that with SAP Build Process Automation, you can tackle both. You can start simple and reach a complex state or even start to improve a complex scenario in an automated process.

Figure 3.28 illustrates an invoice approval process with posting into SAP S/4HANA after the required approvals are received.

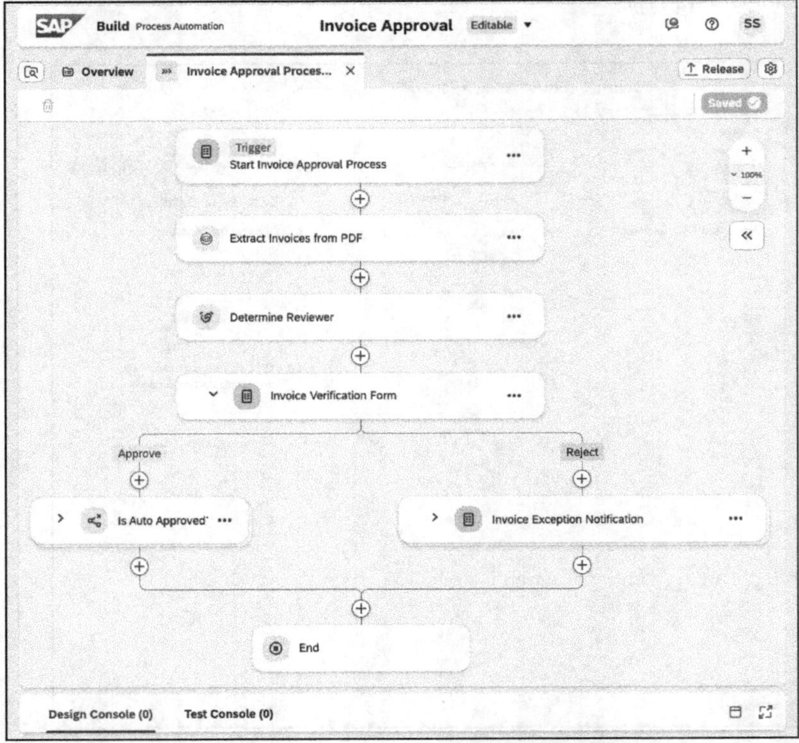

Figure 3.28 Invoice Approval Process with Posting in SAP S/4HANA

This is a typical extension of a standard business process. Before you post the data of the invoice for further processing in SAP S/4HANA, you want to perform some checks and approvals. In this process model, you'll find a trigger to start the process, this could be an API trigger, an event, or a form. The first step is then a bot or task automation to extract data from a document in PDF format. (Other formats can be used as well, but this is just an example.) Next, you need to identify who should review the extracted data before moving to the next step, where the actual review is taking place, in a form. This helps to ensure proper execution moving forward and also that all important information is available.

Depending on the result, approved or rejected, there are now two possible flows. If the review leads to rejection, another form will be used to inform the requester of the identified issues and ask for corrections to be made and the invoice to be resubmitted. If the invoice review comes to a positive result and is approved, it goes another route with further steps, forms, decisions, and so on, as you can see on the expanded process view in Figure 3.29.

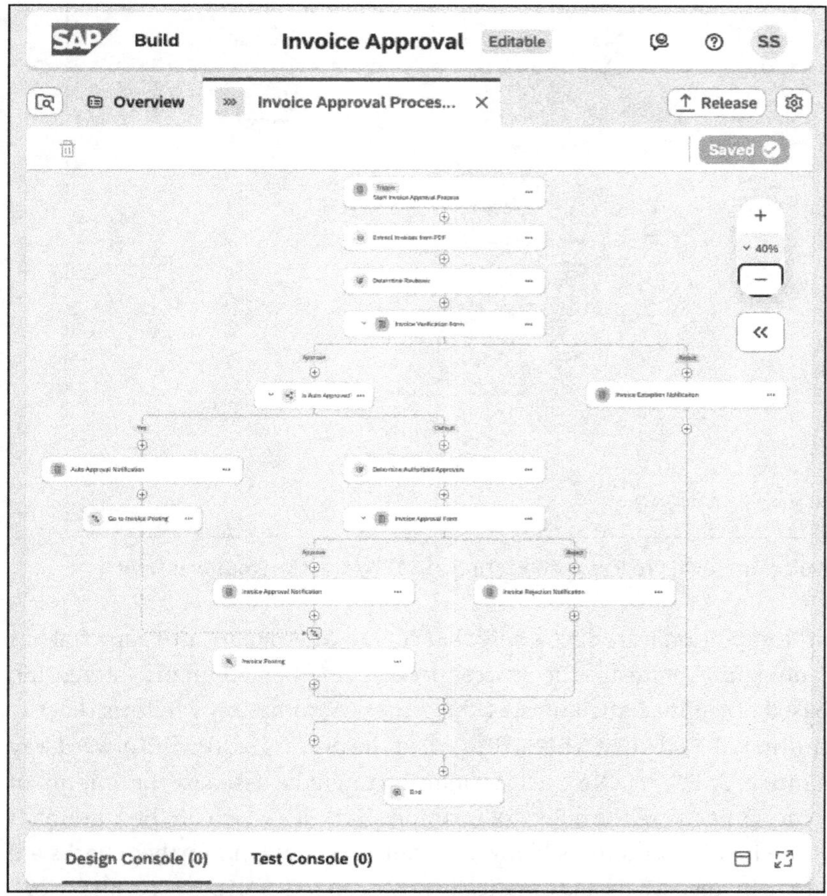

Figure 3.29 Invoice Approval Process with Posting in SAP S/4HANA: Complete View

In this view, we've now expanded the flow that comes after the approved invoice. It's getting a little more complex now, which is normal. SAP Build Process Automation provides several measures to handle increased complexity, which you'll learn about in Chapter 5.

Let's now follow the flow after the approval in Figure 3.30. Here again, we have two possible flows.

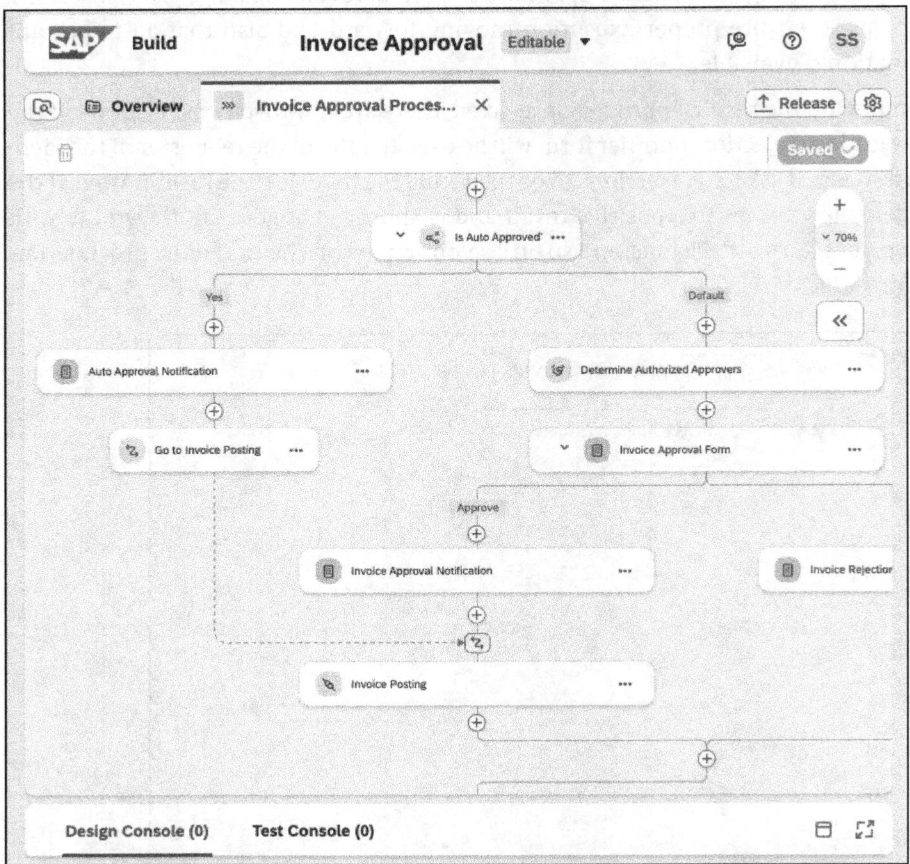

Figure 3.30 Invoice Approval Process with Posting in SAP S/4HANA: Approval Flow

This is due to the condition introduced here that checks whether an auto approval has occurred. An auto approval means the process decides whether a human intervention is necessary based on defined attributes, or the process automatically follows the way until the posting in SAP S/4HANA. This is the goal of the process: a straightforward way to post the invoice in SAP S/4HANA. Such a condition might be based on the amount of the invoice or the company of the biller, for example. Here, it's based on the amount of the invoice: if the invoice amount is below a certain threshold, no further checks are necessary, and the invoice will be automatically posted in SAP S/4HANA via an action.

If this isn't the case, a manual approval is required. To include different parameters to assign the right person for the approval, we again use a decision. If you're wondering where all the data is coming from, it's stored in the process context, extracted from the invoice at the beginning and available everywhere in the process, including the data types, which are part of the process context.

Once the approver has been selected, a form is now used to display the approval data with the action to either approve or reject this invoice. With rejection, a notification is sent out, and the process ends. In our example, we are using a form for this, but it could also be an email notification. If the invoice is approved, a more cheerful notification is sent: the approval notification. Now it's time for the final step when the process goes this route: the posting in SAP S/4HANA.

To summarize, the process started and extracted data from an invoice. This data got checked for potential errors and reworked if errors were found. If everything is fine, the process checks whether approvals are necessary, depending on the amount of the invoice. If no approvals are needed, then the direct posting in SAP S/4HANA occurs. If approvals are needed, then the system checks who could be an approver for this invoice and sends this information. In addition, the user can decide to approve or reject. If the user rejects, the process ends and a notification of the rejected invoice is sent. If the invoice is approved, a notification is also sent, and the posting in SAP S/4HANA occurs. With this, the process ends, and we move back from a process extension to the standard business process in the business application: SAP S/4HANA.

In this invoice approval process, we've seen several artifacts without really explaining what they are good for. We'll now provide you with an overview of each, although the details are reserved for upcoming chapters, starting with Chapter 5.

3.4.4 Decisions

The decisions are what gives processes the extra something. Here, you can maintain any kind of policies, regulations, laws, special offers, specific customer handling, and much more. You'll answer questions such as which customer gets which discount, who should approve any additional discounts, and whether further offerings should be sent to this customer. You can transfer these to any other scenario or process. Acting dynamically and based on the business data makes the difference here.

Decisions are collections of one or more business rules that use input parameters and provide a result at the end that is then processed further. By the way, for input and output, you're using data types. Let's revisit one of the decisions from the invoice approval process in Figure 3.31.

You can embed a decision anywhere in the process where it's needed. In the decision editor in Figure 3.32, you'll see a decision diagram, a graphical representation of the flow of your business rules within the decision artifact.

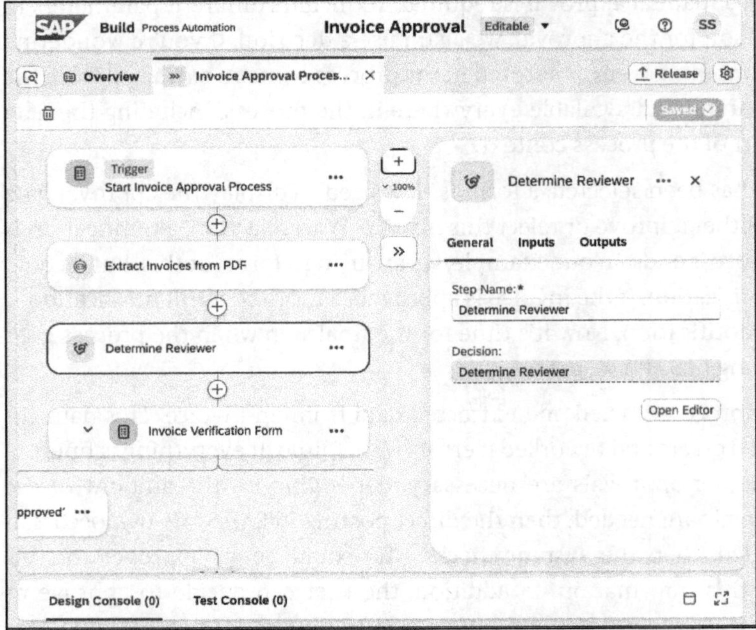

Figure 3.31 Decision with Input and Output Parameters in the Invoice Approval Process

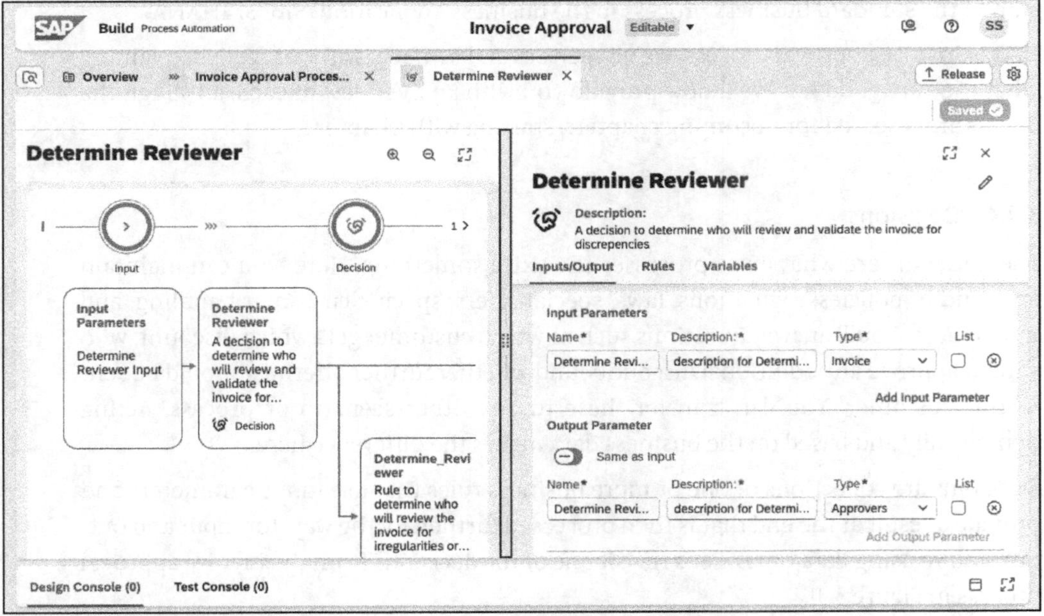

Figure 3.32 Editor to Define a Decision and Add Rules

In this case, the diagram isn't that complex, as we only have one rule here, though it could be more. In general, you have two possibilities to create a business rule:

- **Text Rule**
  You'll use this when you want to express rules in the if-then format.
- **Decision Table**
  This is the place where you maintain multiple rules expressions, as in Figure 3.33. Several input attributes, starting top-down, contribute to different possible results, which will then be stored in the data type for the output.

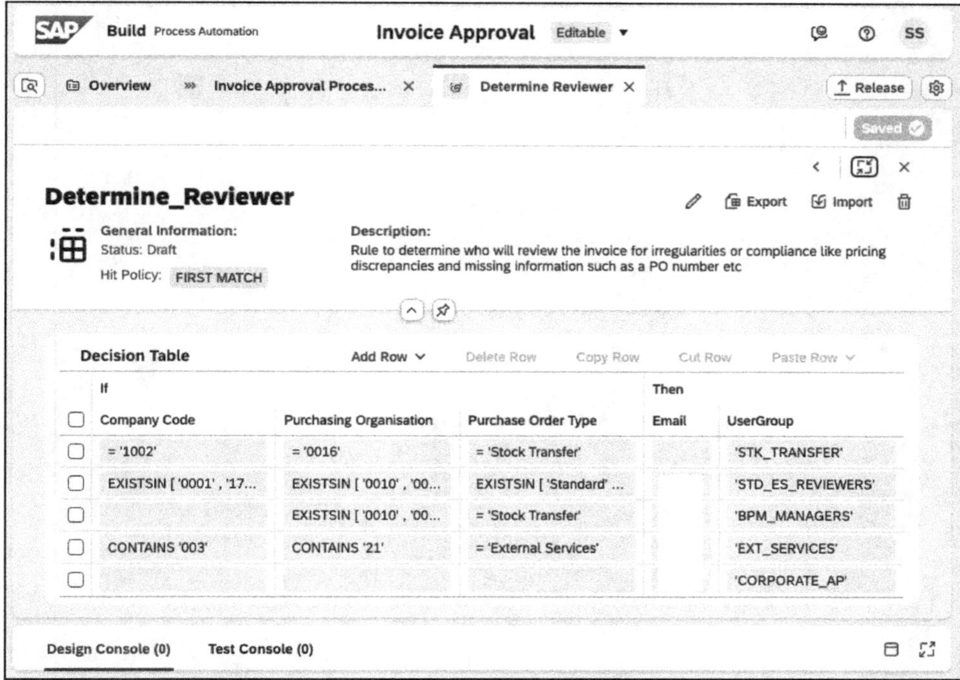

Figure 3.33 Decision Table to Define Who Should Review the Invoice Based on Different Attributes

You can also export such tables as spreadsheets in Microsoft Excel format and import them. In Chapter 8, we'll discuss this and other topics regarding decisions.

3.4.5 Forms

In our investment approval process, we've often used a form to communicate with the end user either to collect some input data or display information. You've likely recognized two different types of forms used here: the approve/reject form where we've asked the user to specifically approve or reject the request, and the notification form that provides information regarding what has happened so far. There's also a third form you can use to trigger the process. The three different types are outlined in Figure 3.34, but they should all have one thing in common: be easily consumable by the end user with a delightful UI.

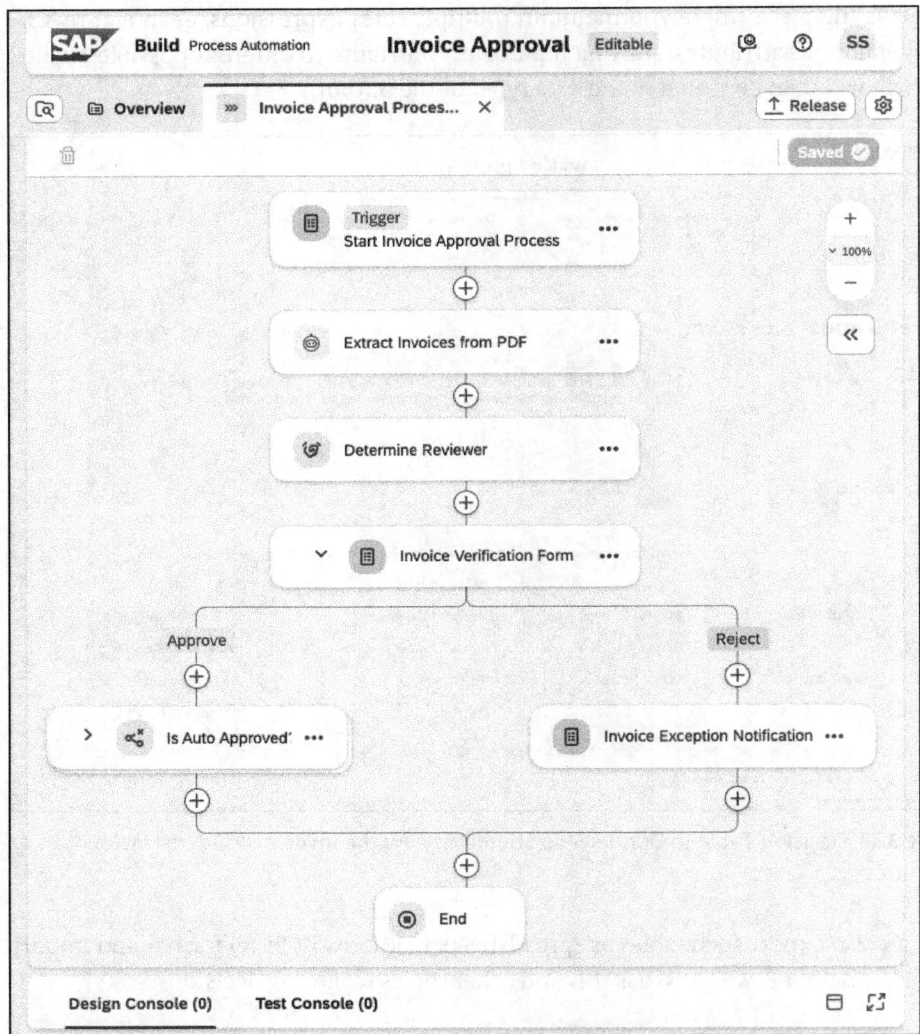

Figure 3.34 Forms in the Process to Interact with the End User

The form to trigger the process and the form to notify about the invoice exception have only one branch. This means, the end user can only confirm the display information or click the button to move on with the next steps in the process. This becomes clearer in the form builder, as shown in Figure 3.35.

In our example, we only have one button assigned, but you have the flexibility to add additional buttons and label them accordingly. This leads to several different branches.

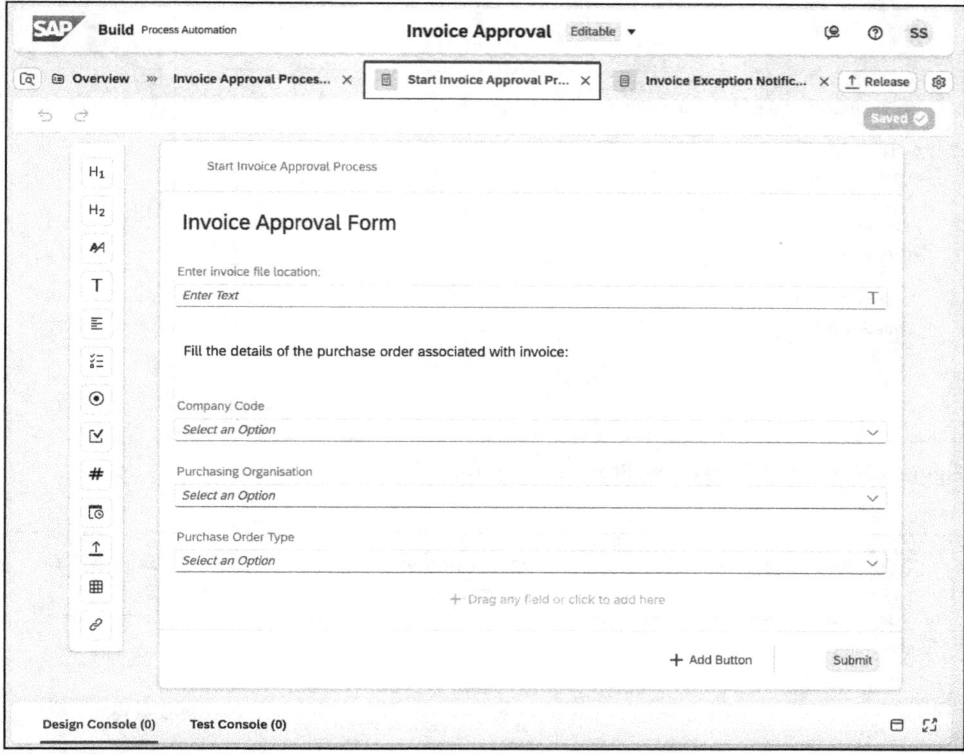

Figure 3.35 Form to Trigger a Process

> **Note**
>
> If you design a form to start the process, make sure to include only one button with the purpose to start the form. You don't want to include several buttons because no matter what is selected, the process will be started.

In addition, at runtime, you can either open the form that triggers the start of the process via a URL, as shown in Figure 3.36, or you could include this as a tile in your personal workspace created with SAP Build Work Zone. Some dropdown lists have been included in this trigger form as well. The values can either be hardcoded or derived from a data source, for example, an SAP S/4HANA system.

On the left-hand side of the forms editor, you see different layout and input elements, as shown in Figure 3.37.

Invoice Approval Form

Enter invoice file location:

Enter text

Fill the details of the purchase order associated with invoice:

Company Code

(Not selected)

Purchasing Organisation

(Not selected)

Purchase Order Type

(Not selected)

Submit

Figure 3.36 Form to Trigger the Process at Runtime

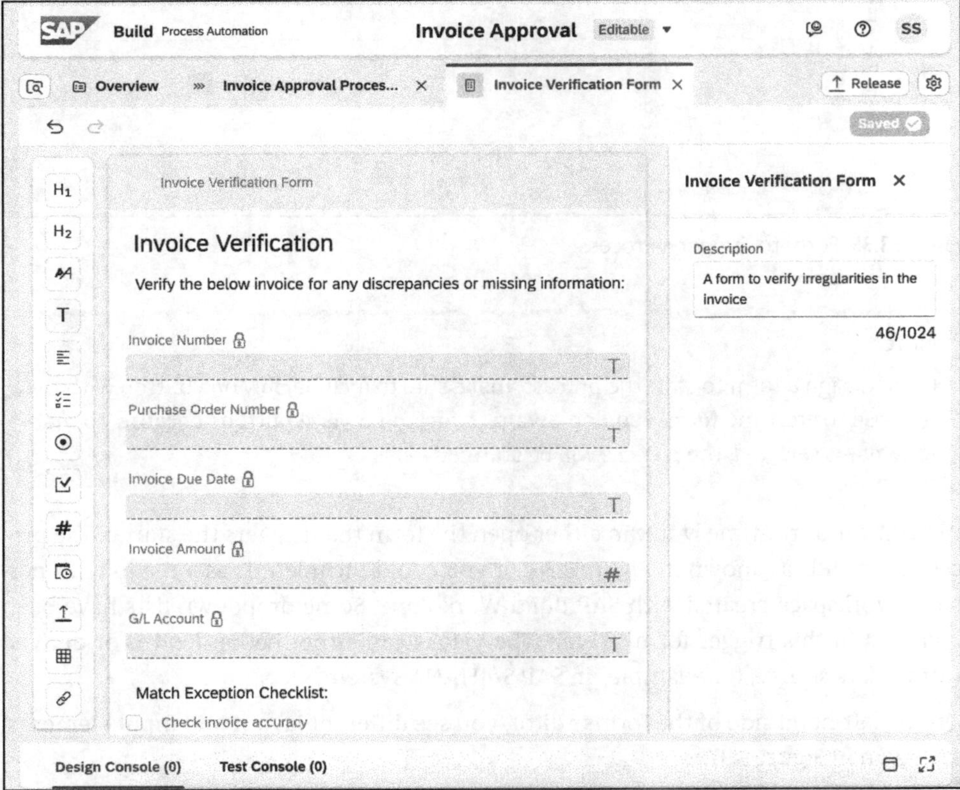

Figure 3.37 Input and Layout Elements in the Forms Editor

How you design your form is completely up to you. There are layout elements, such as headers or paragraphs, as well as a set of input elements: text, text area with more space, dropdown that can be connected to a dynamic data source or include hardcoded entries, choice, checkbox, number, date, file upload and attachment, tables, and link. Some of the fields you see in Figure 3.37 are grayed out and have a padlock icon to signal that these can't be edited by the end user at runtime but will only display information. You also define this in the forms editor for every input field.

Besides the forms editor, in the process editor, you also can define some form attributes for the related process step. The process step is the executable representation of the form, or vice versa, the form is the graphical representation to the end user of the related process step. In the process step, you define potential recipients of this step that can also be dynamically added depending on the entries, as shown in Figure 3.38, where we've chosen **Email** from the process context.

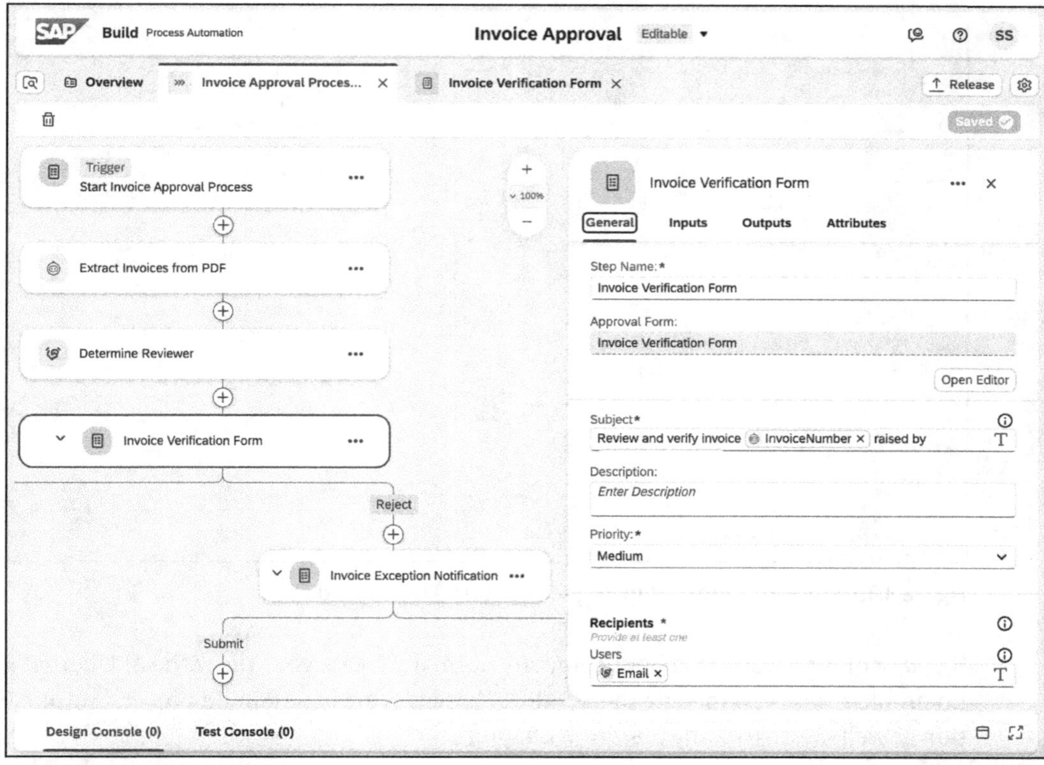

Figure 3.38 Process Step for the Form

Inputs and **Outputs** are related to the definition in the form. In our example, the **G/L Account** field is only there because it has been defined in the form before (compare to Figure 3.37 shown earlier). The same is true for all other fields. We've already discussed the process context, when we explored data types. The process context is also displayed here and includes all data and data types we need.

As shown in Figure 3.39, from the task automation **Extract Invoices from PDF**, we selected the data type **InvoiceDetails** and the entry **GLAccount** and then created the mapping to the **G/L Account** field in our form. We also used the easy drag-and-drop functionality, which is one of the main characteristics of SAP Build Process Automation.

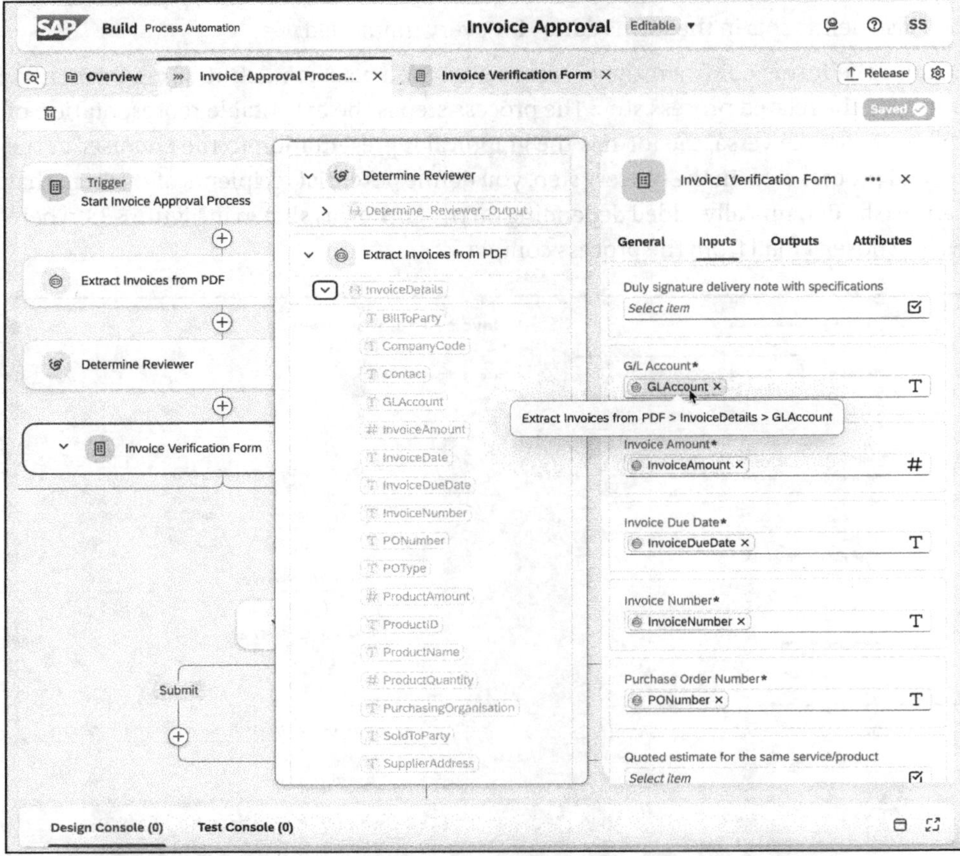

Figure 3.39 Process Context Mapping to the Form's Input Fields

If you or other designers at your company normally create your UIs with SAPUI5, that's no problem—you can include these SAPUI5-based forms in SAP Build Process Automation as well (more to come on this in Chapter 5).

3.4.6 Automations

Remember the boring task described at the beginning of collecting hundreds of individual sales orders in a spreadsheet and maintaining them in your ERP system manually every day? We're talking about repetitive, rule-based tasks, which are time-consuming and, as mentioned, really boring. They also invite errors due to their nature.

For these tasks, you act like a robot, so why not let a bot do the work via task automation?

We come back to our earlier example of the invoice approval process. After the process has been triggered with a form, we immediately use an automation to extract data from the provided invoice (see Figure 3.40), leveraging built-in AI capabilities for document information extraction. We'll cover this in more detail in Chapter 10.

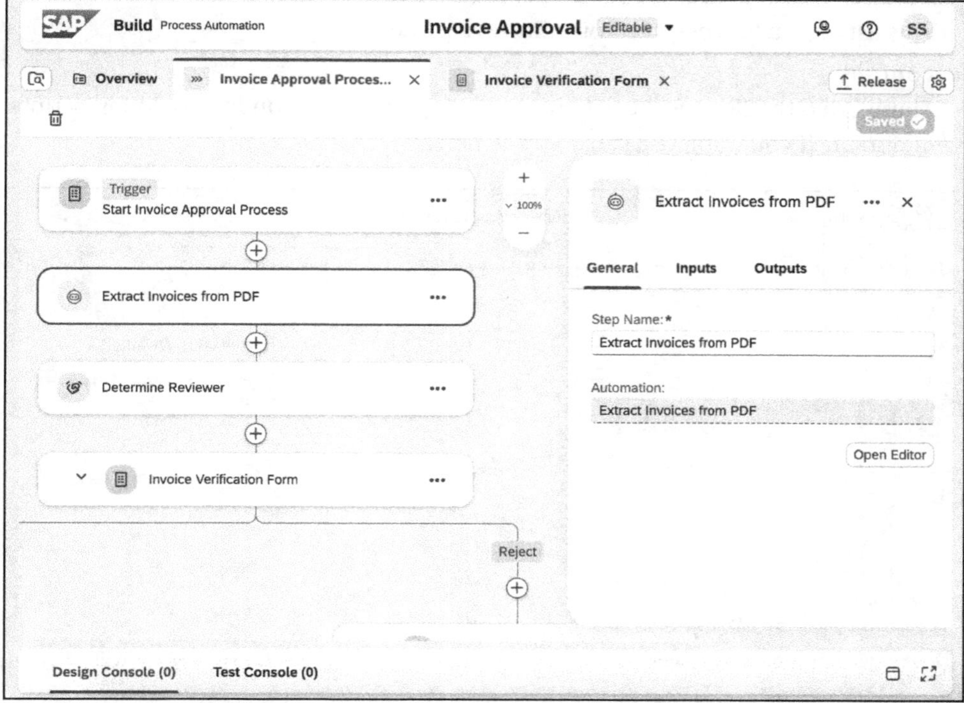

Figure 3.40 Task Automation to Extract Invoice Data from a PDF Document

The inputs you can define here are, again, data types. But let's take a closer look at the automation itself in Figure 3.41.

In our example, the flow is straightforward, although you have the flexibility to include loops, branches, and so on. Therefore, within an automation, you can also create a flow of different steps. Just to get a better understanding, let's browse through the tools available on the right side:

- **Action Groups**
  These allow you to include predefined actions directly in your automation. Actions are used to connect to other applications.
- **Automations**
  These are the already-created automations in the current flow of activities.

- **Activities**
  These come with the SDKs within your project and are predefined; for example, **Open PDF** and **Extract Data** are activities.
- **Processes**
  These are the already-created processes in your project for direct integration in your automation.
- **Data**
  This lists the data types that you'll find everywhere.
- **Controls**
  These drive the flow of the activities, for example, with conditions, loops, scripts, and so on. If you want, you can use code here.

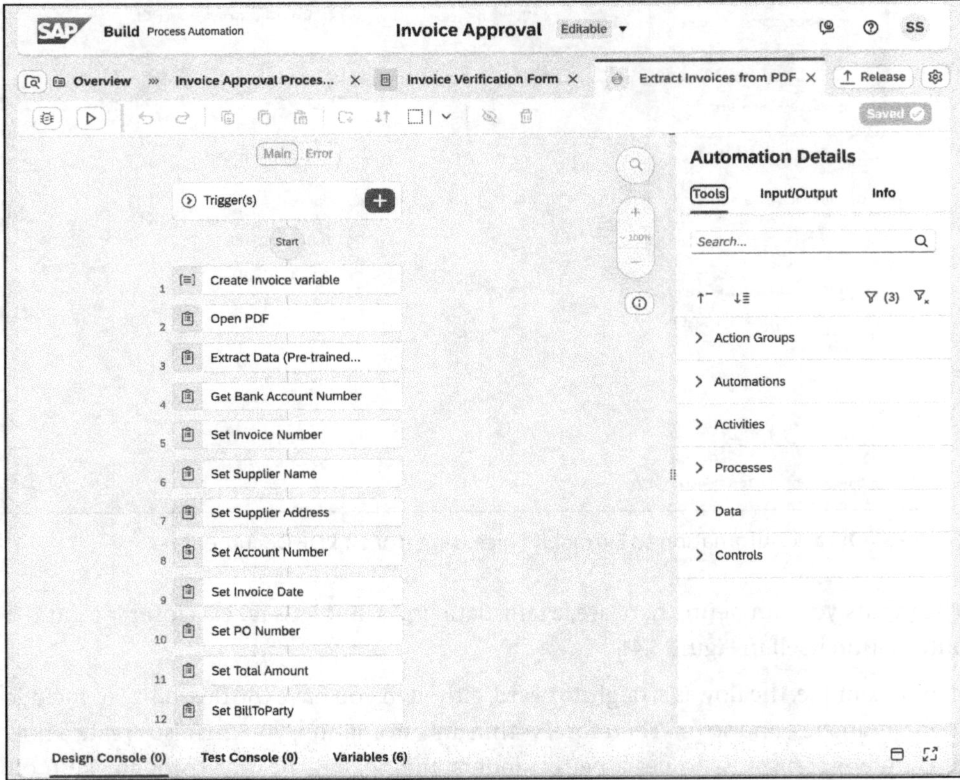

Figure 3.41 Automation Editor

Only the tools where content is already available are listed here. What is missing in our example are screens, where you capture the actions you perform on the screen with any application via screen recording. The automation editor also contains a test environment, which is useful to ensure the bot is doing the right activities and delivers the required results.

Finally, automations are differentiated into attended and unattended. Attended automations are started by a user and are supervised. In contrast, unattended automations are completely autonomous, and no user intervention is needed or required.

Automation is a powerful capability within SAP Build Process Automation. You'll learn more about this in Chapter 7.

3.4.7 Actions

We've discussed actions already to some degree in Section 3.3.5. Our focus there was on creating actions, but now we'll check how to include them in a process. Remember, actions are used to connect to external systems.

Let's revisit our example of the invoice approval process. The goal of this process was to post invoice data into SAP S/4HANA after approvals are done. This is where we use an action, as shown in Figure 3.42.

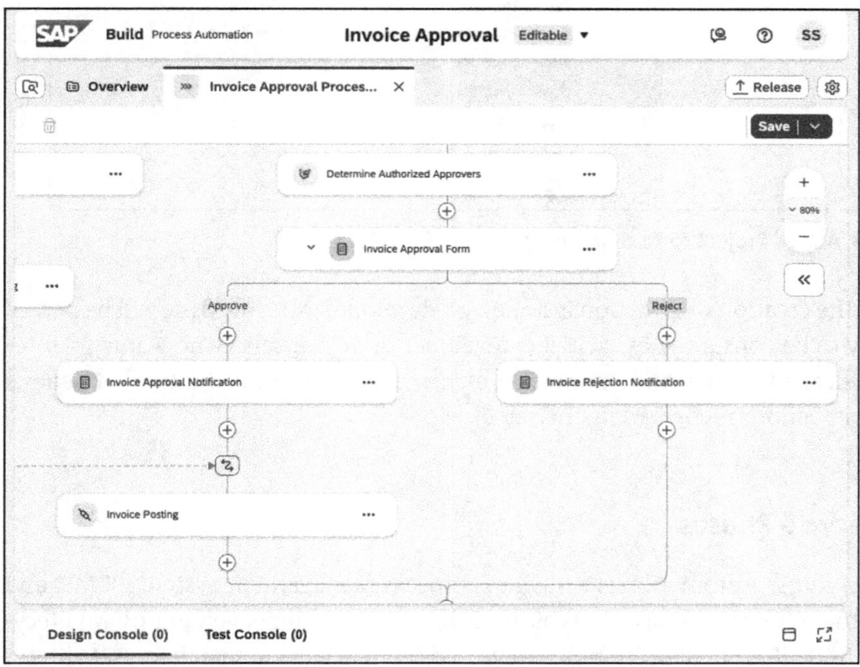

Figure 3.42 Action Invoice Posting as Part of the Invoice Approval Process

The input of this action (**Invoice Posting**) is based on the data types in the process context. The information you post in SAP S/4HANA is up to you, or to be more precise, it also depends on the API. When we open the action project shown in Figure 3.43, we see the information that will be transferred. You define this during the creation of the action project and consume this action wherever you need it, for example, in a process or in an automation.

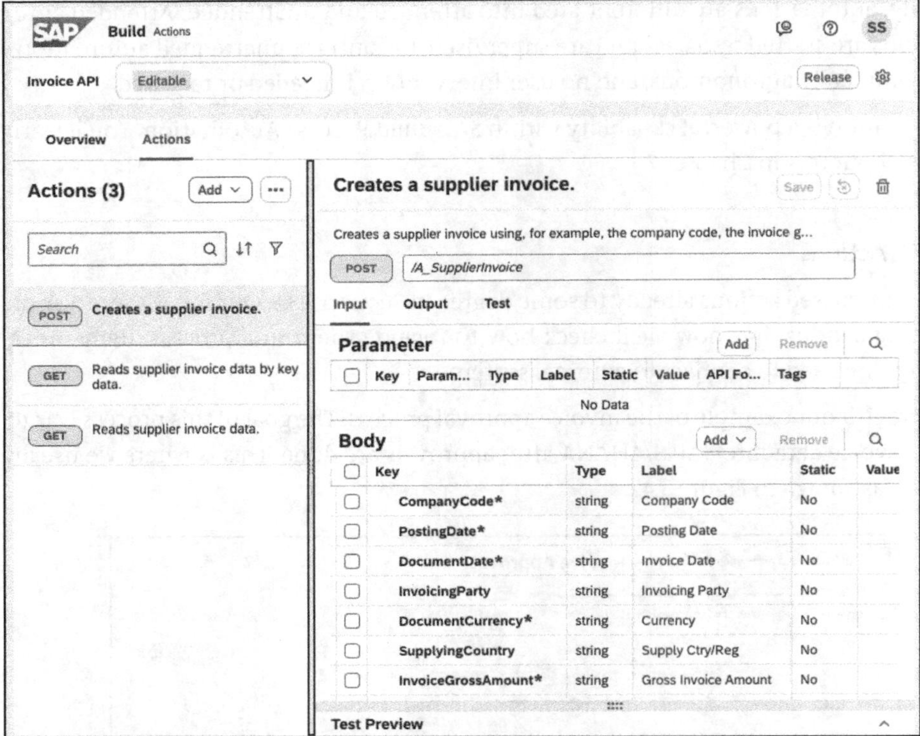

Figure 3.43 Action Project to Post an Invoice in SAP S/4HANA

Typically, the creation of an action is done by a developer. Afterwards, it will be part of the library to be consumed by a citizen developer or anyone else who wants to automate business processes and connect to external systems with the help of actions (more about actions is covered in Chapter 6).

3.5 Lifecycle Phases

SAP Build Process Automation is a business process management system (BPMS) and supports the different phases of a typical business process management (BPM) lifecycle. You'll find different definitions and levels of detail when you search for BPM lifecycle and probably also to some degree slightly different answers when you discuss the topic with BPM experts. However, we most likely can agree on the lifecycle shown in Figure 3.44 and use SAP Build Process Automation to support them. Not shown here is the development lifecycle, so starting in the development system, the processes are transferred to the test system and then executed in the production system.

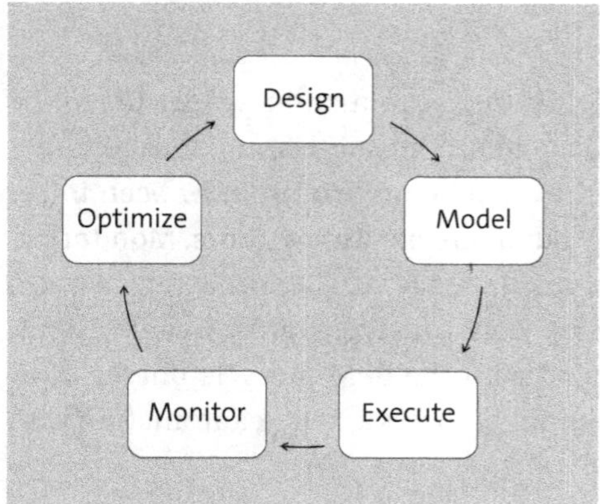

Figure 3.44 BPM Lifecycle

The BPM lifecycle has the following stages:

- **Design**
  You start with the first design of your process, discuss this, and collaborate with different stakeholders based on defined business requirements. In this phase, the actual technical details aren't needed, so sometimes you start with placeholders here.
- **Model**
  Now you get more clarity, add the needed details, and make the process ready for execution. Details such as actions, automations, business rules, and so on will be added, and you complete any activity to make the process ready for the next stage.
- **Execute**
  The process is up and running in the production system. End users are using the process in their day-to-day business and gaining the planned benefits.
- **Monitor**
  Constant technical monitoring and business monitoring are the focus of this phase. You gain valuable information about potential technical issues and also about performance.
- **Optimize**
  You analyze your process. Is it running as expected? Does it fulfill the planned and agreed goals? What feedback are you getting from end users and stakeholders? All of this information will help you optimize your process and get as close as possible to the business needs and confident stakeholders.

Every phase has its own challenges and successes. SAP Build Process Automation helps you master the challenges and come faster to the successes. The flexibility and low entry barrier provided with the low-code/no-code approach will support this. In addition, the openness to include assets from pro developers and include them as well as the business process experts in every phase of the lifecycle is a game-changer.

3.6 Summary

In this overview and introduction to SAP Build Process Automation, we discussed the purpose of this service in SAP Build, outlined the main capabilities, and looked under the hood to have a better understanding of the architecture. You've now been introduced to the different applications on the SAP Build screen—**Lobby**, **Store**, **Monitoring**, **Control Tower**, **Connectors**, and **Inbox**—which you'll see again in later chapters. We also covered the artifacts, such as business process projects, data types, processes, decisions, forms, automations, and actions. We concluded with the BPM lifecycle phases. Now you have a general understanding of how to simplify process and task automation with SAP Build Process Automation.

With this, you're well equipped to dive deep into the details of SAP Build Process Automation, starting with how to create a business process project in Chapter 4.

Chapter 4
Creating a Business Process Project

A business process is a repeatable process that consists of a series of tasks that generally need to be completed in a specific sequence. It refers to any activities that must be performed to complete an end-to-end business scenario. A business process project enables business users to create automations or workflows and then contribute their process expertise to projects to deliver the solutions they need without having to write code. These solutions help organizations close specific automation gaps or create processes that help them differentiate their business in the market.

From this chapter onward, you'll learn about the components of SAP Build Process Automation in more detail by following step-by-step instructions to create and use them. In this chapter, you'll learn to create a business process project and learn about the features of the business process project, which is the wrapper for all artifacts in SAP Build Process Automation. You'll become familiar with the navigation, settings, and configurations involved in the business process project, and usage of the right component in the project.

4.1 What Is a Business Process Project?

The *business process project* in SAP Build Process Automation is a starting point that allows citizen developers to create, configure, and manage automated business processes. It serves as a central place where all the elements needed to automate a business process come together. You can design, test, and deploy automated workflows that streamline and enhance business operations.

The following lists the business process project features:

- Maintains the entire lifecycle of a process from initial design and testing to the deployment and continuous improvement of the process
- Allows collaboration with several team members
- Provides version control where users can test their previously deployed versions of the project

- Publishes the project to the library, which allows the reuse of artifacts of other business process projects by adding them as dependencies
- Publishes the project to SAP Build content store so it can be consumed as a standard template
- Can be exported and imported to different tenants
- Can be managed from the lobby of SAP Build or from the overview of the project in SAP Build Process Automation

4.2 Create a Project

Business process projects are created from the lobby of SAP Build, as shown in Figure 4.1.

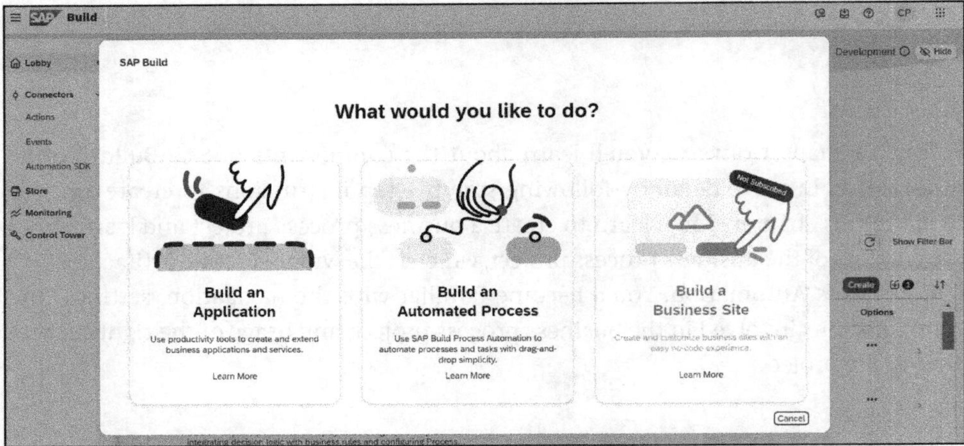

Figure 4.1 Creating a Business Process Project

You can automate the workflows or automations easily via drag and drop. As shown in Figure 4.2, you can create a process or an automation artifact directly from the lobby after providing the project name and description.

You can create process, form, automation, decision, process visibility, and action artifacts, which is explained in Section 4.2.1 to Section 4.2.4, as well as many other components such as document templates, data types, and automation launchers to start your automation, which will be explained in detail in Chapter 7.

The business process project can have one of the following as its only artifact: a process artifact, a decision artifact, an automation artifact, or a process visibility artifact. After successful creation of the business process project, a typical project looks like Figure 4.3.

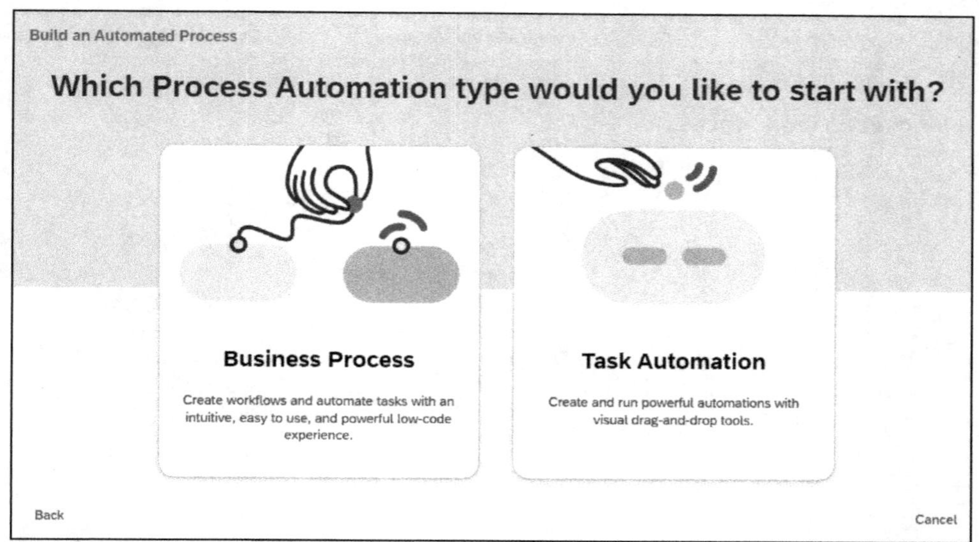

Figure 4.2 Creating a Process or Automation

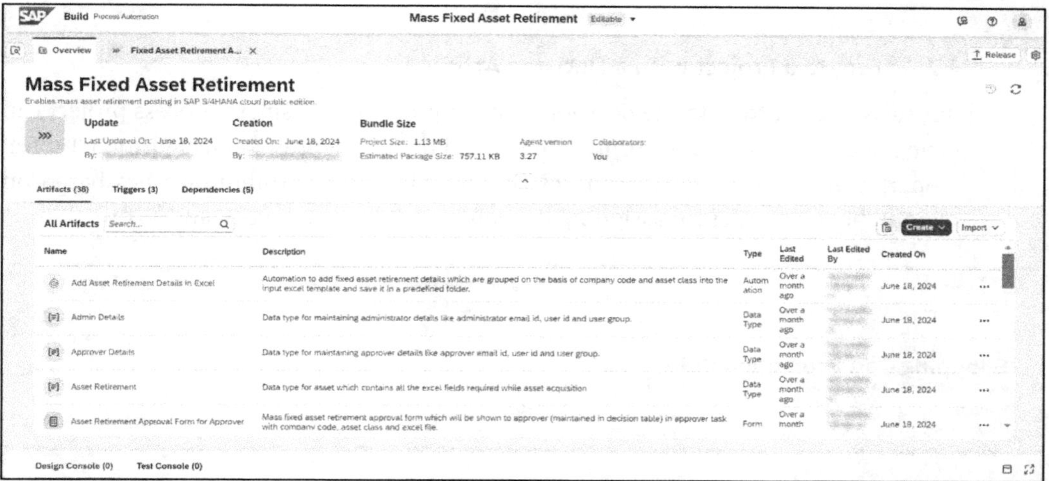

Figure 4.3 Sample Business Process Project

4.2.1 Create a Project with a Process Artifact

Processes are flowchart-like blueprints that define the orchestration of tasks. A business process project can have a process artifact as the only artifact, as shown in Figure 4.4. Typical scenarios include approving a capital expenditure, approving a purchase order, processing a bill payment, hiring an employee, requesting a replacement part, and many more.

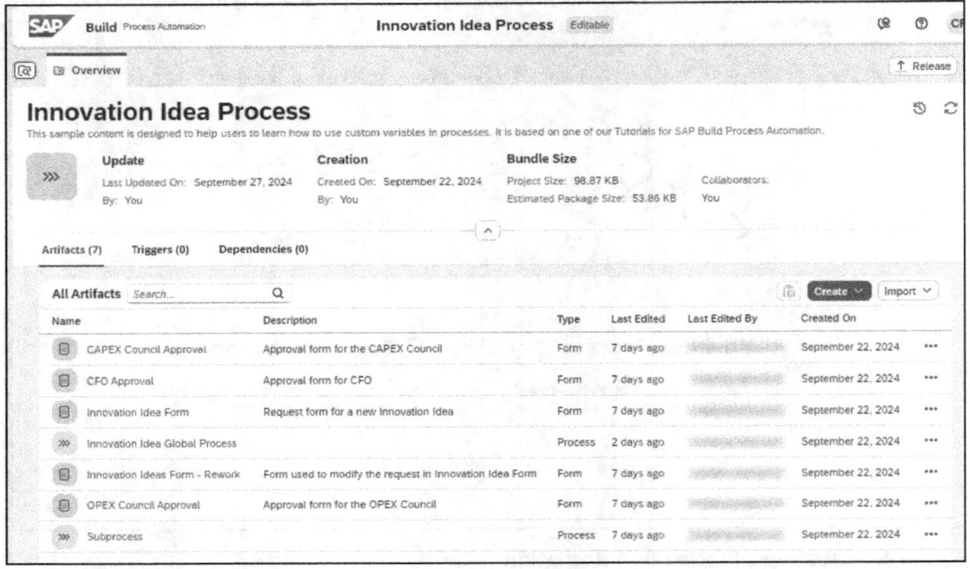

Figure 4.4 Business Process Project with a Process Artifact

4.2.2 Create a Project with a Decision Artifact

Business rules automate the decision-making process. A business process project can have a decision artifact as the only artifact, as shown in Figure 4.5 and Figure 4.6. Typical scenarios are creating business rules for shopping carts, promotion rules based on regions, business rules for assigning the equipment for employees, and so on.

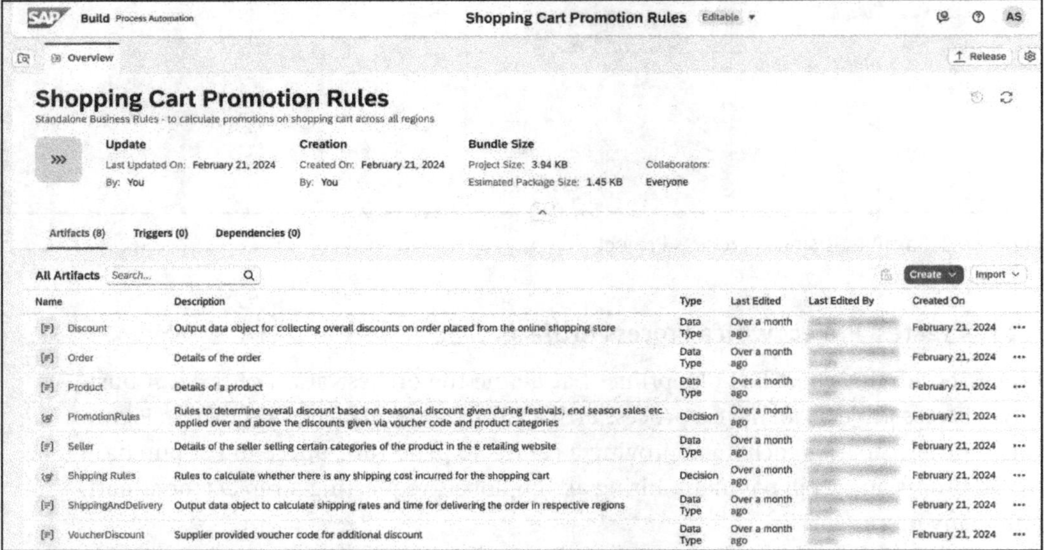

Figure 4.5 Business Process Project with a Decision Artifact

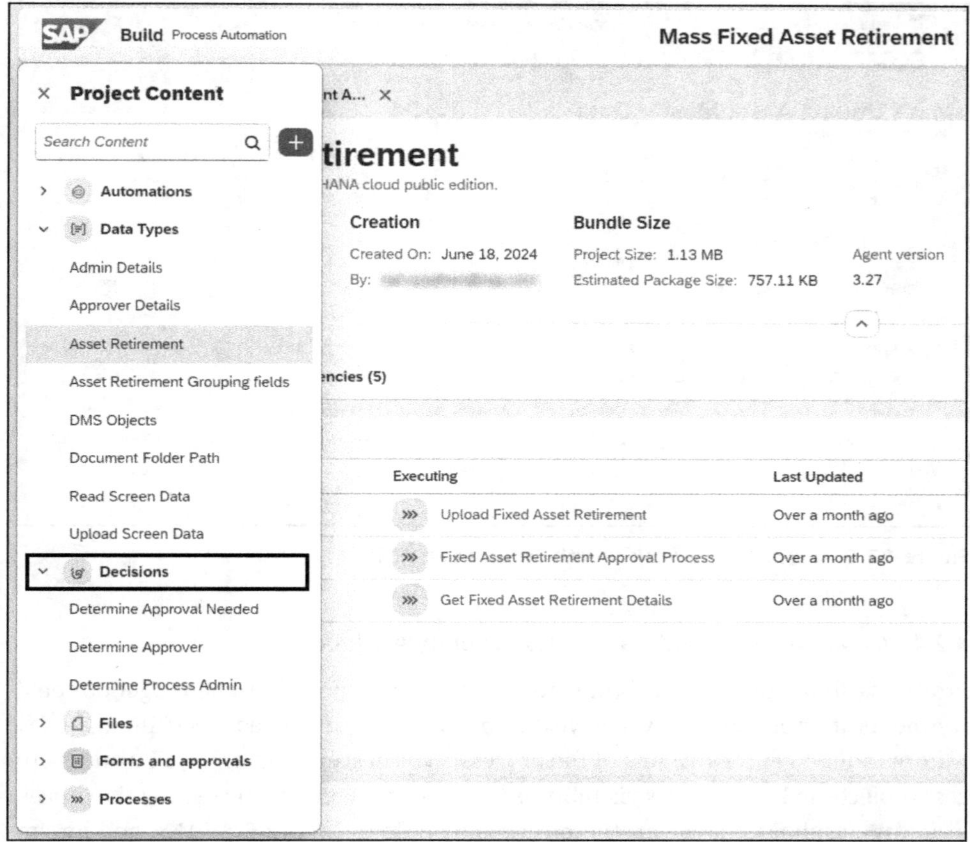

Figure 4.6 Decision Artifacts with Promotion and Shipping Rules

4.2.3 Create a Project with an Automation Artifact

Automation allows you to create bots to reduce operation costs and automate repetitive manual work. A business process project can have an automation artifact as the only artifact, as shown in Figure 4.7. Typical scenarios include automating the mass update of asset master data records in the backend with the asset details available in a Microsoft Excel sheet, automating the process of extracting the data from the document, automating the process of collecting the template-based journal entry from the user's mailbox and uploading them in the application, and many more.

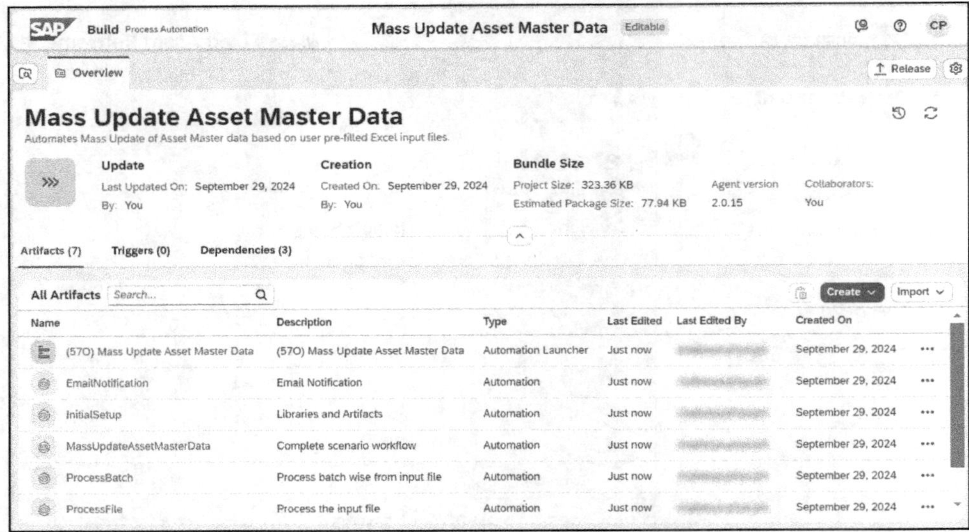

Figure 4.7 Business Process Project with Automation Artifacts

4.2.4 Create a Project with a Process Visibility Artifact

Process visibility provides real-time insights into the process with heterogenous environments. It's not mandatory to have the business process created in SAP Build Process Automation to get the insights of the process. Typical scenarios where a business process project can have process visibility as the only artifact is when creating the process visibility dashboard based on the business workflows in SAP S/4HANA and an employee onboarding process in SAP SuccessFactors that is triggered for a new hire.

4.3 Components of the Business Process Project Overview Page

All the information related to the business process project is available in the **Overview** page of the project screen. In this section, we'll get into the details of each area of the **Overview** page, which is seen as soon as you open the business process project.

4.3.1 Header

The **Header** section gives detailed information about the project such as project creator details, date of creation, project collaborators, and the project and package size (see Figure 4.8).

As a business process project maintains the entire lifecycle of the project, it provides clear visualization of all the project versions, that is, **Editable**, **Deployed**, and **Released**, as shown in Figure 4.9.

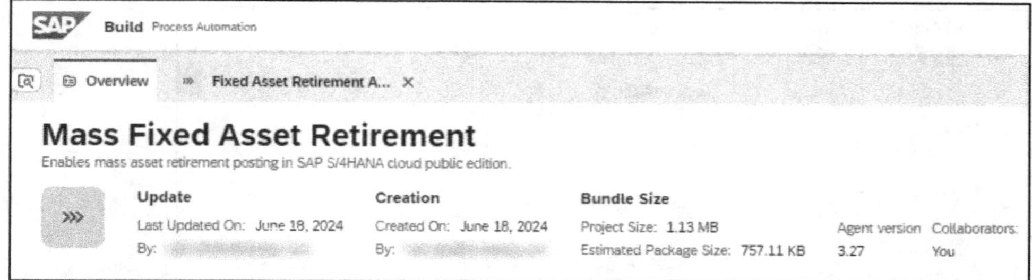

Figure 4.8 Business Process Project Details on the Overview Page

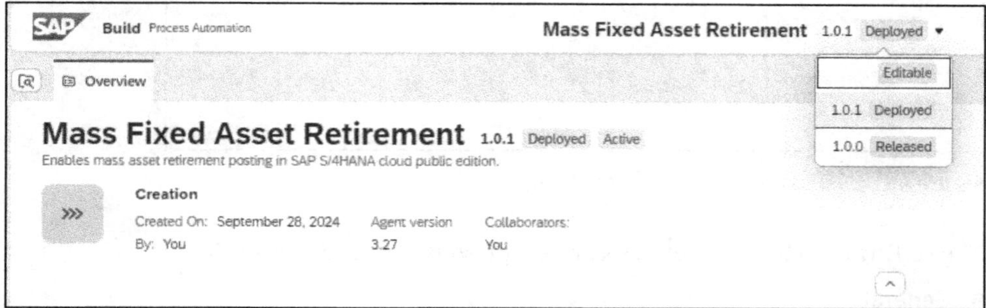

Figure 4.9 Business Process Project Versions

You can always restore your project to the previous version, which provides some flexibility to work on the needed version of the project. Click on the **Restore** icon to view the versions of the project, as shown in Figure 4.10.

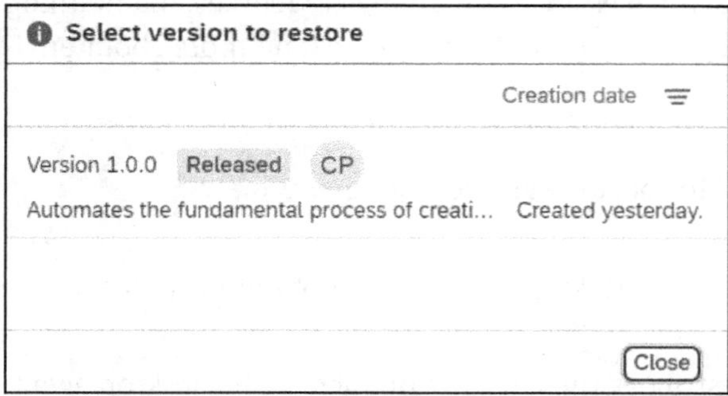

Figure 4.10 Restore Previous Versions of the Project

4.3.2 Settings

Business process project management tasks can be done in the settings, accessed via the **Settings** icon of the project, as shown in Figure 4.11.

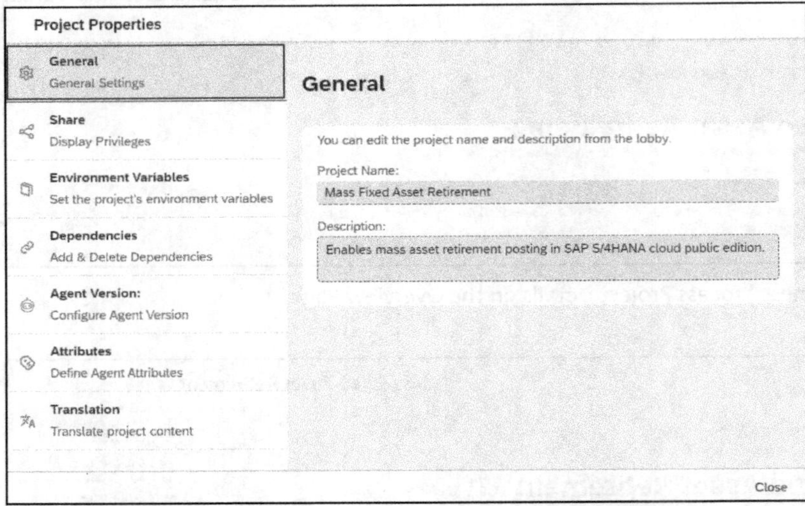

Figure 4.11 Business Process Project Settings

The settings of a business process project provide the following options:

- **General**
  Provides information on the project such as project name and details.
- **Share**
  Displays a list of project collaborators together with their authorizations.
- **Environment Variables**
  Creates and displays a list of environment variables used in the project. Typical use cases for creating environment variables are maintaining credentials, maintaining destinations, and so on. See Chapter 7, Section 7.8, to understand more about environments.
- **Dependencies**
  Adds and updates dependencies to consume artifacts of other projects.
- **Agent Version**
  Maintains the required version of the desktop agent both for Windows and Mac.
- **Attributes**
  Creates agent attributes that distribute the automation jobs to the desktop agent, which matches these attribute values.
- **Translation**
  Translates your project content into many languages, for example, project variables, artifacts, and process steps. As shown in Figure 4.12, you must first export the file to download the translatable file. Maintain the translation in the file without changing the keys, and then import the translated file.

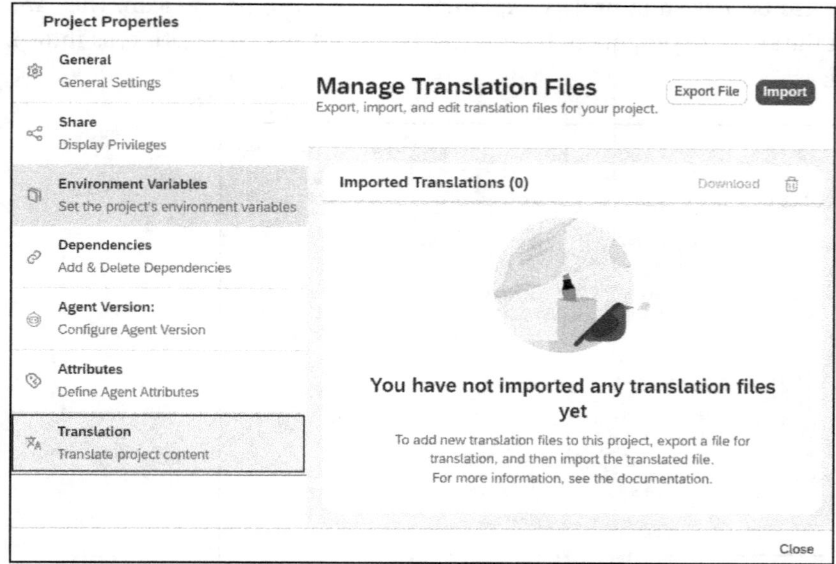

Figure 4.12 Translation in Project Settings

4.3.3 Artifacts

Artifacts represent pieces of business logic so the ordered execution in the process produces a meaningful result. All the artifacts used in the project are available in this section in the **Overview** page, as shown in Figure 4.13. If the project has many artifacts, you can leverage the search option Search... to find your artifacts.

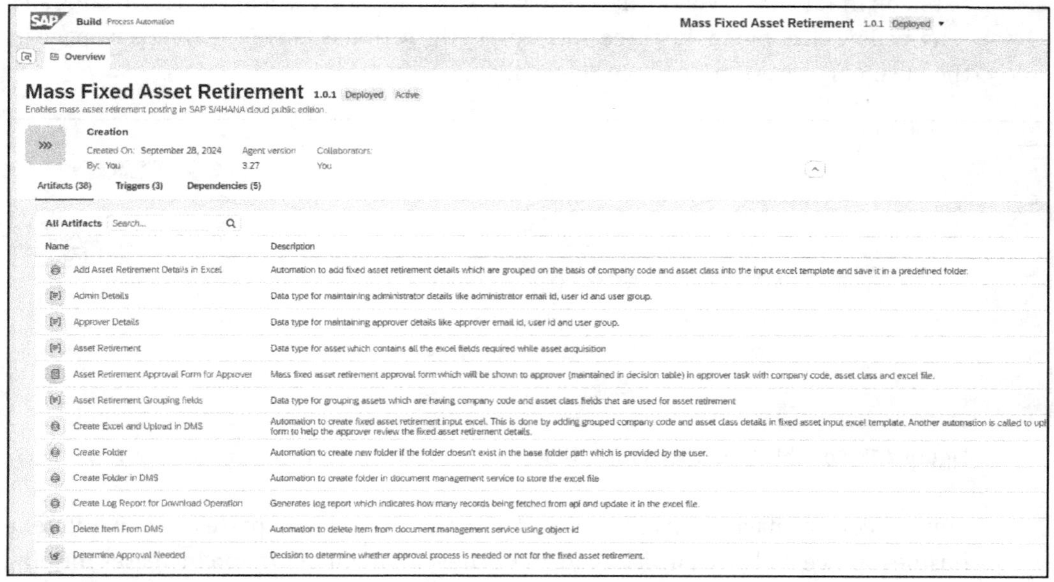

Figure 4.13 Artifacts Overview

Artifacts can also be viewed in project explorer, accessed with the icon. They are segregated at the artifact/component level, which helps with easy navigation, as shown in Figure 4.14.

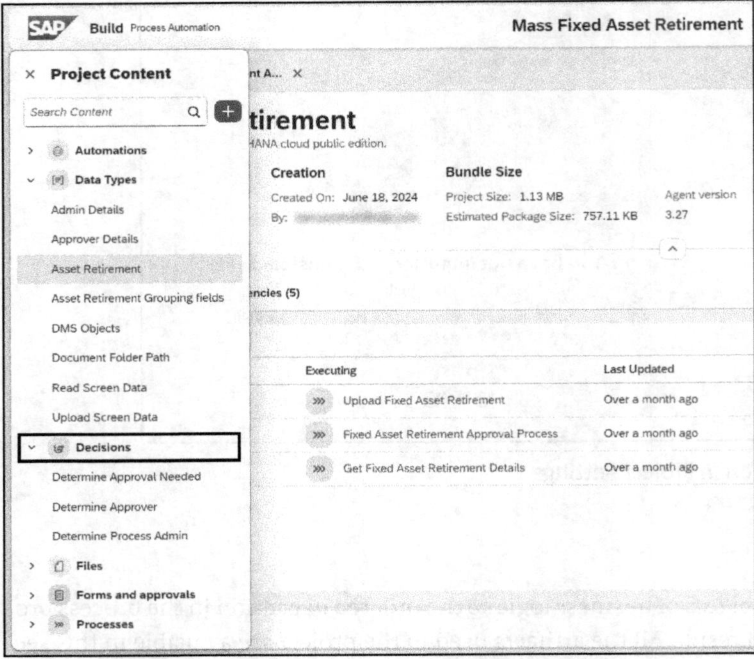

Figure 4.14 Project Content at the Artifact/Component Level

Based on the artifacts type, artifacts can be copied, duplicated, deleted, deactivated, and renamed by clicking on the three dots icon in the **Artifacts** tab of the **Overview** page in the business process project (see Figure 4.15).

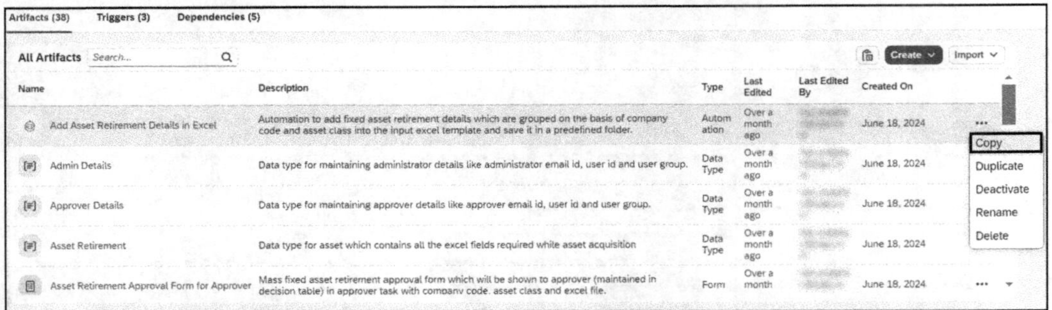

Figure 4.15 Possible Operations for Artifacts in the Business Process Project Overview

You can copy artifacts such as forms and automation and then paste them in different business process objects, which reduces the development effort and increases the efficiency of citizen developers.

4.3.4 Triggers

Business processes can be initiated either through form submission or triggers. All types of triggers (API trigger, event trigger, and scheduled trigger) used in the project are available in the **Triggers** tab of the **Overview** page in the business process project. You can edit, deactivate, and delete the triggers, as shown in Figure 4.16. Refer to Chapter 5 for detailed information on triggers.

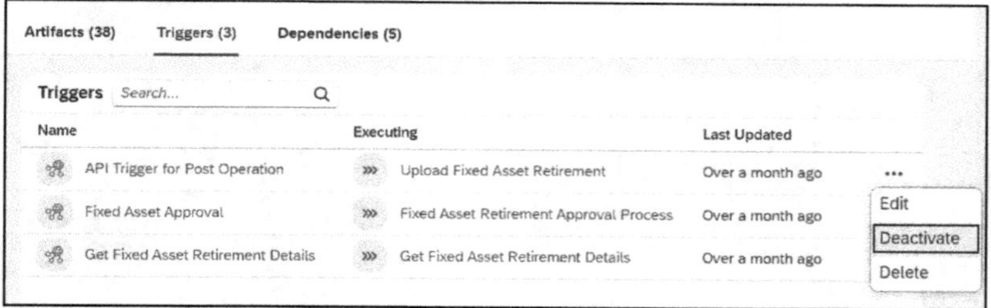

Figure 4.16 Triggers in Business Process Project Overview

4.3.5 Dependencies

Dependencies refer to the relationships and interconnections between artifacts of other projects within a business process project. You can add a dependency to use the artifacts available in SAP Build Process Automation SDK packages or add other business process projects or an action project. All the project dependencies are available in the **Dependencies** tab of the **Overview** page in the business process project, as shown in Figure 4.17.

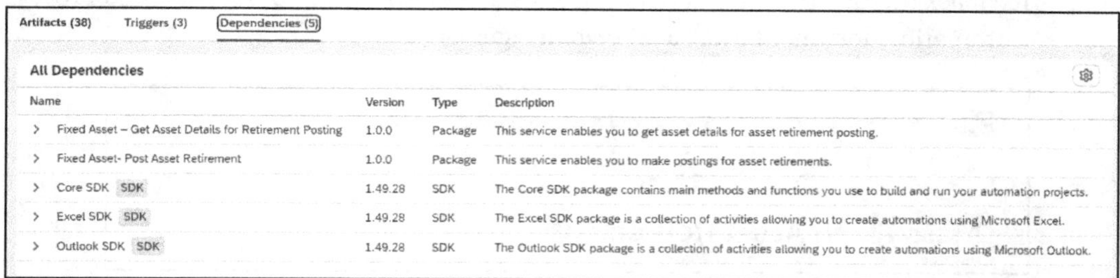

Figure 4.17 Business Process Project Dependencies in the Overview Page

4.3.6 Design Console and Test Console

The *design console* displays errors, warnings, and information during the design time of the business process project. The *test console* displays errors during project execution. When you're testing automation artifacts, you can add watch points that help track the value of the variables during debugging automation debugging. You can watch the values of the expressions used in the custom script or formula editor of an automation. Watch points can be added in multiple ways:

- Right-click on the output parameters of an activity, and select **Watch this**.
- Enable the **Watches** checkbox in the **Variables** area.
- Click the **+** button in the **Watches** area of the **Timeline**, as shown in Figure 4.18.

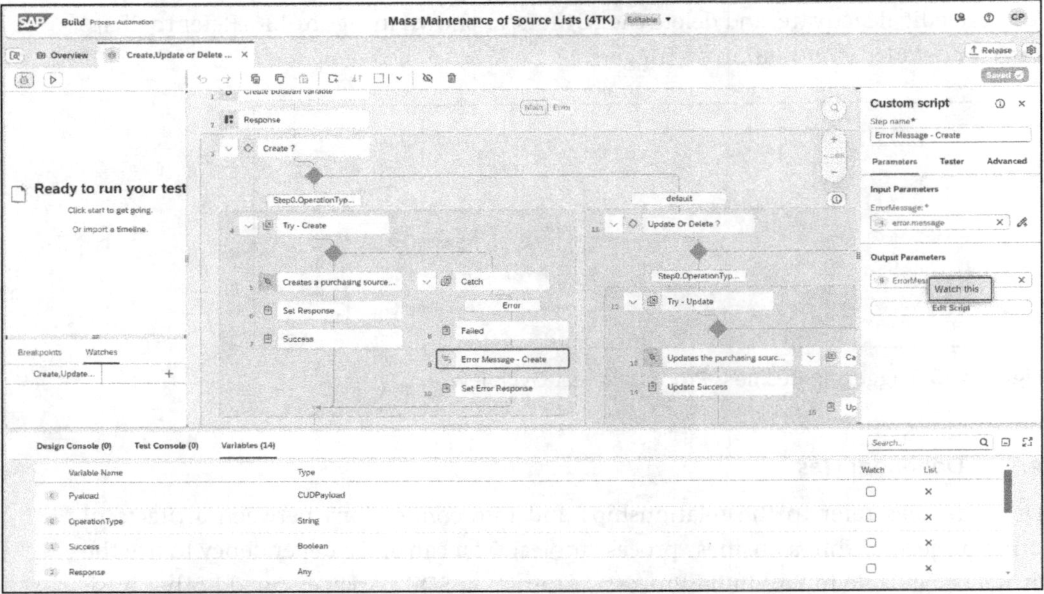

Figure 4.18 Watch Points in Automation

Other interesting options provided in the project are **Feedback** for users to provide feedback on the product and **Help** for users to be directed to the SAP Build Process Automation documentation, as shown in Figure 4.19.

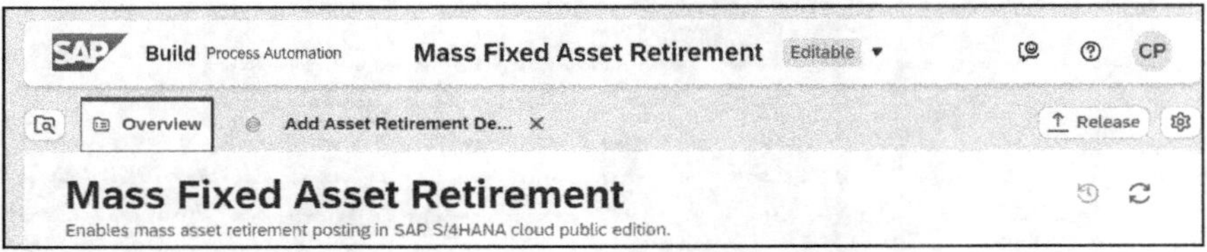

Figure 4.19 Feedback and Help Options

4.4 Summary

You now should have acquired detailed knowledge about business process projects. Additionally, you understand the business process project information available on the Overview page, navigation options, and configuration settings. In the next chapter, you'll learn what a process is, including the role and usage of individual artifacts within the process.

Chapter 5
The Process and Its Artifacts

Processes are flow diagram–like blueprints that define the orchestration of tasks. Every task represents specific business logic that is executed in a specific order to produce a meaningful result. A citizen developer needs to learn how to design the process to complete the business process.

5

Before you design a process, you should first understand what a process is. A *process* is a digital workflow that automates business tasks and activities, orchestrating various steps, decisions, and actions to achieve a specific business outcome. This involves integrating different systems and automating repetitive tasks that enhance efficiency, accuracy, and agility in organizational operations. SAP Build Process Automation provides an intuitive low-code designer to design your process.

This chapter teaches you how to create and configure processes that include different artifacts and activities by using the low-code designer. You'll learn how to use different controls in the process, modularize your process, and initiate the process in different ways. Each activity will be covered in detail.

5.1 Benefits of Low-Code Designer Tools

For application development and automation, SAP Business Technology Platform (SAP BTP) offers low-code and pro-code solutions that enable both business experts and professional developers to drive innovation. SAP Build low-code solutions enable everyone, no matter their skill level, to create and augment enterprise applications, process automations, and business sites with drag-and-drop simplicity. Drag and drop is a simpler and faster way for business users to automate business processes. It combines workflow management, robotic process automation (RPA), decision management, process visibility, and embedded artificial intelligence (AI) capabilities into one intuitive low-code experience. The solution enables business users to create automations and/or contribute their process expertise to projects to deliver the solutions they need without having to write code.

Low-code designer tools offer several advantages for businesses, developers, and nontechnical users. Some of the key benefits are listed here:

- **Speed and efficiency**
  Low-code platforms enable faster development cycles, allowing applications to be

built in days or weeks instead of months. Changes and updates can be implemented quickly, facilitating a more agile development process.

- **Cost savings**
  Less time spent on development translates to lower costs. Companies can save on hiring specialized developers and reduce overall project expenses. Applications built with low-code platforms often require less maintenance, further reducing long-term costs.
- **Accessibility**
  Nondevelopers such as business analysts, managers, and other nontechnical users can create applications without needing in-depth programming knowledge. Teams across the organization can contribute to the development process, fostering innovation and collaboration.
- **User experience**
  Low-code tools often provide drag-and-drop interfaces and visual workflows, making the development process more intuitive.

5.2 Process Editor

The process editor in SAP Build Process Automation is a powerful tool designed to streamline the creation, modification, and management of business processes. Some key features and functionalities of the process editor are provided in the following list:

- **Visual design interface**
  The process editor provides a user-friendly visual interface where users can design processes by dragging and dropping predefined activities and elements into a workflow. It offers a flowchart-like view of the process, making it easy to visualize the sequence of tasks and decision points.
- **Activity libraries**
  The editor includes a library of prebuilt activities that can be easily integrated into processes. These activities can range from simple tasks such as sending an email to complex integrations with other systems. Users can create custom activities to meet specific business requirements, enhancing the flexibility of the tool.
- **Conditional logic and decision-making**
  Users can incorporate decision points within the process to handle different paths based on conditional logic. The editor supports branching, allowing processes to diverge based on predefined criteria or data inputs.
- **Integration capabilities**
  The process editor allows seamless integration with various SAP and non-SAP systems, enabling data exchange and process automation across different platforms. It supports the use of application programming interfaces (APIs) and connectors to integrate third-party applications and services into the automated processes.

- **Data handling**
  Users can define data inputs and outputs for each activity, ensuring that information flows correctly through the process.
- **Variable management**
  The editor allows for the creation and management of variables to store and manipulate data within the process.

As shown in Figure 5.1, the process editor can be accessed by clicking on the process artifact in the **Overview** area of the business process project as explained previously in Chapter 4.

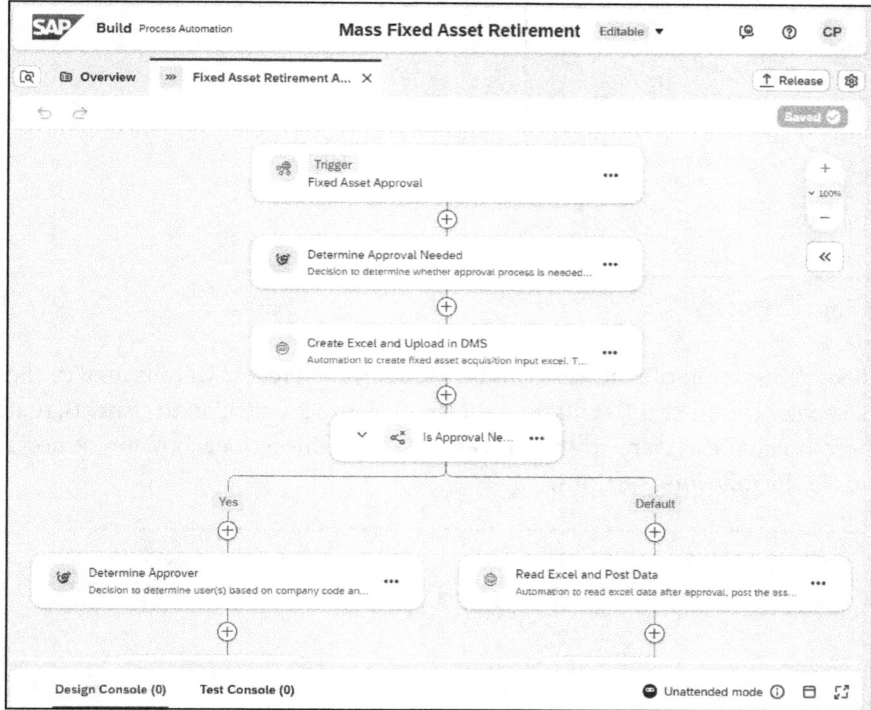

Figure 5.1 Process Editor

5.3 Processes and Artifacts

In this section, you'll learn how to create and design a process using different artifacts and understand different ways to start the process.

5.3.1 Processes

A process is a set of tasks or activities that are completed to achieve a goal or outcome. The prerequisite to create a process is to create a business process project as explained in Chapter 4. A process can be created from the **Overview** area of the project, as mentioned in Chapter 4, Section 4.3.

A simple process can be automating sending an invoice request for approval and posting the invoice in SAP S/4 HANA, as shown in Figure 5.2.

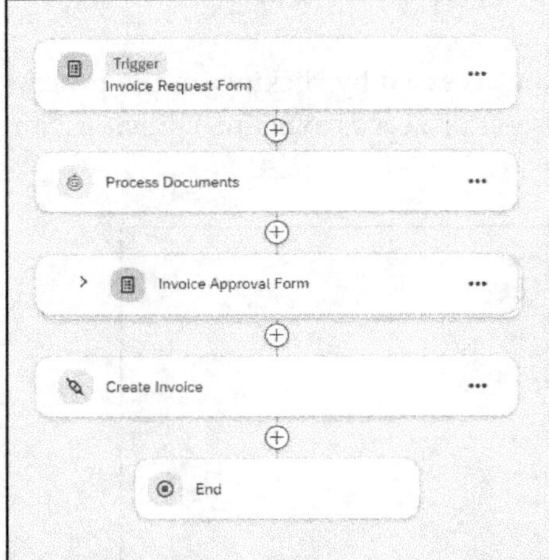

Figure 5.2 Sample Process

The detailed information of a process can be found in the **Process Details** area of the process. To access the **Process Details** area shown in Figure 5.3, click on the canvas, that is, anywhere outside the steps of the process. We'll go through each of the **Process Details** tabs in the following sections.

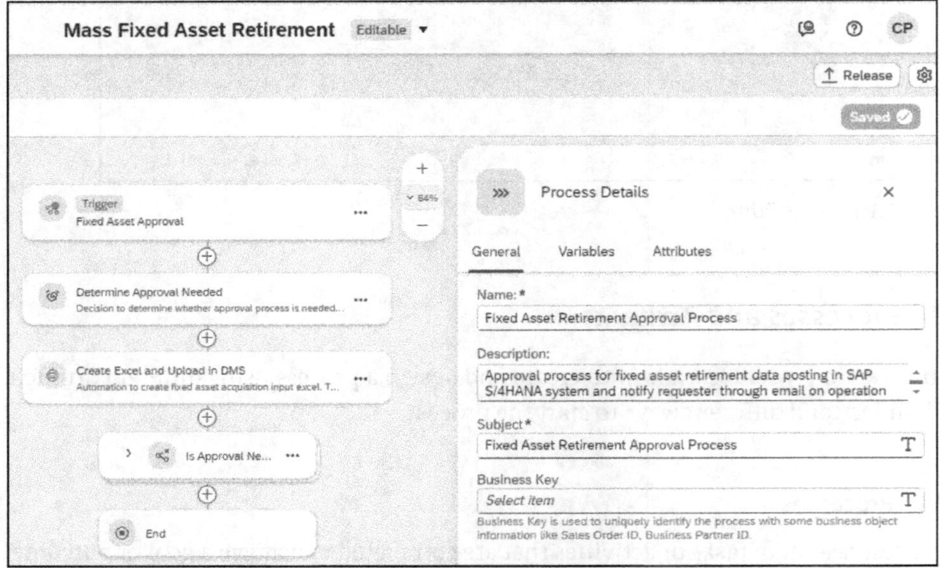

Figure 5.3 Details of a Process

General

The **General** tab of **Process Details** gives an overview of the process such as the **Name** and **Description** of the process, as shown in Figure 5.4. The process is uniquely identified by the **Business Key**. By default, the **Business Key** value is the instance ID of the process. It can even be the Sales Order ID or lead ID of the process because they are unique. This field is typically needed when you're passing the key field to another process or subprocess, and it acts as a unique identifier.

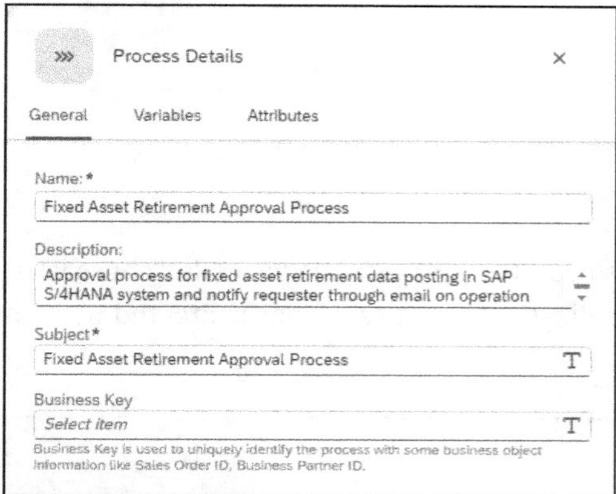

Figure 5.4 General Section of the Process

Variables

The process builder allows for the creation and management of variables to store and manipulate data within the process. **Variables** in process builder are categorized as **Process Inputs**, **Process Outputs**, and **Custom Variables**, as shown in Figure 5.5. The following list describes the different categories of variables in a process:

- **Process Inputs**
  Based on the option, the process is started either through a form trigger, API trigger, or an event trigger, and the corresponding fields of the specific trigger will be automatically added as the input variables for the process. This reduces the manual effort of adding the variables.
- **Process Output**
  In SAP Build Process Automation, process output variables are used to capture and store the results of a process after it has been executed. These variables can then be used for various purposes such as further processing, reporting, integrating with other systems, or informing stakeholders about the outcomes. They facilitate the transfer of data either from one part of the process to another or from the process to external systems and applications.
- **Custom Variables**
  Custom variables can be created at any stage of the process and are accessed at the global level of the process. This option supports the basic data types and the data

types that are created in the business process project. It need not be carry forwarded as input or output of an artifact. Because it's accessible in all the process steps, the value of this field can be updated based on the requirements.

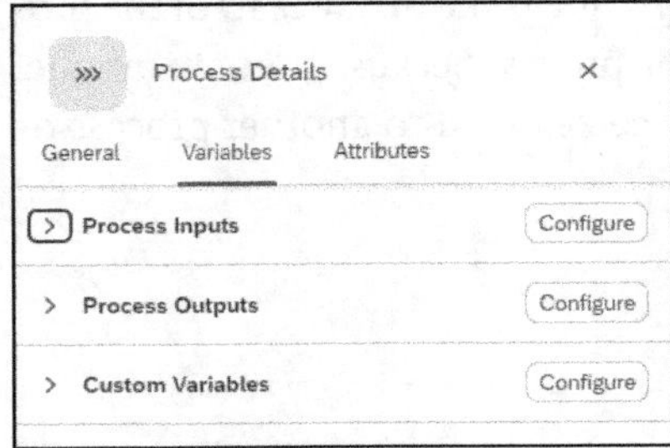

Figure 5.5 Variables Section of the Process

Click on the **Configure** button shown in Figure 5.5 to configure the variables These variables can be declared as basic data types, collections, or a type that is created in data type of the project, as shown in Figure 5.6.

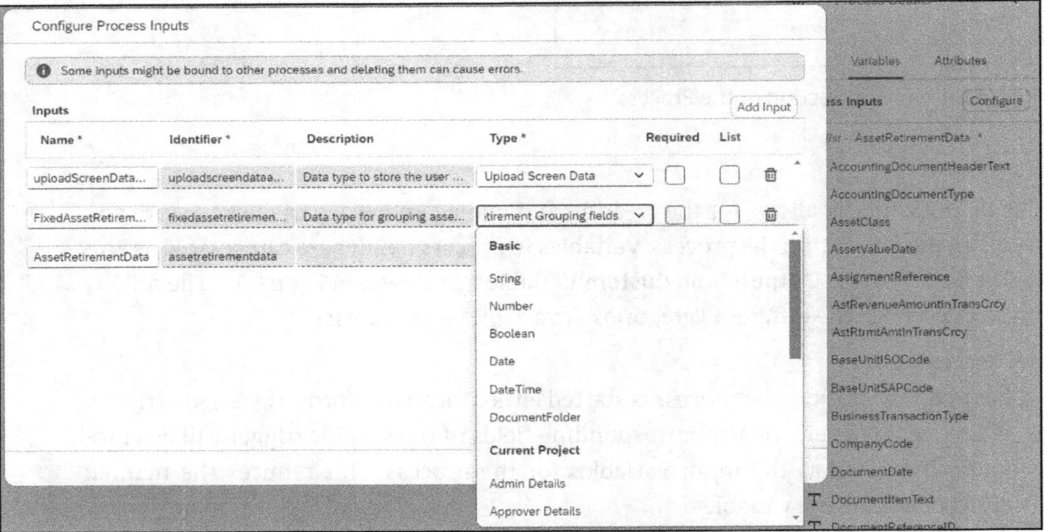

Figure 5.6 Create and Configure Variables in a Process

Let's consider an example scenario in which you need to design the process for employees to submit innovation ideas. The process should by triggered by the employee who fills out the form. Then, the CFO wants it to be divided into capital expenditure (CAPEX) and operating expenses (OPEX) categories, so two different councils can process the ideas and decide whether to approve. If the innovation idea requires additional information, the council will send it back to the requester. If the process is approved by the council, the CFO will receive the case for the final approval.

The requester must receive an email with notification of the decision by the council or CFO.

In Figure 5.7, you can see that **Global cost**, which is the estimated cost for the project, is one of the custom variables.

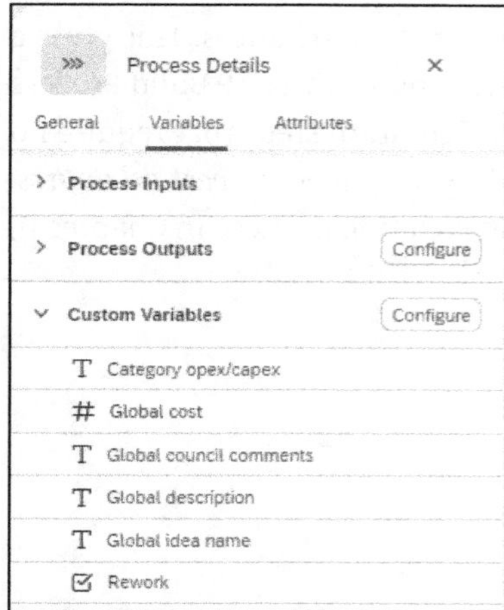

Figure 5.7 Custom Variables in a Process

Based on the conditions, the approval can go either to CAPEX or OPEX. The estimated cost for the project has to be accessible during the entire process to notify the employee of the final cost after the approval. Therefore, **Global cost** is created as a custom variable and bound to both the **Approval** forms; the same custom variable can be updated at all stages of the process. To update the custom variables, click on the step for which you want to update the field, which opens the **Process Content** popup. Update the custom variable with the listed variables, as shown in Figure 5.8.

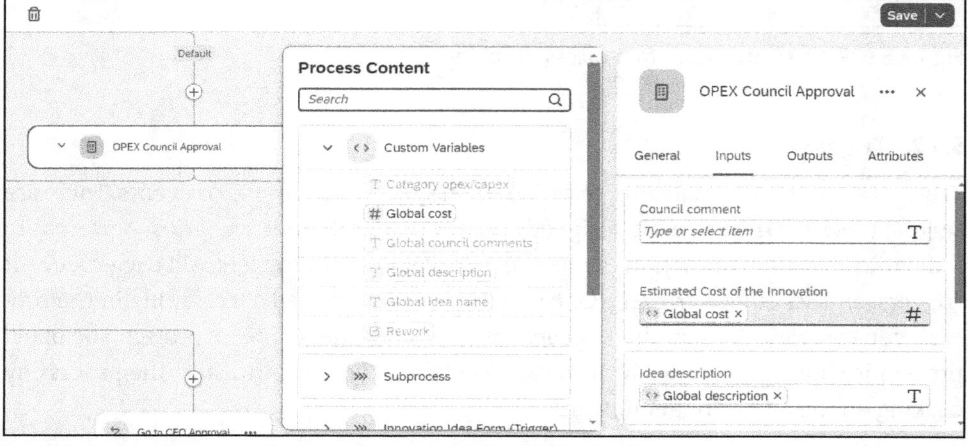

Figure 5.8 Accessing Custom Variables in the Process

Attributes for Process Visibility

The *process visibility* capability in SAP Build Process Automation enables you to build dashboards to monitor processes in and across SAP and third-party applications. Event-driven process views, AI-based cycle-time predictions, milestone tracking, service-level agreements (SLAs), and key performance indicators (KPIs) enable process participants to identify process bottlenecks and at-risk process instances. This gives a complete view of your enterprise processes. For processes built with SAP Build Process Automation, you can create dashboards with a few configuration steps that require no coding expertise. The attributes used in process visibility are called *process attributes*, as shown in Figure 5.9. You can find a more detailed explanation of these in Chapter 11, Section 11.4.

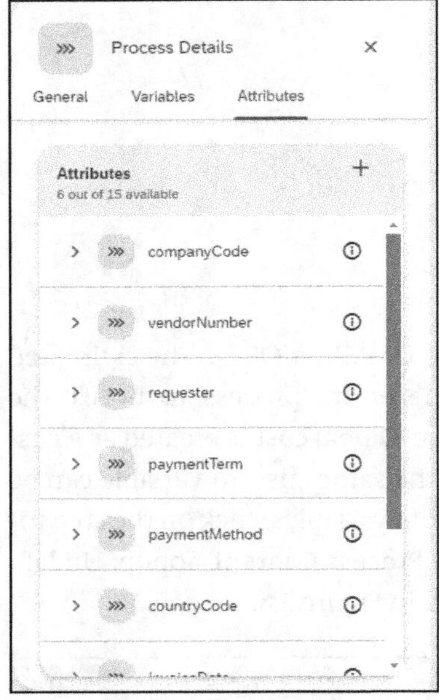

Figure 5.9 Process Attributes for Process Visibility

5.3.2 Triggers

Triggers are used to initiate processes based on specific conditions or events. There are several types of triggers, each designed to respond to different scenarios. A process in SAP Build Process Automation can be started through form triggers, API triggers, event triggers, and scheduled triggers. Once the triggers are created successfully in the process, you can view them in the **Triggers** area of the business process project. The blank process is shown in Figure 5.10. You can decide how you want to start the process by clicking on the **Add a Trigger** button.

In the following sections, we'll walk through the use of each trigger.

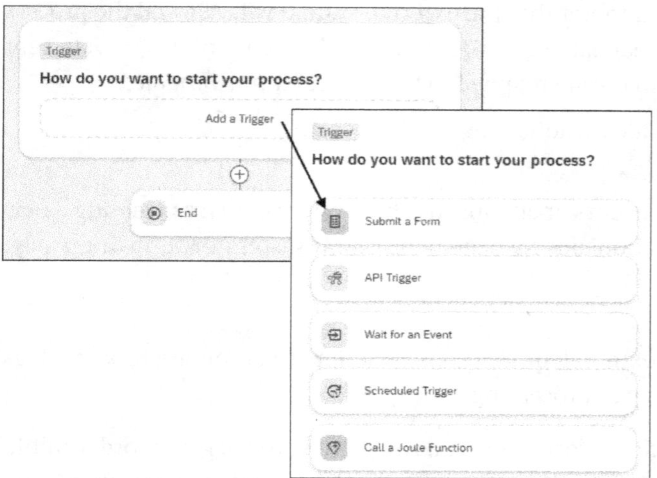

Figure 5.10 Triggers to Start the Process in SAP Build Process Automation

Form

A process starts when a user submits a form, known as a *form trigger*. Forms are typically used to capture user input, which can then be used to drive subsequent actions within the process. The forms are designed using the form builder. The input data from the form is automatically mapped to process input variables, which are then used in the process workflow.

Typical usage scenarios include the following:

- **Request forms**
  Initiates processes such as leave requests, purchase requests, or service requests based on user input.
- **Data collection**
  Gathers information from users, such as feedback forms or survey responses, to start a process.

As an example, consider an employee submitting a travel request form. The submission triggers a travel approval process—capturing details such as destination, dates, and budget—and then creates a travel request in the backend.

Section 5.7 later in the chapter explains the creation and consumption of forms in a process.

API Trigger

An *API trigger* initiates a process when an API call is made to a specific endpoint. This allows external systems or applications to start processes in SAP Build Process Automation.

You need to define the API endpoint that the external system will call, and the process is triggered when an HTTP request (e.g., POST) is made to the endpoint. The data sent with the API request is automatically mapped to the process input variables.

Typical usage scenarios include the following:

- **System integration**
  Starts a process based on events or actions in other systems, such as creating a new record in a customer relationship management (CRM) system or receiving a payment notification.
- **Automation**
  Triggers processes from external applications or scripts to automate tasks such as data synchronization or batch processing.

For example, an e-commerce platform sends an API request to trigger an order fulfillment process whenever a new order is placed. As shown in Figure 5.11, the process is initiated through an API trigger.

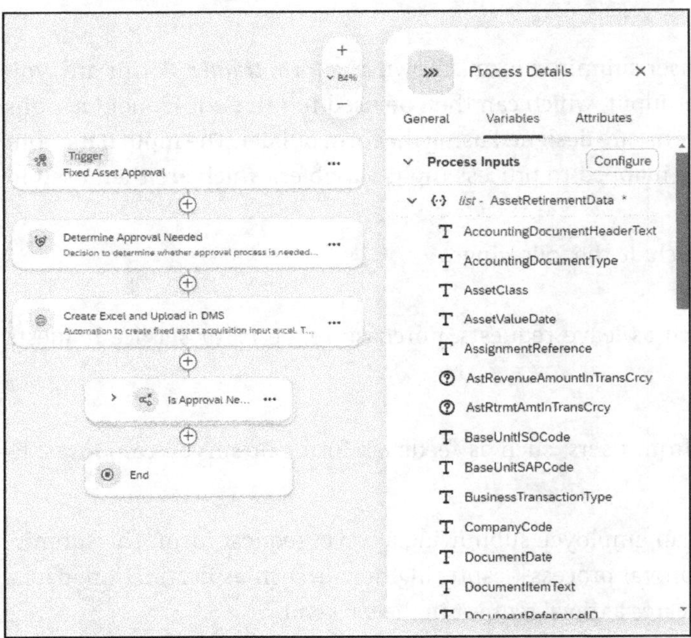

Figure 5.11 API Trigger for an Order Fulfillment Process

Event Trigger

An *event trigger* initiates a process based on specific events raised from SAP S/4HANA, SAP SuccessFactors, Internet of Things (IoT) devices, or third-party business applications. SAP Build Process Automation supports events for publishing in SAP Business Accelerator Hub, or you can upload the specification of the event source of the third-party application.

Usage scenarios include the following:

- **Data changes**
  Triggers processes when specific changes occur in data, such as updates to a database record or changes in a data feed.
- **External events**
  Initiates processes in response to events from external systems or services, such as receiving a message from a message queue or a notification from a monitoring system.
- **Real-time data processing**
  Initiates processes in response to real-time data events from IoT devices, webhooks, or other sources.

One example is triggering the business process to automate the approval of sales orders that are changed and blocked for credit in SAP S/4HANA, as shown in Figure 5.12. Section 5.4 provides detailed information on events.

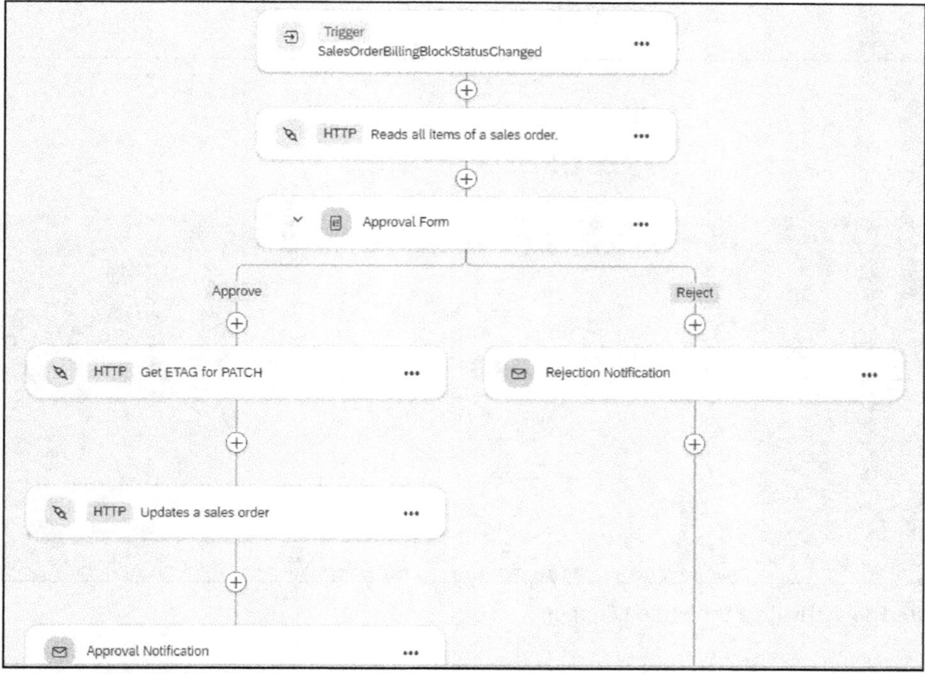

Figure 5.12 Event Trigger in Process

Scheduled Trigger

The processes can be started at scheduled times or intervals, such as daily reports or periodic data refreshes. The user need not start the process manually. To schedule the triggering of the process, create a scheduled trigger in the process, and then release and deploy the process. Figure 5.13 shows the sample process with a scheduled trigger.

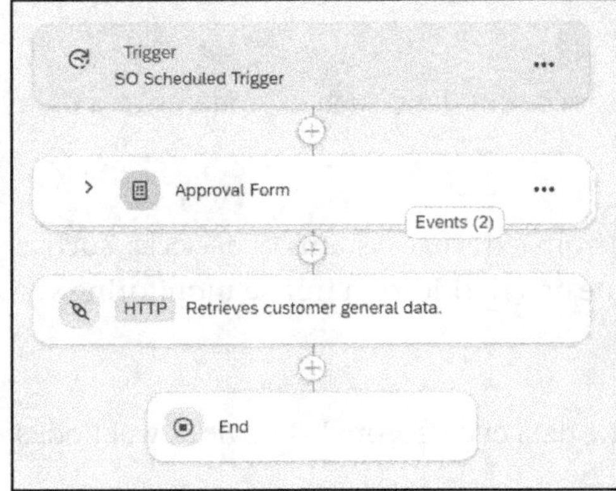

Figure 5.13 Process with a Scheduled Trigger

The scheduled trigger needs to be activated and configured from the SAP Build homepage. Navigate to **Control Tower • Environments • Public • Triggers**, as shown in Figure 5.14.

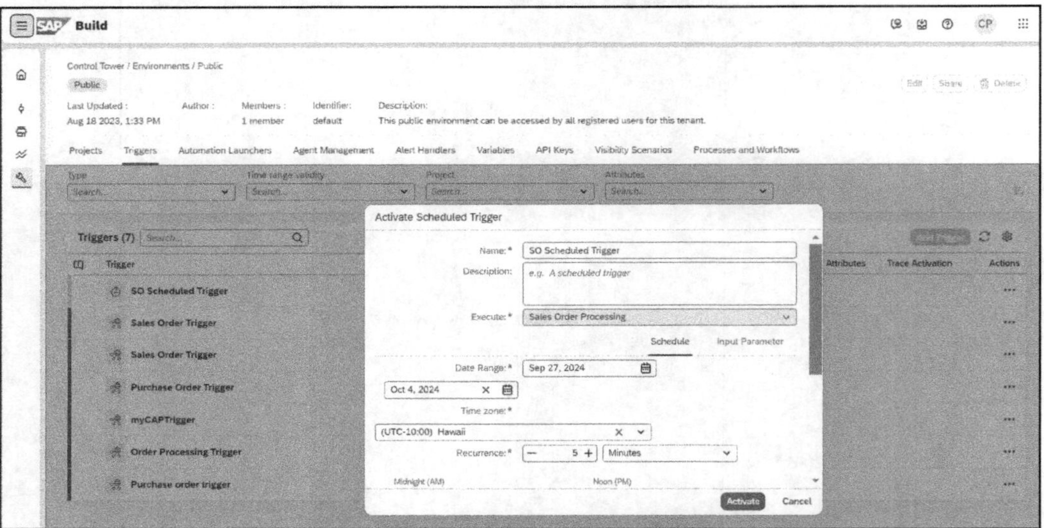

Figure 5.14 Activate a Scheduled Trigger

Each type of trigger allows for flexible and responsive initiation of the business process, catering to a wide range of business needs in SAP Build Process Automation.

5.3.3 Artifacts and Activities

Artifacts are the fundamental building blocks used to define the steps or tasks within a business process. They represent individual units of work that need to be performed as part of the overall process. The usage and sequence of artifacts depend on the process being developed. All artifacts you create are housed within your business process project and displayed in the project overview. This section gives an overview of all the available artifacts and activities. Detailed explanations are provided in the following subsections.

Artifacts can be created either from the **Overview** are of the project or during design time when creating the process by clicking on the **+** button, as shown in Figure 5.15. Artifacts in the process include **Form**, **Action**, **Automation**, and **Decision**.

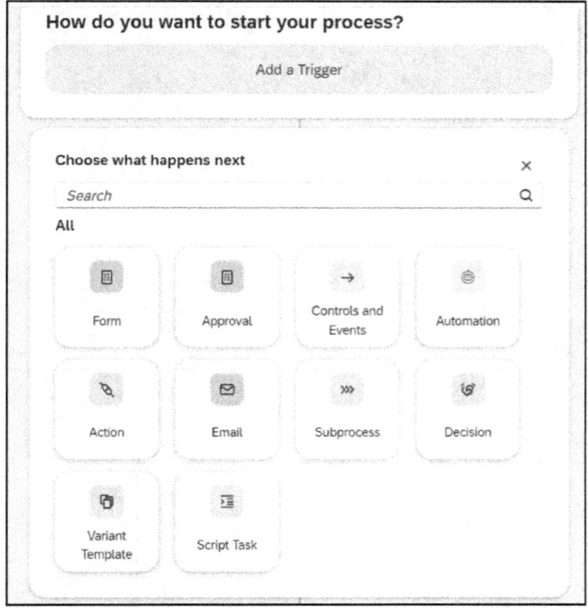

Figure 5.15 Activities and Artifacts in a Process

Form

Forms in SAP Build Process Automation refer to user interfaces (UIs) that are designed for data input and interaction within automated workflows. These forms are integral components that enable end users to provide necessary information, make decisions, or trigger specific actions within the automation process.

You can design forms for requests and approvals. Forms are explained in more detail in Section 5.7. Figure 5.16 shows a process with a **Form** artifact, which is represented with the icon.

Action

The action feature in SAP Build Process Automation connects processes with both SAP and non-SAP external systems. This is an important piece of the puzzle, especially if you want to automate or extend your business processes for any available processes such as in SAP S/4HANA, SAP Ariba, SAP SuccessFactors, and so on. These extensions can be easily built using SAP Build Process Automation. By using actions, you can connect to your given systems for any kind of GET, POST, PATCH, or other calls. Figure 5.17 shows a process with an **Action** artifact, which is represented with the icon. You'll learn more about actions in Chapter 6.

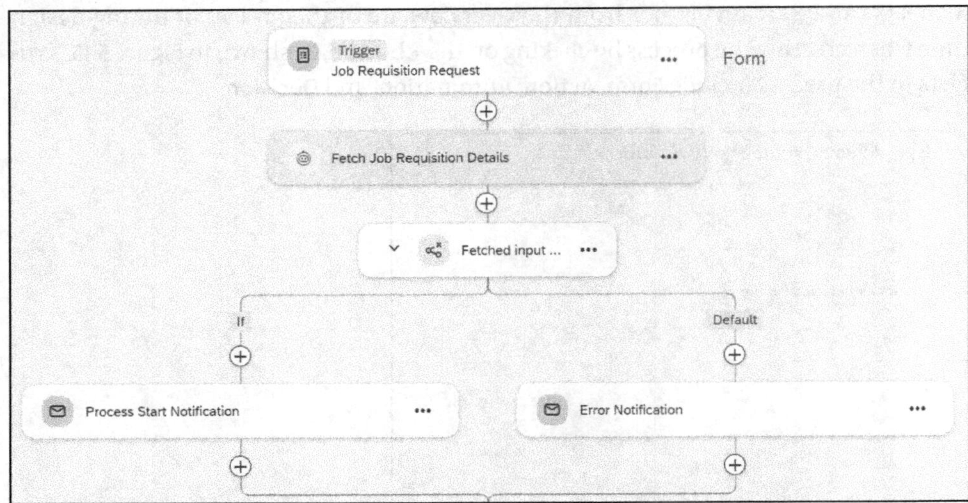

Figure 5.16 Form in a Process

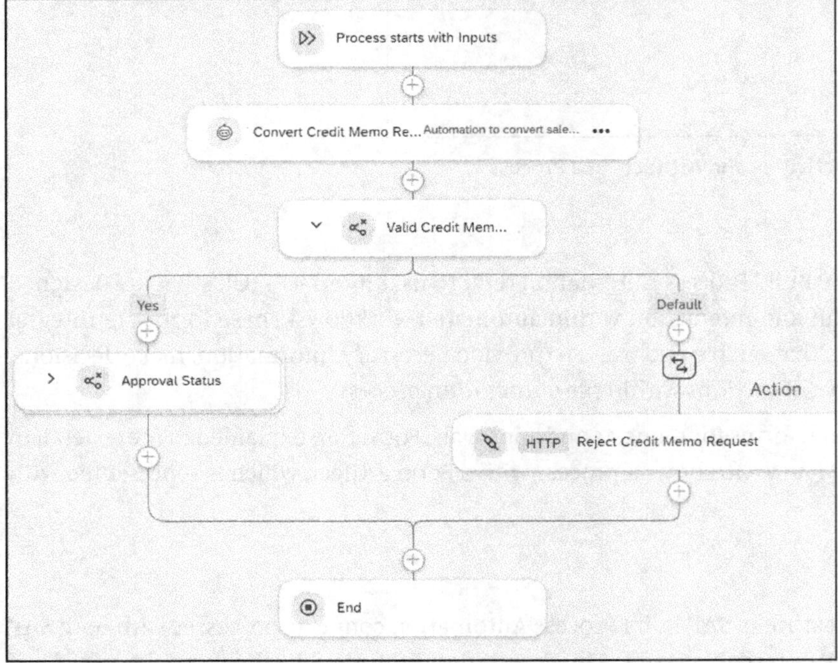

Figure 5.17 An Action in a Process

Automation

Task automation, also known as robotic process automation (RPA), automates the routine tasks with RPA capabilities. It performs repetitive tasks for you, thus reducing the human effort involved. Automation saves time, reduces error, and allows you and your colleagues to focus on work that really adds value.

Depending on the complexity of the automation, users can use either the no-code editor or—when the automation requires the use of JavaScript—switch to the low-code experience. You can capture fields, drag and drop automation steps into the bot flow, and connect steps using the building blocks of standard activities enabled by the solution. UI-based automations from both SAP and third-party applications can be created fast and easy using the built-in automation recorder. Bots orchestrate multiple activities on different applications and screens. Instead of dragging and dropping numerous activities into the editor canvas, you can simply click the record button and step through your typical process for any task. The solution records each step and exports it into an automation flow. You can then adjust the automation flow using intuitive editing features. Figure 5.18 shows a process with an **Automation** artifact, which is represented with the icon. Chapter 7 explains automation capabilities in SAP Build Process Automation in much more detail.

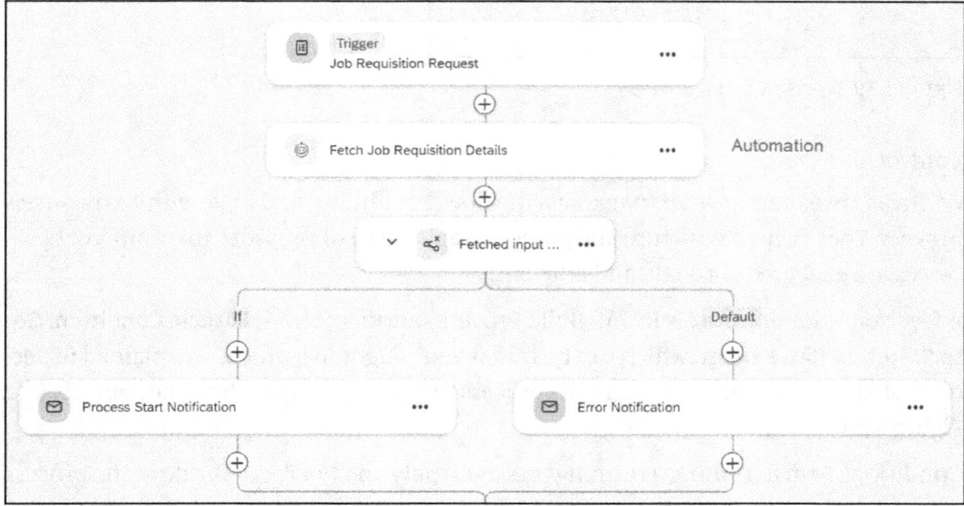

Figure 5.18 Automation in a Process

Decisions

Decision management enables you to codify corporate policies and business rules into repeatable processes. The business rules editor enables everyone to create spreadsheet-like decision tables or text-based rules. Other features include validation and error highlighting, the ability to create and manage rule versions, orchestration to order rules execution, and a rich expression library of the functions and operators needed to model complex rules. Business users, who best understand the organization's processes, can create, maintain, and manage rule versions without relying on IT for additional development effort. Figure 5.19 shows a process with a **Decision** artifact, which is represented with the icon. Chapter 8 provides much more information on business rules capabilities in SAP Build Process Automation.

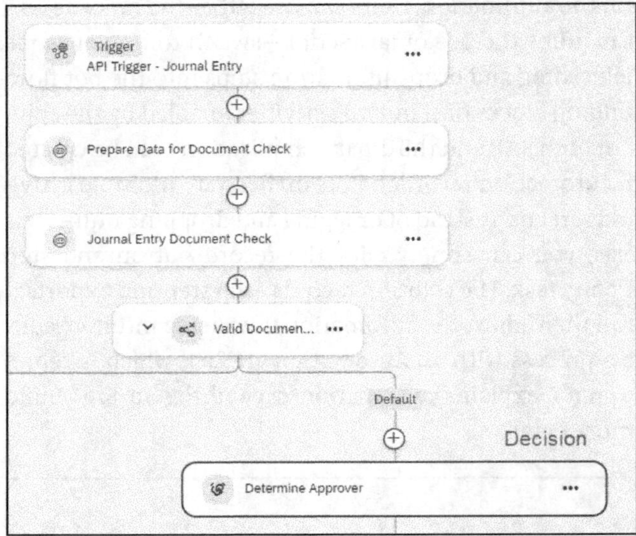

Figure 5.19 Decisions in a Process

Controls

Controls are elements used to manage the flow, conditions, and logic within a business process. They help to structure and guide the execution of activities to ensure that processes are executed correctly and efficiently.

A few examples of controls in SAP Build Process Automation are **Branch**, **Condition**, **Go-to Step**, and **Raise Event**, which can be raised and caught in a process (explained in Section 5.4.2). Click on the **+** button in the process to open a popup with all the controls (see Figure 5.20).

Conditions help in routing a running process. Apply an `If` or `Default` rule to the process context, and define the next steps based on the condition. You can use branches to run multiple tasks at the same time. The tasks in the branches run concurrently, and the process will progress once all the tasks are completed. Click on **Condition** and **Branch** in **Control** to open the **Controls and Events** popup shown previously in Figure 5.20. Your process will look like Figure 5.21 after successfully adding and configuring the conditions and branches. In Figure 5.21, ❶ shows a conditional branch and ❷ shows a parallel branch. See Section 5.5 to learn more about branches.

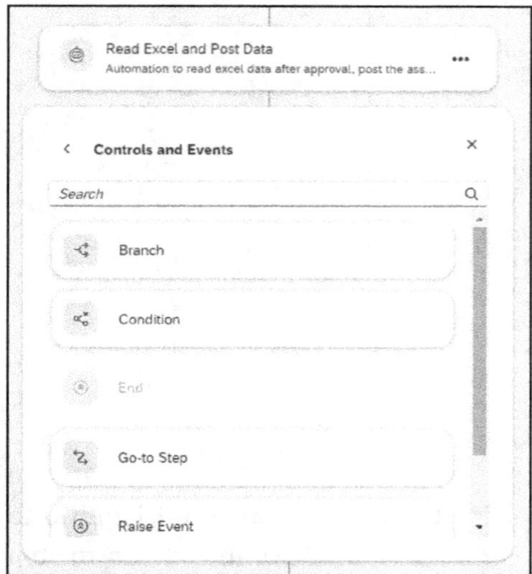

Figure 5.20 Controls and Events

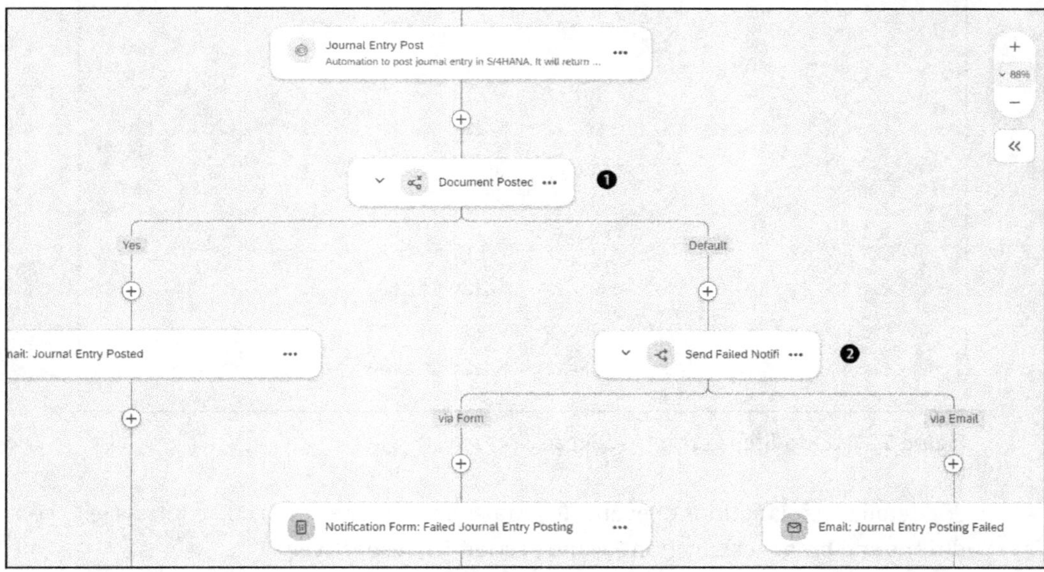

Figure 5.21 Conditional and Parallel Branches in a Process

Wait for duration control is like a timed wait option that controls how long the process pauses. You can configure different intervals of minutes, hours, days, or months, as well as provide any dynamic duration or reference dates based on the context information. You can add multiple timed wait events. Click on **Wait for Duration** in the **Control and Events** popup shown earlier in Figure 5.20, and configure it as shown in Figure 5.22.

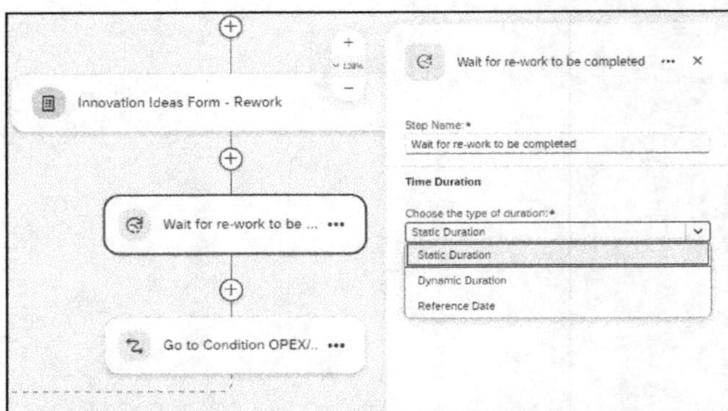

Figure 5.22 Wait for Duration in a Process

You can navigate to any step in the process using **Go-to Step** control. Click on **Go-to Step** in the **Control and Events** popup shown earlier in Figure 5.20, and add **Go-to Step** in the process, as shown in Figure 5.23.

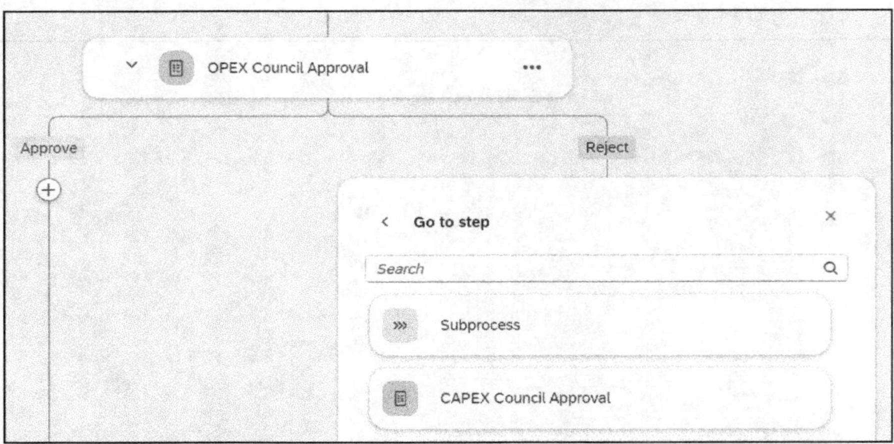

Figure 5.23 Go-to-Step Control in a Process

For detailed explanations of events (e.g., raise and timer events) and controls (e.g., conditions and branches), Section 5.4 and Section 5.5, respectively.

Mail

With the mail feature, you can define and send preconfigured emails and notify the relevant stakeholders. The mail body can include text, process context information (e.g., values generated as an output of an action after successfully posting data into the backend such as a sales order ID, field values such as comments from the approval form, etc.). You need to configure the email header fields (**To**, **Subject**, **CC**, **BCC**).The prerequisite to use the mail feature is to configure Simple Mail Transfer Protocol (SMTP) mail destinations. Click on the **+** button in the process, and select **Email** from the popup, as shown earlier in Figure 5.15. A sample process with the **Mail** activity is shown in Figure 5.24. For more details on the mail feature, Section 5.9.1.

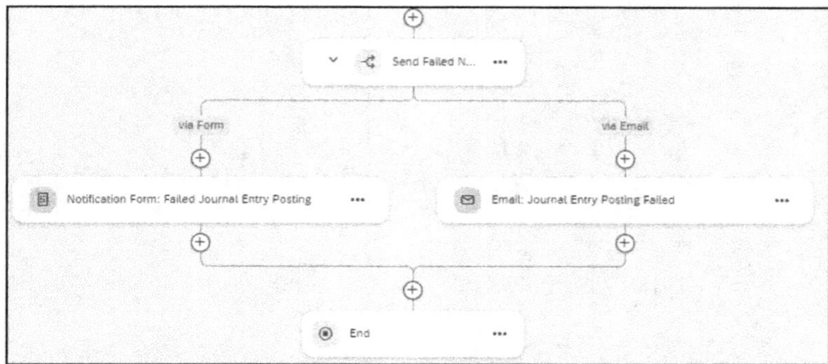

Figure 5.24 Mail Notification in a Process

Subprocess

Subprocesses are referenced subflows that help you modularize your process automation projects, create and reuse a set of common process activities in a subflow, and leverage these activities in one or across multiple master processes without duplicating efforts. You can make use of workflows created in SAP Business Application Studio and a dedicated step to call the workflow by maintaining the workflow definition ID and input and output parameters. Click on the **+** button in the process, and select **Subprocess** from the popup that appears as an extension of the Controls and Events popup shown earlier in Figure 5.15. If a subprocess is already created, a list of subprocess is displayed. You also can create a blank subprocess, as shown in Figure 5.25. See Section 5.9.2 to learn more about subprocesses.

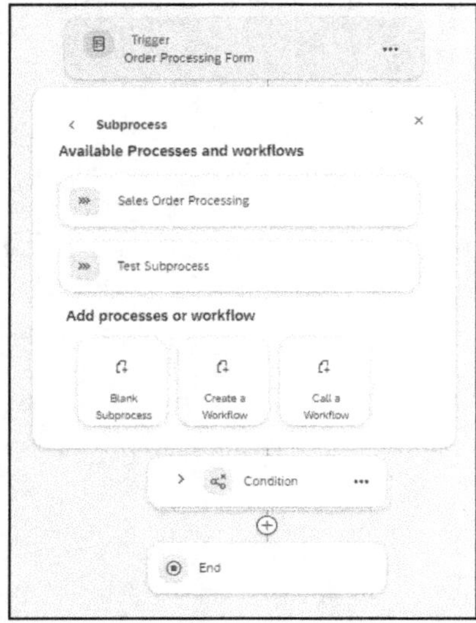

Figure 5.25 Create a Subprocess in a Process

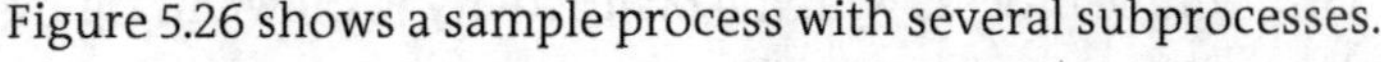

Figure 5.26 shows a sample process with several subprocesses.

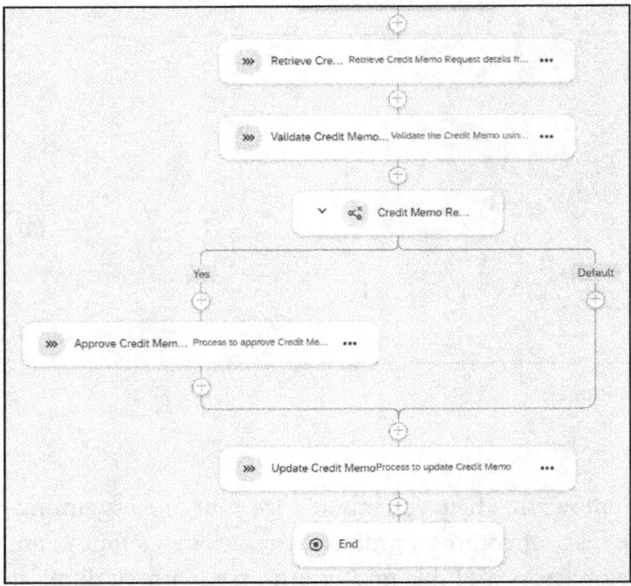

Figure 5.26 Several Subprocesses in a Process

Variant Template

A variant template consists of process steps and a preconfigured process variant, which represents a minimum viable process. The delivered process steps can be used to configure custom process variants from the preconfigured process variant. Click on the **+** button in the process, and select **Variant Template** to create a template and add in a sample process, as shown in Figure 5.27. See Section 5.8 for a more detailed understanding of variant templates.

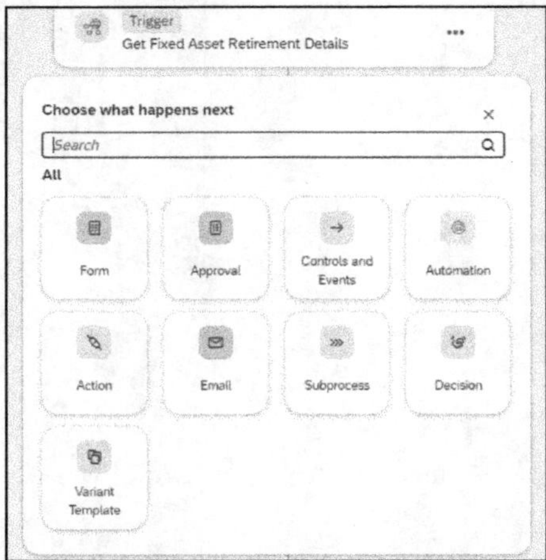

Figure 5.27 Create a Variant Template

5.3.4 Configurations

In this section, you'll learn how to transfer your data in the process, how you can reduce the development effort by reusing the artifacts of other projects by adding them as dependencies in your business project, and how to receive notifications of a process in SAP Build Process Automation via configurations.

Mappings

In SAP Build Process Automation, mappings are crucial for linking and transferring data between different artifacts of a workflow. Mappings ensure that the correct information is passed between tasks, subprocesses, and other components within an automated process. Click on any step in the process; for example, in Figure 5.28, the CFO approval step is selected, which displays detailed information and configurations for the step. Navigate to the **Inputs** tab, and click on any variable to open the popup to map the fields to the context of the process (see Figure 5.28). See Section 5.6 for a detailed discussion of mappings.

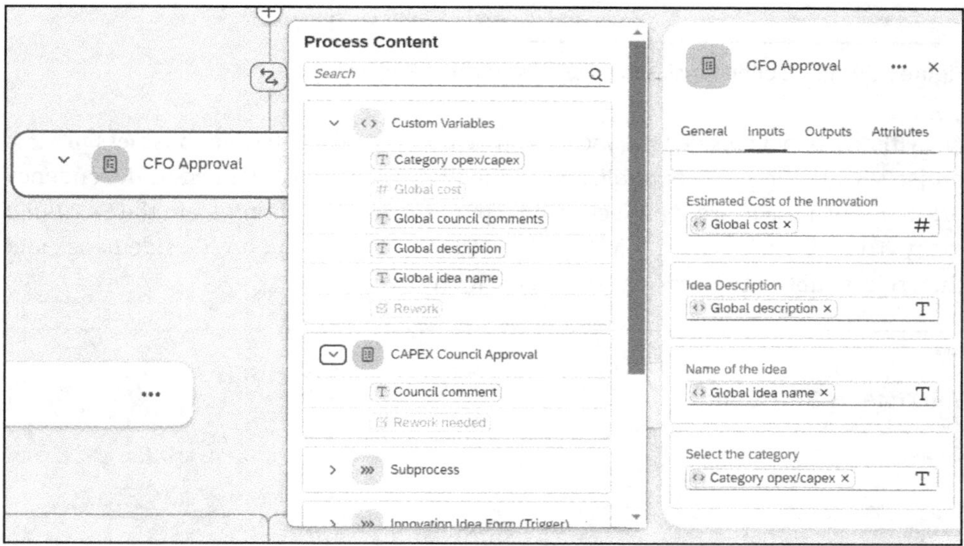

Figure 5.28 Mappings in a Process

Dependencies

Dependencies refer to the relationships and interconnections between artifacts of other projects within a business process project. You can add a dependency to use the artifacts available in SAP Build Process Automation software development kit (SDK) packages or add another business process project or an action project. This allows you to leverage existing resources, ensure consistency, and save development time.

Dependencies are added from the **Settings** area of the business process project as explained in Chapter 4. SDK dependencies are the dependencies related to the task automation artifact. For example, if you want to use Outlook activities in your automation, you need to add the Outlook SDK as dependency in your project using the **Add** button shown in Figure 5.29.

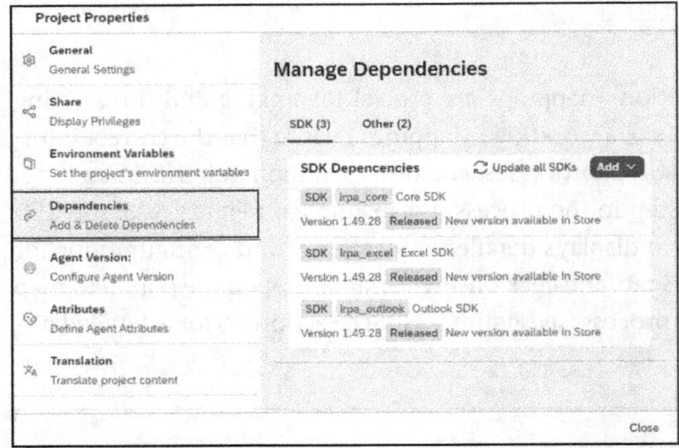

Figure 5.29 SDK Dependencies in a Business Process Project

The **Other** tab includes dependencies from action projects and other business process projects. Only the projects published to the library can be consumed as dependency projects. Business process projects can be published to the library from the SAP Build lobby. To do so, select the project, click on the three dots icon on the side panel, and then select **Publish to Library**, as shown in Figure 5.30.

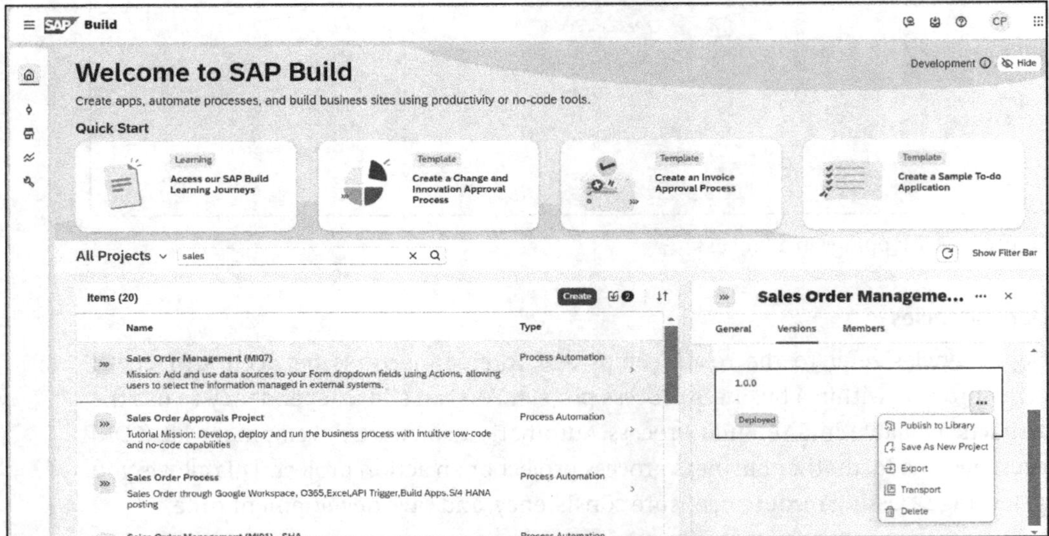

Figure 5.30 Publish Business Process Projects to the Library

Action projects can be published from the action editor, which is explained in detail in Chapter 6.

When you add these projects as dependencies by clicking on the **Add** button shown in Figure 5.31, you can consume all artifacts of those projects. This saves time and effort and further helps in modularizing the process.

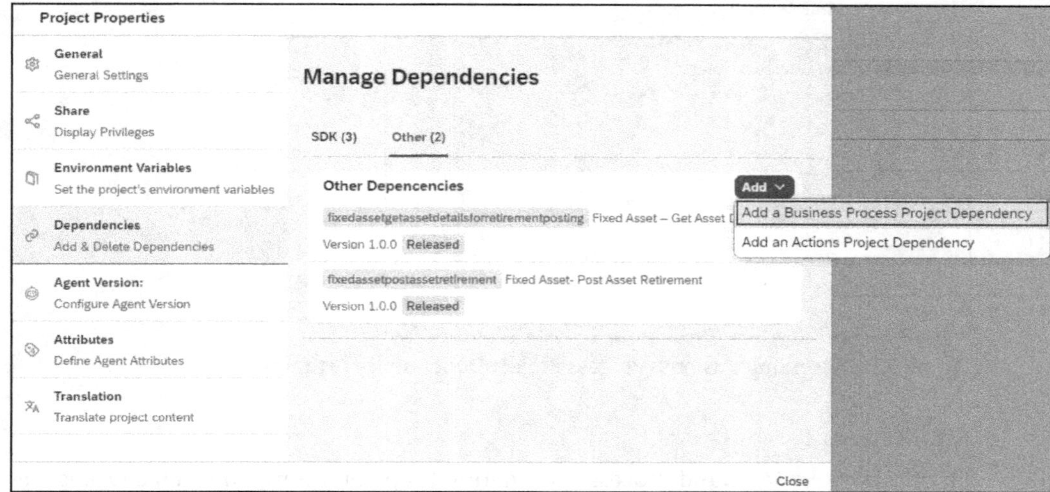

Figure 5.31 Other Project Dependencies in Business Process Projects

Business process project dependency refers to adding other business process projects into your existing business process project. The artifacts of the dependency project can be consumed in your existing project.

As shown earlier in Figure 5.30, the **Sales Order Management (MI07)** project is published to the library and is available to be added as a dependency project.

The artifacts can be accessed in other business process projects as well. Open any business project from the SAP Build lobby. Navigate to the process artifact, and click on the **+** button in the process. It displays all the artifacts and activities, such as the **Form** artifact. The **Available Forms** list for Sales Order Manager (MI07) are available for consumption in the process, as shown in Figure 5.32.

When you add an action project as a dependency, you can consume all actions created in the project (for more details, see Chapter 6, Section 6.5).

The added dependencies can also be viewed in the **Overview** area of the project. When a newer version of the dependency becomes available, the user can update the dependency to the latest version from the **Settings** area of the project.

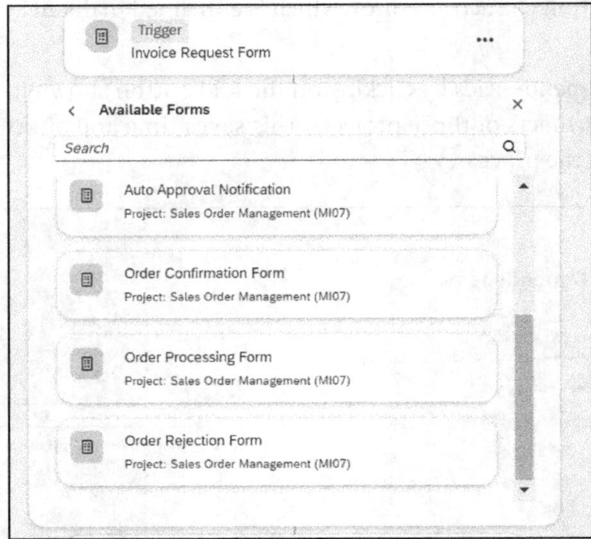

Figure 5.32 Accessing Business Process Project Dependency Artifacts

Notifications

Notifications in SAP Build Process Automation keep you informed when a task is created in SAP Task Center or the My Inbox app of SAP Build Work Zone, standard edition. By receiving real-time alerts, users can respond to the task quickly before it passes the due date to ensure business continuity without delays.

The prerequisite to receive notifications is to configure SAP Build Work Zone, standard edition, with SAP Task Center enabled or to configure the My Inbox app. Refer to SAP Help Portal for detailed information on enabling notifications in SAP Build Process Automation. Notifications are available by clicking the bell icon, as shown in Figure 5.33.

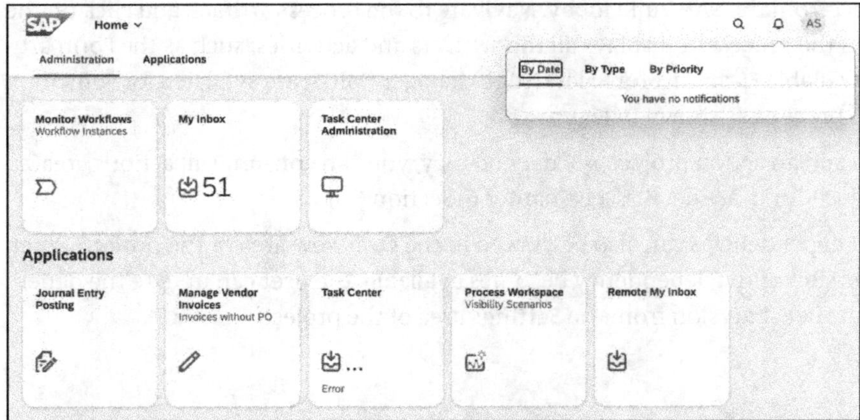

Figure 5.33 Notifications in SAP Task Center

5.4 Events

An event is a significant change in state. Events are originated from the event source to inform users of the change. Event sources can be SAP S/4HANA systems, SAP SuccessFactors, IoT devices, or third-party business applications. Event brokers typically distribute the events and send the event descriptions. Event descriptions generally hold minimal data, such as just the information related to the change. Event consumers registered with the event broker get informed about these significant changes.

In this section, you'll learn about events, understand the importance of event-driven architecture, discover different types of events available in SAP Build Process Automation, and learn how to use events effectively.

5.4.1 Enabling Event-Driven Architecture

Event-driven architecture provides significant benefits from both a technology and business perspective. It allows businesses to improve their situational awareness by getting information in real time. Businesses can base their business decisions on up-to-date information. Organizations can hyperautomate their business processes for improved reaction time and lower costs. From a technology perspective, event-driven architecture allows loose coupling, resulting in flexibility and scalability. It provides incremental growth by adding event consumers and event sources step-by-step to enable better operations and improved quality. The architecture allows for a new kind of real-time technical integrations and extensions and allows businesses to build responsive, decoupled, and scalable systems by leveraging asynchronous event communication. This architecture is particularly beneficial for modern, distributed, and real-time systems.

Events come in two categories: notification events and data events. *Notification events* alert users, administrators, or managers about noteworthy occurrences in a computer network. These occurrences can range from problems or failures in network devices to crossing threshold values or even informational notifications such as upgrades. *Data events* refer to changes in the data in the backend.

SAP Build Process Automation integrates the event-driven approach by allowing workflows to be triggered by various types of events. This enables organizations to build scalable, automated processes that respond dynamically to business needs and system events.

5.4.2 Types of Events

SAP Build Process Automation supports events that are raised from external sources such as SAP S/4HANA systems or any third-party applications, as well as events raised and handled within the process. In this section, we'll describe the different types of events: business, escalation, and timer.

Business Events

Business events are data events where relevant state transition is defined in an event schema. SAP Build Process Automation supports the events that are published in SAP Business Accelerator Hub and also the event sources from third-party applications by uploading an event specification file.

Here's how it works:

1. SAP Event Mesh is configured in the SAP BTP subaccount to listen for events from systems.
2. A business event is created in any third-party system.
3. The SAP Event Mesh instance listening to this event gets the event details and triggers the corresponding process in SAP Build Process Automation.

The prerequisite to consume events in SAP Business Accelerator Hub is to configure SAP Integration Suite, advanced event mesh, or the SAP Event Mesh service in SAP BTP. Refer to SAP Help Portal for how to configure the service and connect to SAP Build Process Automation.

Let's consider the scenario of updating the billing block of a sales order in the order-to-cash process. A billing block is applied to a sales order by internal sales representatives to prevent an invoice from being created if there is a sales return. These billing blocks need to be updated or deleted in random or on-demand fashion, and it's a time-critical activity in the order-to-cash process. In many organizations, this process is still manual, where internal sales representatives get billing block change requests in the form of a Microsoft Excel or email attachment. Sales representatives download these attachments, and then sales orders are manually updated in the SAP S/4HANA system based on data in the Microsoft Excel files or attachments.

By leveraging events in SAP Build Process Automation, you can modify or extend the standard business process from SAP S/4HANA. Because all the SAP S/4HANA events are available in SAP Business Accelerator Hub, you can directly consume all events of the sales order. To create business events, navigate to the SAP Build homepage, and select **Events** under the **Connectors** area, which provides you with options to create events from SAP Business Accelerator Hub or by uploading an API specification, as shown in Figure 5.34.

In our scenario, we need the **Billing Block Status Changed** event, which is triggered when the billing block of the sales order is modified. The billing block status changed event schema sends the information regarding the required sales order and billing block status details, sales order number, distribution channel, and organization division, as shown in Figure 5.35.

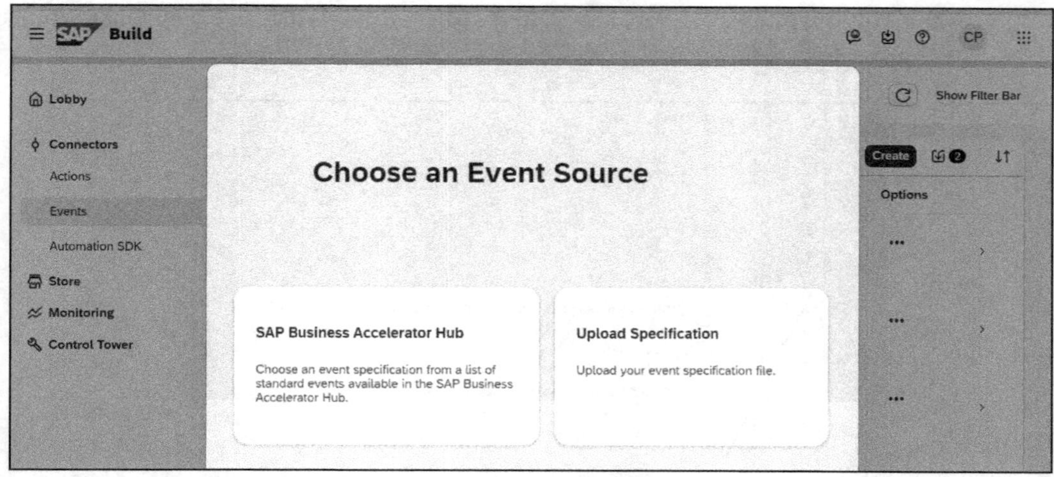

Figure 5.34 Creation of Events in SAP Build Process Automation

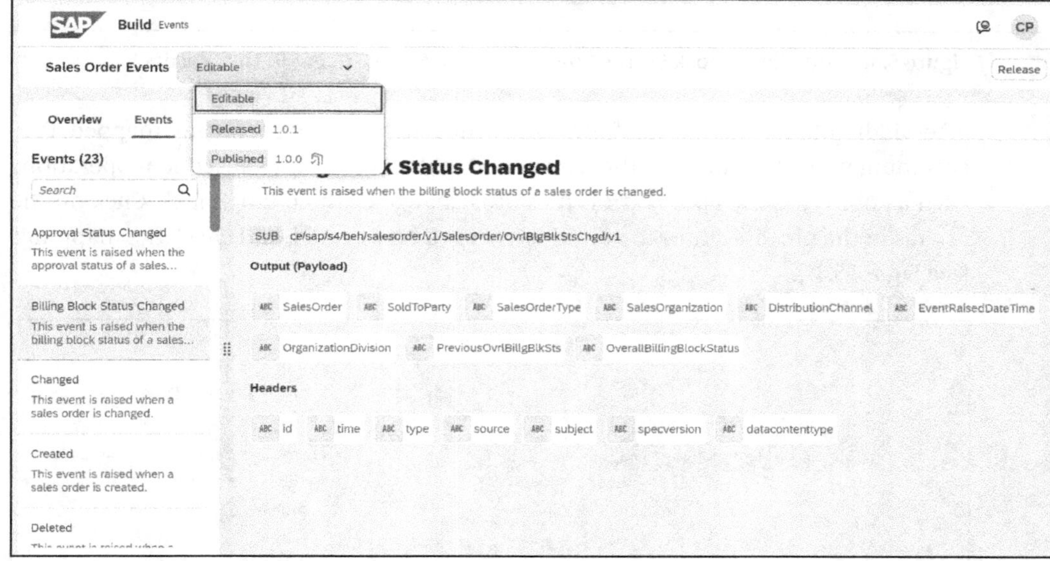

Figure 5.35 Sales Order Event: Billing Block Status Changed

After successful creation of the events, the project needs to be released and published to the library, which is available for consumption in the business process project. The standard business process from SAP S/4HANA is modified by creating a business process with the event trigger to trigger the process when the billing block status is changed in SAP S/4HANA. To consume the published events in a business process, navigate to the desired business process project from the SAP Build lobby, open the process artifact, and click on **Add a Trigger**. A popup appears to select the ways to initiate your process. Because your process must be started by an event, select **Wait for an**

Event. This opens a list of published events. Select the **SalesOrder Billing Block Status Changed** event, as shown in Figure 5.36.

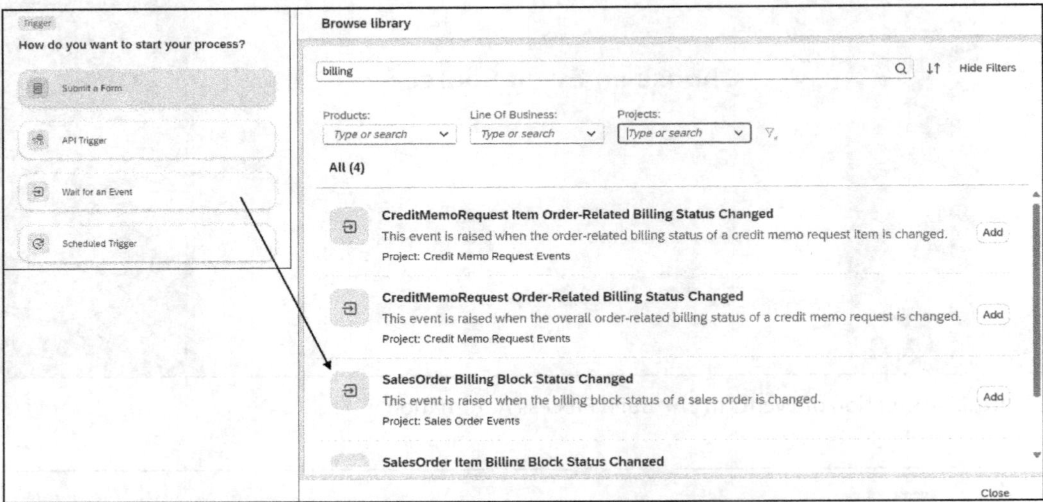

Figure 5.36 Add Billing Block Status Changed as the Event Trigger in the Process

After adding the event successfully, the event schema is automatically mapped. Process input variables can be further consumed in the process for subsequent operations such as notifying the approver of the status change of the billing block. Click on the canvas in the process editor to view the **Process Details** panel, and check the mappings (see Figure 5.37).

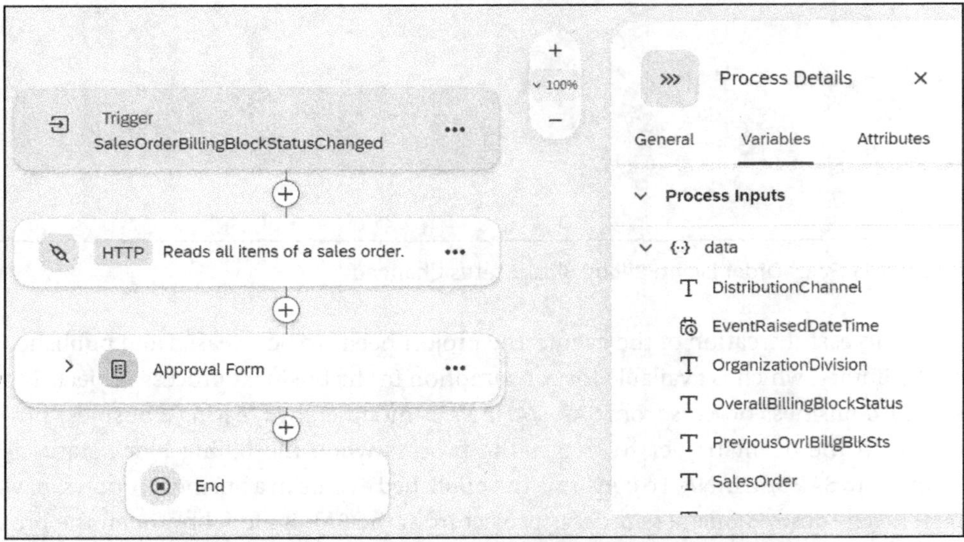

Figure 5.37 Extended Standard Business Process from SAP S/4HANA with SAP Build Process Automation

The process is triggered when there is any change in the billing block in the SAP S/4HANA system. The status of the process and event information can be viewed in the **Monitoring** area of the SAP Build homepage, as shown in Figure 5.38. Chapter 11, Section 11.4, explains in detail the technical monitoring of triggers.

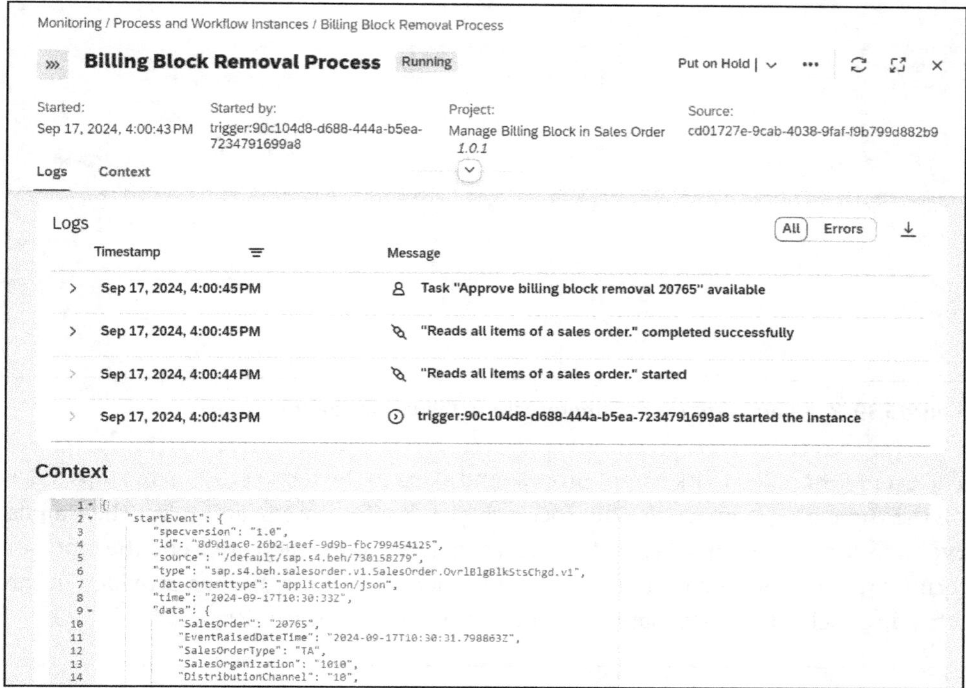

Figure 5.38 Monitor the Process Triggered by the Event

Escalation Event

The escalation event in SAP Build Process Automation communicates to a higher flow scope. These events can be raised in the subprocess or a workflow and can be handled in its parent process. Escalations are noncritical and noninterrupting events where the outgoing sequence flows of `Throw` and `Catch` are both taken. The subprocess throws an event, and this throws the escalation to a higher flow scope. The escalation event is caught by the `Catch` event in the parent process.

For example, in the process flow, the forms need to be validated for errors before being sent for approval. If a form contain errors, an event can be raised to notify the requestor, which is caught in the main process. In the subprocess, click on the **+** button, and select **Controls and Events**. Click on the **Raise Event** option to create an event, as shown in Figure 5.39.

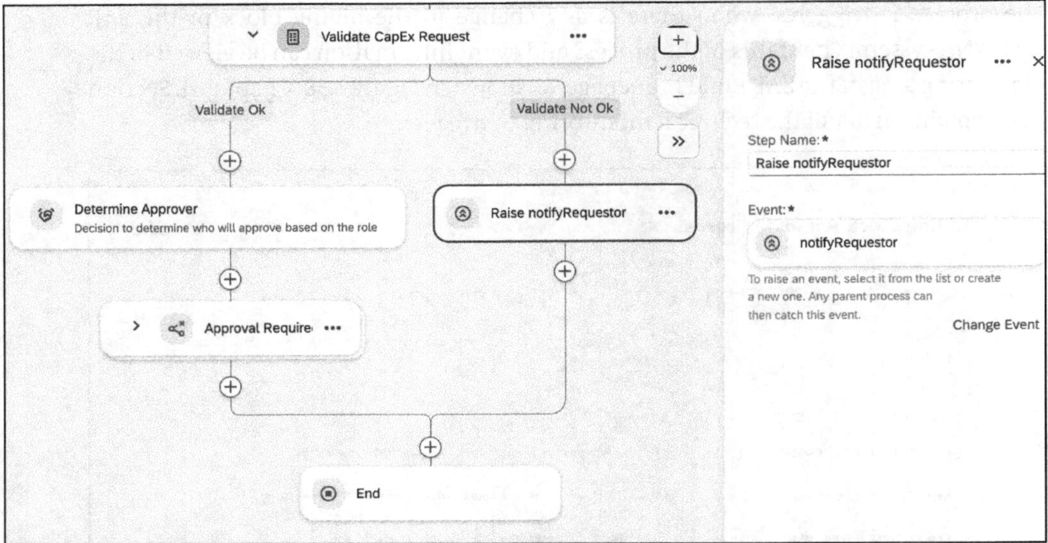

Figure 5.39 Raise an Event in the Subprocess on Validation Failure

The event is handled in its parent process and skips to the next subsequent steps. Click on the three dots icon on the subprocess in your main process, and select the **Handle Event** option, as shown in Figure 5.40. You can move to the next step, end the block, or terminate the process based on your requirements from the options provided in the left side panel of the event handling block, as shown in Figure 5.40.

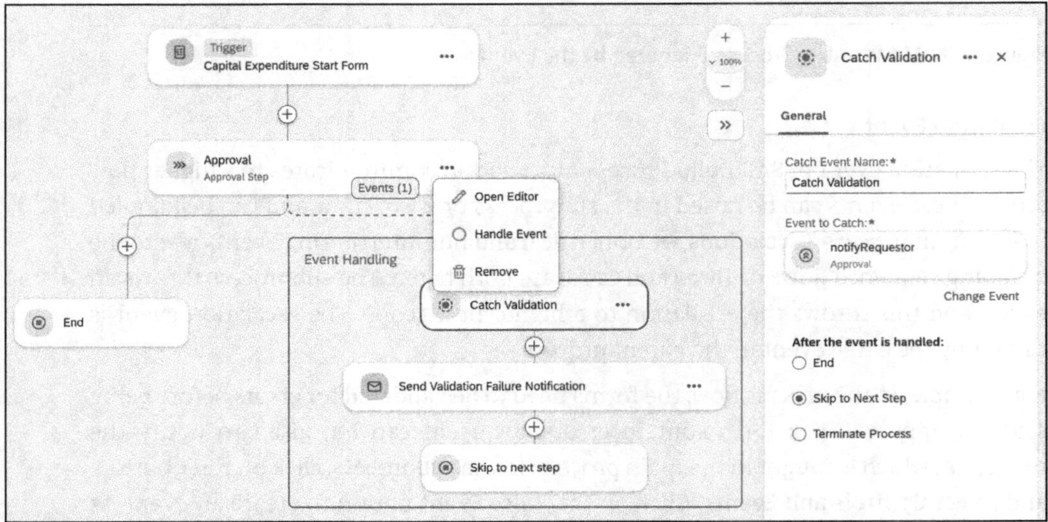

Figure 5.40 Handling the Escalation Event in the Main Process

Timer Event

Timer events in SAP Build Process Automation are triggered by a defined timer. This event type can be raised for a form or approval form. A process can have one or more timer events.

A typical scenario of for a timer event is a purchase order that is submitted for approval. If the designated approver doesn't approve the task within 48 hours, you want to send an automated reminder email to approve the task. If the task is still not approved in the designated time, the task should be automatically forwarded to another approver.

To achieve this use case, set the due date of the approval form as 48 hours by clicking on the **Approval Form** artifact, which opens a side panel. Maintain the **Due Date** for the approval form in the **General** tab, as shown in Figure 5.41.

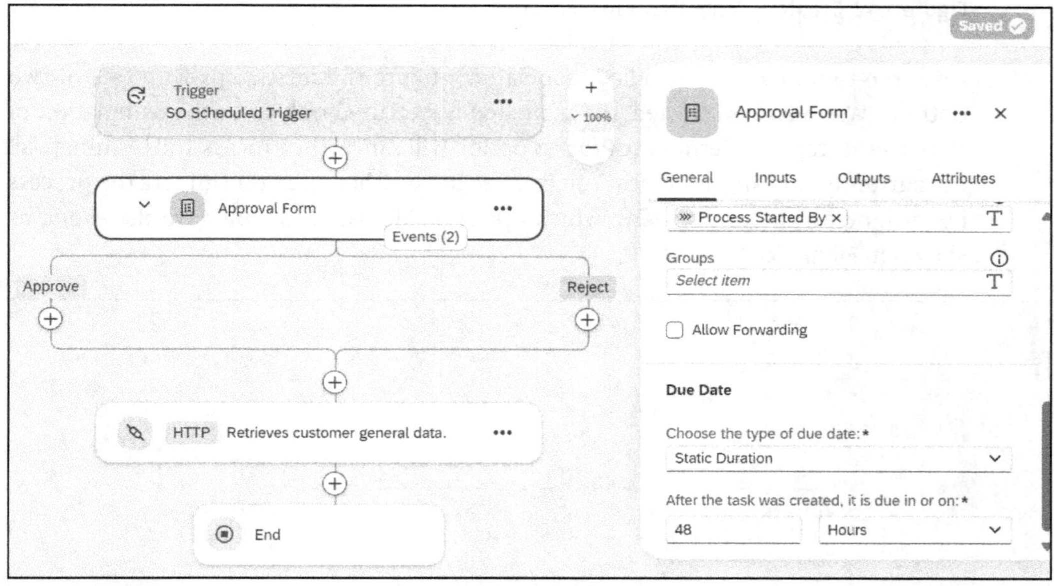

Figure 5.41 Setting the Due Date Configuration for an Approval Form

Create two timer events: one timer event is to send a reminder email to the approver and set the timer till the due date is received. The other timer event will forward the task to another approver. The steps to assign/forward the task to another approver based on predefined rules can be done in a subprocess. These timers will be raised after the task is created. To create the timer events, click on the three dots icon on the **Approval Form** artifact, and select **Handle Event**, as shown in Figure 5.42. An **Event Handling** branch is added. Select the **Timer** event to create the events. As shown in Figure 5.42, two timer events are created for our scenario as explained previously.

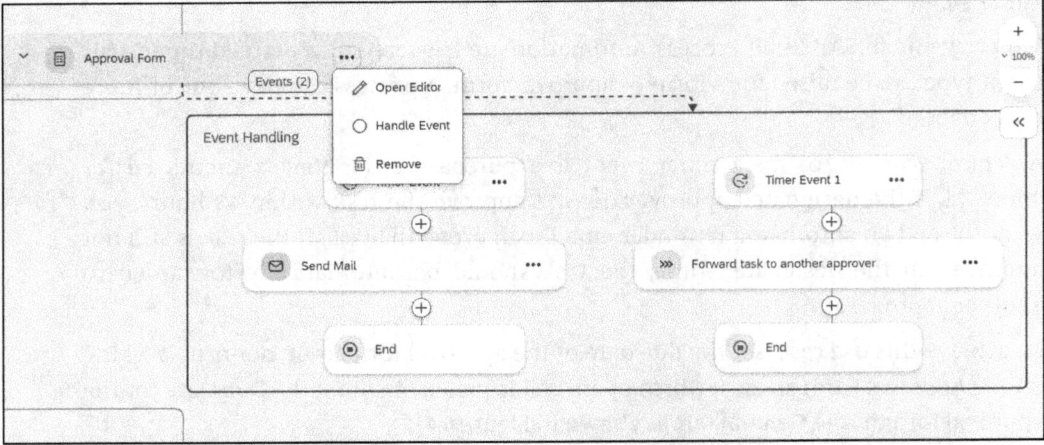

Figure 5.42 Create a Timer Event in a Process

After the timer event is handled, you can configure the next steps with one of two options. When **End** is selected, the next step is executed only after the completion of the current step. The **Terminate Process** option will cancel the process and terminate all the currently running activities. You can decide to either end or terminate the process by clicking on the timer event, which opens a side panel to configure the event, as shown in Figure 5.43.

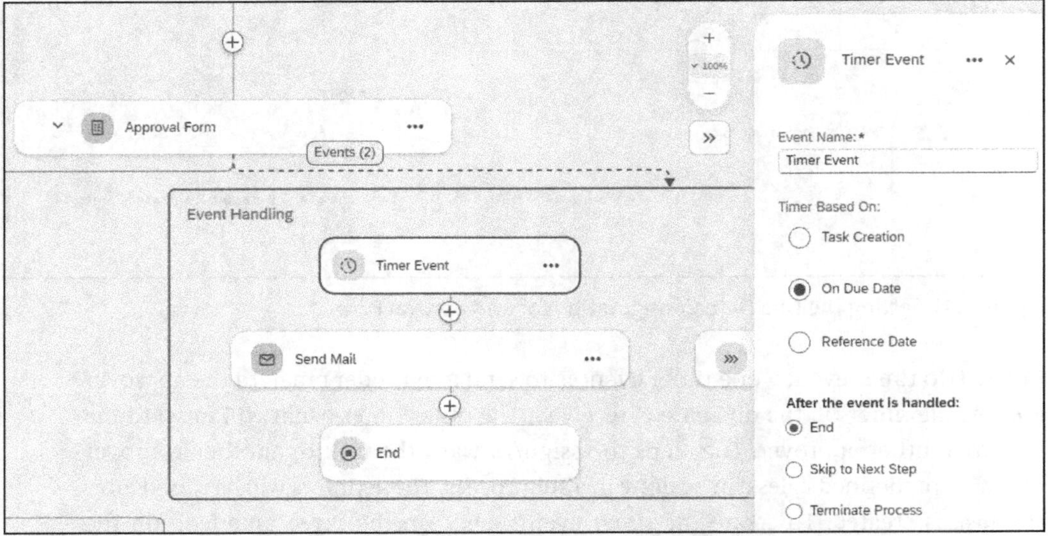

Figure 5.43 Timer Event Handling Configurations

5.5 Branches

Branches are used to create decision points within workflows, allowing the process to take different paths based on specific conditions or criteria. This enables more dynamic and responsive process flows. The conditions are usually based on data values, user inputs, or other criteria relevant to the process. Branches allow flexibility and scalability of the process, eventually increasing the efficiency of the process. The SAP Build process supports conditional and parallel branching, which we'll describe in the following sections.

5.5.1 Conditional Branch

Conditional branches execute different paths based on whether a specified condition is true or false. If the condition is satisfied, it navigates to the **If** branch; otherwise, it executes the default branch. You can add and modify the names of the branches for easy understanding in the side panel by selecting the artifact, as shown in Figure 5.44.

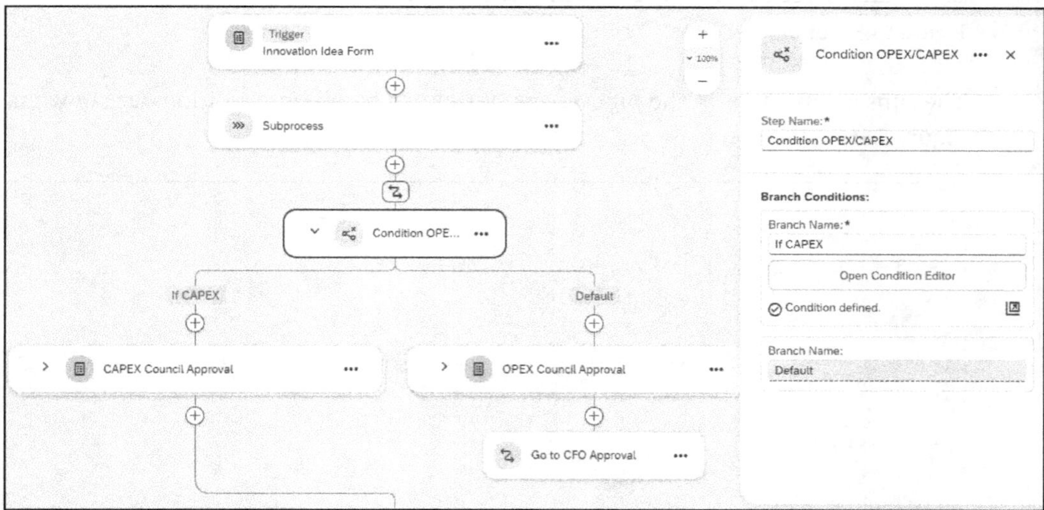

Figure 5.44 Renaming Branches in Conditional Branching

Using the condition editor, you can add conditions and group them with AND/OR logic. The conditions added in the editor can be viewed in the summary section. Click on **Open Condition Editor**, as shown in Figure 5.44 to configure the conditions shown in Figure 5.45. Typical examples where we use conditions are sending a request for manager approval or auto-approval based on the order amount or sending approval requests based on the shipping country.

Figure 5.45 Condition Editor

The subsequent steps in the process can be defined based on conditions, as shown in Figure 5.46.

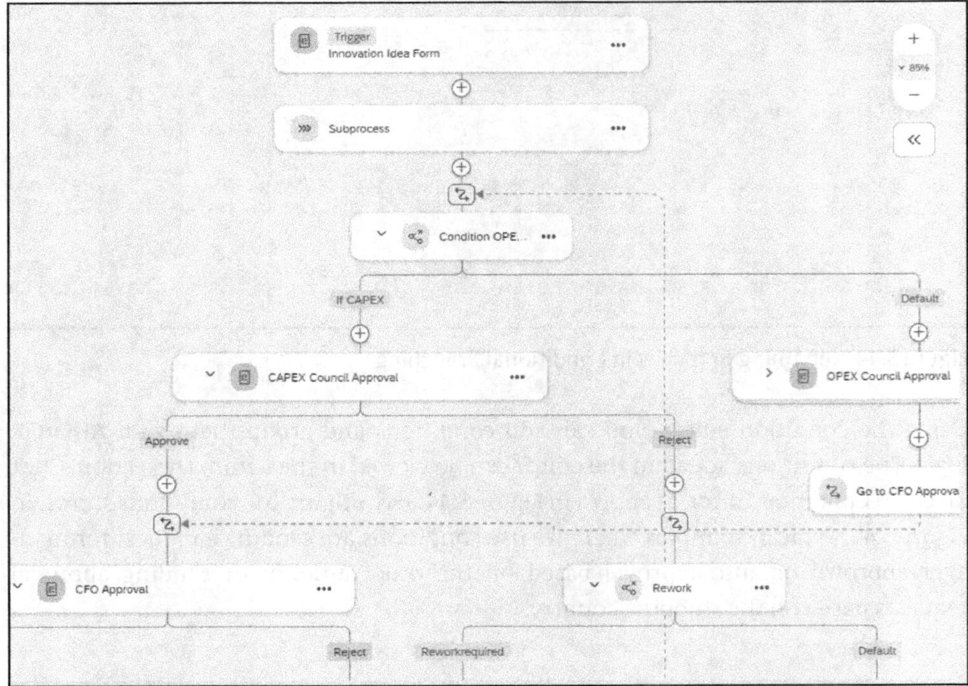

Figure 5.46 Conditional Branching in a Process

5.5.2 Parallel Branch

Parallel branches execute multiple paths simultaneously, converging back into a single path upon completion. The process is terminated if the execution of one of the branches fails.

Let's consider the scenario of determining the equipment and training for a new employee and send both the equipment and training details for a manager approval. In this case, you can create a parallel branch and, by leveraging the business rule capability, create two decision artifacts that will determine equipment and training details based on the employee grade level. This will enable the determination of equipment and training to occur simultaneously, improving processing speed as there is no interdependency between the two decisions. After successful execution of the branches, the branch is merged, and both the equipment and training details are available in the approval form. Parallel branches can be created by clicking on the **+** button and selecting **Controls and Events**. Click on **Parallel Branch** to add a parallel branch. You can configure the number of branches by selecting the activity, which opens up a side panel to configure the parallel branch. Figure 5.47 shows a sample process with parallel branches.

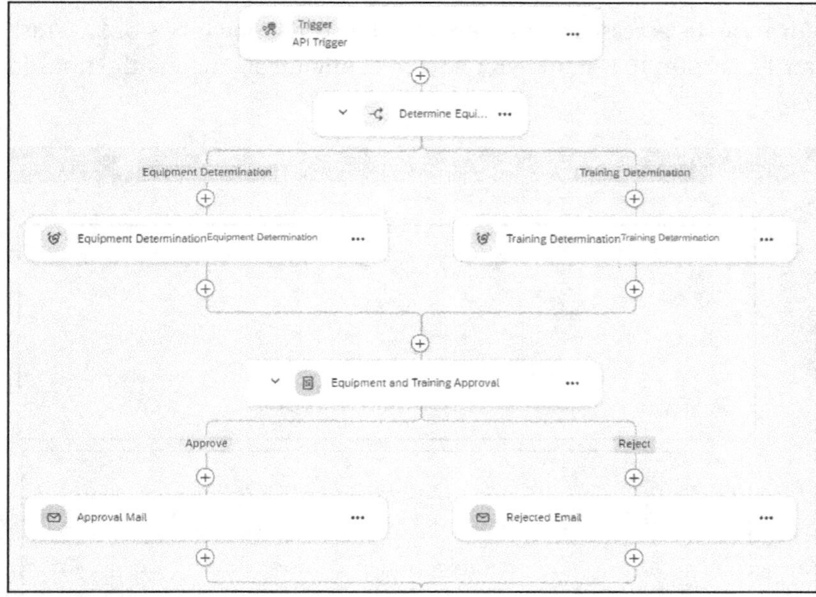

Figure 5.47 Parallel Branches in a Process

You can check the execution of the branches in the **Monitoring** area of the SAP Build homepage. (For more details on technical monitoring of process and workflows, see Chapter 11, Section 11.4.1.) After successful execution of the branches, the branch is merged, and both the equipment and training details are available in the approval form, as shown in Figure 5.48.

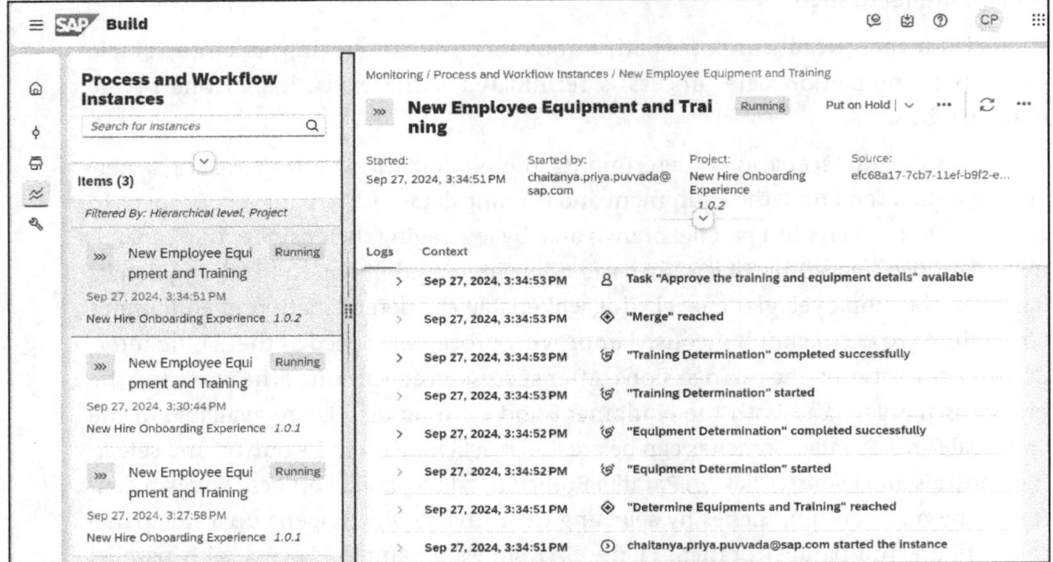

Figure 5.48 Parallel Branches Execution in the Monitoring Area

The approval form can be accessed from the SAP Build Work Zone inbox or SAP Task Center. (Chapter 16, Section 16.1, shows you how to configure and access the tasks in SAP Task Center.)

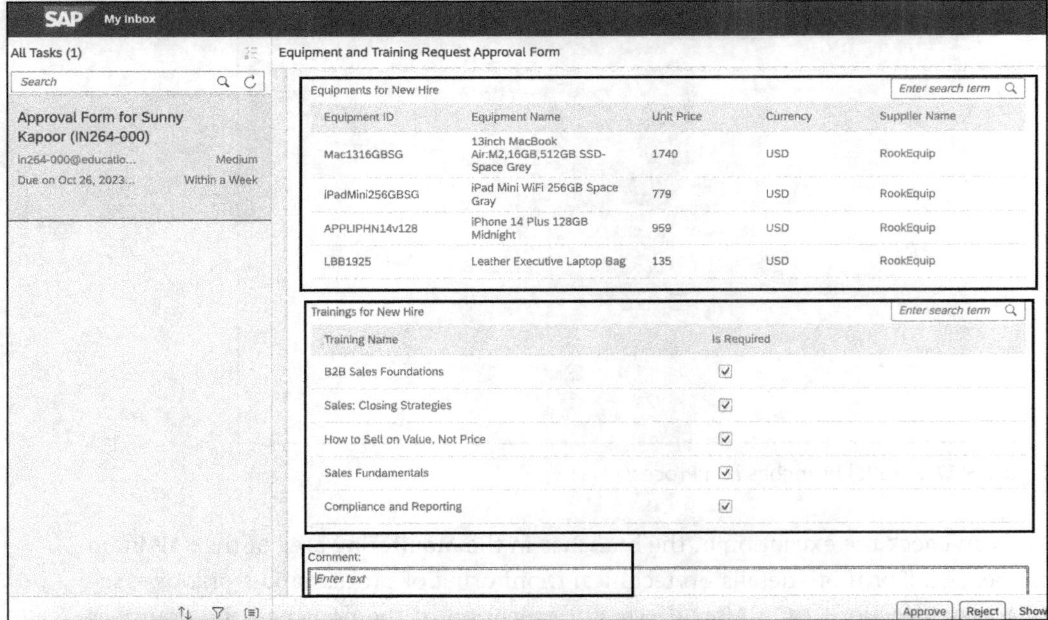

Figure 5.49 Approval Form with Training and Equipment Details

As shown in Figure 5.49, after successful execution of branches, the approval form consists of details about both the branches.

5.6 Mappings

Mappings define the data transfer between different components in a workflow and ensures that the right data flows between the components and maintains data consistency across different stages of a workflow. One of the key components involved in the mappings is context data, which refers to the data that is used in the process and flows between different process elements. Data binding refers to how the data in different artifacts is connected to the variables. Variables in process are mapped to the corresponding inputs and outputs across activities.

Benefits of Mapping Processes

Mapping ensures that the data collected or generated in the process is accurately passed to another activity reducing the risk of data loss or errors. Mapping makes it easier to integrate with external systems and APIs by defining how data is mapped in internal workflows and external applications while ensuring data exchange happens efficiently. You'll experience better data management and clear visibility into data paths, which makes it easier to monitor and optimize the workflows as needed. Well-defined mappings lead to simplified maintenance as changes to one part of the workflow can be easily managed without effecting the entire process.

Next, we'll discuss the bounding process context for artifact properties. In SAP Build Process Automation, the context data of a process is available in process context and data flows within different artifacts such as forms, decisions, automations, and actions by mapping to the corresponding input and output variables across the artifacts or activities. A process can be initiated through forms, triggers, or events. The form data or trigger and event parameters are automatically mapped as process input variables, as explained in Section 5.4.2. The subsequent steps of the process require a manual mapping between the input and output variables of the artifacts or activities.

Let's understand mappings in SAP Build Process Automation with an example. Imagine a business process where users need to submit a purchase request form. The data from the form, such as item name, quantity, and price, needs to be passed to a decision artifact that identifies the right approver based on certain rules. Once the right approver is identified, the data is passed to the **Approval Form** artifact, as shown in Figure 5.50, and then transferred to action artifact to create a purchase order if it's approved.

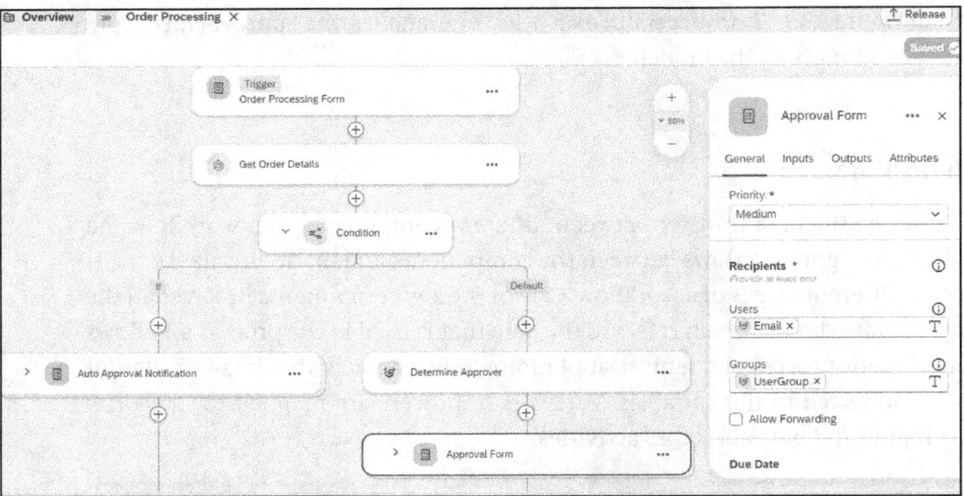

Figure 5.50 Mappings of Decision Artifact Output Variables to Approval Form Data

5.7 Forms

The form builder in SAP Build Process Automation allows you to design interactive forms to capture and share the information. By using the intuitive drag-and-drop interface, you can add input controls such as text, dropdown, checkbox, file controls, or tables. The fields can be validated to minimize errors during the submission process. These forms can be duplicated to speed up the creation of new forms. It can be added as a start trigger to initiate the process or used as a step later in the same process. Apart from data collection, Forms can be used to send notifications to the users. SAP Build Process Automation supports several types of forms such as a basic form, approval form, and SAPUI5-based form.

This section provides a comprehensive overview of forms in SAP Build Process Automation, including how to create a form using the form builder, the various types of forms available, and the best practices for using each type of form.

5.7.1 Need for Forms

First, let's start by understanding why forms are necessary. Forms serve as a primary interface for collecting input from end users. They enable users to provide necessary information, make selections, or enter data to initiate a process. This is particularly useful in purchase requisitions, customer feedback collection, and many more scenarios. It helps to automate routine tasks such as approvals, data validation, and document submission, thereby speeding up processes and enhancing overall efficiency. Using forms also facilitates collaboration among multiple stakeholders by enabling structured information exchange and decision-making within automated workflows.

For example, let's consider a scenario of initiating a new product development process in a manufacturing company that involves multiple stakeholders across various departments such as R&D, marketing, finance, procurement, and sales. The goal is to collaborate efficiently to develop, approve, and launch the product. This process has steps for cross-departmental approval and for providing feedback. The approval forms can be simultaneously sent to various departments, have a sequential flow of approvals, or have approvals based on conditions.

By embedding forms within automated business processes, organizations can streamline their processes, minimize errors by data validation, gather approvals, and coordinate tasks among multiple stakeholders.

5.7.2 Form Builder

The *form builder* is an editor in SAP Build Process Automation to create forms with an intuitive interface by dragging and dropping. Forms can be created either from the **Overview** area shown in Figure 5.51 ❶ or when needed while building a process flow in the process editor by clicking the **+** button in the workflow, which opens a view to select the **Form** artifact ❷.

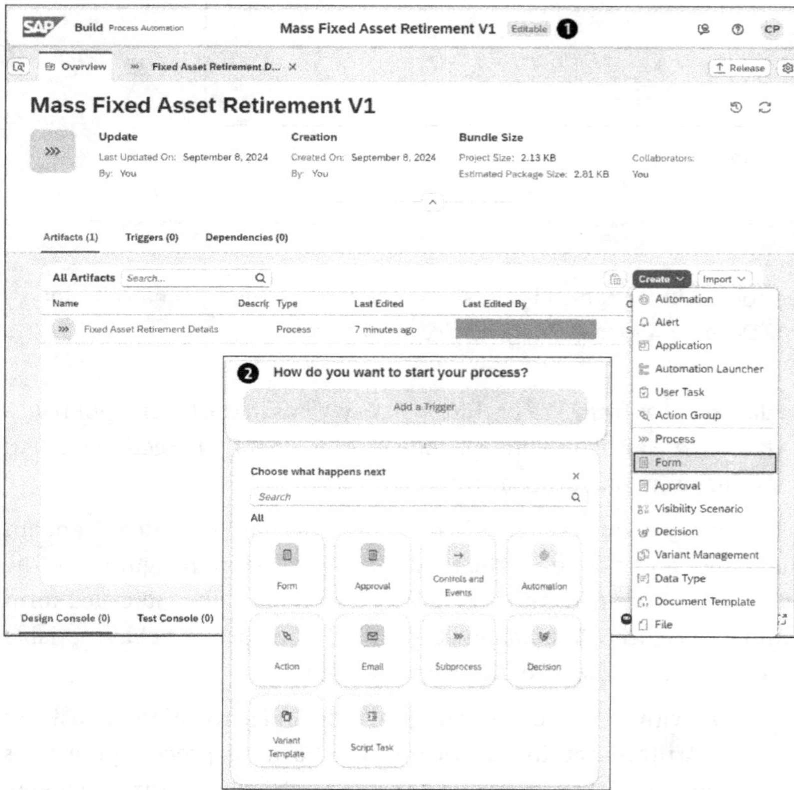

Figure 5.51 Different Ways to Create a Form

Users can create clean and easy-to-consume layouts such as headlines and paragraph simply by dragging and dropping controls onto the canvas. You can incorporate text, number and date fields, dropdown menus, tables, links, file uploads, and checkboxes, as shown in Figure 5.52.

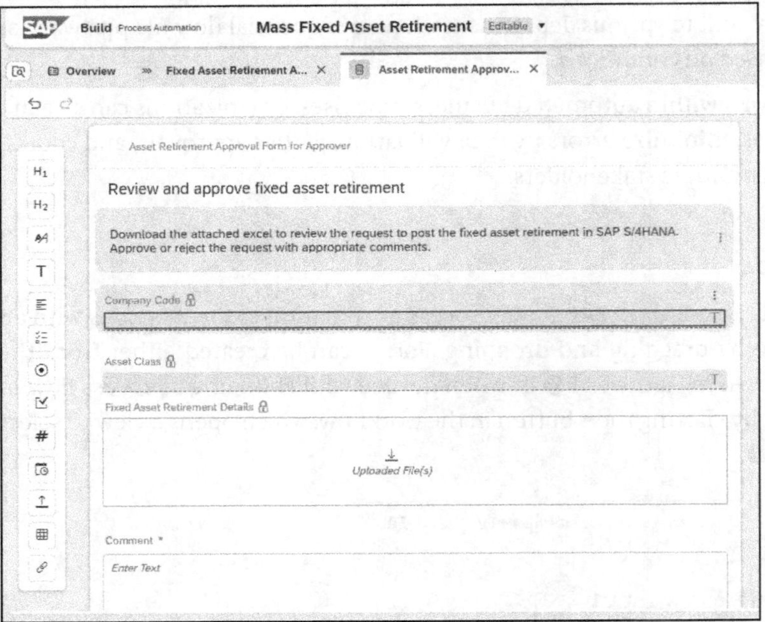

Figure 5.52 Form Builder

> **Note**
> The **File Upload** control isn't enabled by default in the form builder. To enable it, subscribe to the SAP Document Management service in SAP BTP.

You can also add data sources from SAP or third-party systems in the form input fields. The prerequisite is to create an action project (explained in Chapter 6, Section 6.3.2) to mark the array of values as main output array.

Fields can be formatted, configured as required or read-only, and validated. Standard and custom validations can be done for the input fields, as shown in Figure 5.53. The labels of the form controls are generated as variables automatically and called form outputs. For example, in Figure 5.52, **Company Code** is a UI control text field. A variable named **Company Code** is created automatically by the framework.

Forms can be duplicated within the process and can also be copied to different business process project in the **Artifacts** tab in **Overview** of the business process project, as shown in Figure 5.54. This reduces effort and increases the efficiency of citizen developers.

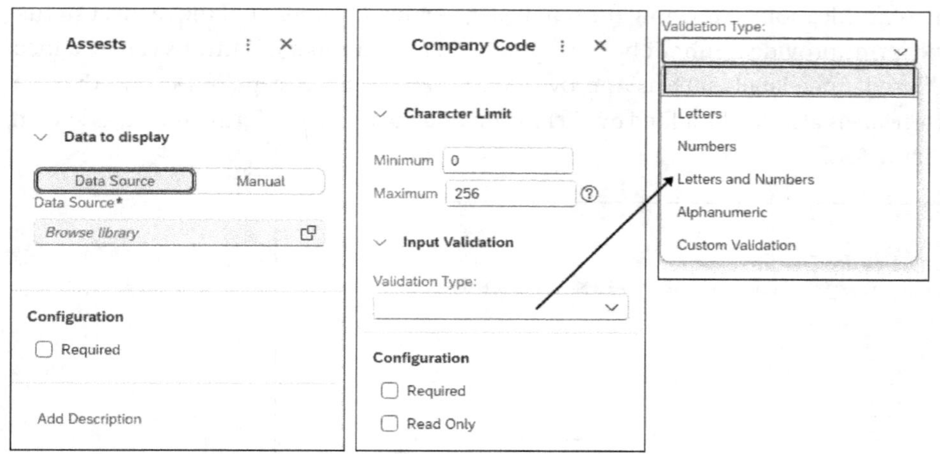

Figure 5.53 Features in Form Builder

Figure 5.54 Copy and Duplicate Forms

5.7.3 Types of Forms

SAP Build Process Automation supports three types of forms, each suited to different scenarios: basic form, approval form, and SAPUI5-based form. We'll discuss each in the following sections.

Basic Forms

The simple basic form is best suited for straightforward data-collections tasks or basic user actions. Some common use cases include purchase requisition requests, asset requests, and customer feedback forms. Apart from data collection, these forms can be sent as notifications informing the purchaser, vendor, or supplier of the request status. Basic forms provide a **Submit** button by default. Any number of buttons can be added with predefined labels such as Approve, Reject, Next, Send, Accept, or Decline. Custom button labels also can be added by clicking on **Add Button** in the form builder, as shown in Figure 5.55.

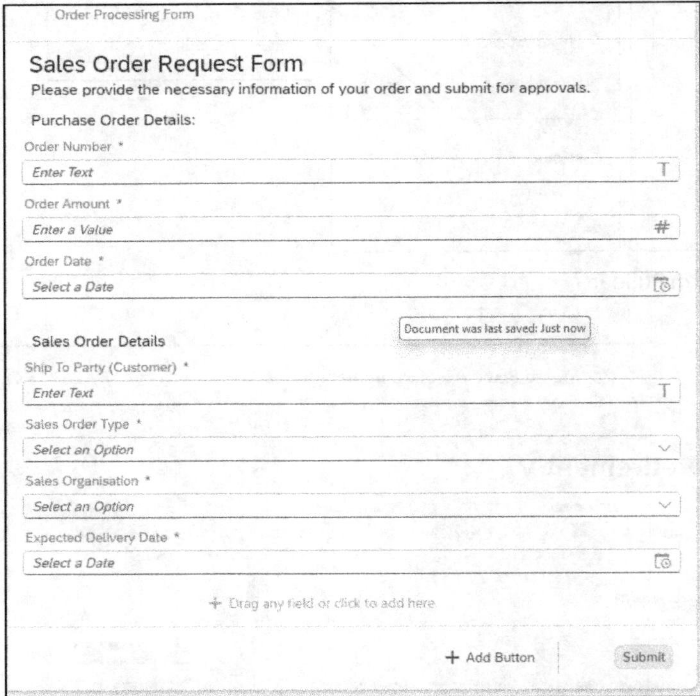

Figure 5.55 A Basic Form in Form Builder

Approval Forms

Approval forms are required when there is decision-making in the automated workflows. Based on the decision, that is, the outcome of the approval form, the subsequent actions are triggered. Approval forms can be routed to the right approvers based on predefined business rules. This routing ensures that the approval process is efficient and timely, reducing delays and manual follow-ups. Approval forms provide **Approve** and **Reject** buttons by default. You can add any number of buttons and label them accordingly just like with basic forms.

Basic forms can also have customized buttons such as **Approve** and **Reject**, which can perform similar functionality as approval forms, but the main difference between basic forms and approval forms is that the former can only be used to initiate the process. The process editor provides an option to create an approval form during the design of the process. By clicking on the **+** button and selecting **Approval** artifact, you can create the approval form and get a template for a typical approval form, containing **Approval** and **Reject** buttons by default, as shown in Figure 5.56. Required UI controls can be added as well to make form building faster.

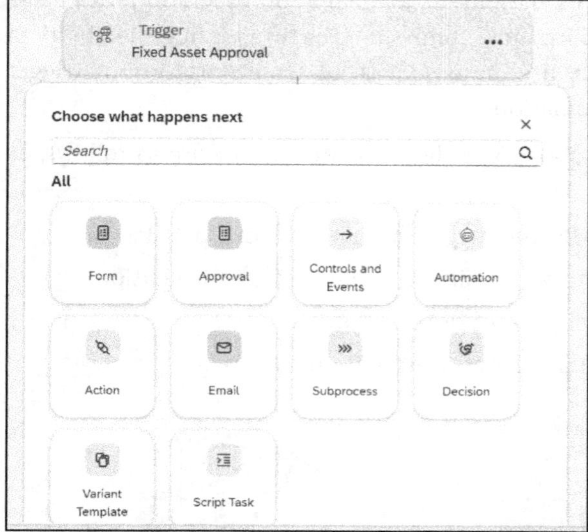

Figure 5.56 Creating an Approval Form from the Process Editor

Typical use case scenarios for approval forms are sending order details for approval before a sales order is created in the backend and approving the new joiner's equipment and training details before they are assigned to the employee, as shown in Figure 5.57.

Asset Retirement Approval Form for Approver

Review and approve fixed asset retirement

Download the attached excel to review the request to post the fixed asset retirement in SAP S/4HANA. Approve or reject the request with appropriate comments.

Company Code

Asset Class

Fixed Asset Retirement Details

Uploaded File(s)

Comment *

Enter Text

+ Drag any field or click to add here

+ Add Button Approve Reject

Figure 5.57 Approval Form

SAPUI5-Based Forms

Form builder supports integration of SAPUI5 applications into basic forms for when you need to incorporate complex validations, data lookups, or integration with back-end systems for real-time data fetching and submission. The result is huge flexibility to use custom code. These forms will be used as user tasks but not as start forms to trigger a process in SAP Build Process Automation.

The steps involved in integrating SAPUI5 applications into forms are as follows, as shown in Figure 5.58:

1. Create an SAPUI5 app using SAP Business Application Studio in SAP BTP.
2. Configure the *Manifest.json* file to be used in SAP Build Process Automation.
3. Push the SAPUI5 app to the HTML5 application repository.
4. Import the SAPUI5 app in SAP Build Process Automation.

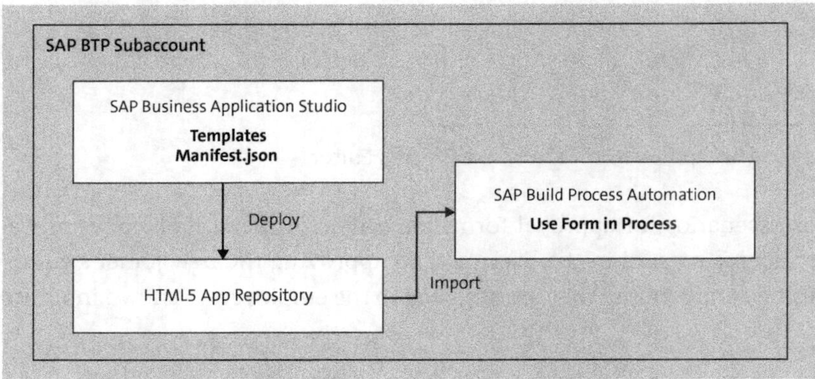

Figure 5.58 SAPUI5 App Integration in Forms of SAP Build Process Automation

After successfully creating the SAPUI5 application and deploying it in the HTML5 application repository of SAP BTP, you can import the SAPUI5 app in SAP Build Process Automation, as shown in Figure 5.59.

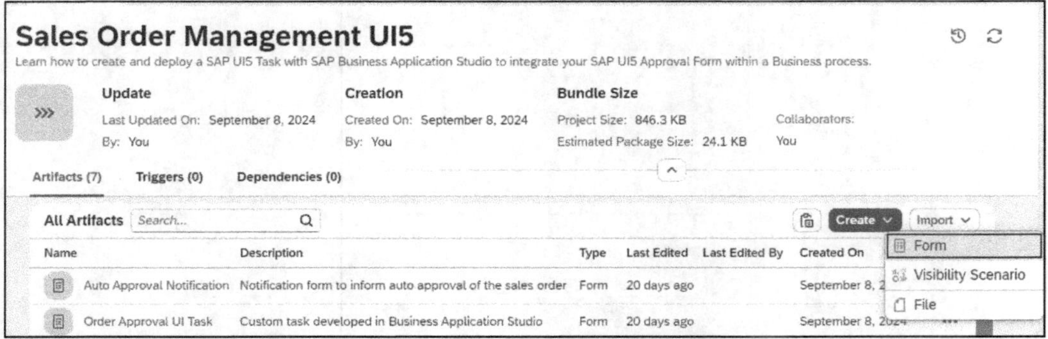

Figure 5.59 Importing the SAPUI5 App to the Business Process Project

Once the SAPUI5 app is imported, you can open the form from the **Artifacts** area to see the details, as shown in Figure 5.60.

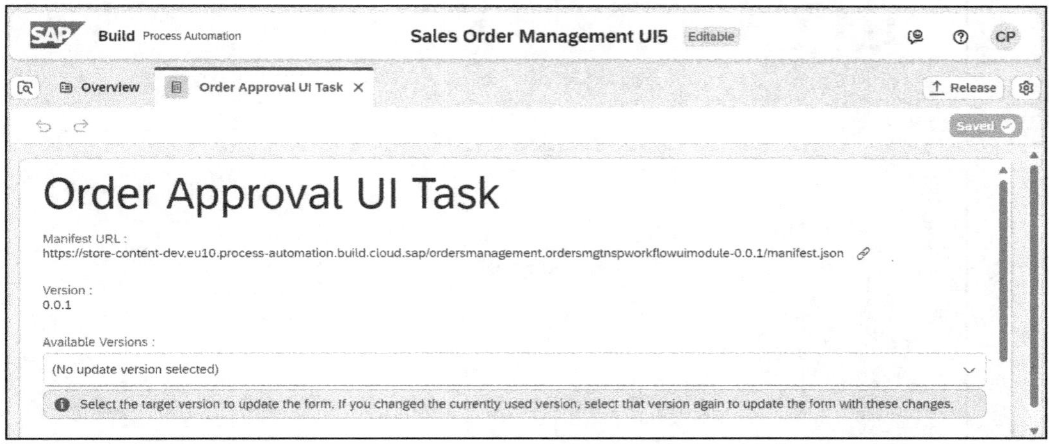

Figure 5.60 SAPUI5 App Details in a Form

By understanding the different types of forms and their appropriate use cases, you can now create forms to meet specific business requirements.

5.7.4 Form Configuration in Processes

Now let's discuss how forms are configured in a business process project. Here's an overview of the steps involved in specific areas of the screen:

- **General**

  As shown in Figure 5.61, the **General** area allows you to configure the form that is accessed by the business user who views and edits the form (typically approvers) and the users who submit the forms. The form, which is available in the SAP Build Work Zone inbox or in SAP Task Center for an action from business users, is called a task. (For more details on SAP Build Work Zone and SAP Task Center, refer to Chapter 16, Section 16.1.) The `ProcessAutomationParticipant` role must be assigned to the business user who processes the task. (See Chapter 13, Section 13.1, which details authorizations in SAP Build Process Automation.)

  You can configure the subject and set the priority of the task as **Low**, **Medium**, or **High**, which indicates how quickly the user should take action. You can assign a specific email address to the task by mapping to the context data or assign groups of users to the task by mapping the role collection maintained in SAP BTP. The task can be forwarded to another user via the **Allow Forwarding** feature. One of the most important configuration options is the **Due Date**, which informs the user of the deadline to take an action on the task. You can create timer events based on the due dates, as explained earlier in Section 5.4.2.

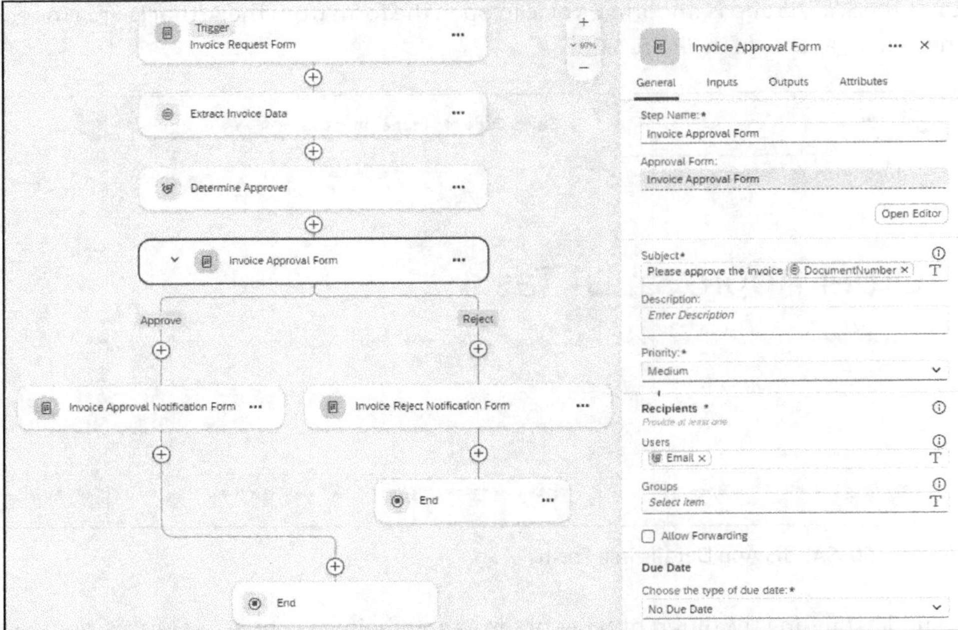

Figure 5.61 Form Configuration in the Process

Figure 5.62 shows an example of the task received in the inbox based on the form configuration settings shown in Figure 5.61.

Figure 5.62 Task in the Inbox of SAP Build Work Zone

- **Inputs/Outputs**
  Configuration of the form in the process also involves mapping form data to the variables in the process context using the **Inputs** and **Outputs** tabs. All the editable

fields of the form are automatically added as inputs and outputs of the form in the process. Read-only fields are available only as inputs to the form. If the form is used to initiate the process, the form variables are automatically mapped as process input variables as shown in Figure 5.63. The **Inputs** and **Outputs** tabs are shown in Figure 5.64.

Figure 5.63 Mapping of Form Variables for a Form as a Start Trigger

If the form is used in any other step in the process, form data must be mapped manually to the process context. As shown in the first screen in Figure 5.64, the UI controls used in an approval form in the form builder are available as form variables. You can map the forms by selecting the **Form** artifact in the process (as shown in the second screen) to open a side panel and then select the **Inputs** tab. Select the variables that require mapping. The **Process Content** popup appears where you can access all the variables of the process, as shown in Figure 5.64.

In a typical approval scenario, most of the fields in the approval form are read-only and are either mapped from the request form in the process or mapped by fetching data from the backend system. The editable fields (e.g., approver comments) are shown in the form builder in the first screen of Figure 5.64. Therefore, the **Comments** variable is available in the **Outputs** variables tab in the form of a process, as shown in Figure 5.65.

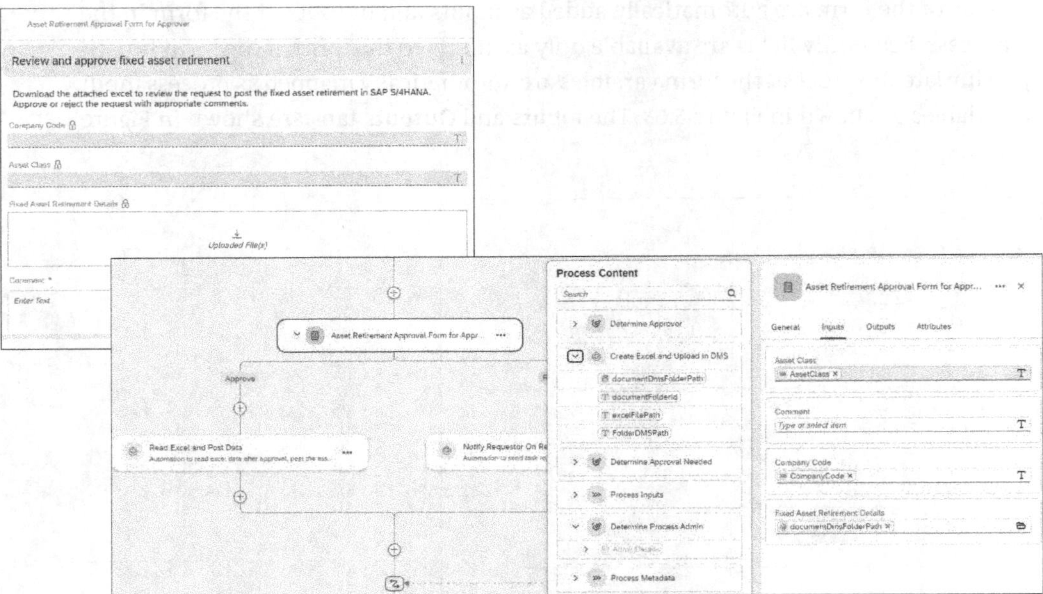

Figure 5.64 Approval Form Mapping

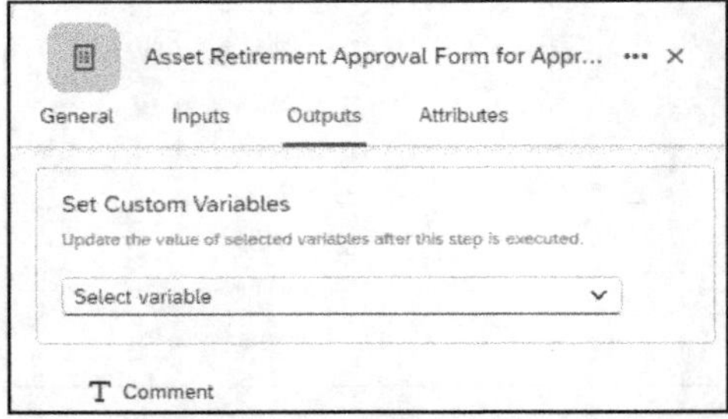

Figure 5.65 Mapping of Form Read-Only Fields as Output Variables in a Form

Based on the number of buttons in the forms, the branching in the process is determined accordingly.

Approval forms in a process can be customized to support various types of approval flows, such as sequential approvals (where each approver acts one after the other), parallel approvals (where multiple approvers can act simultaneously), or conditional approvals (where approvals depend on specific criteria). Refer to Section 5.5, which explains branching in a process.

After successful release and deployment of the process, if the form is used as a start trigger to initiate the process, it generates a **Form Link** and **Launchpad Configuration Parameter** in the **General** tab of the form artifact, as shown in Figure 5.66.

Clicking the copy icon beside **Form Link** generates the URL that can be used to start the process from a browser, as shown in Figure 5.67.

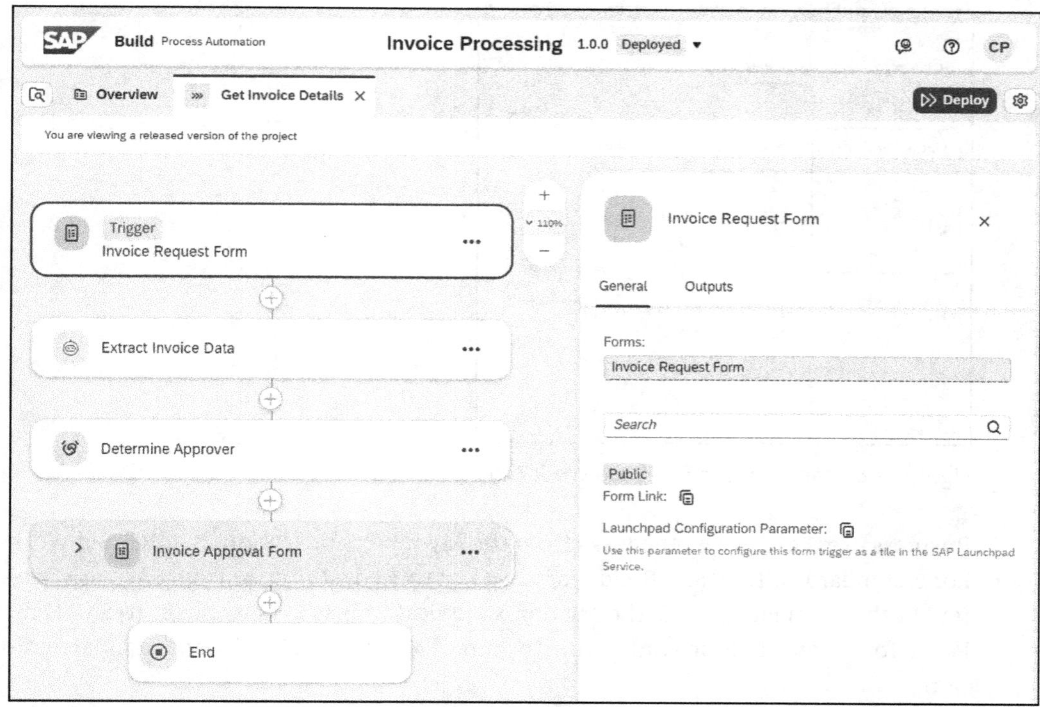

Figure 5.66 Deployed Process with a Form Trigger

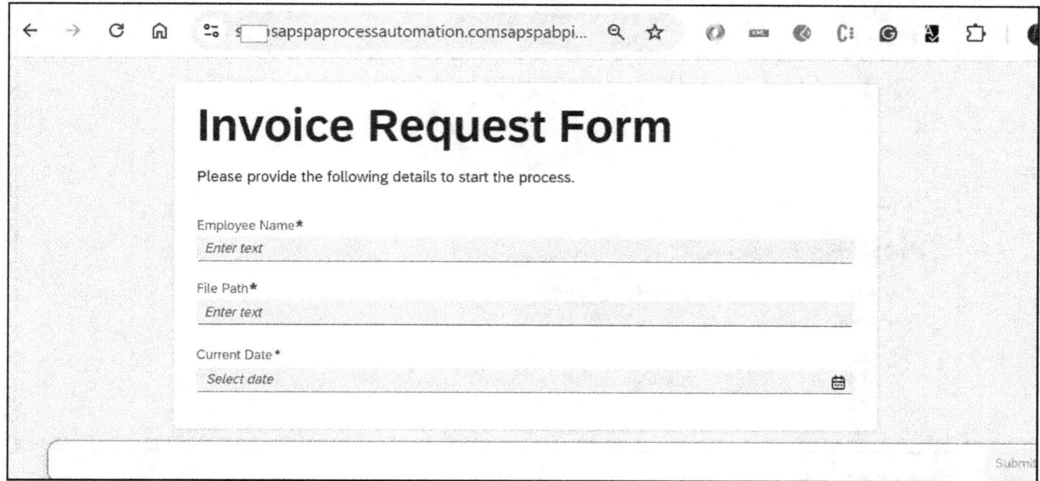

Figure 5.67 Start Process with a Form Trigger from a Browser

A process can also be started from SAP Build Work Zone, standard edition, as shown in Figure 5.68, by configuring the tile that uses **Launchpad Configuration Parameter**, which is generated after deploying the process as explained earlier. Refer to SAP Help Portal for more details on configuration in SAP Build Work Zone, standard edition.

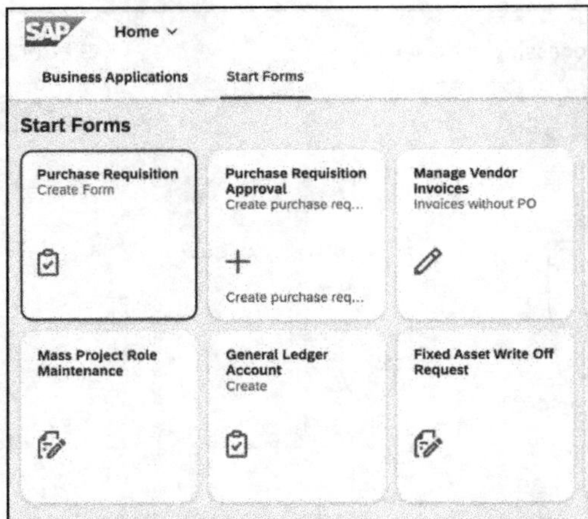

Figure 5.68 Start Forms in SAP Build Work Zone

Business users can access the tasks from the My Inbox application in SAP Build Work Zone, standard edition; SAP Build Work Zone, advanced edition; or from SAP Task Center. As shown in Figure 5.69, the My Inbox application has 51 tasks. Refer to SAP Help Portal for more details on configuring the inbox of SAP Build Work Zone, standard edition.

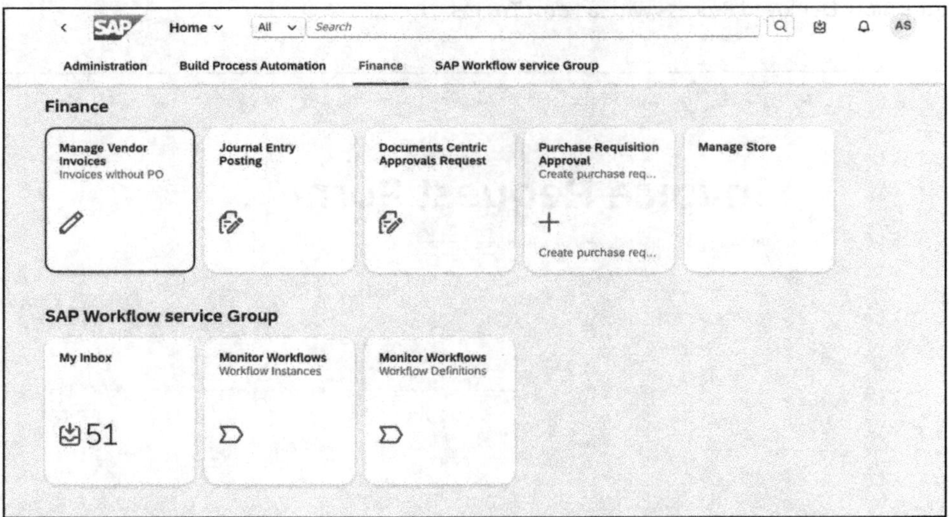

Figure 5.69 Inbox of SAP Build Work Zone, Standard Edition

Click on the **My Inbox** tile to access your tasks that require further action. Figure 5.70 shows the list of tasks in My Inbox application.

Business users must be assigned to the process participant role to access the form URL or the task in My Inbox. (See Chapter 13 to learn more about authorizations.)

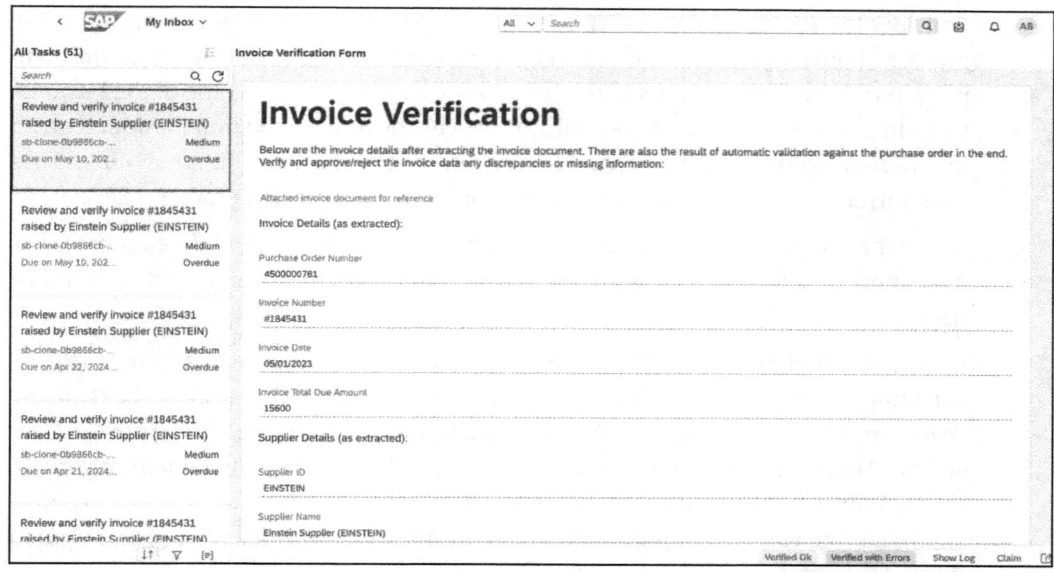

Figure 5.70 Tasks in the My Inbox Application of SAP Build Work Zone, Standard Edition

5.8 Variant Templates

Before we get into the details about variant templates, let's discuss the need for variants templates, as shown in Figure 5.71.

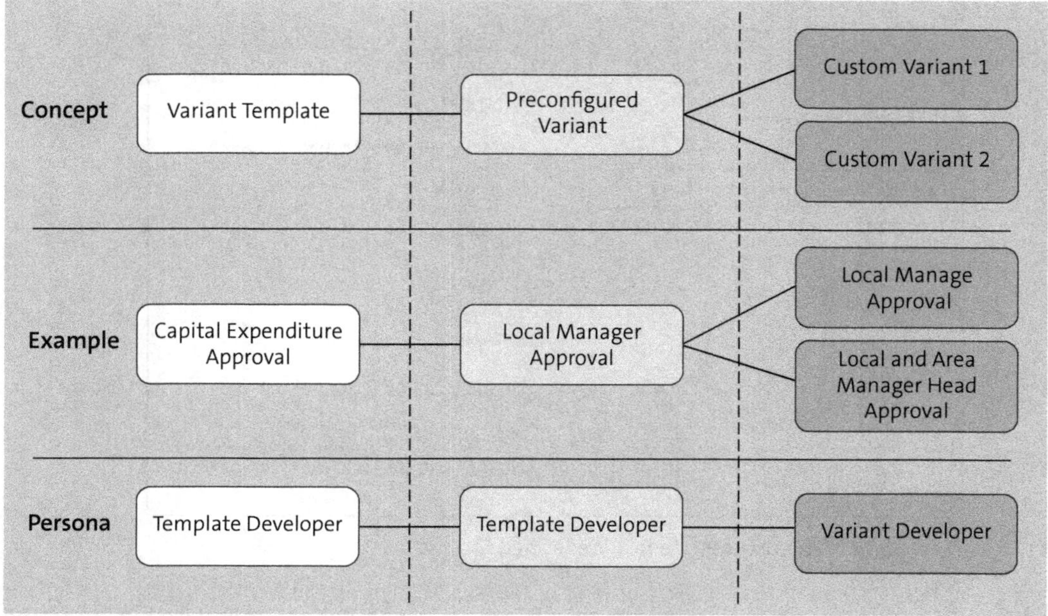

Figure 5.71 Variant Template Concept

Consider the scenario of a purchase requisition and approval in an organization. The process of ordering the equipment and getting approval is mandatory across the company, but there are variations in the process for scenarios where some of the business units in a company require two-level approval or the limit of the order amount varies in different regions of the company that requires auto-approval. *Variant template* has been introduced to create those configurations in the predefined process.

Another typical scenario is the CAPEX approval process where a predefined process has only approval from the local manager but has multilevel approval for different industries.

A variant template consists of process steps and a preconfigured variant called the *blueprint variant*. The persona who creates the variant template is called the *template developer*. As shown in Figure 5.72, a variant template can be configured from the **General** tab. Under **Configuration**, you can create the **Default** variant and **Blueprint** variant. The default variant executes when none of the conditions for variants are executed. The blueprint variant simplifies variant creation, which includes the preconfigured steps of a variant.

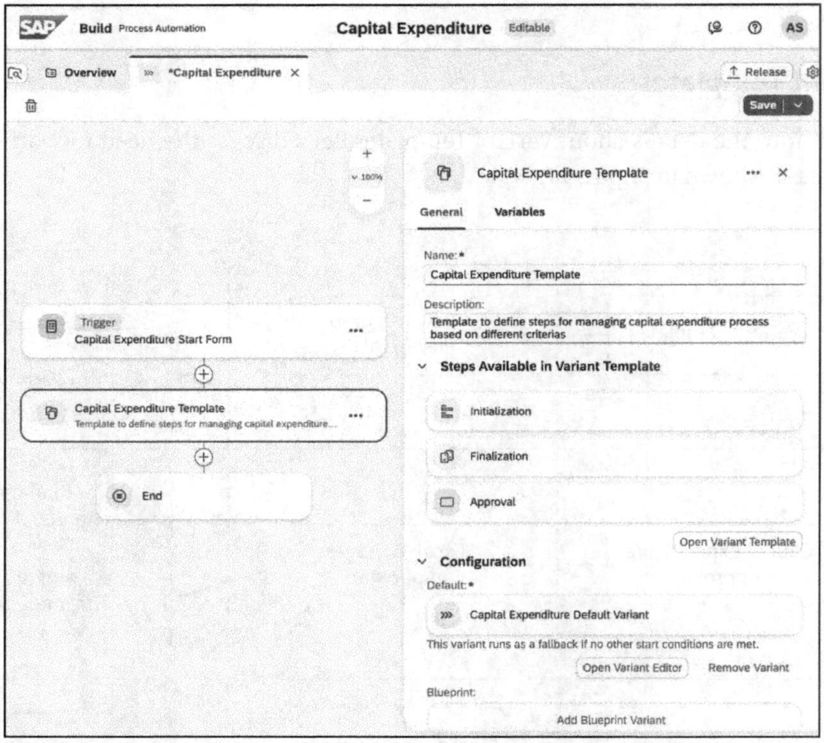

Figure 5.72 Variant Template in a Predefined Process

To add steps in a variant template, click on the **Open Variant Template** button in the **General** tab, as shown in Figure 5.72. You'll be taken to the variant editor, as shown in

Figure 5.73. A variant template can have steps containing artifacts such as forms, decisions, email tasks, and many more. The CAPEX approval process contains a preconfigured variant with only the **Approval Step** (see Figure 5.73).

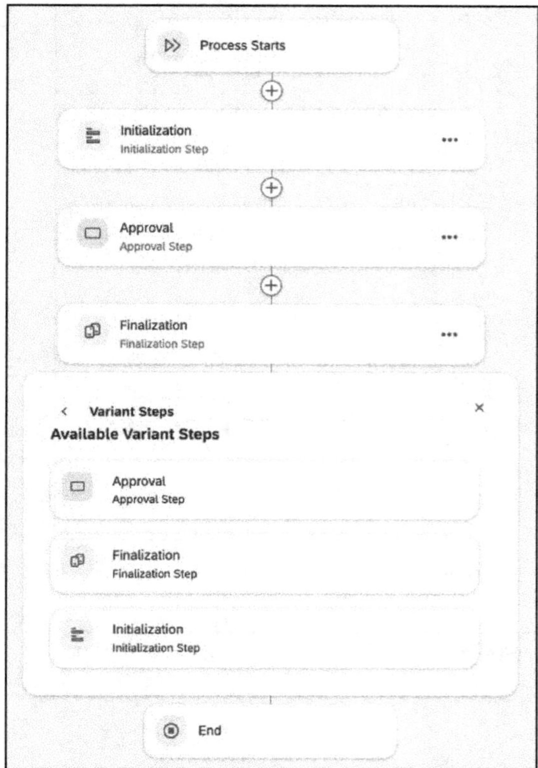

Figure 5.73 Steps in the Variant Template

The steps in the variant template are configurable, so you can maintain the usage of each step, for example, making a step mandatory. This can be achieved by clicking on the step that you want to configure, which opens a side panel with all the **General**, **Inputs**, **Outputs**, and **Variant** tabs, as shown in Figure 5.74. You can even provide value help to the step inputs by enabling the **Value Help** button, as shown in Figure 5.74.

Via the delivered process steps, it's possible to configure custom variants that contains additional approval steps in *variant management editor*. The persona who configures the variant is called *variant developer* and in organization they are typically called *business process configuration expert*.

Variants can be configured in variant management application, whose tile can be accessed from SAP Build Work Zone, standard edition. Variants are executed based on the priority maintained in variant management. You can change the order of variants, deactivate the variants, or delete them by clicking on the three dots icon on the variant in the **Variants** area, as shown in Figure 5.75.

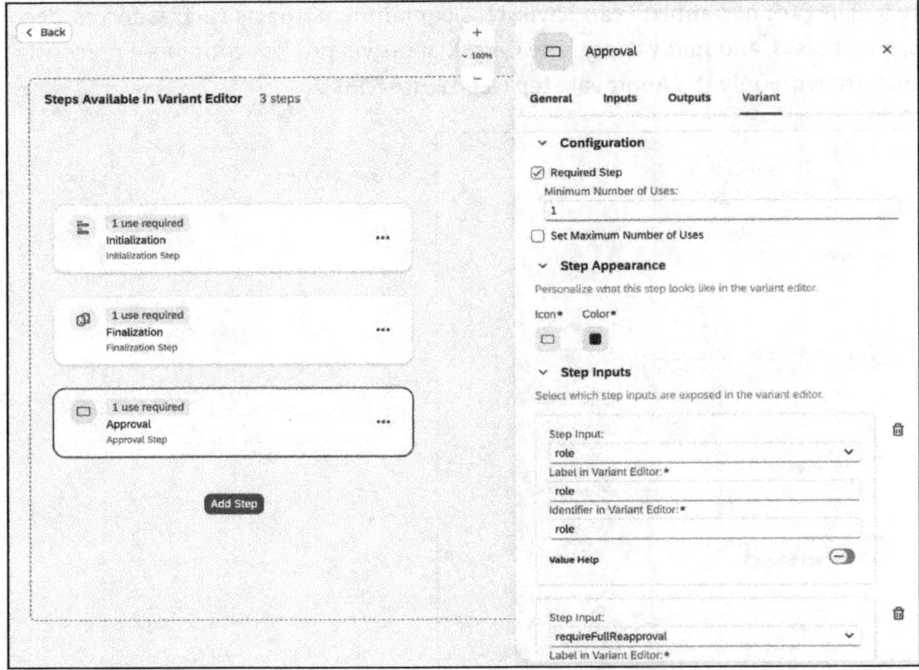

Figure 5.74 Step Configuration in a Variant Template

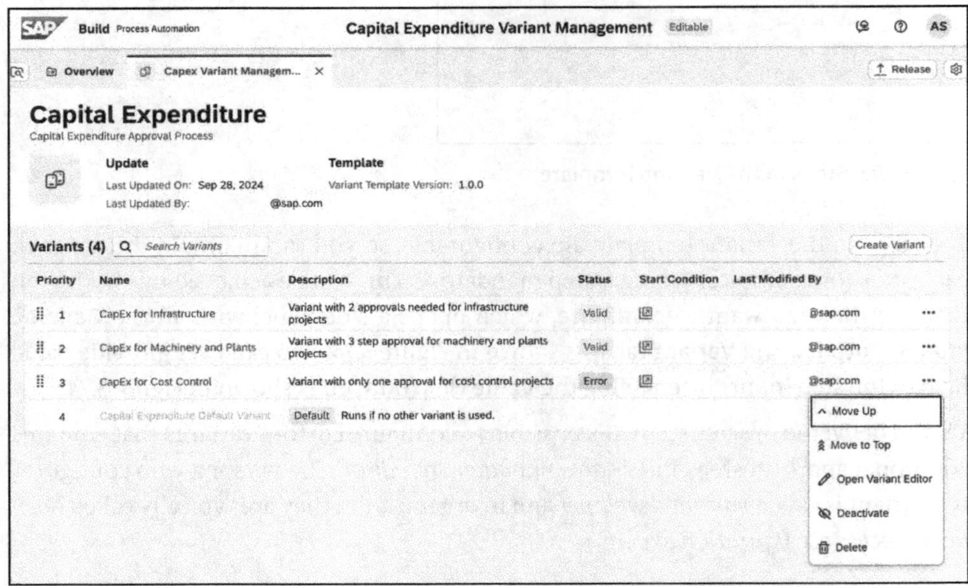

Figure 5.75 Variant Management Application

For the CAPEX approval process, which requires configuration of a two-level approval, a new variant must be created by clicking on the **Create Variant** button shown in Figure

5.75. You'll be taken to the variant editor shown in Figure 5.76 where you can configure additional approval steps by clicking on the + button.

Figure 5.76 Variant with Two-Level Approval in the CAPEX Approval Process

Any number of variants can be configured within the specified conditions. The default variant is executed if none of the conditions are satisfied. Both personas will have a skillset mix of business and IT/development. It's more likely that someone with solid development expertise prepares the variant template and process steps, while someone with less or no development expertise configures the variants.

Variant templates and the configuration of variants provide a framework for business process experts within which they can flexibly adapt to changing business requirements with no code. It also hides the complexity of the underlying processes, which might contain many more steps, conditions, or flow controls that are visible in the modular variants.

5.9 Other Activities

In this section, you'll learn about the frequently used activities such as mail, subprocesses, and script tasks in detail. A brief introduction of mail and and its subprocesses was already covered in Section 5.3.3.

5.9.1 Mail

The mail feature in SAP Build Process Automation allows you to send automated email notifications within the process. Emails can be triggered for any notifications such as task completion, error notification or rejections, and they can be sent at different stages of the process notifying different stakeholders as needed. Email can be configured with the sender, recipient(s), subject, and body of the email directly within the process editor. The recipient can be a fixed email address or dynamically selected based on process data in the process context. All of these features can be achieved by selecting the mail artifact in the process and configuring the notifications per your requirements, as shown in Figure 5.77.

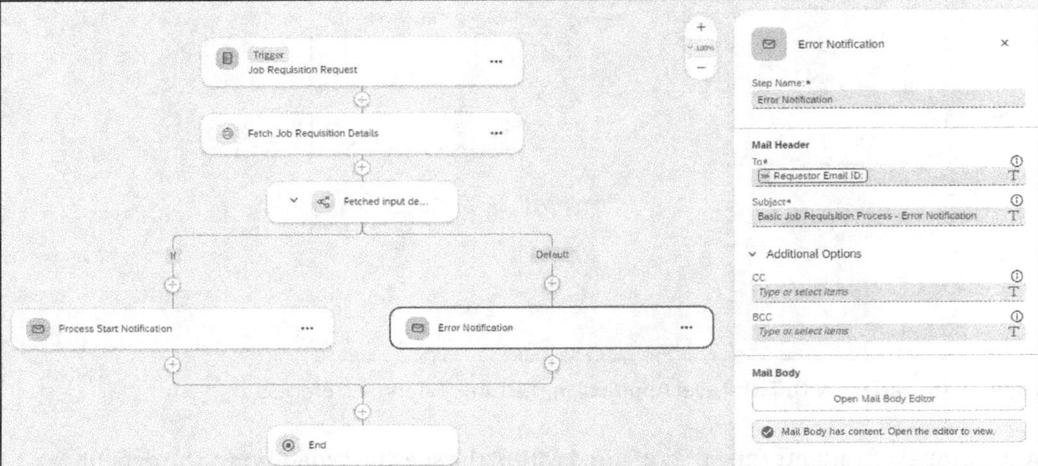

Figure 5.77 Mail Notifications in a Process

The email body can be customized using HTML content for formatting options that include text, links, and process context information (e.g., values generated from the output of any artifact). Click on the **Open Mail Body Editor** button in Figure 5.77, which opens a popup to add the mail body (see Figure 5.78).

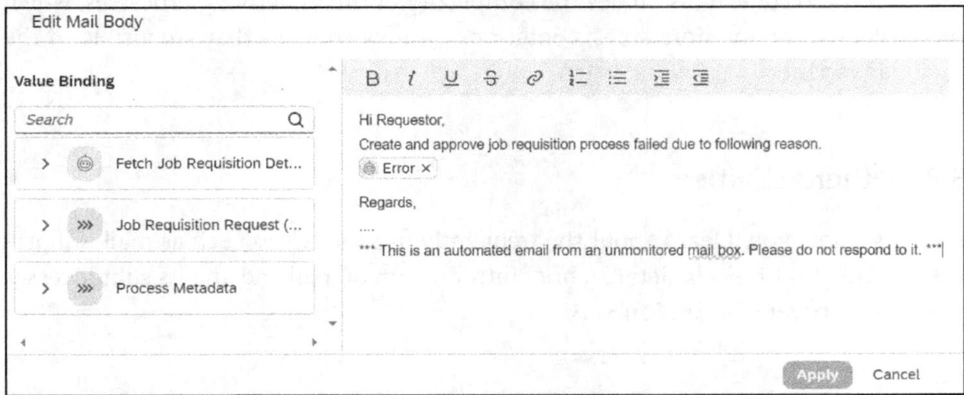

Figure 5.78 Mail Body Configuration

5.9.2 Subprocesses

Subprocesses are referenced subflows in SAP Build Process Automation that enable users to create and reuse a common set of process automation steps (workflows, forms and approval tasks, bot automations, business rules) across multiple processes. With subprocesses, you can streamline your process automations, reduce the risk of errors, and simplify maintenance.

Figure 5.79 shows a **Validate Invoice** subprocess, where validation of invoice is done in a subprocess. Figure 5.80 shows how you call this subprocess from the main process.

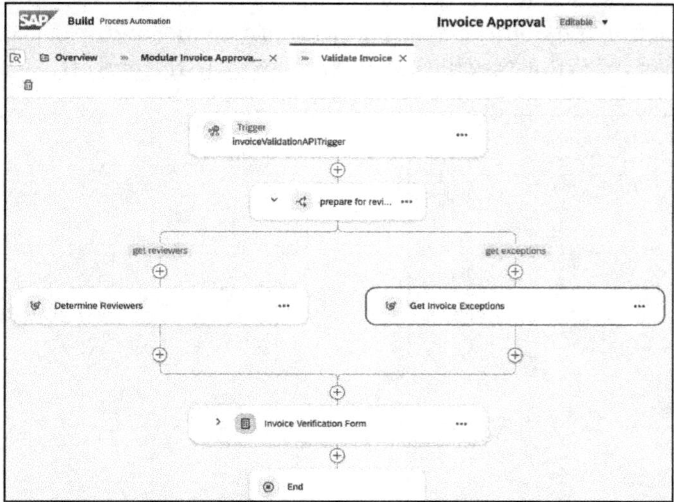

Figure 5.79 Validation of Invoice Subprocess

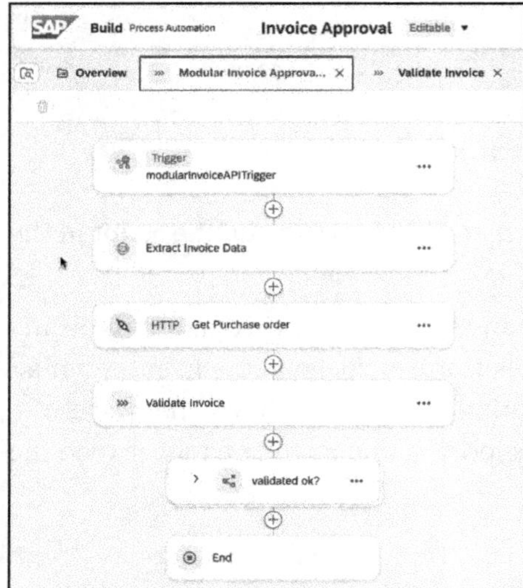

Figure 5.80 Calling the Subprocess in the Main Process

Subprocesses in a process artifact work synchronously. Subprocesses can be called in an automation artifact as well, but it works asynchronously. You can access subprocesses of other projects when they're added as business process project dependencies, as explained earlier in Section 5.3.4.

5.9.3 Script Tasks

A script task in a process refers to the user-defined code that allows for handling operations that can't be achieved using the standard built-in activities provided by SAP Build Process Automation. These scripts are typically written in JavaScript. To add a script task in the process, click on the **+** button to display a list of activities and artifacts, as shown in Figure 5.81. Select the **Script Task** tile to add the activity to the process.

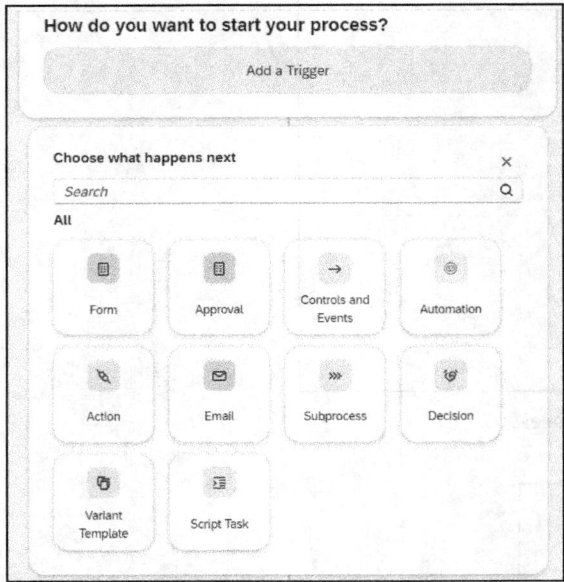

Figure 5.81 Script Task in the List of Artifacts and Activities

Click on **Open Editor** in the side panel, as shown in Figure 5.82, to add the script in the script editor.

Process context is available in the script editor, and you can perform operations on any of the variables. The script task editor provides features such as IntelliSense, syntax highlighting, and compile-time error highlighting. Figure 5.83 shows the usage of a script task, which performs operations such as looping and passing data between the variables.

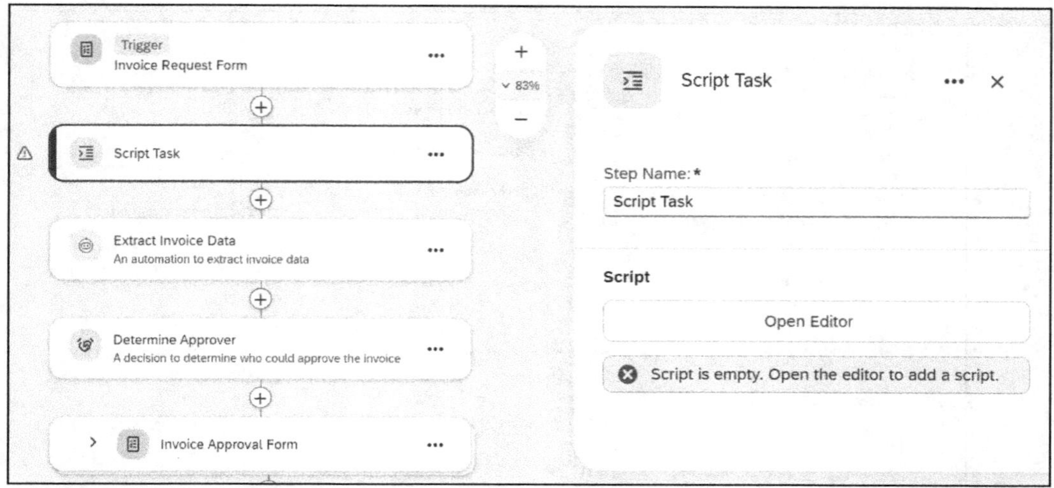

Figure 5.82 Script Task in the Process

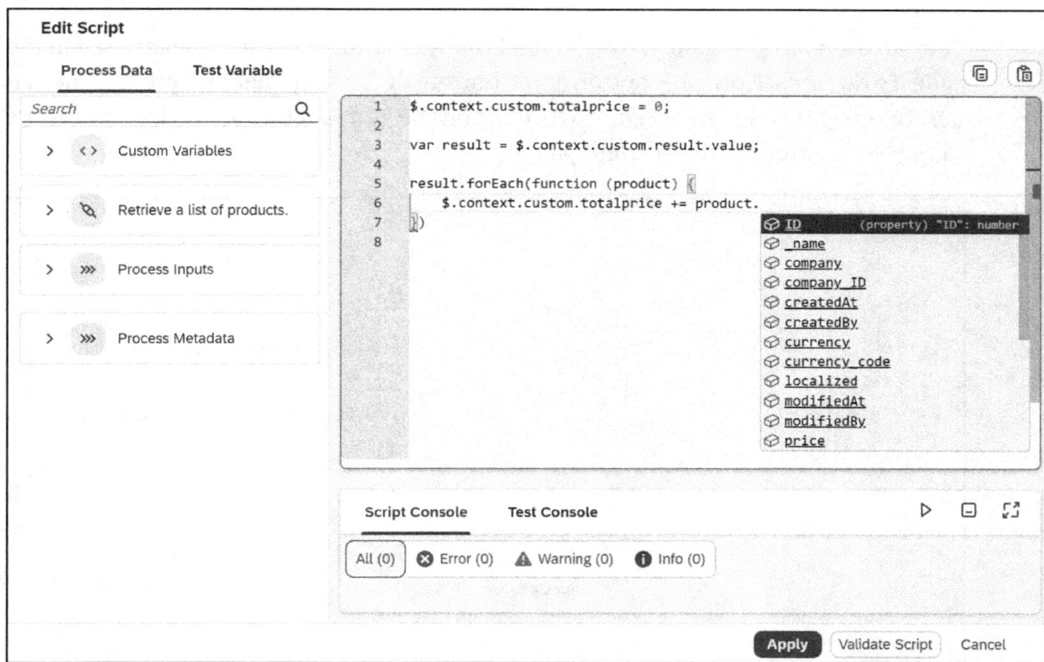

Figure 5.83 Script Editor

The code can be validated using the **Apply** button at the bottom of the screen, as shown in Figure 5.83, which highlights the errors in the script (see Figure 5.84).

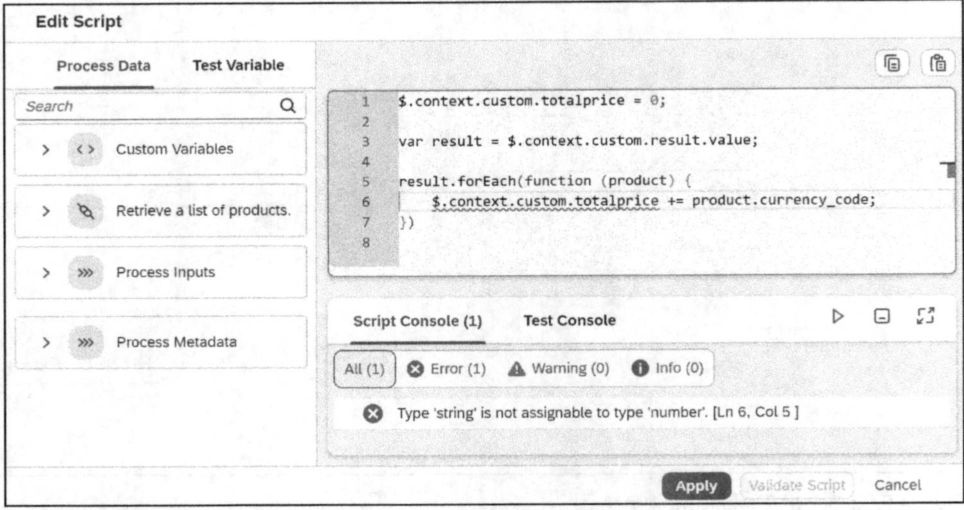

Figure 5.84 Code Validation in the Script Editor

You can even test the code in the script editor by creating the test variables used in the code. To do so, click on the **+** button at the bottom of the left panel, and provide the values for testing, as shown in Figure 5.85. Click on the **Run Test** button to view the results in the **Test Console** tab in the right panel.

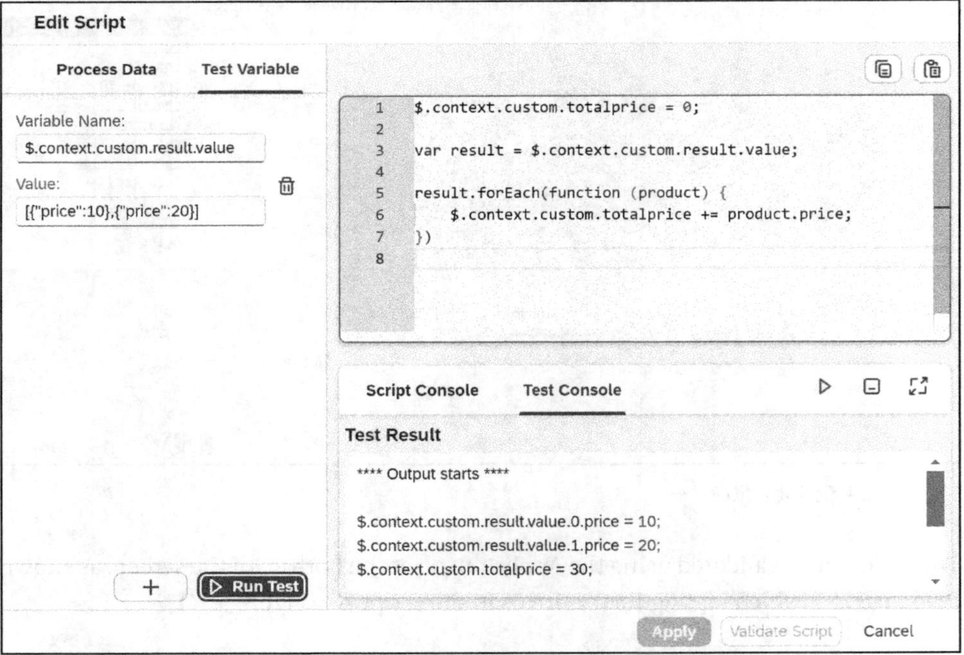

Figure 5.85 Test the Code in the Script Editor

5.10 Summary

In this chapter, you've gained a comprehensive understanding of the business process creation journey. This includes insights into the overall process flow, the use of key artifacts and activities, and the various features of the process editor. You've learned when to use specific components and have been introduced to best practices that ensure effective process design. With this knowledge, you're now well-equipped to start building a business process from scratch. In the next chapter, you learn about the action artifact in detail, which is used to connect processes to external systems.

Chapter 6
Action Artifacts

In SAP Build, an action is used to connect to external systems. In this chapter, we explain what actions are, their benefits, and how they can be configured. We'll also clarify FAQs on actions and APIs, including how actions are used to connect to external systems via SAP Build Process Automation.

The action project is one of the core building blocks of the SAP Build portfolio, which offers a central platform for cross consumption of application programming interfaces (APIs) and artifacts. Action play a crucial role in cross-application integration, ensuring a seamless user experience for leveraging several services across SAP and non-SAP systems. It represents a collection of create, read, update, delete (CRUD) operations with a defined customizable signature that acts like a thin layer of abstraction over an API simplifying its consumption. In this chapter, you'll get a complete view of the actions concept and how to use it.

6.1 What Is an Action, and How Does It Differ from an Application Programming Interface?

Imagine you have different tools and systems in your business, such as SAP S/4HANA for managing finances, SAP Ariba for procurement, and SAP SuccessFactors for HR. Each of these systems has its own set of functions and data. Now, let's say you want to make these systems work together more efficiently or add new capabilities to them. With actions in SAP Build Process Automation, you can easily build extensions to tailor your business needs by connecting to a given SAP S/4HANA, SAP Ariba, or other SAP line of business (LoB) systems for any kind of `GET`, `POST`, `PATCH`, and other calls.

Now, let's understand how an action differs from an API. An *application programming interface (API)* is a set of definitions and protocols for building and integrating application software. An API is used to connect different systems.

As mentioned earlier, an *action* is a core building block used to connect the business process with an external system—both SAP and non-SAP applications—by consuming the APIs. This single operation or method references an API and can be consumed in a reusable manner.

To help understand the difference, let's consider a practical example of posting of an invoice to an SAP S/4HANA system after approval of the invoice via a business process. As shown in Figure 6.1, actions are used to post the data in the backend system based on the process outcomes (approval, validation, etc.), that is, the standard business processes are tailored to the process outcomes before posting data to the SAP S/4HANA system, and the actions reuse the standard SAP S/4HANA APIs to perform POST operations.

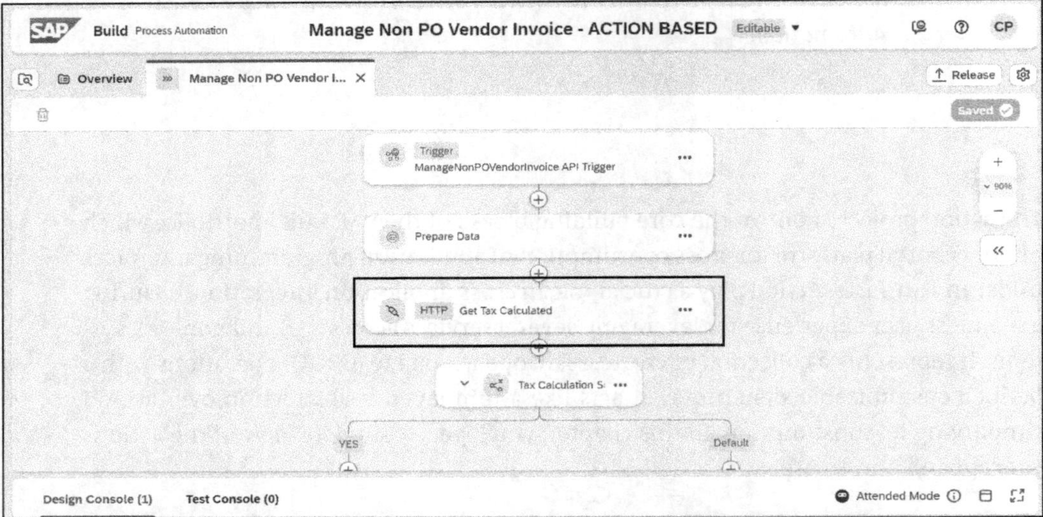

Figure 6.1 An Action Project in an SAP Build Process Automation Business Process

Other real-world examples where actions can be used are for getting the employee details from an SAP SuccessFactors system for the off-boarding process, updating purchase requisition details for order changes in the SAP Ariba system, and getting asset details from the asset management system before sending it for approval for asset depreciation.

6.2 Connectivity Options

In the dynamic world of business process automation, integrating various systems and reducing manual effort are crucial. The connectivity options provided by SAP Build Process Automation enables you to streamline your business workflows by connecting multiple systems effortlessly, eliminating the need to switch between different applications.

You can use various API sources and SAP systems to consume them as actions in your business processes. The connectivity options in SAP Build have been categorized based on two key factors:

- **API availability**
  This refers to whether the API you need for the integration is already available and ready to be used. If the API exists and is accessible, certain connectivity options will be appropriate for linking SAP Build Process Automation. This option is referred to as **Live API**.
- **API configuration/definition**
  This refers to how the API is defined or set up, including its specifications, endpoints, authentication methods, and other technical details. Depending on how well-defined and configured the API is, different connectivity options may be more suitable. This option is referred to as **API Specification**.

An action project can be created from the **Connectors** area of the SAP Build homepage with either of the connectivity options, as shown in Figure 6.2.

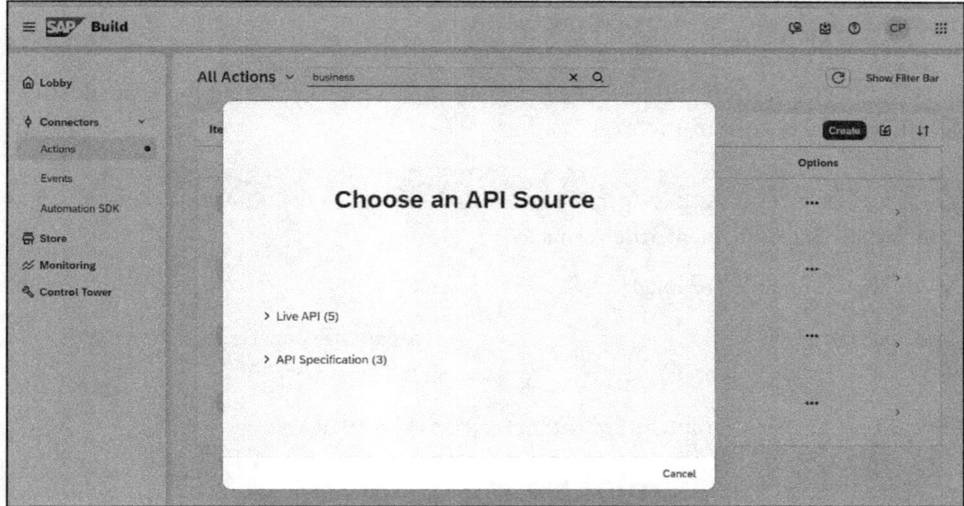

Figure 6.2 Connectivity Options for Actions in SAP Build

In the following sections, we'll delve into the options that make it easier to consume APIs.

6.2.1 Live Application Programming Interface

This option is used when an API is available, making it easier to consume the API through a destination-driven approach. Specifically, you can now use SAP Cloud Application Programming Model, ABAP RESTful application programming model, and any SAP Business Technology Platform (SAP BTP) destination when creating action projects, as shown in Figure 6.3.

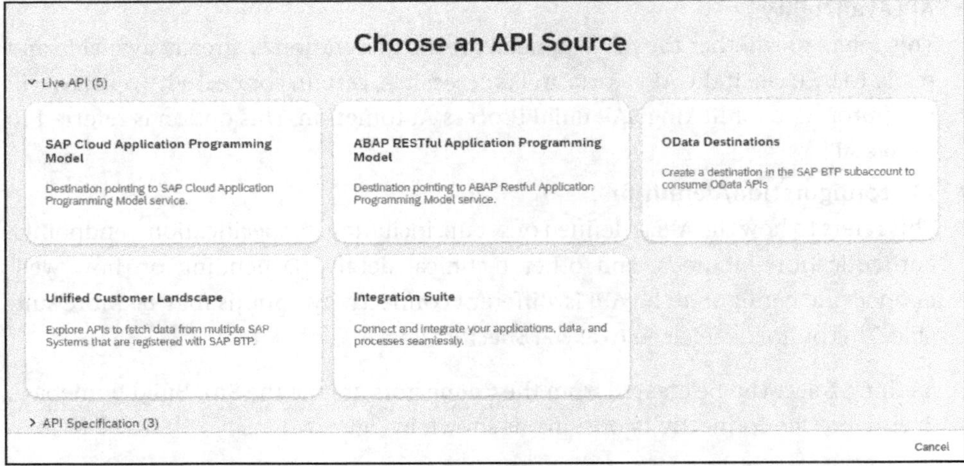

Figure 6.3 Live API Actions: Connectivity Options

To connect your actions project, the destination in SAP BTP should have the destination properties shown in Table 6.1.

Name of the Parameter	Value
sap.applicationdevelopment.actions.enabled	true
sap.processautomation.enabled	true
sap.build.usage	Bases on the connectivity option that is chosen, e.g., **CAP**, **RAP**

Table 6.1 Destination Parameters of Action Projects in SAP BTP

As shown in Figure 6.4, you can create destinations and add the destination parameters under **Destinations** in the **Connectivity** area in SAP BTP.

The following list details the connectivity options in **Live API**, as shown earlier in Figure 6.3:

- **SAP Cloud Application Programming Model**
  SAP Cloud Application Programming Model simplifies the development of enterprise-grade services and applications by providing a set of languages, libraries, and tools. By integrating SAP Cloud Application Programming Model with action projects, you can easily consume SAP Cloud Application Programming Model–based APIs to automate and streamline your business processes. Along with the parameters maintained in the destinations of SAP BTP mentioned in Table 6.1, maintain parameter **sap.build.usage** as **CAP**.

Figure 6.4 SAP BTP Destination Configuration for Action Projects in SAP Build Process Automation

- **ABAP RESTful Application Programming Model**
  ABAP RESTful application programming model facilitates the creation of scalable and robust RESTful services using ABAP. With ABAP RESTful application programming model support in action projects, you can efficiently leverage ABAP-based APIs, enhancing the automation of business tasks and processes.

  Along with the parameters maintained in the destinations of SAP BTP mentioned in Table 6.1, maintain parameter **sap.build.usage** as **RAP**.

- **OData Destinations**
  You can consume OData APIs in SAP Build by creating destination in SAP BTP. Along with the parameters maintained in the destinations of SAP BTP mentioned in Table 6.1, maintain parameter **sap.build.usage** as **odata_gen**.

- **Unified Customer Landscape**
  You can connect to standard and custom APIs by registering the SAP system to the system landscape list and maintaining the formation type as **Integration** with SAP Build in SAP BTP. All the APIs can be viewed only after the system is registered successfully. A few examples of the SAP systems are SAP S/4HANA Cloud, SAP SuccessFactors, and SAP Commerce Cloud.

- **Integration Suite**
  You can create an action project that consumes business data from a single semantically connected data graph using the graph destination in SAP BTP. Maintain the parameter **sap.build.usage** as **graph** with the parameters mentioned in Table 6.1. You can connect to the iFlows of SAP Integration Suite and also to SAP Business Accelerator Hub to consume APIs from a centralized API catalog managed by your organization. Maintain the parameter **sap.build.usage** as **apihub_enterprise** with the parameters mentioned in Table 6.1.

6.2.2 Application Programming Interface Specification

This option is used in scenarios where the definition of API exists and the actual API may or may not exist. There are three options to build an action project by consuming your API specification or actual API, as shown in Figure 6.5.

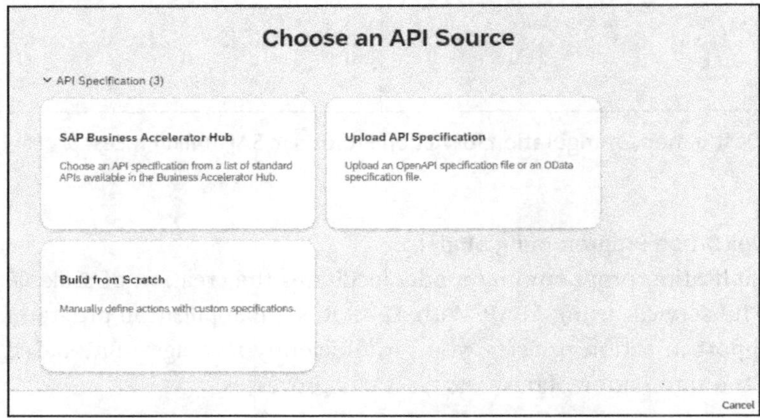

Figure 6.5 Connectivity Options for API Specification

Let's dive into the details of each option:

- **SAP Business Accelerator Hub**
  Standard APIs from various SAP products (e.g., SAP S/4HANA, SAP SuccessFactors, SAP Customer Experience, SAP BTP, SAP Ariba, SAP Fieldglass, and SAP Concur) that are available in SAP Business Accelerator Hub can be easily consumed to create action projects in SAP Build. Maintain the destination parameters as mentioned in Table 6.1. As shown in Figure 6.6, select the product from **SAP Business Accelerator Hub**, for example, **SAP Ariba**.
  After the SAP Ariba product is selected, a list of associated packages is displayed, as shown in Figure 6.7.

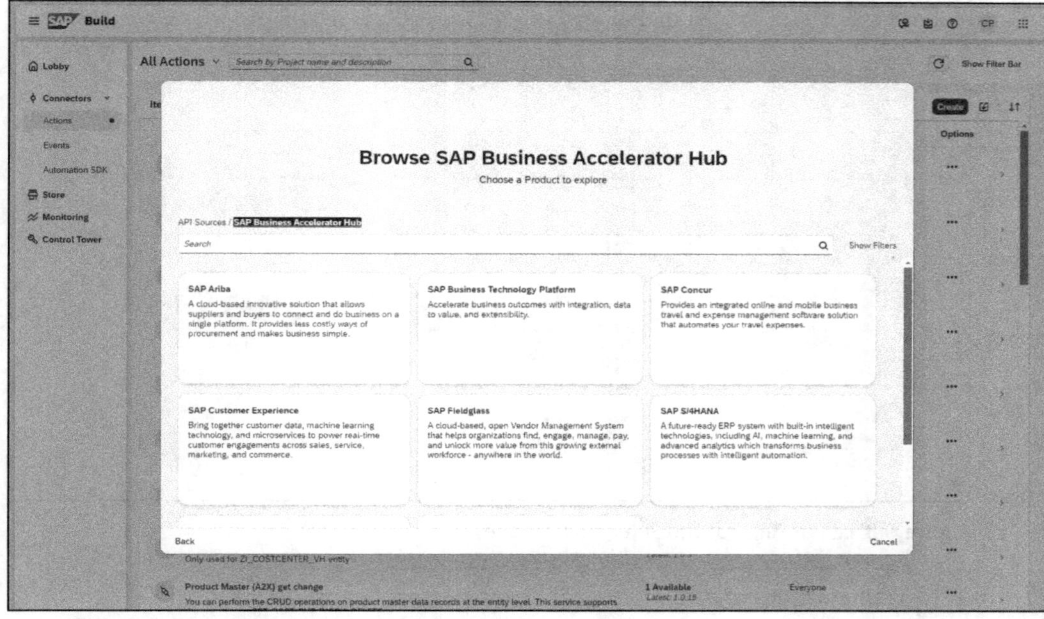

Figure 6.6 List of Products from SAP Business Accelerator Hub

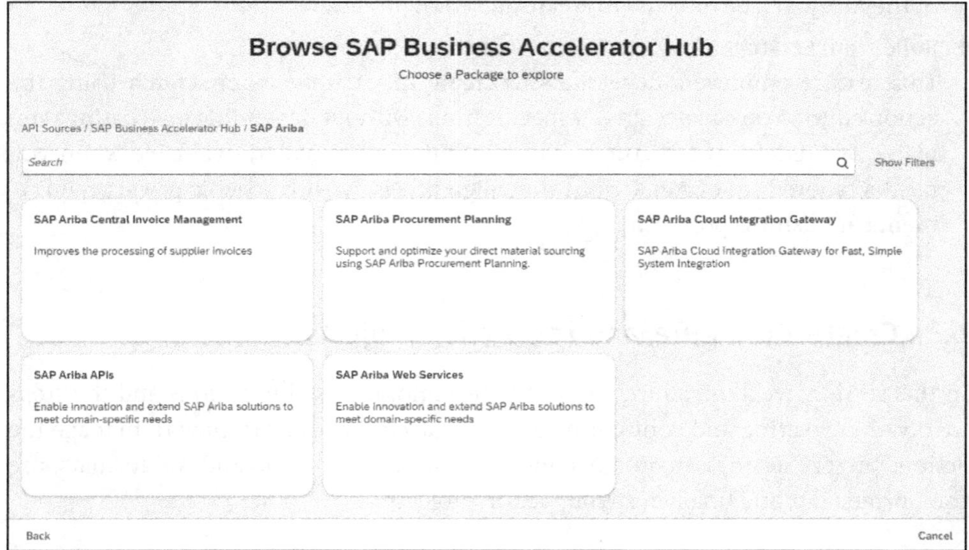

Figure 6.7 List of Packages for SAP Ariba

Finally, select the API from the package for your use case, as shown in Figure 6.8.

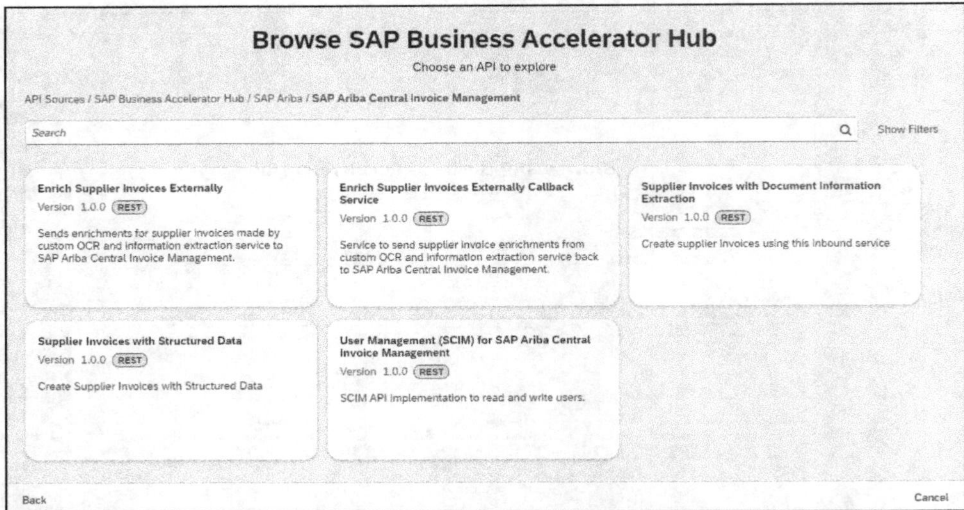

Figure 6.8 List of APIs for the Selected Package

- **Upload API Specification**
  You can upload an API specification or an OData specification file by selecting the **Upload API Specification** tile, as shown earlier in Figure 6.2. After successful upload of the API, you'll be directed to create an action project.
- **Build from Scratch**
  This feature empowers developers to create API actions from scratch using the action editor. You can design API specifications without depending on existing templates or predefined structures. This capability is particularly valuable when you need a tailored set of API actions that align precisely with specific project requirements or business processes.

6.3 Create, Configure, and Test Action Projects

In this section, we'll introduce you to the terminology used in actions and the steps involved in creating and configuring action projects. You'll learn how to manage the actions project settings, maintain standard/custom parameters and API formats, use the Outputs tab, and finally test your action projects.

6.3.1 Terminology

Before we get started, let's get to know the terminology used in actions, as listed in Table 6.2.

Term	Definition
Action	A single operation or a method that references an API can be consumed in a reusable manner. It simplifies the experience of citizen developers as they can consume business functions/APIs from the library of the curated content.
Action project	This is a collection of customizable actions that is implemented using the OpenAPI specification by citizen developers.
Action editor	This editor defines and maintains the action project and the actions for subsequent cross-consumption.
Action library	This enable citizen developers to quickly compose processes, automations, and applications by consuming from the library of curated content.

Table 6.2 Terminology in Actions

6.3.2 Create and Configure an Action Project

Action projects are created from the SAP Build lobby using any of the connectivity actions, as shown in Figure 6.9 and explained earlier in Section 6.2.

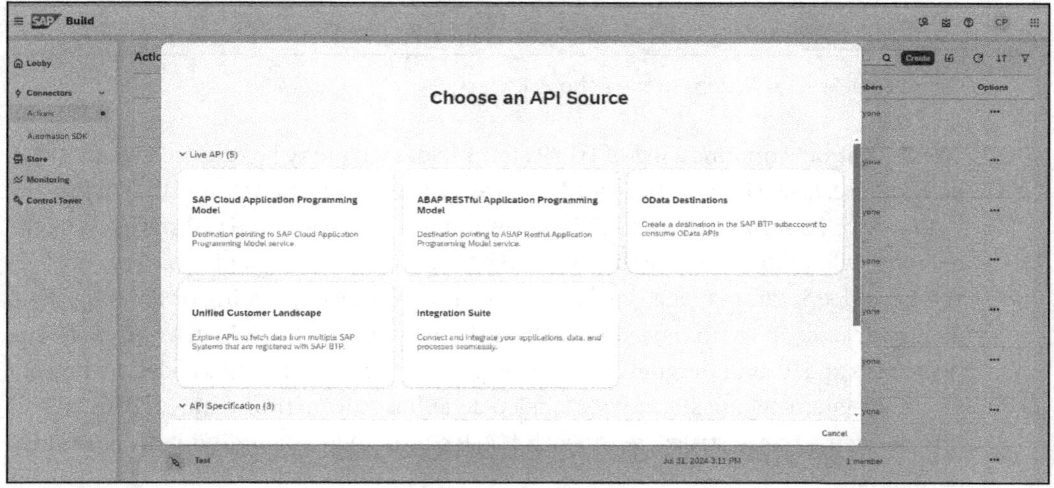

Figure 6.9 Create an Action Project in SAP Build

Let's understand how action projects can be configured with various features available in the action editor after successful creation of the action project, as shown in Figure 6.10. You can add the required methods, actions, or APIs to the action project based on your requirements. In our example, we've added several actions such as posting data to create a lead in SAP Sales Cloud, so we've added the POST method, retrieving the change history of the lead using the GET method, and so on. Select one of the actions (e.g., **Create a new lead**), which displays the input and output parameters of the action, as shown in Figure 6.10.

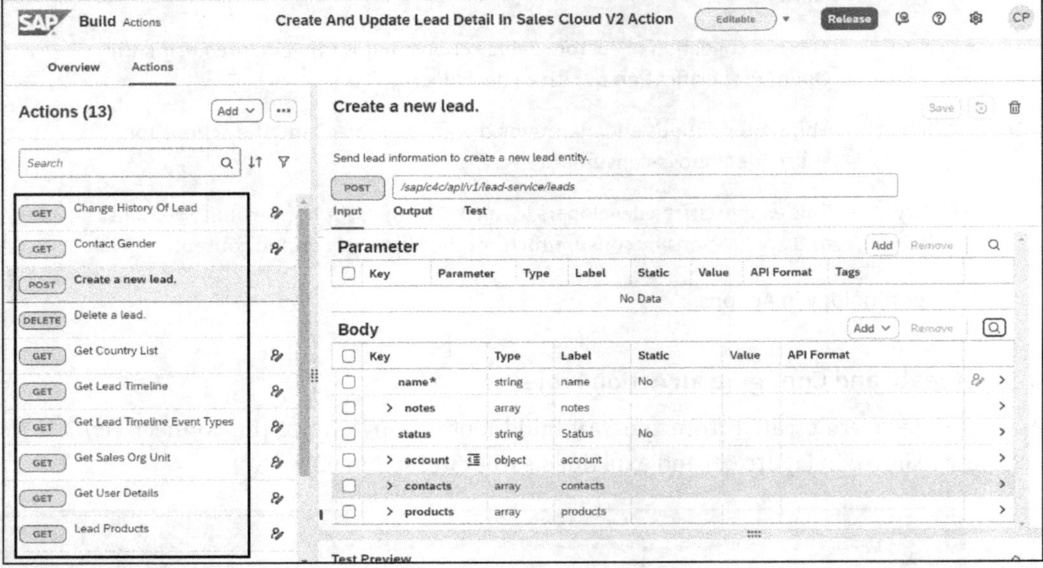

Figure 6.10 Several Actions in the Actions Project

The **Output** tab contains a list of HTTP status codes that may be returned by an API in its response. Each HTTP status code has an associated list of parameters that define the structure of its response. When the API responds with a specific HTTP status code, the response will follow the structure defined by the citizen developer in this section. **Citizen** developers can manage the list of HTTP status codes by adding or deleting them using a dropdown list in the Output tab, as shown in Figure 6.11. When a new HTTP status code is added, it undergoes a validation process. If no HTTP status codes are present in the schema, a default status code labeled default is automatically added. You can customize the response structure for each HTTP status code by modifying its associated parameters.

Note

To avoid cluttering the business process with unnecessary fields, use a minimalist set of input and output fields. This ensures a cleaner process design and execution. Additional fields can be added or updated later as requirements evolve.

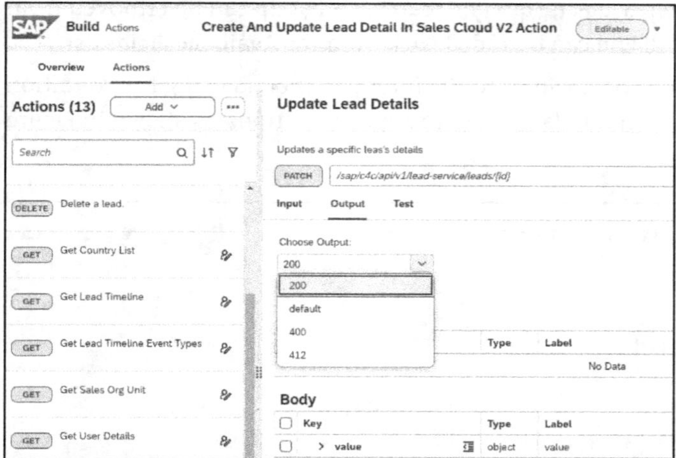

Figure 6.11 Output Tab for an Action

Now, let's discuss frequently asked questions in configuring action projects:

- **How can I add a filter parameter to my API to retrieve particular data?**
  Standard input, output, and parameters of the body can be deleted and added back according to the requirements. The removed fields can be accessed from the **Available Fields** tab, as shown in Figure 6.12.

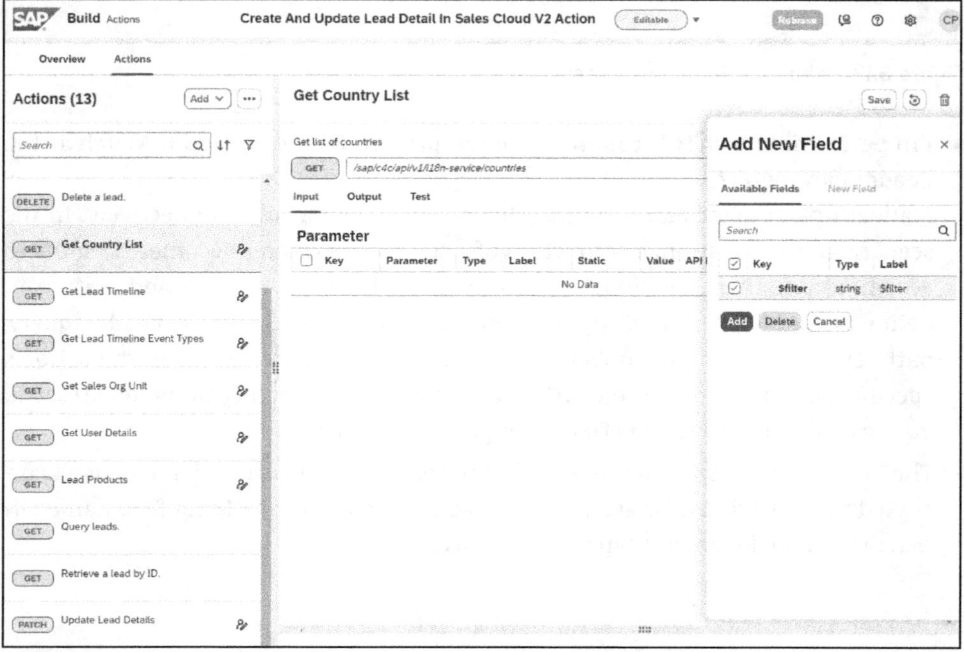

Figure 6.12 Standard Parameters: Adding Filters to an Action

Fields such as **$top**, **$filter**, **$select**, **$query**, **$search**, and **$skip** can be removed from the **Input** tab of an API call and can be added back from the **Available Fields** list.

You can even apply conditional values to the filter parameters using the condition editor. Select the icon in the **Value** field to apply conditions, as shown in Figure 6.13.

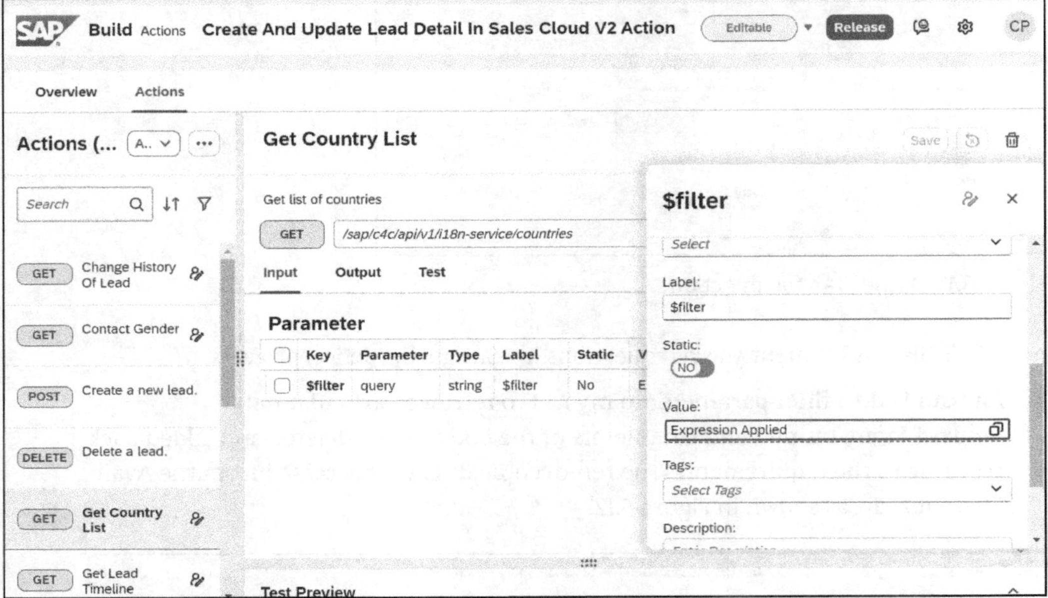

Figure 6.13 Values for Filter Parameters

- **I'm performing a PATCH call in my action project. How can I add If-Match as the header parameter?**
  Standard parameters alone are insufficient for invoking an API effectively. In the scenario in the question, introduction of a custom parameter becomes essential to accommodate these requirements. An action editor can use the **Add** feature to define and add customized input parameters for use cases such as **header**, **query**, **path**, or **custom** body parameters that aren't initially defined within the action's specification in an action project. Click on the **Add** button in the **Parameter** area, and you're taken to the **Add New Field** popup, as shown in Figure 6.14.

  The `If-Match` request header is added in the `PATCH` call to ensure that data isn't corrupted while patching. This can be achieved by adding the new **Input** field named **If-Match** with the **Parameter** type as **header** (see Figure 6.15).

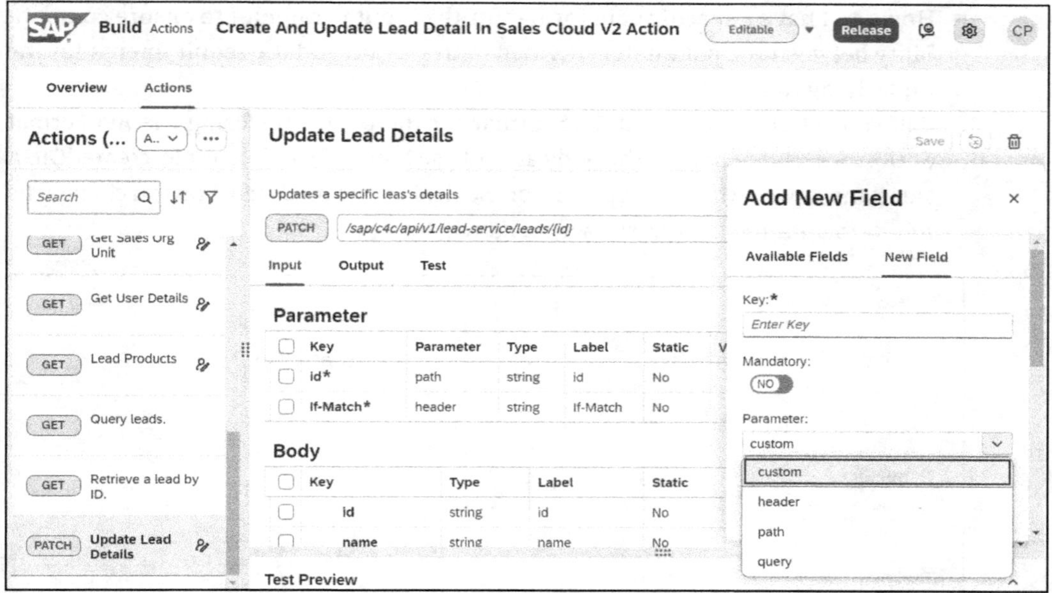

Figure 6.14 Custom Parameters in Action

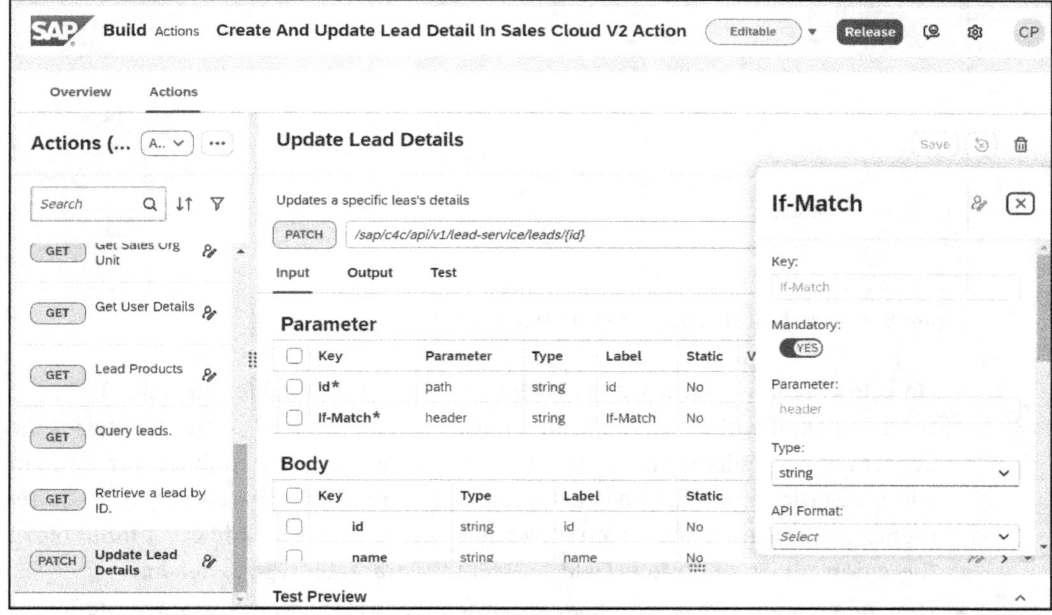

Figure 6.15 Adding a Header Parameter in Custom Input Parameters

- **How can I have a specific date format for the input parameter to ensure compatibility between the parameters defined in the action and the requirements for the underlying API?**
  You can select the required date format for the parameter using the **API Format** option. Select any field in the **Body** area of the **Input** tab, for example, **createdOn**. A side panel opens where the options for the field can be modified. You can change the API format for the field, as shown in Figure 6.16.

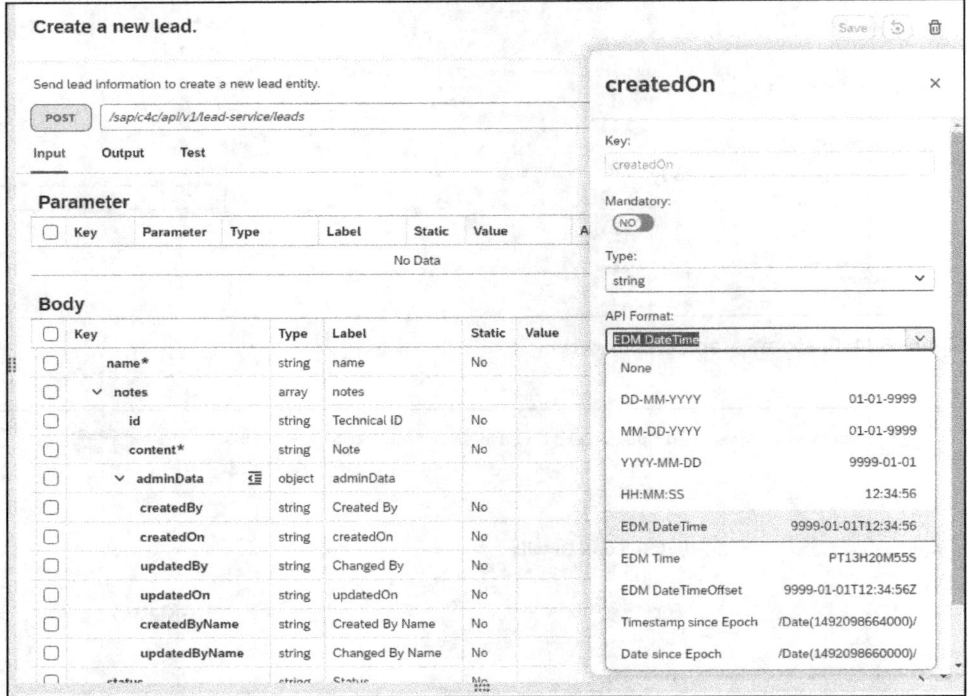

Figure 6.16 Maintain Specific Date Parameters in an Action

- **How do I maintain static values for the parameters defined for an action?**
  We'll explain it with an example. In requests such as POST or PUT, the client uses the **Content-Type** header to specify the type of content being sent to the server. You can add the **Static** value to the header parameter **Content-Type** based on your requirements. To navigate to the screen shown in Figure 6.17, click on **Add** in the **Input** tab of the action editor. Select **New Field**, and set the **Parameter** type as **header**.
- **How do I enable a** Cross-Site Request Forgery **(CSRF) token for create, update, and delete operations in an action?**
  This can be done in the **Project Settings** (accessed by clicking the ⚙ icon) of the action editor, as shown in Figure 6.18.

Figure 6.17 Maintain Static Values for Parameters in an Action

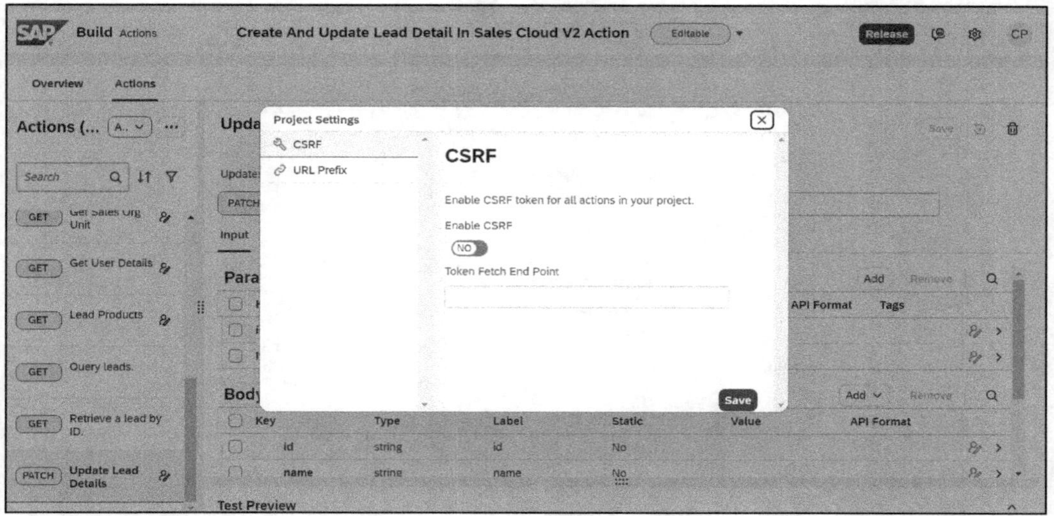

Figure 6.18 Enable the CSRF Token in Action

- **How is a custom action used in an action project?**
  The **Custom Action** option is required when you want to create an API from scratch. This option can be used when you want to add a custom action into an existing action project where the API already exists or when you build the action project

from scratch, as explained in Section 6.2.2. In the action editor, click on **Add** in the **Actions** tab, and then choose **Custom Action** to add a custom action, as shown in Figure 6.19.

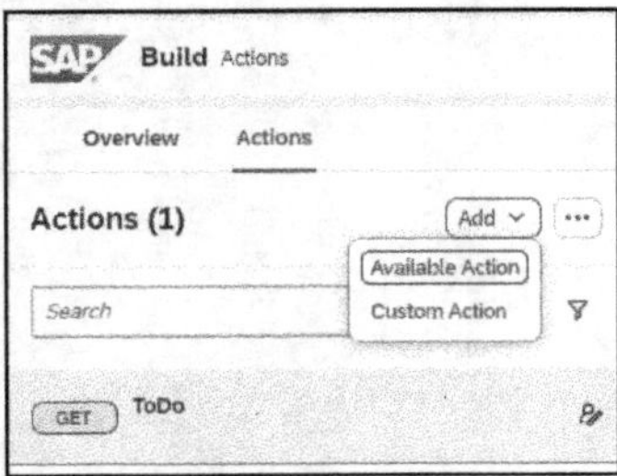

Figure 6.19 Custom Action in Action Projects

- **How do we trim down the parameters of specific output HTTP codes that might not be relevant to the business processes that are part of the API response?**
  Users can make modifications to the structure of the selected code by first selecting the output code using the **Choose Output** option, as shown in Figure 6.20.

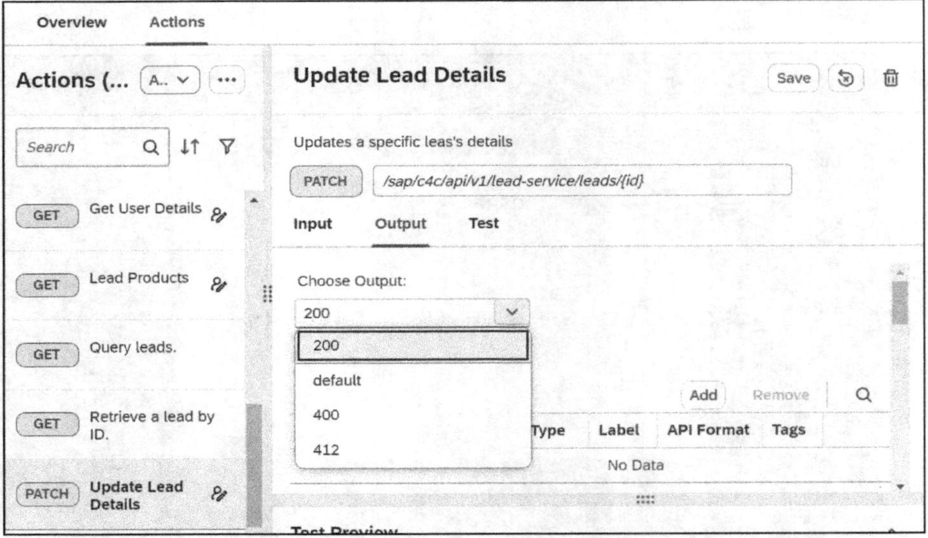

Figure 6.20 Customizing Output Parameters

For easy parameter navigation, users can sort the field by selecting the **Key** column header and can search for the parameter to be deleted using the **Search** icon (see Figure 6.21). Click on the checkbox beside the parameter that you want to delete, and click on the **Remove** button, as shown in Figure 6.22. The deleted parameters are soft deleted from the table. You can add them using the **Add** button. To delete the fields permanently, you can delete them in the available field section when you click on the **Add** button.

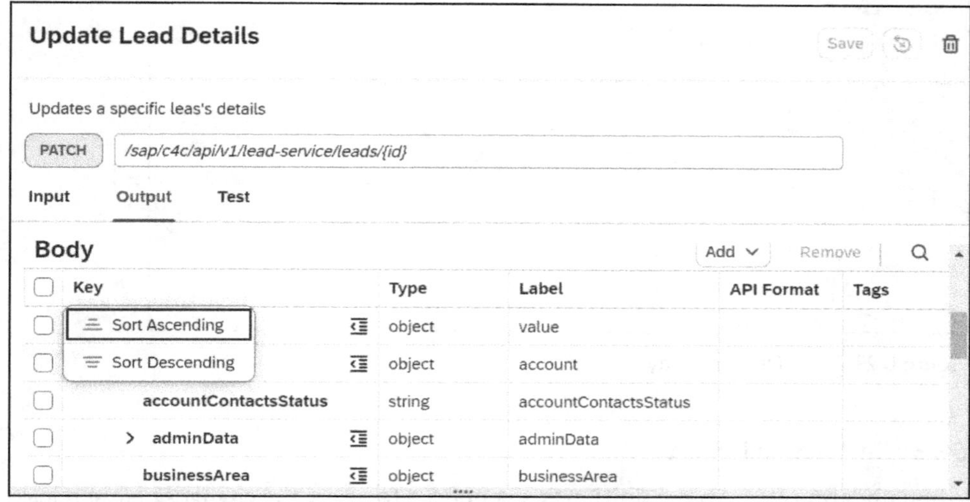

Figure 6.21 Sorting the Parameters

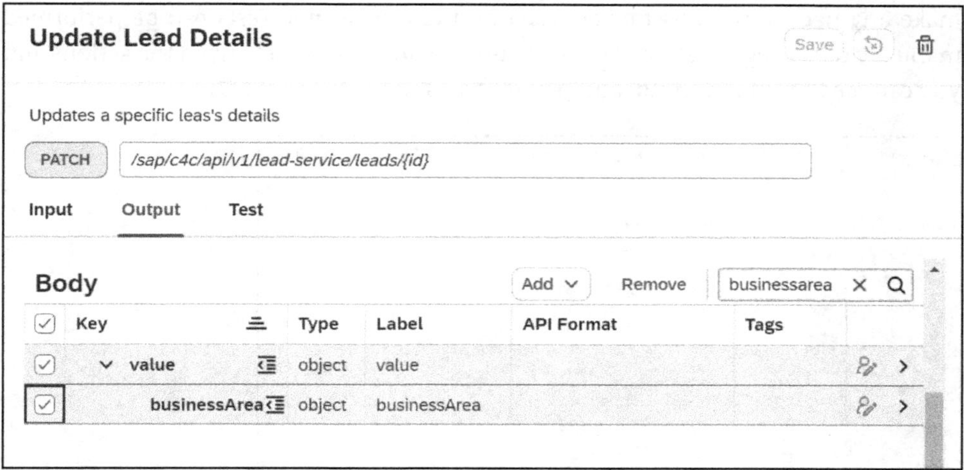

Figure 6.22 Removing the Selected Parameters

- **How do I make output of the GET request as a data source for a dropdown in the form builder?**
  The output of the GET request in an Action project must be marked as **Main Output Array**, as shown in Figure 6.23. This feature allows users to select the data in the drop-downs from an external system. This allows citizen developer to dynamically add the values in the dropdown of the form.

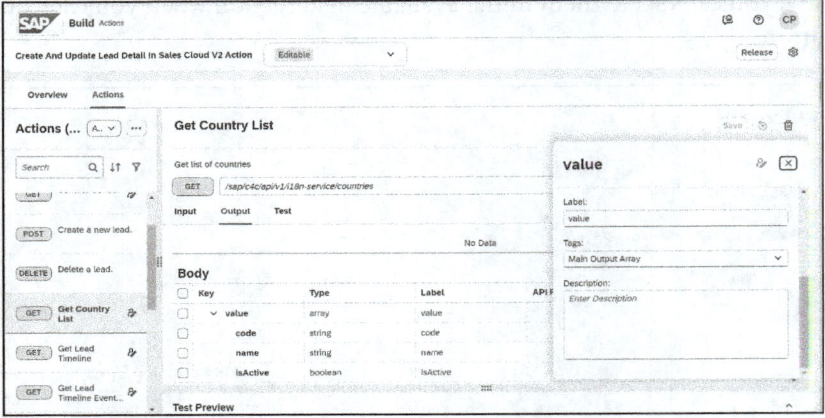

Figure 6.23 Main Output Array

6.3.3 Test Action Project

When updating an action definition, it's beneficial to verify the changes before releasing the project. The **Test** tab is designed for this purpose. It allows you to gather the required configuration and test data, enabling you to assess the modifications and make any necessary adjustments based on the test results. Tests can be performed manually or through destinations configured in SAP BTP. Select any of the actions and you can see the **Input**, **Output**, and **Test** tabs, as shown in Figure 6.24.

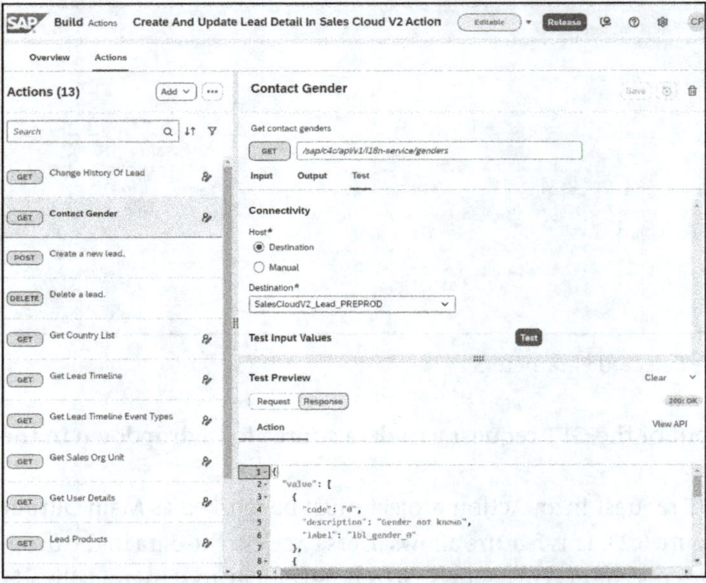

Figure 6.24 Test Action Project

Now you've successfully created, configured, and tested your action project.

6.4 Release and Publish Actions

To make your action project available for cross-consumption, you need to *release* and *publish* the actions of the action project in the *action library*. When the project is released using the **Release** button shown in Figure 6.25, it creates a new package, and the packages are versioned and stored in the registry.

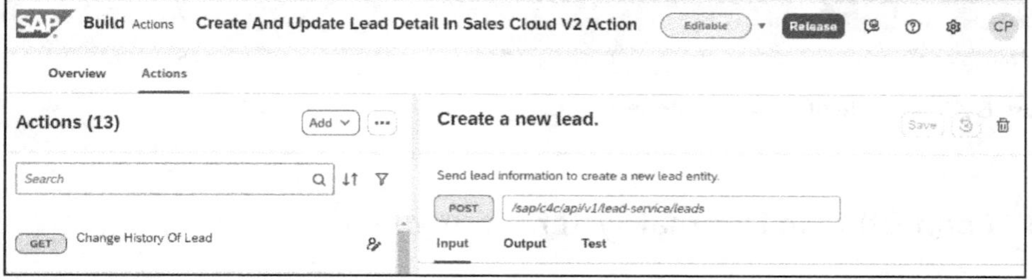

Figure 6.25 Release Action Project in the Action Editor

Once the project is released successfully, the package must be published to the action library using the **Publish to Library** button, as shown in Figure 6.26. The published packages are available in the SAP Build lobby. Actions in the published packages are reusable artifacts that can be consumed in a business process project.

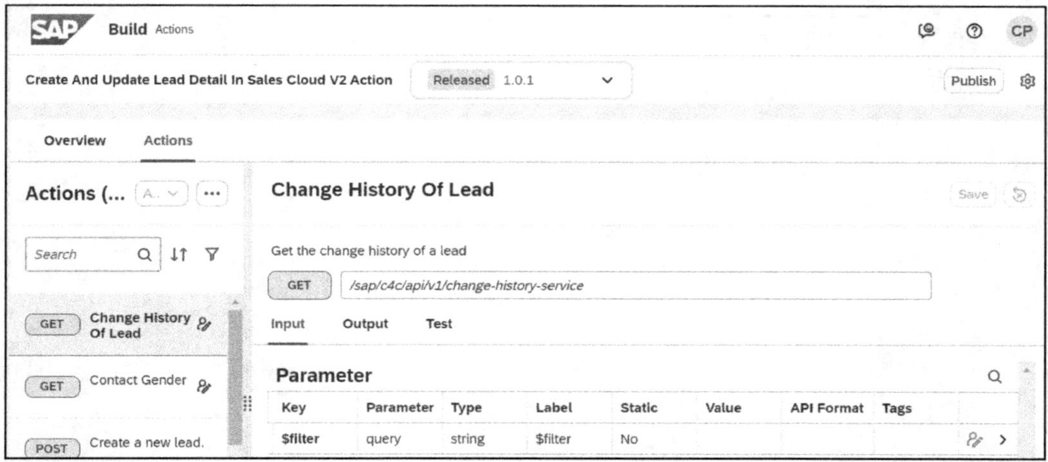

Figure 6.26 Publish Action Projects to the Action Library

You've created an action project from the SAP Build homepage. After successful creation of an action project, you'll be taken to the action editor to create and configure actions. To consume an action project in a business process, you need to release and publish the action project to the library, which contains action projects and their artifacts. Figure 6.27 provides an overview of the typical lifecycle of an action project.

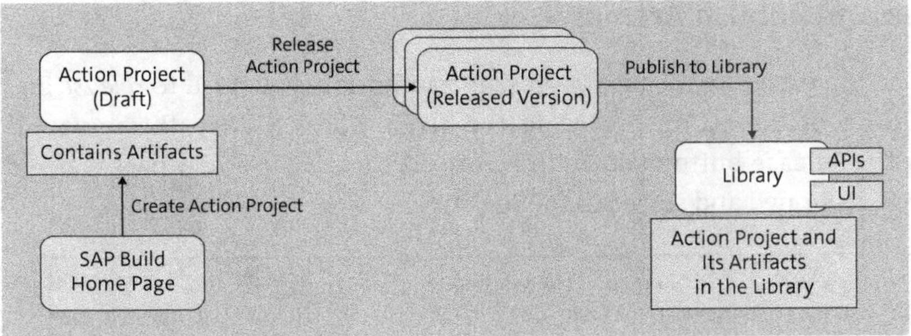

Figure 6.27 Lifecycle of an Action Project

6.5 Connecting to External Systems using Actions

The published action project can now be consumed in your business process project to post data to the backend. You'll add the action in the business process to connect to the backend system. Navigate to the SAP Build lobby, and open the business process project. Navigate to the process of the project, click on the ⊕ button, and select the Action artifact, as shown in Figure 6.28.

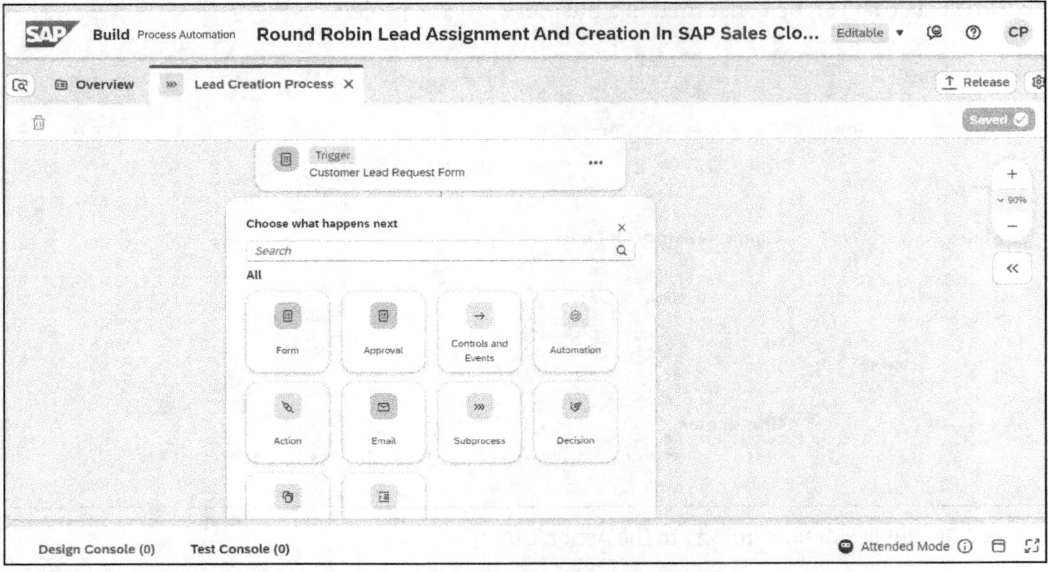

Figure 6.28 Adding the Action Artifact in the Business Process

You'll then be directed to choose the relevant action from the Actions library, as shown in Figure 6.29.

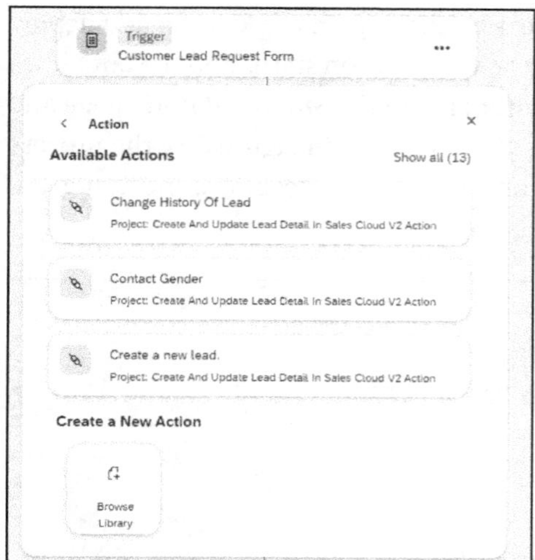

Figure 6.29 Available Actions from the Action Project to the Add-In Business Process

After successfully adding the action project, the business process looks like Figure 6.30. You can configure the action's input, output, and destination parameters on the right-side panel of Figure 6.30. You need to create an environment variable to access the destination created in SAP BTP as explained in Section 6.2. A destination is needed to connect to any backend system to execute the APIs. The destination-based environment variable will be used to configure the action of the business process.

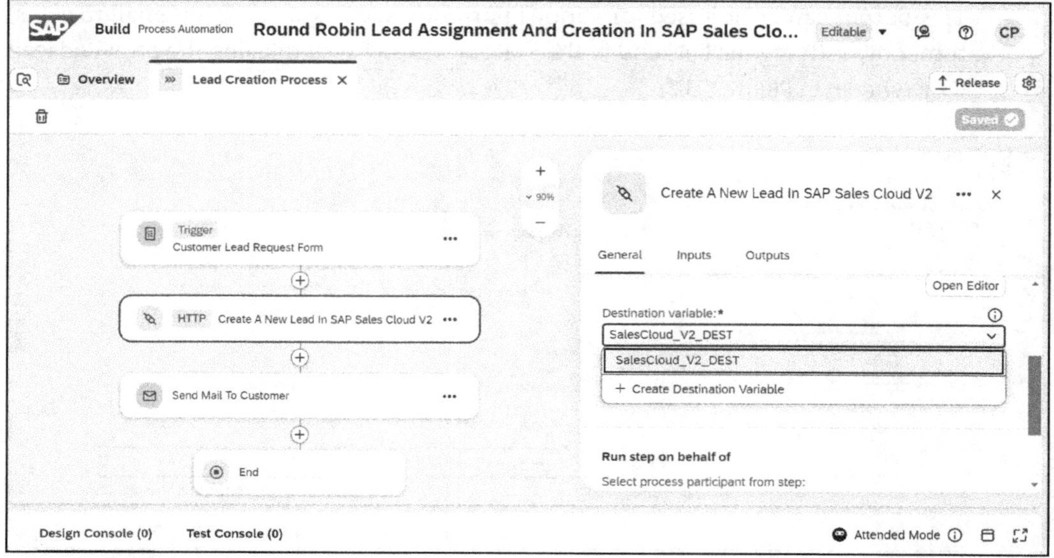

Figure 6.30 Configuring Destination Parameters for an Action in the Business Process

Finally, release and deploy the business process by clicking on the **Release** and deploy buttons in the business process project. Monitor its execution status and the outcomes of action execution by navigating to the **Monitor** area in the SAP Build homepage and selecting the **Process and Workflow Instances** tile. As shown in Figure 6.31, the process context will be updated with the output from the action execution, including the newly created lead ID.

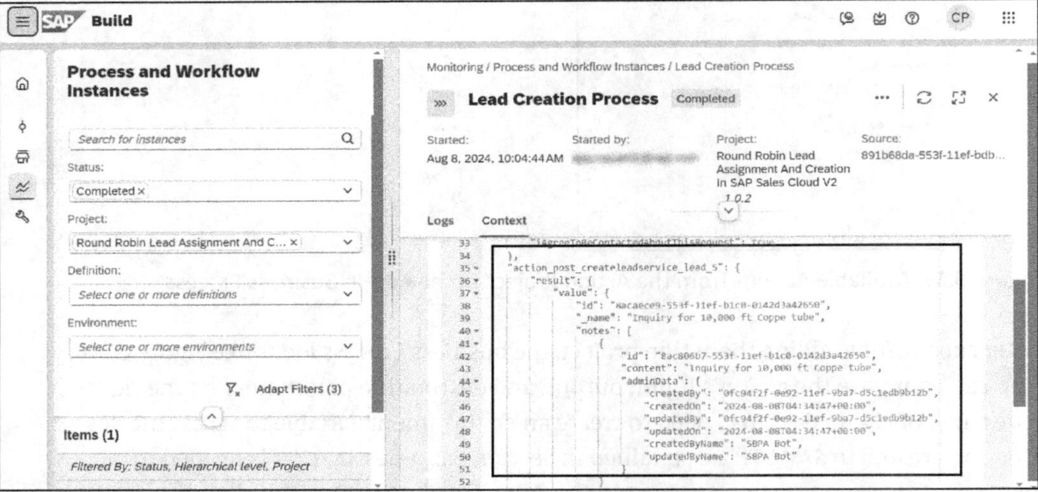

Figure 6.31 Monitor Status and Outcomes of an Action

This output can be accessed for various purposes, such as sending an email to a customer or verifying the backend in the SAP Sales Cloud system using the provided lead ID, as shown in Figure 6.32.

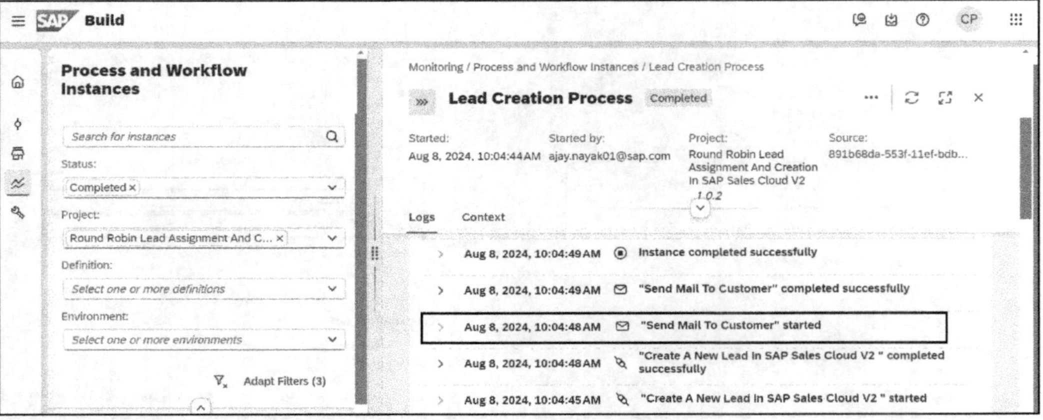

Figure 6.32 Consuming the Output of an Action

In Figure 6.33, the left side shows an overview of all steps to create and configure an action project in the action editor, and the right side shows a list of all steps to consume an action project in a business process project.

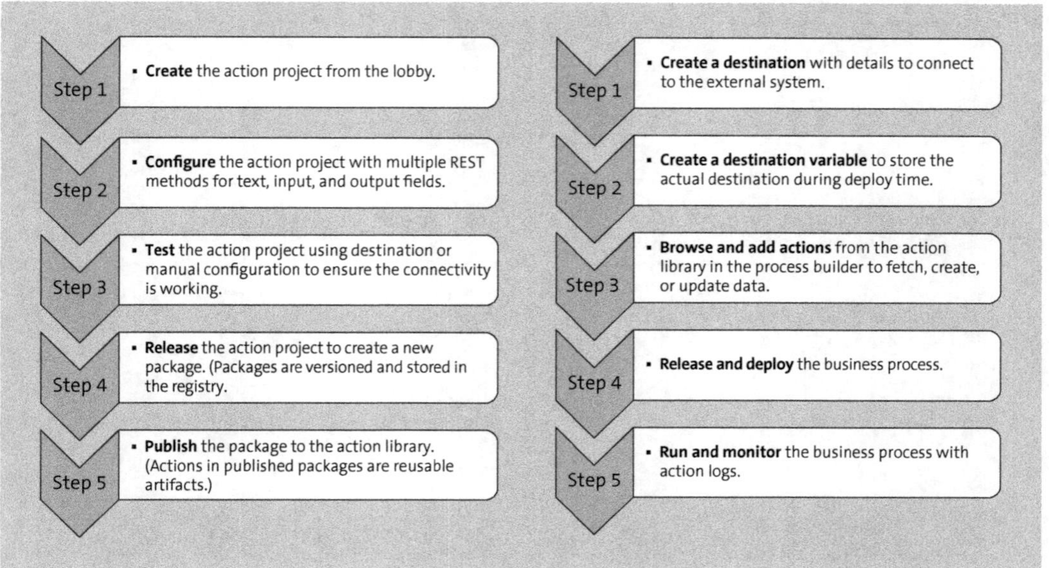

Figure 6.33 Steps to Create and Consume an Action Project in a Business Process in SAP Build Process Automation

6.6 Summary

You've learned how to create and consume action projects and how to build and manage business processes that involve connectivity using APIs from SAP and non-SAP systems. You've also understood how easily you can consume APIs of SAP products in SAP Business Accelerator Hub using destinations configured in SAP BTP. Based on the outcome of the action project in a business process project, you can perform further steps in the workflow. You learned how to monitor the action logs in a business process, including the status of the process. This powerful feature allows for streamlined integration and enhanced control over your business workflows, ensuring seamless communication between various systems and services.

In the next chapter, you'll learn about the automation artifact and how a bot can automate the repeated manual operations in a business process. By leveraging action projects, you can efficiently define, customize, and execute API-based actions, making it easier to automate complex processes and improve overall efficiency. This functionality is a game-changer for anyone looking to optimize their business operations and leverage the full potential of API connectivity.

Chapter 7
Automation Artifacts

In today's complex IT landscape, the ability to automate across different systems with agility and precision requires intricate automation capabilities. This chapter teaches you how to create and configure automation artifacts. You'll learn about software development kits (SDKs) that can be used to create automations for SAP and non-SAP platforms with complex flow designs. By the end of this chapter, you'll have a comprehensive understanding of how these concepts can be applied to automate entire processes fluidly and effectively between applications.

Automation artifacts, also known as robotic process automation (RPA) bots, refer to software that allows users to automate repetitive, rule-based tasks. Often, these typically mundane and time-consuming tasks are prone to human error yet are necessary for business operation. RPA bots mimic human interaction with different systems, executing tasks such as data entry and transaction processing. They are essential to businesses looking to free up human resources for higher value activities while reducing human error and creating a more efficient, scalable, and compliant enterprise.

Automation artifacts are just another of such artifacts to help accomplish comprehensive automation across different systems and applications. Inside an automation artifact, you can do the following:

- Pilot applications and operate on top of their user interface (UI).
- Complete web service calls.
- Interact with file systems (local and network drives).
- Use Microsoft Excel, Microsoft Outlook, and so on, as well as cloud-based applications.
- Execute document-processing tasks.
- Monitor, verify, and manipulate data.
- Create a complex, conditional, and iterative flow of activities.

You can see what an automation artifact looks like inside an SAP Build Process Automation project in Figure 7.1.

Figure 7.1 Automation Artifact Created in a Project

Automation artifacts are designed and implemented from within the automation editor, which we'll discuss in Section 7.1 Once inside the automation editor, you can add steps to the automation flow that fall under five different categorizations:

- Screens
- Activities
- Controls
- Data
- Already-existing artifacts (automations, actions, and processes)

As we make our way through the chapter, we'll discuss each of these categorizations in its own subsection. However, the most powerful use cases are those that combine the functionalities from many of these subsections to efficiently and accurately model automation flows. In fact, as we discuss each categorization, you may notice that functionality from another is included in some explanations and examples. You should always be thinking about how you can combine these functionalities to create the most powerful, robust, and scalable automation.

7.1 Automation Editor

When we talk about automation artifacts, we're referring to automations created in the automation editor. In this section, we'll discuss how to use the automation editor to create an automation artifact. We'll first explain the different features available on the automation editor's screen and then how to begin using these features to design an automation flow. We'll also discuss how to test and debug an automation with the automation editor.

7.1.1 Overview

The automation editor has the following features, as shown in Figure 7.2:

❶ **Automation editor canvas**
The automation editor canvas is where you design your automation. Here, you can drag and drop activities in the sequence needed. You can also click on specific activities to get more information about the activities or to change their parameters.

❷ **Automation Details (right-hand panel)**
If an activity from the canvas is selected, the activity details and parameters will be shown here. However, if no activity is selected (or you click the automation editor canvas), the **Automation Details** area appears (as shown). This is where you can add any of the activities or options from ❹ through ❿ to the automation editor canvas. You can also toggle this pane to the **Input/Output** option. Doing this will allow you to set inputs or outputs for your automation artifact. You can also select the **Info** option in the pane to display and adjust the automation artifact information, such

as its name or timeout duration (for both specific activities or the entire automation).

❸ **Search bar**
You can use this search bar to search for specific activities or options from ❹ through ❿ that you may want to drag into your automation editor canvas.

❹ **Action Groups**
You can call configured actions from your automation artifact. Any actions inside the project (can be added as dependencies in the project settings) will be available to add from here (discussed later in Section 7.6).

❺ **Automations**
You can call already-created automations from your automation artifact. Any automations inside the project (or those added as dependencies in the project settings) will be available to add from here (discussed in Section 7.6).

❻ **Screens**
You can automate on top of the UIs of different applications. You first need to capture the applications and their corresponding screens and elements inside the application editor. Once you've captured these items, you can drag and drop the screens onto your automation editor canvas and add activities to them (discussed in Section 7.2).

❼ **Activities**
Activities are predefined by the SDKs inside your project. You can easily add SDKs as dependencies to your project based on the types of activities you want to use. You can then drag and drop their associated activities onto the automation editor canvas (discussed in Section 7.3).

❽ **Processes**
You can call already-created processes from your automation artifact. Any process inside the project will be available to add from here (discussed in Section 7.6).

❾ **Data**
Data types and data management activities can be added to your automation artifact to manage data and variables throughout the automation flow (discussed in Section 7.5).

❿ **Controls**
Controls can be used to create complex automation flows with conditions (if... else), iterations (loops), and error handling. You can drag and drop different controls into your automation editor canvas (discussed in Section 7.4).

⓫ **Design Console**
The **Design Console** will alert you to any errors, warnings, and information that arises while designing your automation artifact. For example, if you add an activity but forget to fill out one of its required parameters, an error will occur in the **Design Console**.

⓬ **Test Console**

The **Test Console** will alert you to any errors, warnings, or information that arises while testing your automation artifact. For example, if your automation fails during execution because you improperly used an activity, an error will occur in the **Test Console**. You can also print messages (and variables) to the **Test Console** using the **Log Message** activity.

⓭ **Variables**

You can keep track of all variables associated with your automation artifact here. You can see the variable name, associated step or activity, type, and whether the variable is a list. You can also set the variable as a *watch variable* so you can monitor its values throughout the automation flow during testing.

⓮ **Breakpoints/Watches**

You can keep track of the different *breakpoints* and *watches* (or watch variables) throughout the automation flow during testing. This is part of the *tester panel.*

⓯ **Timeline**

The timeline shows the event logs of the automation flow during testing. You can also interact with the timeline by clicking on specific events to get more information. This is also part of the tester panel.

⓰ **Tester (debug) panel**

This icon opens the tester panel.

⓱ **Start test (debug)**

This icon starts the test (debug) session for your automation artifact. If you have input variables, you'll be asked to provide values.

⓲ **Automation toolbar**

The automation toolbar contains multiple tools to help design and edit your automation artifact. You can also access most of these tools by right-clicking activities or groups of activities and using the corresponding dropdown list.

⓳ **Settings**

This icon opens the **Settings** popup, where you can edit many aspects of your project. This is where you can add dependencies to your project, such as SDKs, to increase the functionality and types of activities you can add to the automation editor canvas. You can also look at general settings, share your project, edit environment variables, change the agent version, and add agent attributes.

Note

The **Screens, Automations, Action Groups**, and **Processes** dropdown options in the right-hand panel won't be available if they are empty. If they haven't been created yet, there's nothing to be shown.

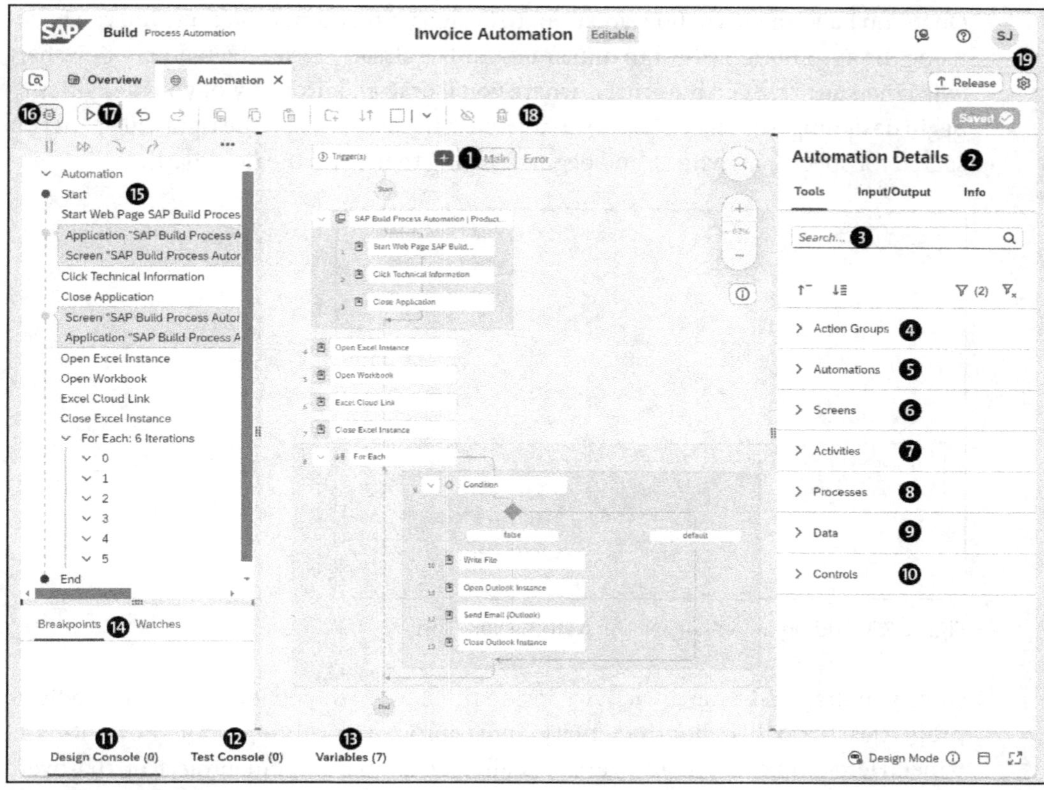

Figure 7.2 Automation Editor with an Example Automation

To create an automation artifact and get started inside the automation editor, you can navigate to the **Overview** section, click **Create**, and choose **Automation**. You can give it a **Name, Identifier** (automatically chosen, but you can edit it), and **Description.**

7.1.2 Using the Automation Editor

Now that you're familiar with the automation editor, let's take a step inside and see how to begin using it to accomplish your automation. Prior to creating anything inside your automation editor, architect your solution and create a flowchart (pen and paper will do) to understand the basic flow of activities you plan on automating. This will help you better understand the activities you want to use, the SDKs you want to add, and the applications you want to capture. Of course, your preliminary flowchart doesn't need to be perfect and may not map 1:1 with activities available in the automation editor. As you become more experienced with automation, your preliminary flowcharts will naturally begin to map more closely with what is available. Over the course of the implementation, you should test, validate, and iterate many times to find the correct flow of activities to accurately execute your automation.

Once you have an idea of how you want to design your automation flow, you can begin using the automation editor to implement it. The already-created but empty flowchart (with the **Start** and **End** buttons) is where you'll drag and drop any of your activities to begin designing (see Figure 7.3). To do this, click the activities on the right-side panel (or search for specific activities), and begin dragging them into the flowchart.

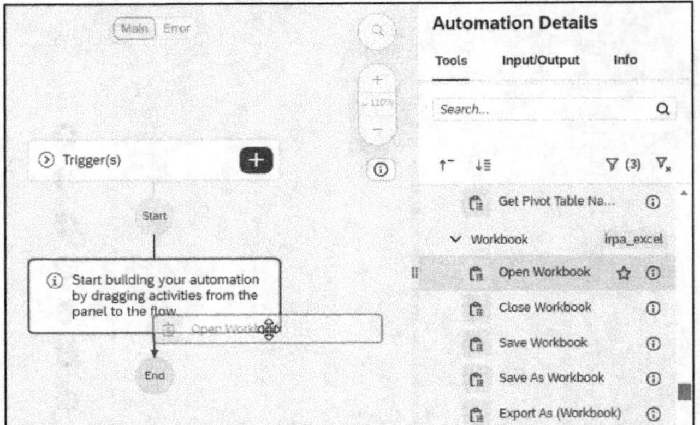

Figure 7.3 Adding an Activity to the Automation Flow

Once you drag and drop an activity inside the flowchart, you'll likely want to edit its properties. To do this, click the activity, and its properties will appear on the right-hand panel. Here, you can use any variables already created in your project or that were brought into the project as inputs or environment variables. You'll see the recommended list of available variables to be used by clicking inside one of the activity parameters, which will produce a dropdown list of the variables. You may notice that some variables in this dropdown have a number or letter next to them, as shown in Figure 7.4.

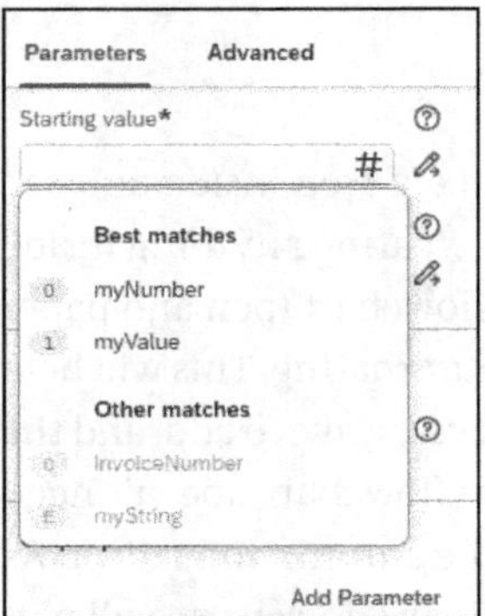

Figure 7.4 Editing Activity Parameters

This number represents the specific step that the variable was created from. This step number corresponds to the number to the left of each of your activities. Input variables will receive a **O** designation, while environment variables will receive an **E** designation. Some variables may also be grayed out. This represents that the variable is of the wrong type (e.g., the parameter is looking for a number, but the variable is a string).

You'll also see the ✎ icon, which allows you to open the expression editor to designate your own values inside the activity parameters. For some activities, you can even create more complex logic or expressions that may not be possible from outside the expression editor.

You should also be aware that when there are errors with your automation from a design-time perspective, they will show in the **Design Console** at the bottom of your screen, as shown in Figure 7.5. You can usually just click the error, and it will automatically take you to the activity and its associated error. You'll need to solve all **Design Console** errors before you can test your automation.

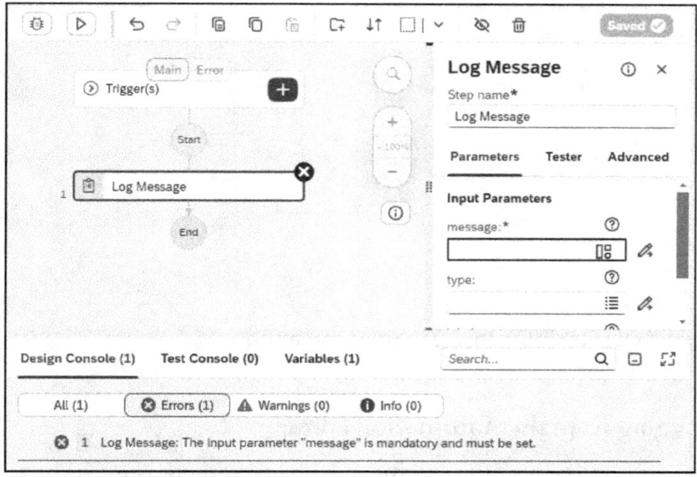

Figure 7.5 Using the Design Console to Evaluate Errors

Another benefit of outlining the flow of your automation prior to building it is that you can have a general view of what data you'll be using. While this can always change over the course of your implementation, it's good to have a general understanding of what data you'll need, what data you'll create, and what data you'll output. Often, an automation artifact is only one part of the larger process. Sometimes, multiple automation artifacts can be chained together, and an automation is often combined with the other artifacts described in this book. To make use of this powerful, modular approach, you need a way of passing data between your different artifacts and you need a good way of using data available from previous artifacts inside your automation artifact.

Let's say, for example, we wanted to use an invoice number that we collected from a form artifact earlier in our process. We can set that as an input to our automation to be used by the automation's activities. Then, during the automation, we used our document understanding capabilities to read that specific invoice and extracted all relevant information. We could then set this information as an output of our automation to be used later in the process, such as in an approval artifact. To do this, we need to click the **Input/Output** toggle on the right-side panel from within the automation, as shown in Figure 7.6. From here, we can define the variable names and types to which the data will be assigned (and later assign them by clicking the **End** node in our automation flow). We'll then be able to use all inputs over the course of the automation and all outputs from outside the automation in other artifacts.

Figure 7.6 Creating Inputs and Outputs from Inside Your Automation

7.1.3 Testing and Debugging from the Automation Editor

Understanding how to test and debug your automations is extremely important to implementing your automation flows. You should strive to test and validate early and often, iterating as necessary. You'll find over the course of implementation that, especially with complex functionality and flows, activities may not perform exactly as you expect them to. Perhaps you might notice that an activity requires case-sensitive strings or that an application doesn't respond quickly enough, requiring you to add a delay. There are a countless number of unpredictable issues you may run into that are best caught early so that they don't compound and create complex issues that are more difficult to solve later. When you test, you want to make sure you're considering possible corner cases that may cause your execution to fail. When thinking of corner cases, it's often helpful to consider what an end user with no knowledge of how the automation is designed might use that could cause an issue. What input might they provide that unknowingly causes the automation to fail? Remember, you may plan on deploying your automation to many users. And, while many users will use it exactly as intended, some may not. Thus, you want to make your automation flow as dynamic as possible and consider all reasonable corner cases when testing. Further, when debugging, you want to reduce the number of things that may cause an issue. To do this, again, it's best to test early and often so that you can narrow down the list of troublesome cases as much as possible.

To test your automation, click the ▷ button inside the automation editor. This will launch a popup that allows you to fill any fields relating to environment variables or input variables that may be used during the automation flow. Once you click **Test**, your automation will connect to the locally installed desktop agent where it will begin executing the automation.

When testing your automation, it may also be helpful to use breakpoints to pause your test at specific points during execution. This may be because you need to configure something to test a specific functionality during execution or because you want to inspect specific variables or elements prior to or immediately after specific steps. To inspect variables, you can set watch variables or use the **Log Message** activity to log a variable or other message.

In Figure 7.7, we've created a simple automation where we loop three times, logging a message every time. We've set a breakpoint on the **Log Message** activity, where the execution will pause during every iteration (you can **Resume** ▷▷, **Step Over** ▷▷, **Step Into** ↘, and **Step Out** ↗ of each activity/module with the buttons at the top of the debugging pane). You can also see a timeline of activities in the debugging pane, where you can click into any of the specific activities and get more information, as shown in Figure 7.8. This is particularly helpful if a specific activity has an error during execution. That activity will show as red in the timeline. You can then click the activity and receive information on what the error is related to.

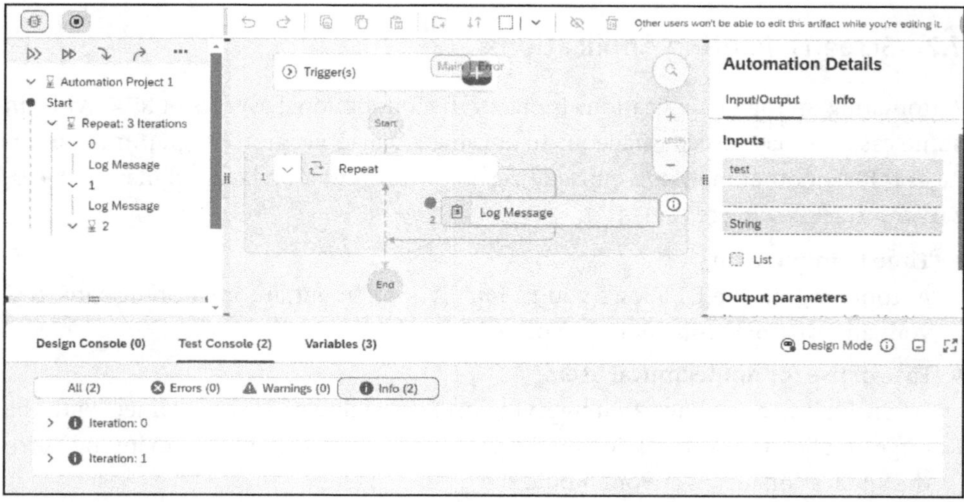

Figure 7.7 Debugging a Simple Automation

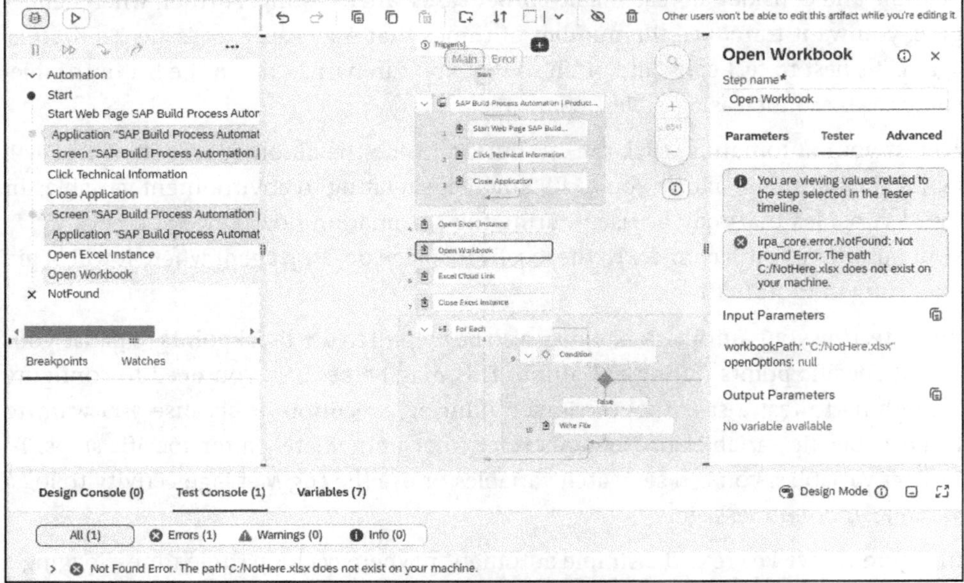

Figure 7.8 Debugging an Error in an Automation

Testing and debugging automations is a very important skill to develop during the design of your automations. It not only helps you to create robust and powerful automations but also allows you to iterate with new functionality more quickly. Ultimately, being able to quickly test and iterate allows you to create better projects more efficiently.

7.2 Screens: Piloting Applications

Automating on top of applications is one of the foundational aspects of RPA. While in some cases, it may be advantageous to automate via application programming interfaces (APIs) or other methods, automating on the presentation layer/UI may be necessary for the following reasons:

- **Time to implement**
  Automating on the UI allows you to rapidly implement automation designs, test your automations, and iterate as you see fit.
- **Ease of use for nontechnical users**
  Users don't need to have an underlying understanding of how to connect with the target application. Automation on the presentation layer allows you to interact with the application just as anyone would.
- **Noninvasiveness**
  You don't need to change anything about your source application.

- **Ability to automate on legacy applications**
  Most legacy applications don't offer the ability to automate through APIs or other methods. Users must automate on top of the UI.

SAP Build Process Automation can capture several application technologies, including the following:

- Web
- Full client (using UI automation)
- SAP GUI
- Java automation
- High Level Language API (HLLAPI)

In the following sections, we'll explore how to pilot applications on top of the presentation layer. You'll start out by learning to use the application editor to capture your applications and the applicable features of their UI that you plan to automate with. Then, we'll discuss some best practices for capturing web-based applications and full client applications.

7.2.1 Using the Application Editor to Capture and Declare Applications

In this section, we'll discuss how to use the application editor to capture the applications that you plan to automate. You'll learn the basics of how SAP Build Process Automation sees your applications and how to navigate within the application editor, as well as how to capture, declare, and automate the different aspects of your applications.

Applications, Screens, and Elements

When you pilot applications using automation artifacts, you often look at the most efficient method through which a person would interact with the application to complete the process. When piloting an application, there are three main components you must first understand though: applications, screens, and elements.

If you were to create a flowchart of the overall process, you would start with the larger *applications* being automated. From there, you would consider the different *screens* inside those applications that the user may touch. Then, within each of these screens, you would detail every *element* that a person would interact with. This includes every button, every search bar, and every piece of information on the screen that a person would need to perform the actions of the overall process. To understand these concepts, let's use Facebook as an example. In this example, Facebook is the application, encompassing all the other screens within it. From there, all the different people's profiles are considered different screens within the larger Facebook application. Then, within each person's profile (or screen), you have elements where you can perform actions: the profile picture you can click on to enlarge, the status bar you can type in to write something on someone's wall, or the **See all photos** hyperlink you can click on to navigate to their photos. Applications, screens, and elements work in a hierarchal fashion. All elements belong to a specific screen, and all screens belong to a specific application in the same way every element in a person's Facebook profile belongs to that person's profile (screen), and every profile (screen) belongs to Facebook (the application).

In Figure 7.9, under the **Declared Application** area, you can see the highest part of the hierarchy, the application titled **SAP Build Process Automation Home Page**. Under this application, you see its associated screens, titled **SAP Build Process Automation | Overview**, **SAP Build Process Automation | Product Features**, and **SAP Build Process Automation | Technical Product Information**. If you look further inside the **SAP Build Process Automation | Overview** screen, under **Declared Elements**, you can see three different elements captured. One of which is the **Features** button, which corresponds to the button on the screen that helps you navigate elsewhere.

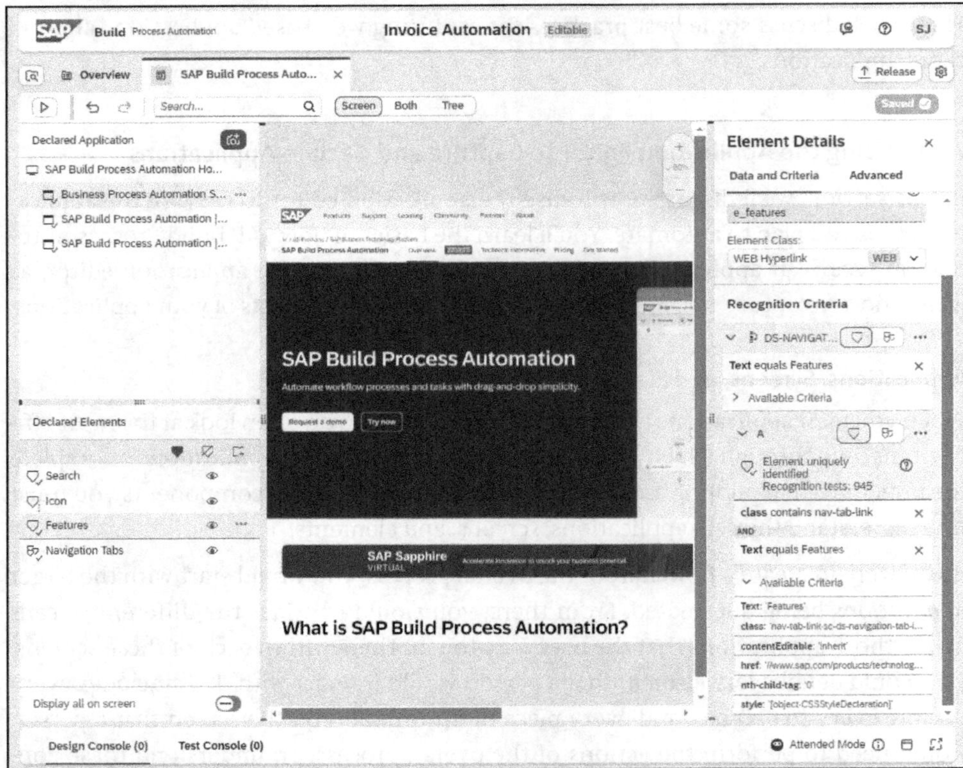

Figure 7.9 Application Capture of the SAP Build Process Automation Homepage

Application Editor

When piloting applications, there are two main activities you must perform: capture and declaration. *Capturing* refers to the process of taking a snapshot of the application and associated screens that are to be automated, while *declaring* refers to the process of creating unique recognition criteria for each application, screen, and element so that the bot knows which items to automate on.

To capture applications, click the [+] button, select the **Create** option, and then choose **Application**. You'll be given the option to add an application name and description. Once you do this, you'll see a new popup screen, as shown in Figure 7.10.

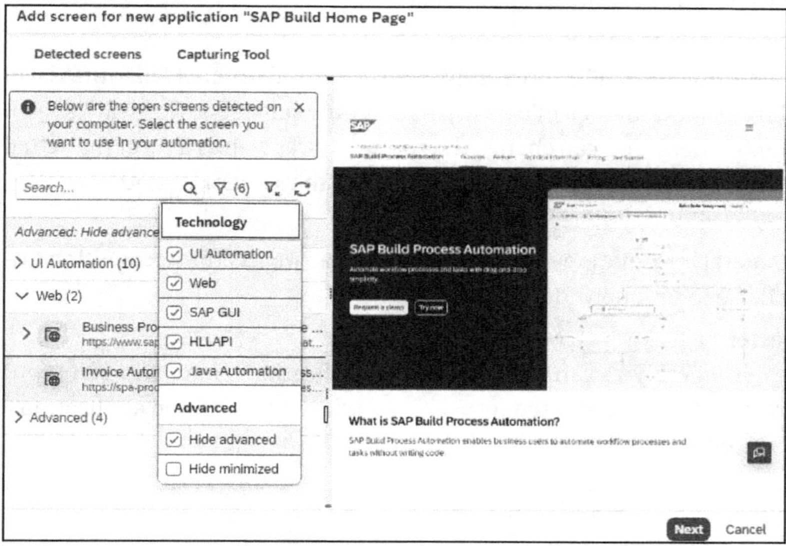

Figure 7.10 Application Capture Page Showing All Available Technologies to Be Captured

After choosing the correct application from the list provided, you can then capture the application, as well as associated screens and elements, which you'll see later. Applications, screens, and elements can either be captured manually or with an automatic recorder (discussed later).

Note

Before capturing an application, ensure that SAP Build Process Automation and the application to be captured are running in two separate browser windows.

Tip

The window zoom factor may affect how applications, screens, and elements are captured. Typically, you'll want both your computer zoom factor and the browser or application zoom factor to be 100%.

Once you've captured your application and associated screens, navigate to the *application editor* where your new application artifact was created, as shown in Figure 7.11:

❶ **Declared Application**
Displays your captured application, as well as any of its associated screens. Click on any of those items to view their respective contents. Clicking on the icon allows you to add more screens to the application. Clicking on specific screens allows you to see the declared elements in that screen.

❷ **Declared Elements**
Displays your captured elements in each respective screen.

❸ **Screen view**
Displays the captured screen. Elements actively selected and uniquely identified will have a green box around them, while elements not uniquely identified will have an orange box around them. You can change the zoom level, as well as look at the actual display of the screen, the tree (underlying structure of the screen), or both.

❹ **Element Details**
Displays the **General Information** of the selected item as well as the **Recognition Criteria** that you'll use to uniquely identify the item.

❺ **Application tester**
Launches a tester where you can test pilot your application. Here, you can see if items are correctly identified and see if specific activities work on the applications, screens, and elements.

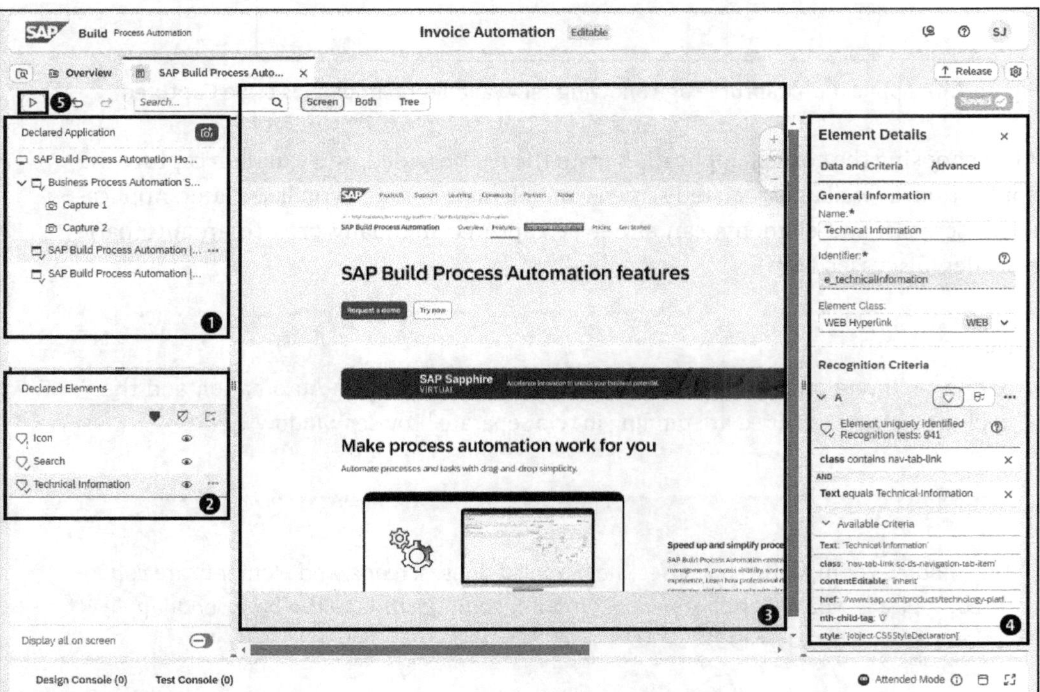

Figure 7.11 Application Editor: Declare Applications, Screens, and Elements, as well as Test Pilot Your Applications

Selecting Unique Recognition Criteria

When you capture applications, screens, and elements, you must also consider how you're expecting the computer to recognize and uniquely identify each feature. As a simplified example, let's say we have two blue buttons, one labeled PRESS ME and the other labeled DO NOT PRESS ME. How would the computer know which button to press if we only told it to press the blue button? In this scenario, we must make sure to recognize the proper button by its PRESS ME label. This way, the computer is absolutely sure it's pressing the correct button as we've provided it with uniquely identifiable criteria.

This notion is particularly powerful when identifying screens and elements in your application. You can future-proof your automations by choosing criteria that you know won't change. Therefore, if an application's UI update alters the location of a button, it won't matter if you've chosen sufficiently robust recognition criteria. When you record applications, SAP Build Process Automation automatically chooses recognition criteria it thinks is best suited for the application. However, the user should check the automatically chosen recognition criteria and adapt it if they think it can be more robust. This may also depend on the specific technology of the application being captured. We'll cover some best practices later in this section.

When you select unique recognition criteria for applications, screens, and elements, SAP Build Process Automation provides you with several operators to adequately select your criteria. Criteria will typically take the form of either a Boolean (i.e., checkboxes that are checked or not checked), a string, or a number. Booleans can be set to either true or false, while strings and numbers have operators, as shown in Table 7.1 and Table 7.2.

String Operator	Description
equals	The property value is strictly equal to the argument.
not equal to	The property value is strictly *not* equal to the argument.
contains	The property value contains the argument.
doesn't contain	The property value does *not* contain the argument.
starts with	The property value starts with the argument.
doesn't start with	The property value does *not* start with the argument.
ends with	The property value ends with the argument.
doesn't end with	The property value does *not* end with the argument.

Table 7.1 String Operators

String Operator	Description
is empty	The property is empty—no argument.
is not empty	The property is *not* empty—no argument.
matches	The property value matches the argument.
doesn't match	The property value does *not* match the argument.

Table 7.1 String Operators (Cont.)

Number Operators	Description
=	The property value is strictly equal to the argument.
≠	The property value is strictly *not* equal to the argument.
>	The property value is strictly greater than the argument.
<	The property value is strictly less than the argument.
≥	The property value is greater than or equal to the argument.
≤	The property value is less than or equal to the argument.
≤	The property value is less than or equal to the argument.
≯	The property value is *not* strictly greater than the argument.
≮	The property value is *not* strictly less than the argument.
≱	The property value is *not* greater or equal than the argument.
≰	The property value is *not* less or equal than the argument.

Table 7.2 Number Operators

The **Match** operator allows you to locate a string on a screen. You must enter a regular expression in the **Value** field that respects the POSIX Basic Regular Expression syntax and that is case-insensitive.

A quick example will help you understand the **Match** operator better. Let's say you were looking at a screen and wanted to capture a specific set of elements that consisted of a digit. You could set the **Operator** to **Matches** with the **Value** set to **\d** to find every element with a digit, as shown in Figure 7.12.

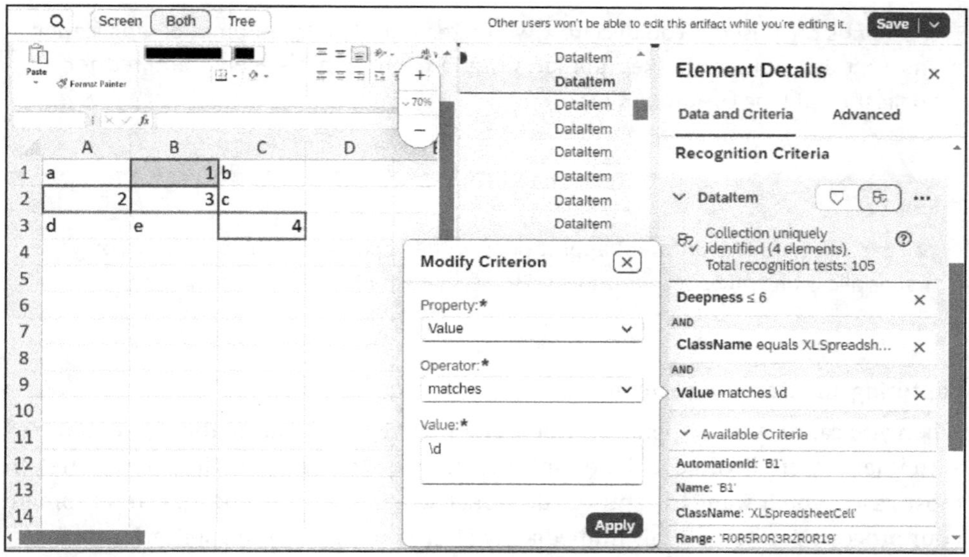

Figure 7.12 Using the Match Operator to Find All Elements with a Digit

Capturing and Declaring Applications

As described earlier, to capture applications, you'll click the + button, select **Create**, and then choose **Application**. You can then add an application name and description. Once you do this, you'll see a new popup screen. From here, capturing the application is relatively straightforward, as you'll only need to choose the applicable screen from the application to be captured, and the associated application will be automatically captured. There are advanced capture options you can change based on the specific technology you're capturing. Details regarding these options can be found in the documentation, although you rarely need to change them.

Declaring applications is also relatively straightforward, as the **Recognition Criteria** is selected automatically during capture. However, in some cases, you may want to change or add additional criteria. For example, the **Domain** property will often be automatically selected for web-based applications. While this is often adequate for web-based applications, there may be situations where you need additional criteria, such as when you have a development environment application and a production environment application. Here, you may want to select **contains** "development" or "production" as part of the URL for an additional recognition criterion.

Note

UI automation technology captures will provide the **EXE** (the name of the application) and **TITLE** (title of the main associated window) properties in the application data. Typically, **EXE** will be set under **Recognition Criteria** while **TITLE** will be ignored. The **TITLE**

criterion is only used if you declare a web screen inside a UI automation application and isn't available in the **Recognition Criteria** of the other screens captured for an application of type UI automation.

Note

You can't declare two different applications with the same criteria; otherwise, only the first application will be recognized, and the second application will be hidden.

Capturing and Declaring Screens

When you capture your application, at least one screen will automatically be captured with it (and more if you decide to capture multiple screens at the time of capture). In most cases, you'll want to capture additional screens of an application to automate your process. You can do this immediately after you capture the application and first screen using the **Capture Another Screen** button or after you navigate to the application using the **Go to Application** button. From the application artifact, under **Declared Applications**, you can click the icon to capture and add another screen. In this case, you can follow the same steps you completed when you originally captured the application to capture the new screen.

Warning

When you capture new screens, you need to ensure they are still part of the same application you originally captured. This means that whatever recognition criteria you assigned to the application must also apply to any new screens you're capturing. For example, if you used the **Domain** of the application as **Recognition Criteria**, all screens captured for that application must have the same domain or they won't be recognized.

Tip

When capturing a large screen, you may notice that the capture may scroll throughout the page to capture all elements. If you want to capture only part of the screen (as the bottom of the page may be unnecessary to your automation), or if you want to capture a specific dropdown **On Hover**, which may disappear (discussed later in this section), you can select **Disable Auto Scrolling** to force SAP Build Process Automation to only capture the part of the screen visible at time of capture.

You can also create multiple captures of the same screen. This is particularly helpful when you want to access multiple versions of the same screen, such as with dropdown menus. In some situations, you can capture these as different screens, but typically two captures of the same screen with only small differences won't have different enough recognition criteria to recognize them separately. For example, in the case of web applications, a screen with a dropdown menu collapsed and the same screen with the dropdown menu expanded will have the same *domain*, *title*, and URL. In these cases, it's helpful to create a second capture of the same screen. To do this, you can click the ••• (options) icon next to the screen you want to take a second capture of, and then click **Add Capture to Current Screen**. From here, you can follow the same process you used to capture other screens.

In Figure 7.13, you can see the **SAP Build Process Automation Home Page** application with three different pages: **Overview**, **Features**, and **Technical Product Information**. However, on the **Overview** page screen, we wanted to expand the **About** dropdown menu to be able to identify and eventually click the **Company Information** button. To do this, we created a second capture of the **Overview** page so that we could access this dropdown menu. Therefore, in the first capture of the screen, we can click the items underneath the dropdown menu, and in the second capture, we can click the items contained in the dropdown menu. While these are two different captures, they are still considered to be the same screen, as they share the same recognition criteria (they both use the **Overview** URL in this case).

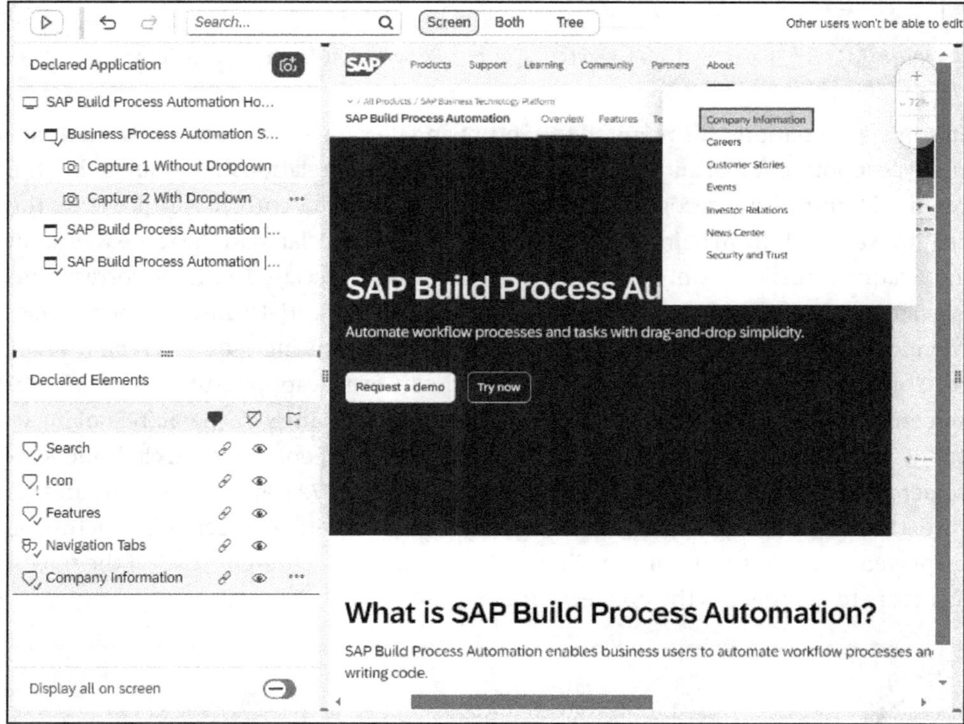

Figure 7.13 Application with Multiple Screens, As Well As Multiple Captures of the Overview Page Screen

Tip

Some dropdown menus disappear when you click outside of them or when you stop hovering over them. To allow you to still capture these menus, SAP Build Process Automation gives you the **On Hover** option. To access this capability, begin by clicking the icon to add a new screen or click the ∘∘∘ icon and select **Add Capture to Current Screen** to add a new capture. This will launch either the **Add Screen** or **Add Capture** popup, which look almost identical. From there, you'll click the **Capturing Tool** at the top of the popup, and then click the **Launch Capturing Tool** button. This will launch the capturing tool, which allows you to hover over the screen or parts of a screen that you want to capture, and hold [Ctrl] (can also be set to [Shift]) to capture the screen (or parts of the screen), as shown in Figure 7.14.

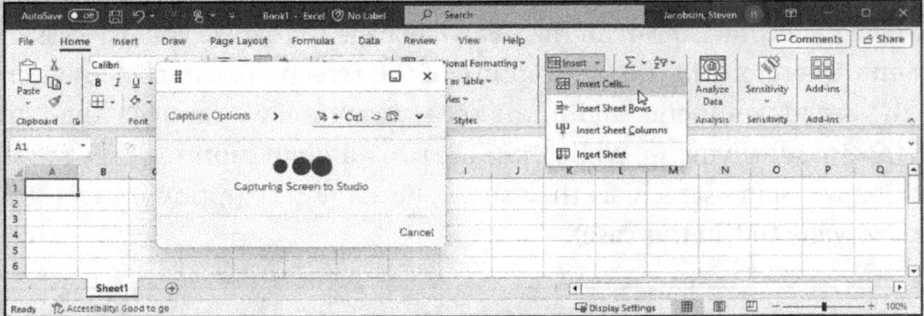

Figure 7.14 On Hover Feature Used to Capture a Dropdown List in Microsoft Excel

In some automations, it's required to capture and declare a subscreen. Subscreen capture is supported for UI automation only. This subscreen relationship ensures that the required controls are recognized. However, if not declared correctly, it prevents the required controls from being recognized. A subscreen is declared in the same way as an application screen. The only difference is that it must be declared with the corresponding parent screen. Capturing subscreens may be necessary with UI automation if an element can't be properly recognized/accessed without capturing the subscreen it's part of. You can capture multiple subscreens and can even capture subscreens of subscreens. To do so, you need to click the ∘∘∘ button next to the screen you're looking to capture a subscreen for (often referred to as the parent screen), and then click **Add Subscreen**, where you should see a popup, as shown in Figure 7.15. From here, you can capture the subscreen just like capturing any other screen. The screen structure is displayed on the left-side panel in Figure 7.15. You can click different parts of the screen structure to see them in the preview pane.

Note

Typically, UI automation screens will have **Deepness** automatically assigned as a recognition criterion (in addition to **Name**). *Deepness* refers to the level in the screen structure that the screen you're trying to capture exists.

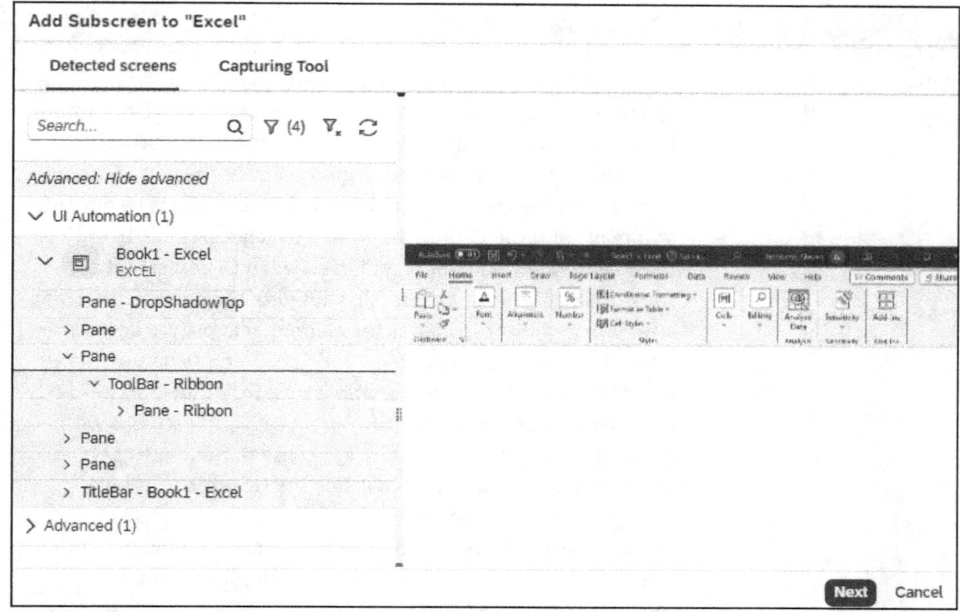

Figure 7.15 Add Subscreen Popup to Add the Subscreen

There's often confusion around when to create multiple captures of a screen, when to capture a subscreen, and when to capture an entirely different screen altogether. While it's very difficult to generalize across all different applications that may be captured, you can certainly work with some advice. Often in development, you must be able to work in an iterative fashion, where you capture a screen (and its associated elements), quickly test the screen (using the **Screen Tester**, discussed later), and adapt your solution design if needed. That being said, you can use some general guidelines as given in Table 7.3.

Type of Screen Capture	Reason to Use
Multiple captures of the same screen	You often want to create multiple captures of the same screen when there are different aspects of a screen that you can't access in one capture, such as a dropdown menu. When you create multiple captures of the same screen, you must ensure that—based on the recognition criteria selected—it's the same screen. In other words, the recognition criteria used to select the first capture must apply to all captures of that screen. For example, if you use the URL to recognize a screen, a subsequent capture of the screen with a dropdown menu deployed will, almost always, maintain the same URL. There's usually no significant risk in taking multiple captures of the same screen if the recognition criteria still apply.

Table 7.3 Guidelines on When to Use the Different Types of Screen Capture

Type of Screen Capture	Reason to Use
Subscreens	You want to use subscreens when there's a clear structure of the screen being captured, which can be seen when you go to capture the screen. You'll always want to capture the parent screen first, and then capture any necessary subscreens of the parent screen after. Capturing the proper subscreen can sometimes be difficult as it may not be obvious to determine which subscreen to capture for the intended element or action you plan to implement. In these cases, it's recommended to start with the outermost parent screen that still encapsulates the element you plan to use. If you haven't captured far enough "down" the screen structure, the element will either be difficult or impossible to capture, or the element won't work during testing. When capturing screens with complex screen structures, it makes sense to frequently test. If the element can't be captured inside the screen, or if it fails during testing, you can then move "down" the screen structure and capture subsequent subscreens. This can be repeated until the proper subscreen is captured. Err on the side of caution regarding capturing too many subscreens or capturing unnecessary subscreens.
Different screens	You want to capture different screens when the screens to be captured have distinctly different recognition criteria.

Table 7.3 Guidelines on When to Use the Different Types of Screen Capture (Cont.)

Warning

You can manage the order of screens, which helps to recognize the screens in the right order during execution. The order of the screens (from top to bottom) is the order in which they will be recognized. Therefore, if recognition criteria define the screens as *not* mutually exclusive, the first screen listed will be the one recognized. While it's intuitive to order the screens in the chronology in which you use them (ok for organizational purposes if there's no specific reason to change the order), you must always consider whether the order matters regarding how SAP Build Process Automation recognizes your screens. If it does, you must arrange the screens accordingly. To do this, click on one of the screens and drag it above or below another screen. All associated items, such as elements, will be moved with the screen automatically.

Note

Surface automation can be used to automate a screen based on the images shown on that screen. This may be helpful if the screen's document object model (DOM) isn't available at either design time (when you won't be able to capture specific elements) or runtime (the technology won't be able to recognize captured elements). Because surface automation uses the screens images, it isn't as robust, but it can often be helpful if there's no other way to automate the application or screen.

To declare a screen, you need to set recognition criteria, which will be used to identify the screen correctly when your automation runs. This will entail selecting specific recognition criteria that uniquely recognize your screen from the other screens in your application. You can do this by hovering over items in **Available Criteria** under the **Recognition Criteria** section on the right-side panel of the application editor, and then clicking the + icon. Doing so will add a criterion to your **Recognition Criteria**. You can then click on the criterion and edit it using the string operators and number operators discussed earlier.

When uniquely identifying screens, you also have the concept of **Must Exist** and **Must Not Exist** elements. Once an element is captured and declared (discussed shortly), you can set them as **Must Exist** or **Must Not Exist** elements of the screen. While this isn't always necessary, it can be a powerful feature that allows you to identify screens that may be difficult to declare and identify otherwise. This may happen with web pages that can change dynamically without needing to navigate to another page, or if there are widgets on the screen (although the need for **Must Exist** and **Must Not Exist** elements to identify screens comes up in several other cases too).

As a quick example, let's look at the SAP Build Process Automation **Lobby** shown in Figure 7.16.

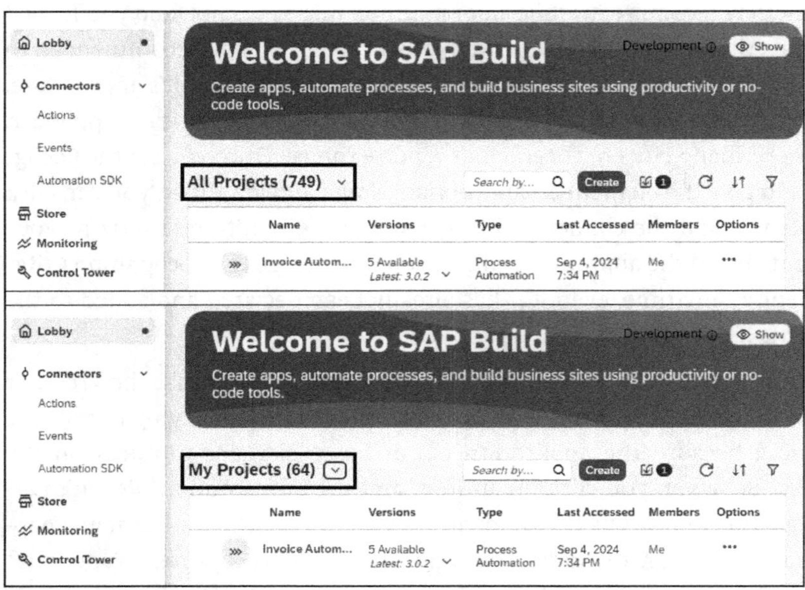

Figure 7.16 SAP Build Lobby with Nearly Identical Screens: One Showing "All Projects" and the Other Showing "My Projects"

Let's say you wanted to iterate through all projects created by only the current user. In the lobby, you can filter **All Projects**, **My Projects**, **Projects to Work On**, and **Projects to Deploy**. However, when you change this filter, nothing about the available criteria to recognize the screen changes. The **Domain**, **URL**, and **Title** all remain identical. In this case, you can capture the filter header and recognize it as an element, entering "My Projects" as its criteria. You can then mark this element as **Must Exist**, which uniquely identifies the screen. This recognition technique is also useful when there are insufficient recognition criteria in general (i.e., if the URL is consistently changing and can't be relied on to recognize the screen every time).

Capturing and Declaring Elements

Now that you've learned how to capture and declare applications and screens, we can move on to what most of the activities in your automation will be implemented on: elements. *Elements* refer to the components of a screen that you interact with when using a UI. Elements consist of the buttons, input fields, list and table items, and many other types of controls on the screen. Capturing and declaring elements occurs in the same step too, as compared to applications and screens where you must first capture them and then declare them. There's no popup screen, and there's no need to access the actual application anymore. Once the application and screen are captured, you'll work entirely off the **Screen** view in your application editor to capture and declare elements. To begin declaring your elements, you can click on specific items on the screen. When you click on items, the item will either show a green or an orange box around it. A green box represents that the criteria automatically assigned to the item selected is sufficient to uniquely recognize it, while an orange box means it's not, and you'll need to add additional criteria. However, even if the box is green, it's still recommended to look at the **Recognition Criteria** automatically assigned and make sure it's appropriate. For example, if the element is dynamic, and a text criterion is automatically applied, it may make sense to change this to criteria that is more robust. Once you set the **Recognition Criteria** and give the element an appropriate **Name** and **Identifier**, you can click Declare Element. This will add the element with its assigned recognition criteria to your **Declared Elements** list of the application editor. You can change the **Recognition Criteria** of the element at any time, even if it has already been declared and added to the **Declared Elements** list.

When clicking on items in the screen, you may sometimes notice that the green or orange box around it doesn't perfectly encapsulate the item you're trying to capture. This is happening because the application editor bases element captures on the underlying DOM. So, when you click on an item and the box around it doesn't perfectly encapsulate the item, it's usually referring to a different DOM item that just happens to be located at, near, or under the item you want to capture. If you can't accurately identify the item with a click, you should refer to the DOM to accurately identify the item by selecting either **Tree** or **Both** on the Screen Both Tree toggle. This will allow you to precisely choose the exact element you're looking to capture. In fact, even if you can identify the item from the **Screen** view, it's still advisable to look at the DOM and ensure you're capturing the correct item.

In Figure 7.17, you can see that the **Features** navigation button on the screen has been selected with a green box around it. However, you can confirm the correct item is selected by consulting the DOM, and making sure the **Navigation Tabs** item is selected.

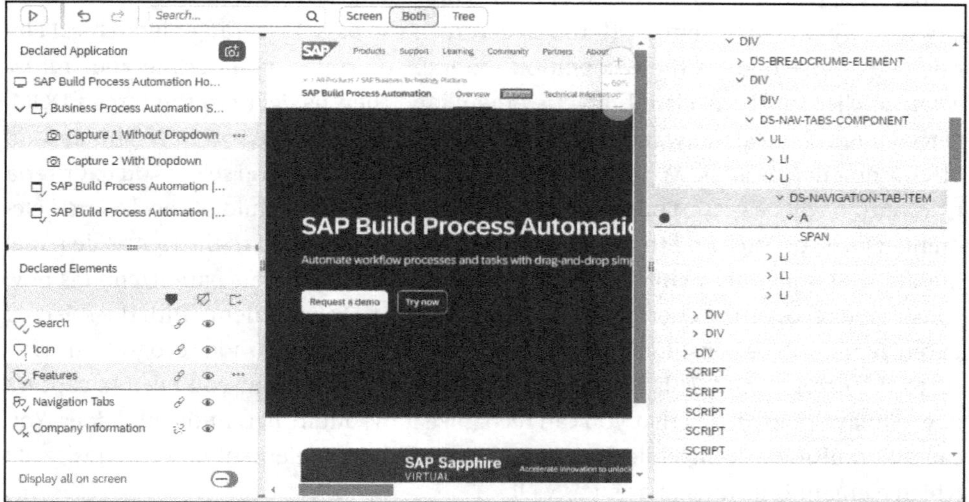

Figure 7.17 Application Editor with Both the Screen UI and Screen DOM Displayed

Tip

There are often multiple elements that you can capture to accomplish the intended activity. For example, when capturing the **Navigation Tabs** item in Figure 7.17, we could have captured the **A** element or the **SPAN** element (assuming we wanted to click on the element). For this reason, it's always suggested to test early and test often when declaring elements.

Note

The **Element Class** is automatically detected by the application editor, although you can override this and manually select the class if you need to.

When declaring an element, often the available criteria provided is sufficient. You need to add + the appropriate criteria to the **Recognition Criteria**, and then, if needed, edit it using the string operators and number operators we discussed earlier. When you consider recognition criteria, you typically want to make the criteria specific enough to uniquely identify it compared to the other elements on the page, but make it general enough so that any small changes don't affect recognition. This could be important, for example, when capturing a table element where the text may change (and thus you don't want to use the text in the capture as your criteria). You always want to consider what is dynamic and how you can capture elements to make them as robust as possible.

Sometimes, you can't identify elements with only their own criteria. You can additionally identify these elements with the structure of the page by adding hierarchical elements (parent/child) to the recognition criteria. To do this, you declare your *target element*, click on the declared target element, and then navigate to the screen DOM. Once you find the already declared target element in the DOM, you can right-click on other elements in the DOM (in the target element's hierarchy) and select **Add to Criteria** (this option will be grayed out if the element isn't a parent or child of the declared element). Once you click **Add to Criteria**, the element will be added to the right-side panel hierarchy as a parent or child element of the target element. This parent/child is now considered recognition criteria for the target element. Only elements with the selected parent/child will be considered when identifying it. You can also add recognition criteria to the parent/child. This is particularly helpful if the element you need to capture doesn't have many fields that you can recognize it by, but its parent or child does. You can add multiple parent/child elements to a single target element and can designate which element is its direct parent as well.

Note

You can change the target element even after the element has been declared by clicking the ∘∘∘ icon next to it (under **Recognition Criteria**) and clicking **Set as Target Element**.

Note

Sometimes, you may need to capture dynamic elements or elements with too few recognition criteria. If you know that this target element will remain in the same element position relative to another static element, you can use the **Labeled by** parameter as recognition criteria. To do this, select your target element, click on the ∘∘∘ option of the other element, and then click **Labeled by**. This can be used for all technologies other than web technology.

Once you've selected recognition criteria, you'll see either an orange or green box. An orange box, which represents that your element isn't uniquely identified, will tell you how many elements are identified with the recognition criteria you provided. This may help you narrow down what additional recognition criteria you need to uniquely recognize your element. If the box is green, it means your element has been uniquely recognized. However, this box will also provide you another piece of information: recognition tests. *Recognition tests* are performance indicators used to identify the number of elements, screens, and subscreens associated with a captured application screen. If you notice performance is low (i.e., speed of identifying the element during automation is too slow), it may be caused by too many recognition tests. You can often change the number of recognition tests by changing the recognition criteria or the specific target element. For more information on recognition tests, see the documentation.

Often, you don't want to capture just a single element, but instead multiple similar elements. To do this, you can use the *collection* feature. A collection consists of multiple elements that all conform to the recognition criteria selected. To do this, toggle the icon to the **Set as Collection** option. Once you do this, assuming **Recognition Criteria** is set correctly, all elements in the collection will have a green box around them on the screen. You can also toggle between the different elements in the collection in the DOM (**Tree** view) by clicking the up arrow and down arrow at the bottom of the viewport (the green dot will correspond to the initial element that was selected when recognizing the collection). Let's consider two examples to illustrate the collections concept and use some advanced recognition techniques to adequately recognize your elements.

Let's say on the SAP Build Process Automation Overview page (see Figure 7.17), you wanted to capture all the **Navigation Tabs** items so that you could iterate through each page. How might you do this? While there are several ways to do this, the easiest is to first click on one of the navigation tabs to find its place in the DOM. You then see that all these navigation tabs are part of a list. You could add the parent, **UL** (unordered list) and then declare the target element (**LI**, or list item) as a collection. Note in Figure 7.18, we only assign **Recognition Criteria** (of type **Text**) to the parent, while the target item has no recognition criteria, other than the fact that it's an **LI** element. Thus, all **LI** elements that are children of the recognized **UL** element (with text: "Overview Features Technical Information Pricing Get Started") will be identified.

As another example, let's say you wanted to capture the first column of a Microsoft Excel table. You could easily do this by using the **AutomationId** criteria and also editing the string operator. If you click on the first cell in Microsoft Excel (in the top-left corner), it will have an **AutomationId** of "A1". You can add this as **Recognition Criteria**, edit the **Operator** criteria to **starts with** and the **Value** to **A**, and set the element to be a **Collection**, as shown in Figure 7.19.

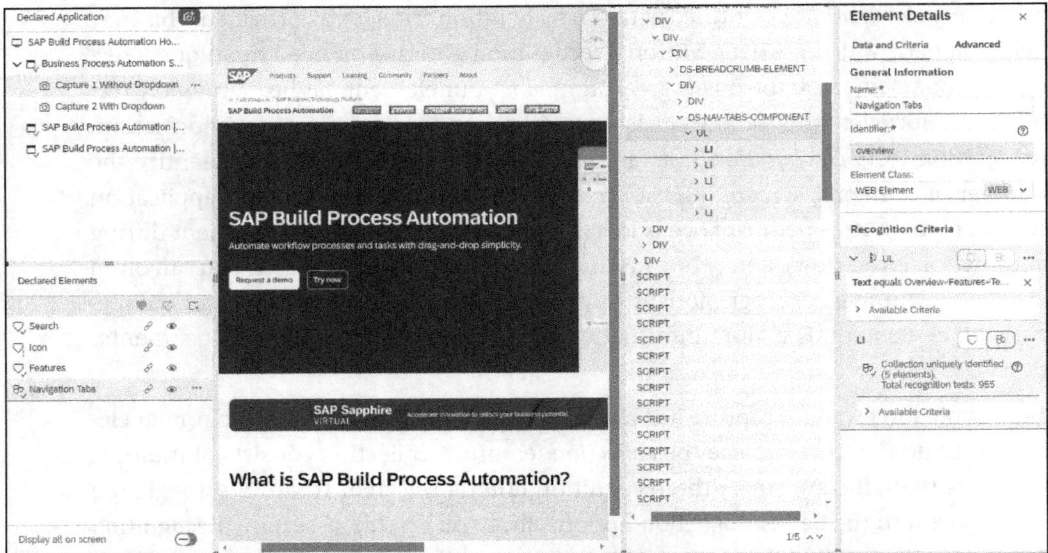

Figure 7.18 Collection of Navigation Tabs Using the Parent/Child Relationship Recognition Technique

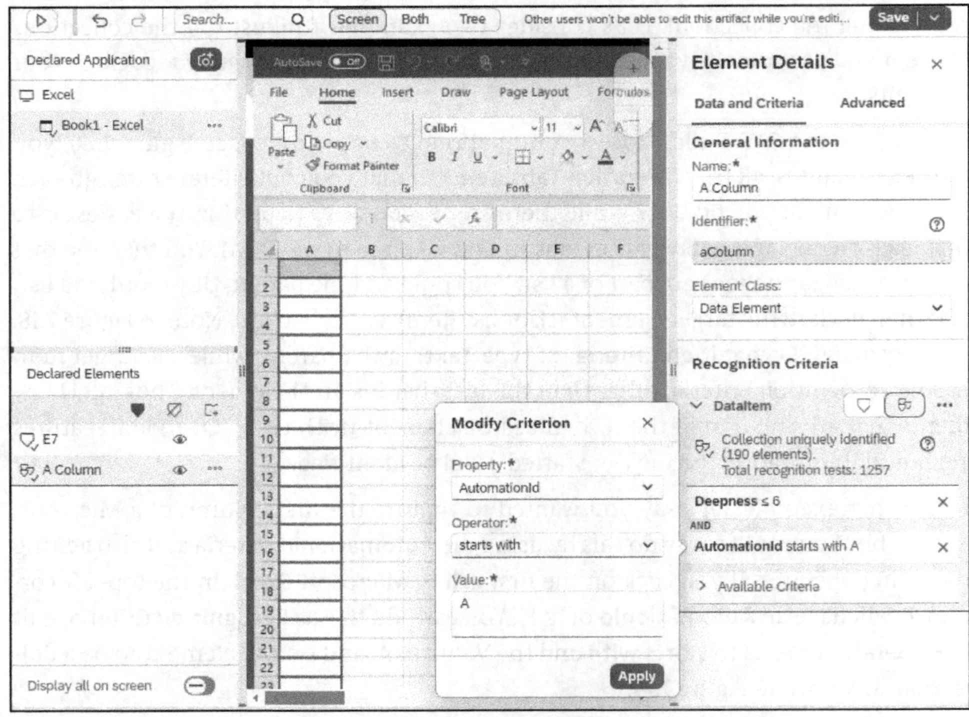

Figure 7.19 Collection of Elements to Capture the First Column in Microsoft Excel Using a String Operator

There are also some additional features in the application editor that allow you to work with elements that don't require too much discussion but are important to know:

- **Moving elements**
  You can drag and drop elements to change the order of the declared elements or move elements to other screens entirely. This may be particularly helpful when using the **Must not exist** screen recognition technique, where you recognize an element on the screen it does exist on, drag it into another screen where it must not exist, and mark it as so.
- **Duplicate**
  You can duplicate any element. This can be combined with the ability to move elements to other screens as previously described.
- **Ignore**
  You can ignore elements on screens that you've already recognized. To do this, click the icon to ignore the element. To stop ignoring an element, click the icon.
- **Linking**
  You can link elements to specific captures of a screen using the icon. Clicking this icon will provide you with the different linking options.
- **Grouping**
  You can group different elements using the icon for organizational purposes. You can set all items in a group to **Must exist** or **Must not exist**, or ignore all elements in a group.

Testing Applications

The application editor provides you with an *application tester* that allows you to test your applications. We've stressed a few times the importance of testing early and testing often while capturing and declaring your screens and elements. The application tester allows you to do this quickly and simply. To do this, click the button in the top-left corner of the application editor. Doing so will launch the application tester, as shown in Figure 7.20. From the application tester, you can jump into any of your declared applications (make sure the application has been launched and is active; otherwise, it won't be detected by the tester). Once inside an application, you can see (based on the **Recognition Criteria** set) if the application editor can identify the screens and elements. Even more powerful, you can select any of your elements and perform sample activities on them, such as simulating a click activity on a button. With collections, you'll see all identified elements in the collection, each with their own numerical index.

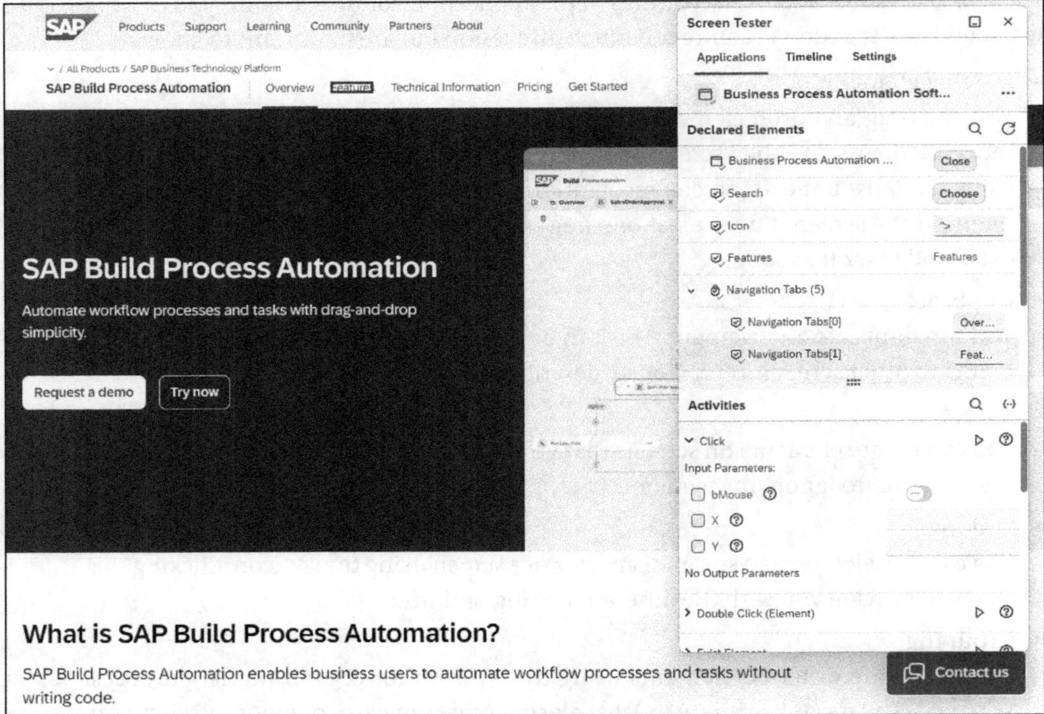

Figure 7.20 Application Tester: Testing a Screen

Adding Activities to an Automation

Now that you've captured your applications, screens, and elements and have tested them thoroughly, you can create your automation and start adding activities to it. You can either create a new automation artifact or add to an existing automation. Once you have the automation artifact open, you can expand the **Screens** option, and drag the screen you're looking to automate on into your automation flow. Next, you can either double-click the screen node you just created or single-click it and choose **Define Screen Activities** in the right-hand panel. This will display the screen, as well as your captured elements with green boxes around them (assuming they have been declared correctly). From here, you can expand the **Activities** option on the right-hand panel or search for specific activities using the search bar. Click the ⓘ icon next to any activity for more details as well as a link to the associated documentation. Figure 7.21 shows an **Automation** artifact with a screen and some associated activities added to the automation flow.

> **Tip**
>
> Once opened, clicking on the screen or elements will limit the activities associated with that specific item. For example, clicking on a declared button will limit the activities to

only applicable ones, such as the **Click** activity. These activities can then be dragged and dropped onto the element or screen and edited at any time by clicking on the added activity in the automation flow.

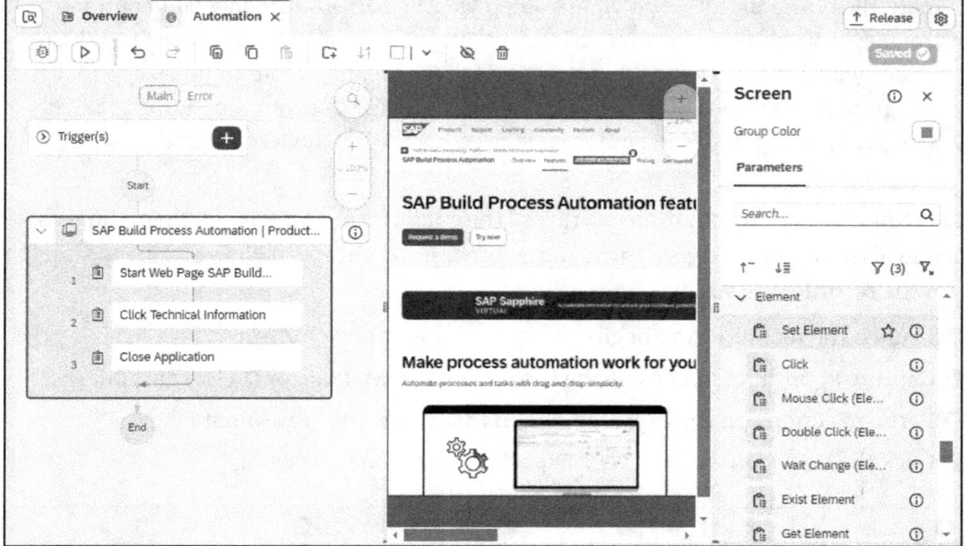

Figure 7.21 Automation Artifact with Screen Activities Added

Using the Recorder

When you begin to create the automation flow, with all associated applications, screens, elements, and activities, you should architect and design the automated process before capturing each of the different aspects. This is most easily achieved by walking through the process and noting step-by-step activities that a person would perform. By walking through the process as a human would, you not only ensure no steps in the process are missed but also adapt and create a more efficient process to be automated. To use Facebook as an example again, when writing down step-by-step instructions, you may notice that instead of clicking on "See all photos" and scrolling through dozens of pictures to find a person's profile picture, you can just click their profile picture on the original profile page.

SAP Build Process Automation makes this process even easier with its own *recorder* functionality. While it's still advisable to detail the different applications, screens, elements, and associated activities while architecting and designing the process, there's no need to capture each characteristic individually (although you can if you like). When capturing an application, the user simply needs to choose **Recorder** under **Type** and click the **Record** button. From here, the user can complete the list of step-by-step activities on the application.

> **Tip**
>
> While the SAP Build Process Automation recorder is a powerful tool that can help automate processes, you still learned how to manually capture and declare applications, screens, and elements for good reason. In some cases, the recorder can nearly automate the entire process, and what you previously learned may only be important to troubleshoot and test. However, in other cases, you may need to use some of the techniques you learned early to augment the recorder, creating more complex and robust recognition criteria and automations. Wherever on this spectrum your process falls, it's always important to understand what may be happening "under the hood."

SAP Build Process Automation can record the applications, screens, elements, and associated activities as you work through the process. As shown in Figure 7.22, the steps to run the recorder are simple:

❶ Launch the recorder, and begin recording using the button.

❷ Capture applicable screens (using the **Automatic**, **Manual**, or **On Hover** modes).

❸ Perform actions on the captured screen that you want to automate.

❹ Click the stop icon, and click **Export** to export the recording.

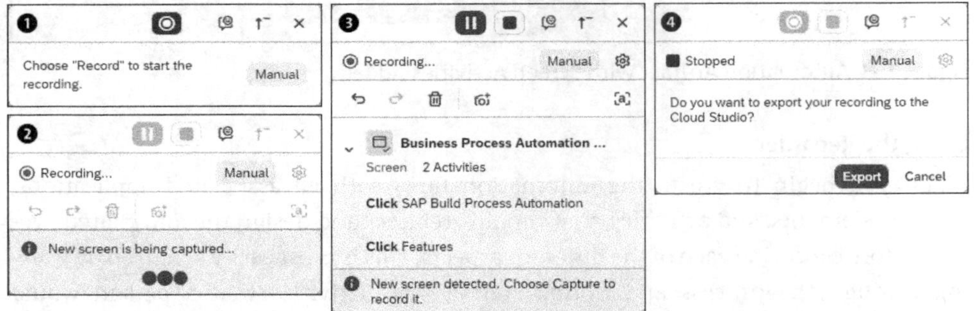

Figure 7.22 Steps to Use the Recorder

> **Note**
>
> While recording activities, elements will automatically be captured along with their associated activities. You can pause the recording to not include activities you want to omit and can stop and export the automation at any time.

The recorder supports three types of screen capture:

- **Automatic**
  When the recorder is set to **Automatic** mode, all new screens and changes to the screen will be captured. The following technologies support the automatic capture mode:

  - SAP GUI for Windows (SAP WinGUI 770 patch 3 onwards)
  - UI automation
  - SAP web applications (SAP WebGUI, SAPUI5, SAP SuccessFactors, SAP Ariba, SAP Business ByDesign, SAP Concur)
- **Manual**
  When the recorder is set to **Manual** mode, new screens and changes to the screen must be captured manually by pressing **New Screen Capture**. Capture hint messages are displayed on the recorder when a new screen appears to indicate that you must manually capture the new screen.

Note

The steps are recorded regardless of whether they are corresponding to the correctly captured screen. If you do not click the button to manually capture new screens, the recorded activities will correspond to the previous, and thus incorrect, screen. When recording in **Manual** mode, you must be careful to capture the new screen before recording the activities on that screen.

- **On Hover**
  Some UI components may not be able to be captured when using the manual capture. For example, submenus, hoverable dropdown menus, and other screen components that must be explicitly clicked on or hovered over to appear may disappear when you go to click the capture button or stop hovering over the item. In such cases, these items can be captured using the **On Hover** mode. This mode enables you to hover over a screen or items and capture them by pressing `Ctrl` or `Shift`.

When you finish, you can stop the recording, and choose whether you want to create a new automation artifact with the recording or add your recording to the start or end of an existing automation. With the SAP Build Process Automation recorder, you can simply and quickly create the foundation for your automation. Of course, the artifacts from the recording can always be edited and adapted as you see fit.

7.2.2 Web-Based Applications

One of the most popular application technologies to automate on is the web-based technology. When capturing web-based applications, you can rely on a few different frameworks to make automation easier. SAP Build Process Automation offers a generic *web framework* that relies on the underlying HTML, but it also offers frameworks from within the SAP ecosystem. Each framework includes framework-specific SDKs to make capturing items and automating activities simple yet flexible.

In the following sections, we'll discuss some specific best practices related to the web technology and its associated applications.

Web

The web technology allows us to capture HTML pages displayed in your web browser. With this technology you can capture the underlying Document Object Model (DOM) of the webpage that you can subsequently use to capture and declare elements you'll automate on. Each new instantiation of the web application, as defined by the application recognition criteria, will be observed as a new instance of the same web application during runtime.

> **Note**
>
> Some web screens are composed of multiple frames which must be captured separately to obtain their DOM. Each of these separately captured screens will require unique recognition criteria so that they can be uniquely identified. Often, **DOMAIN** and **URL** may not be sufficient criteria, as they do not uniquely recognize the screen, or they have a **URL** that changes. It may be helpful to use **FRAMEINDEX** and **MAINFRAME** criteria additionally.

> **Note**
>
> Dynamic iFrames appear on a web page dynamically. There are mainly two types:
>
> - An iFrame injected into a web page after a set of actions is performed.
> - An iFrame already present on the web page but is blank. The HTML contents are added to this blank iFrame after a set of actions is performed.
>
> As of now, only manual capture is supported to automate dynamic iFrames.

When declaring web applications, you must be considerate of the recognition criteria we use. You want to ensure that the criteria is general enough to apply to all web screens in the application but specific enough to ensure it does not apply to other application and other screens in other applications. Usually, the **DOMAIN** is sufficient for this purpose (and it will usually be automatically assigned), but you should always consider if this is the right criterion for your use case.

When declaring web screens, the criteria must either be a subset of the application criteria or not conflict directly with the application criteria. The screen criteria must also uniquely identify the screen from all other screens inside the same application. Typically, using string operators on the **URL** and **TITLE** work well here.

The web technology will first recognize the web application, and then the web screen. Once it has done so, it will search the web screen DOM to find each declared element. When searching the DOM, the technology will target the first DOM component that matches the declared recognition criteria, and if it's a collection, it will target all the DOM components that match the declared recognition criteria. The DOM searched at runtime will be the one accessed at that point in time, which may differ from the DOM we used to design our automation (especially with criteria such as element values or element text). The DOM may also change during future UI updates to the webpage. For this reason, you must ensure that your element recognition criteria are not only uniquely recognizable from other element recognition criteria but also are thoughtfully constructed to robustly withstand possible changes. Usually, you'll set recognition criteria for the properties as listed in Table 7.4.

Property	Best Practice
TAG	An HTML **TAG** of the component (e.g., **INPUT**, **DIV**, **SPAN**, **LI**, etc.) is required by the web technology and will be automatically applied as recognition criteria, which can't be further edited or removed.
id or **name**	Both **id** and **name** may be unique recognition criteria that you can choose to declare your element. However, check the value, and see if it changes on different instantiations of the page (refreshes, recaptures, etc.). If it does change, it's likely not a good recognition criterion. If it remains unchanged, it likely is a good recognition criterion. You also usually will want to use string operators with this criterion, such as **Starts with**, **Contains**, or **Ends with**.
text **class** **title** **aria-label** **type**	You can also use these to uniquely capture your element, although you want to make sure they won't change over time or over the course of your process. The element may also share some of these with their ancestors, making these criteria not completely unique but still helpful as one of multiple recognition criteria.
Advanced capture methods	Sometimes, the preceding criteria aren't sufficient, and you'll need to use advanced capture methods that you learned about earlier, such as the parent/child technique.

Table 7.4 Properties That You Can Use for Recognition Criteria

Note

Often, you may want to use criteria not mentioned in Table 7.4. You must look at each element and determine the best capturing technique for that specific element and use case, as there's no rule you can apply to all declarations.

When adding activities to a web screen, you typically want to add the activities in a fashion that mimics human interaction. However, there are a few tips beyond this general advice:

- Use the **Start Web Page** activity to initiate the screen automation.
- Use the **Wait (Screen)** activity to ensure the web screen has loaded prior to automating, and use the **Wait (Element)** activity prior to initiating an activity on an element. (This isn't always necessary, but helpful if you find the automation is moving too fast and "skipping over" some activities, or if the element takes considerable time to load.)
- Sometimes, the **Keystroke Element** and **Mouse Click** activities can be used instead of the **Set Element** and **Click** activities, respectively, if **Set Element** and **Click** activities aren't working. **Keystroke Element** and **Mouse Click** mimic user interaction with the screen UI more closely. However, they can sometimes be less robust for this reason. If a user is using their mouse or keyboard during **Keystroke Element** and **Mouse Click** activities, it may have an effect. They may also be less effective than **Set Element** and **Click** on VMs where the screen isn't active.

Other helpful guidelines for web technology can be found in the documentation at *http://s-prs.co/v592806*.

Many of the preceding techniques and suggestions (and thought processes behind them) pertain to SAP Ariba, SAPUI5, SAPWebGUI, and SAP SuccessFactors. However, there are still some additional best practices you can typically apply.

SAP Ariba and SAP SuccessFactors

Automating SAP Ariba and SAP SuccessFactors web applications will follow many of the guidelines generally used for all web technology. However, SAP Build Process Automation can recognize the application being capture is SAP Ariba or SAP SuccessFactors and will be able to operate on it accordingly. After capturing the application, make sure that the application and screen are of the correct **Framework** (**Ariba** or **SuccessFactors**) underneath the **General Information** of the application and screen respectively. This should automatically add the SAP Ariba SDK or SAP SuccessFactors SDK to your project, which will be helpful when applying SAP Ariba- or SAP SuccessFactors-specific activities to our applications, screens, and elements in the automation flow.

Helpful guidelines for recording SAP Ariba applications can be found in the documentation at *http://s-prs.co/v592807*. Helpful guidelines for recording SAP SuccessFactors applications can be found in the documentation at *http://s-prs.co/v592808*

SAPUI5

Once again, because SAPUI5 applications are also web applications, they will still follow many of the same guidelines we've previously discussed. You'll also want to make sure you have selected the correct **Framework (SAPUI5)** underneath the **General Information** for the application and screen. The following best practices are recommended to record an SAPUI5 application:

- SAPUI5 applications and screens may have the underlying DOM change based on small interactions with the UI. Things may also time out or stop being available. Thus, it is always helpful to refresh the screen prior to capturing it, and is especially important when using the recorder.
- Try to capture the entire screen when possible. If you're recording, use the scroll navigation bar instead of scrolling the mouse, as they may not operate the same on these applications.
- Try to **Set** or **Keystroke Element** for input boxes instead of capturing additional dropdown menus and clicking the corresponding values or suggestions. You can apply this value to the input by using **Keystroke Element** with the `Tab` or `Enter` key.
- When trying to capture dynamic tables, think about what information from the table you will need to perform activities on before capturing the screen. For example, if you need to capture the first row of the table, make sure to add this row before capturing the screen. When declaring elements, you can always change the value or range of the element to reflect the row you need to capture, but it is always easiest to have the visual representation inside the application editor when designing.
- Elements may change their type after an activity (e.g., some input forms may change to immutable fields). If the element is needed after this, it must be captured again, which may require a second capture of the same screen.

Other helpful guidelines can be found in the documentation at *http://s-prs.co/v592809*.

SAPWebGUI

Automating on SAPWebGUI applications will conform to many previously mentioned guidelines, and similarly you'll want to make sure you have chosen the appropriate **Framework** (**SAPWebGui**) underneath the **General Information** for the application and screen. One main difference between SAPWebGUI and other web frameworks, however, is that it is not recommended to use the **URL** as recognition criteria for screens, and instead we should use the automatically assigned **TITLE** and **MAINFRAME** criteria.

Sometimes SAPWebGUI applications may be embedded in other HTML applications. To capture these instances, follow the procedure in the documentation at *http://s-prs.co/v592810*.

7.2.3 Full Client Applications

SAP Build Process Automation also provides other application capturing technologies to help you automate client-based applications. Specifically, the *SAP GUI* technology allows you to automate on top of the SAP GUI application existing locally on your computer with a framework-specific SDK to easily automate SAP GUI elements and activities. Additionally, *UI automation* technology allows you to automate on top of any full client application. We'll discuss both of these in this section.

For UI automation and SAP GUI element declarations, by default, the **Deepness** criterion is added when you select an element. The **Deepness** criterion can only be used with the less than or equal to (<=) operator.

UI Automation

UI automation technology allows you to capture applications, screens, and elements related to native Windows applications. This allows you to capture a large range of applications that we may need to act on in your automation flow.

When you declare applications using the UI automation technology, the **EXE** property will be automatically set by the application editor as the recognition criteria. This is usually sufficient and appropriate to recognize your application.

To recognize a screen, the UI automation technology searches the DOM tree for the DOM component that matches the criteria set. When declaring UI Automation screens, the recognition criteria must be able to withstand anticipated changes to the screen but also specifically match it to the correct component. You must also decide which screen to capture, as it's not always the outermost component. When in doubt, you should capture the outermost screen, test your elements using the tester, and iterate as needed. You also often want to use **Deepness** as one of the pieces of recognition criteria for screens, as this makes it easier to identify and may increase recognition performance.

When you declare UI automation elements, the general recognition criteria advice that we've already discussed applies here. Specifically with UI automation, you'll typically have an **AutomationID** or **Name** in addition to the **Deepness** that you often can look to use as your **Recognition Criteria**. Advanced recognition techniques may also be needed here.

Other helpful guidelines can be found in the documentation at *http://s-prs.co/v592811.*

SAP GUI

SAP GUI for Windows is a frontend application that you can use to access SAP applications. Most of the advice described in the earlier section (for UI automation) will apply for SAP GUI, but there are some additional guidelines.

When declaring screens, you may often have screens with similar recognition criteria. For this reason, you may use **Must Exist** conditions to uniquely recognize your screens. If possible, you should limit this to one **Must Exist** element per screen to maintain high performance. You should also try to use **Must Exist** instead of **Must Not Exist** (if possible) for performance concerns as well.

Note

The main SAPLOGON screen is not an SAP GUI screen. Do not manually change the technology to select the SAP GUI technology to capture this screen.

When capturing elements, you often want to use the **Id** criterion, if it's stable. If it's not stable, you can look to use the **Name** criterion if it's stable. If neither is stable, you may need to use advanced recognition techniques. You may often look to use **Name** as criterion for collections, such as with **GUITableControl** cells. Frequently with SAP GUI, you'll want to capture larger elements, such as **GUITableControl**, where you can then use the SDK provided to you to assign specific actions to the table, such as specifically selecting one of its nodes (i.e., instead of capturing the node itself).

Tip

When automating on SAP GUI for Windows, scripting must be enabled to access the application. If you find that an automation isn't working, or you are unable to capture a SAP GUI application, this is likely the issue. Scripting must be enabled on both the client and server side. The procedure to do this can be found in the documentation at *http://s-prs.co/v592812* (navigate to the client side or server side procedures respectively on the left hand tree under the **Enabling Scripting** section).

Other helpful guidelines can be found in the documentation at *http://s-prs.co/v592813* and *http://s-prs.co/v592814*.

7.3 Activities: Making Use of Software Development Kits

Whether you're automating on top of application UIs as discussed in Section 7.2 or using activities that can interact directly with applications, SAP Build Process Automation gives you a powerful suite of activities to accomplish your automation needs. Inside automation artifacts, you have several *software development kits* (*SDKs*) available that include drag-and-drop activities, classes, modules, and data types. These SDKs include the following:

- Core SDK
- SAPUI5 SDK
- SAP GUI for HTML SDK
- Excel SDK
- Outlook SDK
- PowerPoint SDK
- Microsoft 365 Cloud SDK
- PDF SDK
- SAP BAPI SDK
- SAP Ariba Solutions SDK
- SAP SuccessFactors SDK
- Document Information Extraction SDK

- Google Workspace SDK, Google Authorization SDK, Google Document AI SDK, Google Vision AI SDK, Google Cloud Storage SDK
- Web SDK
- Java SDK

When you create an automation, you import appropriate SDK packages that provide activities to build the flow of your automation. These activities will be organized into classes and subclasses that you can drag and drop from the activities section of the tools panel into the automation flow. Once these activities are in the automation flow, you can then define their related parameters to make sure they execute as expected. Activities can be added, duplicated, rearranged, deactivated, and deleted at any point from the automation flow.

In this section, we'll discuss how to make use of activities embedded in SAP Build Process Automation. You'll learn how to add provided SDKs to your project and will take a deep dive into some of these SDKs to better understand how they can help you accomplish your automation goals.

7.3.1 Adding a Software Development Kit

The first step to adding activities to your automation flow is to add any SDKs you want to use in your automation. To do this, click the ⚙ icon in the top right of your project screen, and then click the **Dependencies** list option in the resulting popup, as shown in Figure 7.23. From here, you'll notice that two dependencies are already active: **Core SDK** and **Excel SDK**. If you want to add more SDKs, you need to click the **Add ⌄** icon, followed by **Add a Business Process Project Dependency**. From here, you can add any available SDK (if it's not available, you'll need to add it from the embedded store in your lobby). Once you add the necessary SDKs, their related activities will be added to your project, and you can begin to use them in your automation.

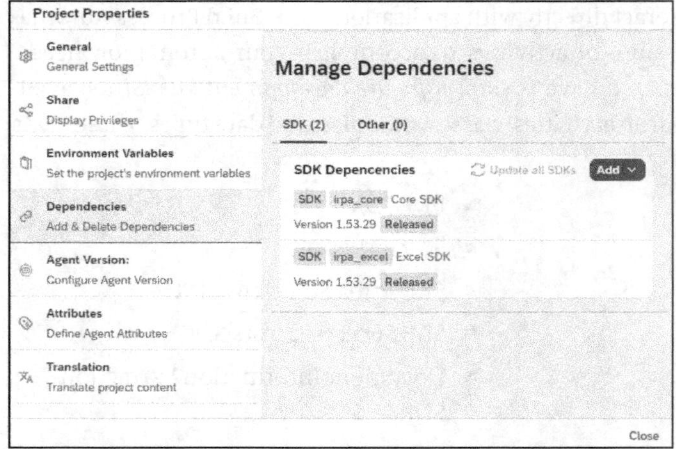

Figure 7.23 Adding SDKs to Your Project

7.3.2 Simple Operations

While SAP Build Process Automation offers complex functionality to allow you to automate across the entire spectrum of business processes, it also offers simple operations to ensure that the most fundamental tasks can be automated. In this section, we'll discuss some simple operations from the Core SDK that are fundamental to automating your processes.

The Core SDK

The Core SDK contains main methods and functions that are fundamental to automation projects. It comes already added as a dependency to your project, as it's essentially necessary to execute an automation flow. This SDK gives you many of the activities needed to pilot applications, as well as interact with your environment. It includes the following collections of activities:

- **Base Core SDK**
  Collection of core functions and activities for accessing applications, screens, and elements.
- **Automation Alerting**
  Collection of activities for sending automation alerts
- **Encoding**
  Collection of functions related to Base64 encoding.
- **Converters**
  Collection of conversion functions between two formats such as JavaScript, JavaScript Object Notation (JSON), and XML.
- **Cryptography**
  Collection of cryptography activities.
- **Data Type Management**
  Contains the functions to work with data types.
- **SAP Document Management System (DMS) Module**
  Collection of functions related to DMS.
- **Errors Management**
  Collection of functions for thrown errors.
- **File System Management**
  Collection of functions for accessing and manipulating file system objects.
- **FTP**
  Collection of functions for File Transfer Protocol (FTP).
- **Information**
  Collection of activities to retrieve environment information.
- **Clipboard**
  Collection of activities related to clipboard management. Use `get` or `set` to retrieve or define the current value of the operating system clipboard.

- **Monitoring**
  Collection of functions to add monitoring activities to your automations.
- **Dialogs**
  Collection of functions to display a dialog for users. Users can receive or supply information.
- **Registry**
  Collection of functions related to Windows Registry. To be used with caution as some registry information might require administrative rights.
- **HTTP Requests**
  Collection of functions related to HTTP requests for retrieving information from websites or web services.
- **Security**
  Collection of activities related to security.
- **Local Agent Variables**
  Collection of activities to retrieve and read variables from the local agent.
- **Local Agent Credentials**
  Collection of activities to retrieve and read credentials from the local agent.
- **Connectors**
  Dedicated activities for different technologies (SAP GUI, UI automation, etc.).

Activities Used to Pilot Applications

Included in much of the Core SDK are activities needed to pilot applications you've already captured, as discussed in Section 7.2. When necessary, SAP Build Process Automation will automatically bring in additional SDKs that relate to the specific application technology you're using.

Once you add a screen to your automation flow inside the automation editor, you can either double-click the screen node you just created or single-click it, and then choose **Define Screen Activities** in the right-hand panel. This will display the screen, as well as your captured elements with green boxes around them (assuming they have been declared correctly). From here, you can expand the **Activities** option on the right-hand panel or search for specific activities using the search bar. Activities will be filtered based on the context of activities you're looking to use. For example, if you click on the general screen, activities will be filtered to those activities that specifically apply to the screen. If instead you click on an element, activities will be filtered accordingly. You can then drag and drop the activity onto the screen or element, and it will automatically be added to the automation flow with the target screen or target element set. You can alternatively drag and drop the activity into the automation flow and select the target on the right-hand panel under the activity's parameters.

Most of these activities tend to mimic human interaction closely, such as with clicks or keystrokes, so we won't discuss the individual activities at length. However, you can click the ⓘ icon next to any activity for more detail and the associated documentation. We also discuss some best practices for specific technologies in Section 7.2.

Working on Files

When you work with files and folders inside an automation artifact, you can use activities to easily mimic human interaction within a file system. The following activities are made possible by dragging and dropping activities into the automation flow, and only represent the most used subset of file system activities. All activities can be found in the product documentation:

- **Move (File/Folder)**
  Move a file (or a folder) from one location to another.
- **Copy (File/Folder)**
  Copy a file (or folder) from one location to another.
- **Remove (File/Folder)**
  Remove a file or a folder from a specified location.
- **Exist (File/Folder)**
  Check whether a file or a folder exists in a specified location.
- **Rename (File/Folder)**
  Rename an existing file or folder.
- **Create Folder**
  Create a folder within a specified path.
- **Get File Collection**
  Return a collection (list) of files belonging to a specified folder.
- **Get Folder Collection**
  Return a collection (list) of folders belonging to a specified folder path.
- **Get Absolute Path Name (File)**
  Return the absolute path name of a file. An absolute path points to the same location in a file system, regardless of the current working directory. To do that, it must include the root directory.
- **Get File Name**
  Return the full name of a specified file (base name and extension).
- **Create File**
  Create an empty file in a specified location.
- **Open File Dialog**
  Open a file explorer dialog to select a file path. This is specific for attended usage (requires user response).
- **Read File**
  Read the content (text) of a specified file based on a supplied encoding (UTF-8, by default). This activity only works for files smaller than 50 MB.

- **Write File**
  Write text content in a file based on a supplied encoding (UTF-8, by default).
- **Append File**
  Append a text content in a file based on a supplied encoding (UTF-8 by default).
- **Unzip File**
  Unzip a ZIP file in a specified folder (without a password).

Note

The **Read File** and **Write File** activities in the preceding list work specifically for text files. However, such activities also exist for comma-separated values (CSV), JSON, and XML).

To emphasize the importance of being able to work with files inside a file system, let's consider an example. Company ABC has three separate divisions: Engineering, Marketing, and HR. Within a larger folder titled Employees, they have three separate text files named Engineering Employees, Marketing Employees, and HR Employees. On a quarterly basis, they need to update each file with the current list of employees for each division. However, because the files are immutable for compliance purposes, they must duplicate the Employees folder, rename it to reflect the current quarter, read through each text file, and update it if the names of the employees have changed. Once the preceding folder has been duplicated, renamed, and updated to reflect the current quarter, the files can be shared via a shared network or emailed to the proper personnel via Microsoft Outlook activities.

Working on File Transfer

With automation artifacts, you also have drag-and-drop activities to manage files using FTP. As a prerequisite to working with FTP, we're assuming that you already have an FTP server established. Once you have this, working with FTP servers is easy with SAP Build Process Automation. You start out by opening a connection to the FTP server using the **FTP Access** activity and filling out the necessary parameters, as shown in Figure 7.24. The different types offered are Secure FTP (SFTP), FTP, and File Transfer Protocol Secure (FTPS).

Once you've opened the connection, you can then interact with the FTP server using the activities listed under the FTP SDK (you can find all such activities by searching for FTP in the activities search bar in the automation editor). Figure 7.25 and Figure 7.26 show examples of how to work with FTP to download files and upload files, respectively.

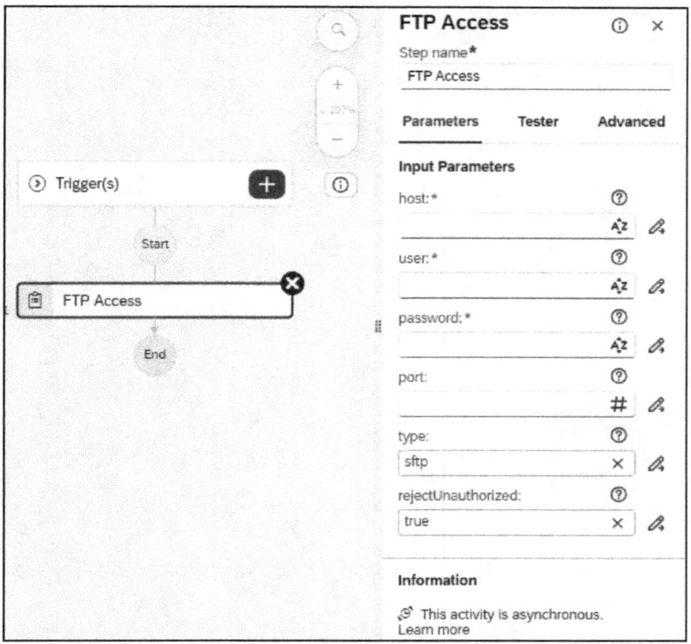

Figure 7.24 FTP Access Activity with Available Parameters

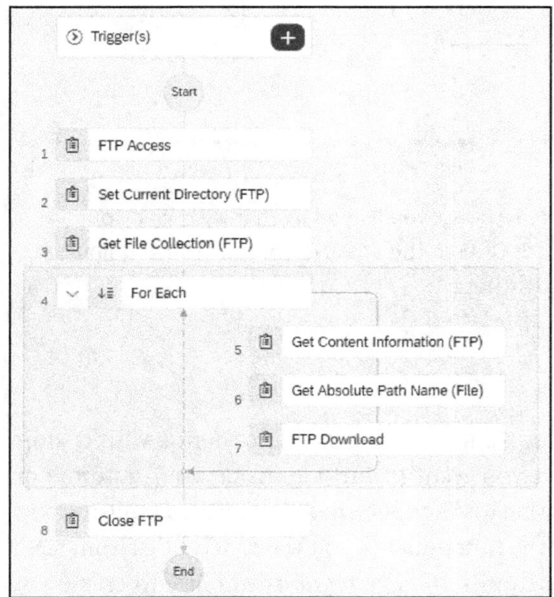

Figure 7.25 FTP Download Automation

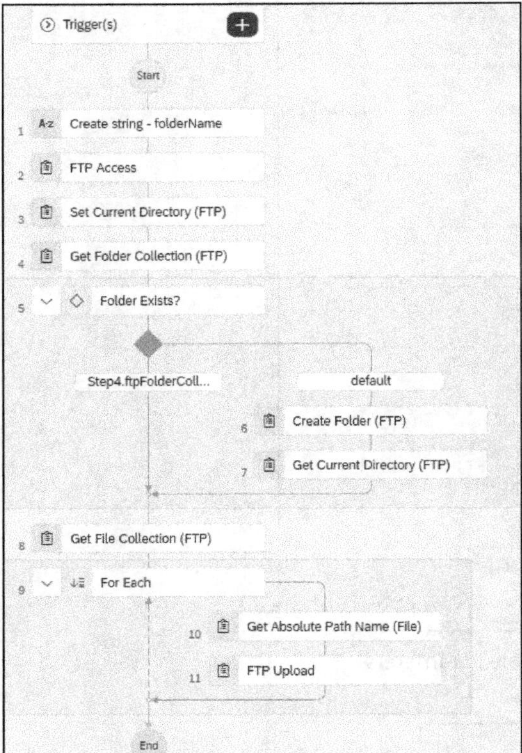

Figure 7.26 FTP Upload Automation

> **Tip**
>
> You can also work with files using DMS. To do so, you'll need to enable DMS in your SAP BTP account and create a destination for DMS in the SAP BTP cockpit. You should then be able to call it from your automation artifact.

7.3.3 Custom Script

The **Custom script** activity, as shown in Figure 7.27, allows you to define with custom JavaScript code any functionality that you want to include inside your automation flow. Typically, it's advised to only use custom scripts when there isn't a built-in activity from one of the provided SDKs for the functionality you want. Activities from SDKs are usually easier to maintain, but custom scripts give you the ability to create any complex functionality that is possible with JavaScript.

Your **Custom script** activity provides an environment to develop code with code completion and recommendations. This feature will recommend modules containing classes, activities, and enumerations, as well as applications, screens, and elements that are already available to or in your automation artifact. You can also set input variables and output variables to be used from inside the custom script.

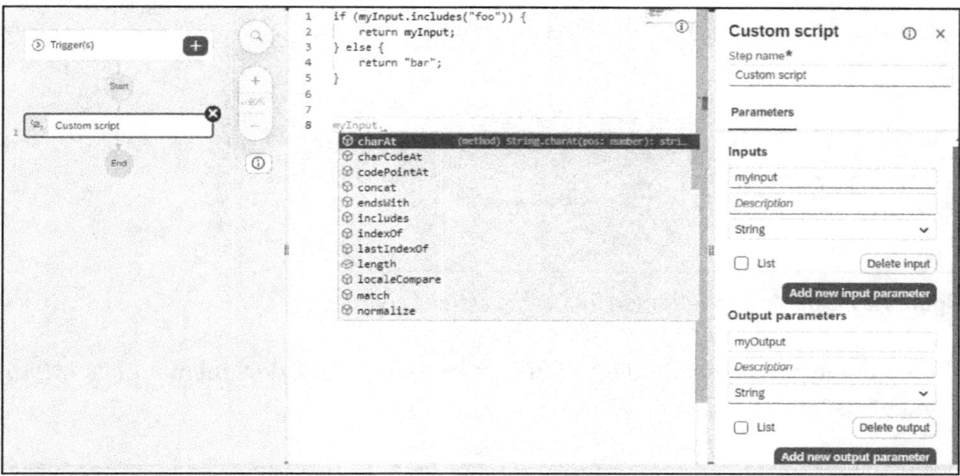

Figure 7.27 Custom Script Activity

7.3.4 Web Service Calls

While you can pilot many applications using the techniques described in Section 7.2, sometimes it may be easier to call a web service to accomplish a task for you (if it's available). You can do this using the **Call Web Service** activity. When using this function, you'll first want to define the parameters of the web service call using a **Custom script** activity, as shown in Figure 7.28. Once you've defined the parameters in your custom script, plug the output of the **Custom script** activity into the input of the **Call Web Service** activity, as shown in Figure 7.29.

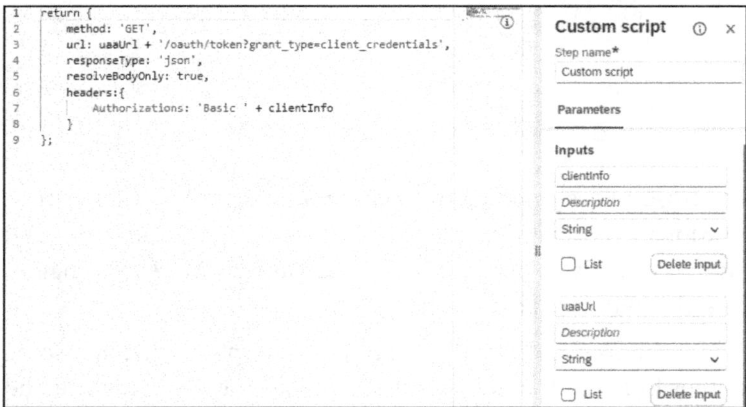

Figure 7.28 Sample Custom Script Parameters for a Web Service Call Used to Retrieve a Session Token from a Server

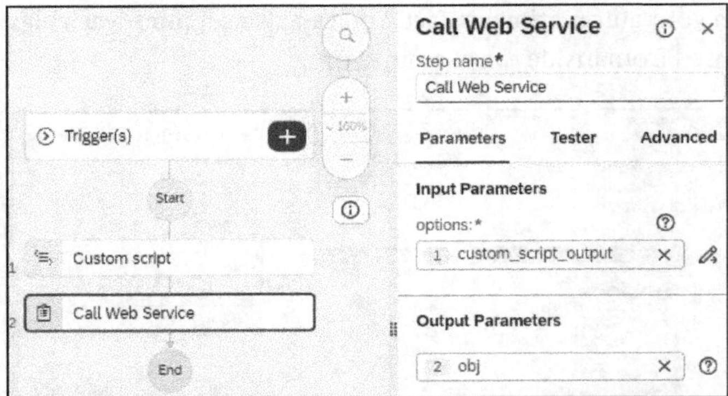

Figure 7.29 Sample Configuration for Calling a Web Service

Some sample parameter configurations can be found in the documentation at *http://s-prs.co/v592815.*

7.3.5 Piloting Microsoft Office Products

Now that you've seen how to pilot an application through automating on the presentation layer as well as how to call web services and use the Core SDK, let's look at how to automate some common applications that are useful in many business processes. SAP Build Process Automation recognized the ubiquity of Microsoft Office products in classic automation use cases and enables you to automate them with drag-and-drop ease through custom SDKs for each product. Of course, you can still capture the application and automate it through the UI layer, but these SDKs allow you to automate these applications even more efficiently. Fetching a table from Microsoft Excel or sending an email through Microsoft Outlook can be implemented in just a few clicks. SAP Build Process Automation provides you with SDKs for the following Microsoft products:

- Microsoft Excel
- Microsoft Outlook
- Microsoft Word
- Microsoft PowerPoint

We'll mostly discuss Microsoft Excel and Microsoft Outlook here, as they are the most used Microsoft applications in automation use cases. However, Microsoft Word and Microsoft PowerPoint can certainly be used in automations as well. More information regarding their SDKs can be found in the documentation.

We'll also mostly discuss the use of these SDKs through examples that can also be found as prebuilt content in the SAP Build tenant store. The activities available and their descriptions can be found in the documentation and are typically straightforward. However, these activities become more powerful when combined to achieve complete use cases. For this purpose, the following subsections will discuss use cases and show you how to construct them using the activities available to you. We'll discuss in each use case the definition of each activity used and some applicable best practices (if available) to aid your understanding of how the activities should be put together in an automation flow.

When opening Microsoft Office applications, attention must be paid to the management of the application instance used during the process. When you initialize an instance, you create an instance for your automation artifact to use. There are two main ways you can initialize an application instance:

- **Create a new application instance.**
  Here, the application isn't yet in use, and you usually want the automation to open it.
- **Use an existing application instance.**
  Here, there's already an instance of the application open on the computer for the automation to use.

Once the application instance is properly initialized, you can begin to automate on it. However, once you're finished automating on the application instance, you need to make sure you release the instance. If you just want your automation to release the instance, you can use the **Release Excel Instance** activity (or the **Release Instance** activity for whichever Microsoft application activity you're using) to release it. Here, the user can continue to work on the application after the automation has finished and can close it whenever they like. If, however, you want to release the application instance and close the instance altogether, you can use the **Close Excel Instance** activity (or the **Close Instance** activity for whichever Microsoft application activity you're using), which will both release and close the application instance.

Note

Application instances aren't always properly closed or released after an automation job. For example, let's say your automation contains the **Open Excel Instance** activity, but it doesn't end with either the **Release Excel Instance** activity or the **Close Excel Instance** activity due to any of the following:

- You haven't added one of these activities at the end of your automation.
- There was an uncaught error in your automation, and the automation stopped before reaching the **Release Excel Instance** activity or the **Close Excel Instance** activity.
- The user has manually closed the Microsoft Excel instance that was opened by the automation.

Any of the preceding situations can cause issues with working with Microsoft Office applications. If you were to run the automation again from these examples, the **Open Excel Instance** activity may use the Microsoft Excel instance that was previously opened during the previous automation, and you would need to enter the current worksheet or workbook used by the Excel SDK activities. This may run correctly, but it also may cause issues as new Microsoft Excel instances may be opened that could cause things to act in unexpected ways. If these situations occur and you want to either prevent the automation from opening the previous Microsoft Excel instance or avoid unexpected behavior due to "hanging" application instances, you can manually end the *Excel.exe* (in the case of Microsoft Excel) task using the Windows Task Manager.

7.3.6 Microsoft Excel Activities

The Excel SDK is a collection of activities allowing you to create automations via Microsoft Excel. In this section, we'll explore some use cases that show the power of this SDK.

Get Range Definition

In this automation, you find the number of rows and columns in the activated worksheet of the workbook located at the file path provided as input. Specifically, you'll find the index of the column, and the index of the row, associated with the last cell in the table range. You'll also convert the column index into the column name and output the range. You can see this automation flow in Figure 7.30 and some of its associated activities in Table 7.5.

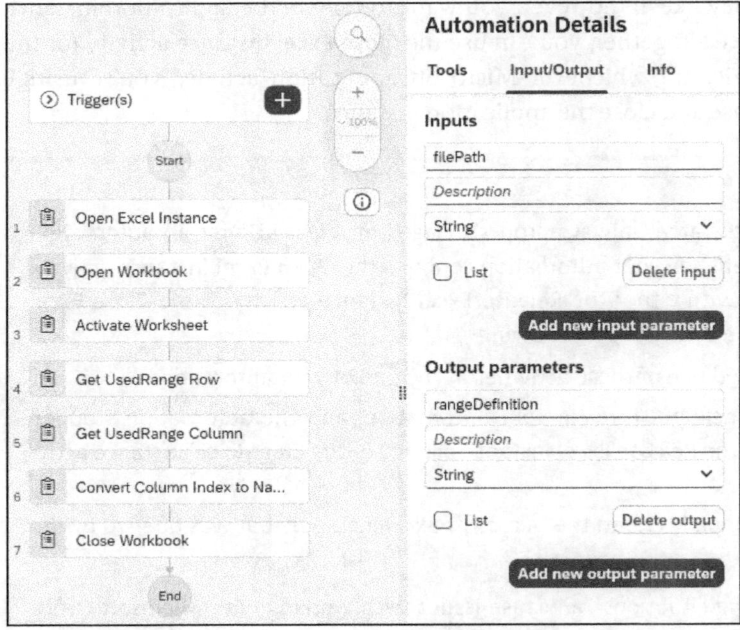

Figure 7.30 Automation for Getting the Range Definition of a Microsoft Excel Table

Activity Name	Description	Best Practice
Open Excel Instance	Mandatory activity that opens an instance of Microsoft Excel, allowing you to use other Microsoft Excel activities	You must always have an open instance of the application running before executing subsequent activities. Make sure that care is taken when managing these instances (as discussed in Section 7.3.5).
Open Workbook	Opens a workbook reference by a workbook path	This usually follows the use of the **Open Excel Instance** activity and precedes the **Activate Worksheet** activity.
Activate Workbook	Sets the active workbook	Although usually not necessary, this can be used to set a previously opened workbook as the active workbook.
Get UsedRange Row	Retrieves the row index of the used range in an active Microsoft Excel worksheet	If you delete the data of the last row used, the activity may still return its index. Test to make sure your worksheet is performing as expected.
Get UsedRange Column	Retrieves the column index of the used range in an active Microsoft Excel worksheet	If you delete the data of the last column used, the activity may still return its index. Test to make sure your worksheet is performing as expected.
Convert Column Index to Name	Converts a column index number into a column name	For example, an index of 1 gives "A", 2 gives "B", and so on.
Close Workbook	Closes the active workbook	–

Table 7.5 Activities Used in the Get Range Definition Automation

Copy Dynamic Table

In this automation, we use a nested automation artifact (**Get Range Definition** automation, as discussed in the previous section) to get the range of the table in the Microsoft Excel file. We then use the **Excel Cloud Link** activity to upload a sample Microsoft Excel file that looks like the file you plan to use during execution. This activity evaluates the file and creates a custom data type that fits the Microsoft Excel table. More specifically, it creates a data type with attributes matching the column headers of the table in the file. When the automation executes, it automatically places the corresponding pieces of data from the Microsoft Excel file into your data structure.

Warning

The column headers used to create the attributes of the data object will be used to extract information from future Microsoft Excel files using this **Excel Cloud Link** activity. Therefore, the column headers must not be changed and must exactly match the data attributes to be extracted correctly. If they don't match, the activity will be unable to extract that column from Microsoft Excel and unable to place it into the data structure.

Once you've loaded the Microsoft Excel table into the automation's context, you then create a new workbook where you copy the table into. You do this by creating a **For each** loop, where you loop around the list of data objects taken from the extracted data, and push the values to the cells in the newly created workbook. You also can use the **Set Values (Cells)** activity, with only a few minor changes within the **For each** loop. At the end of the automation, you format the cells and save the workbook. You can see this automation flow in Figure 7.31 and some of its associated activities in Table 7.6.

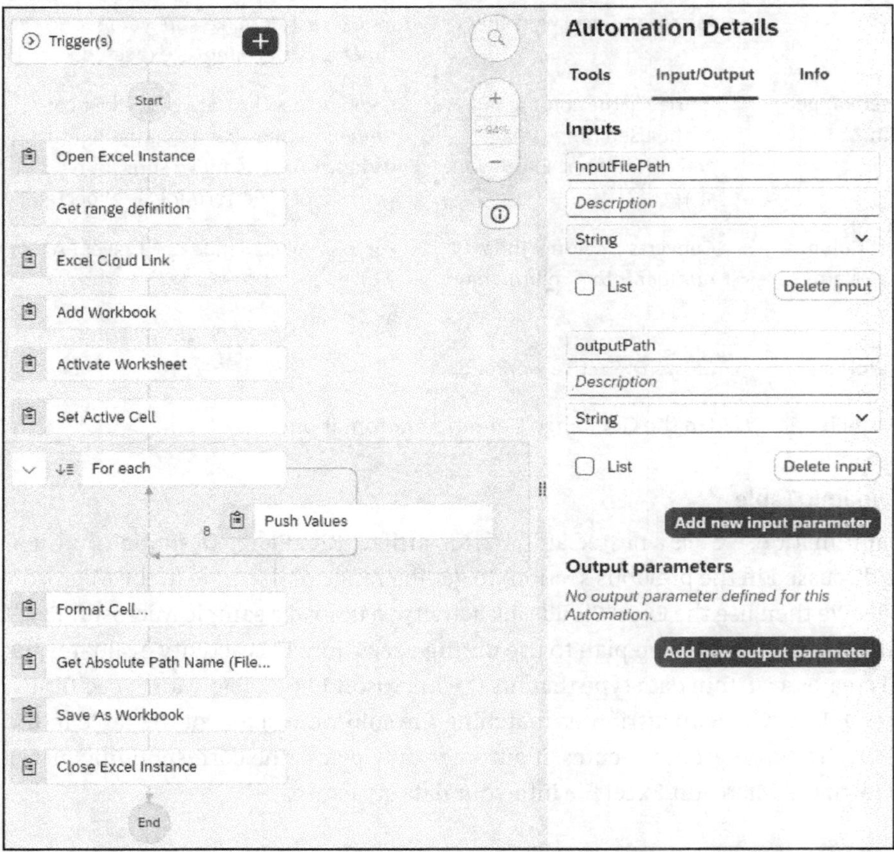

Figure 7.31 Automation for Extracting a Microsoft Excel Table and Copying It to Another File

Activity Name	Description	Best Practice
Excel Cloud Link	Retrieves the cell range values from a specified worksheet. A dedicated configuration screen is available to map the values from the worksheet to an existing data type or a new data type.	The dynamic range detection calculates the area to work on between two given columns. The top row is determined by the first filled cell from the top of the first column. The last row is determined by going down from the top row to the first encountered empty row.
Add Workbook	Adds a new workbook in the active application.	–
Set Active Cell	Sets the active cell in an active worksheet.	–
Push Values	Pushes (writes) the values in the active worksheet. Starts from the active cell, and then the active cell is moved to be ready for the next push.	Can create similar functionality with the **Set Values (Cells)** activity but is implemented slightly differently.
Format Cell	Applies a set of formatting properties to a specified cell range	–
Save As Workbook	Saves the workbook to the given file path.	This will only save the active worksheet and thus must be done multiple times for multiple different active worksheets. Be careful when saving to CSV format as some information may be lost.

Table 7.6 Activities Used in Copy Dynamic Table Automation

Format and Export

In this automation, you trigger a macro called to sort the data decreasing in each column. You then get the number of rows as well as the number of columns, and then convert the number of columns into its name to simplify the use of the Microsoft Excel range for the next steps. Then, you open a new workbook and get the names of the open workbooks to work with. Next, you copy the data from the first Microsoft Excel workbook to the second Microsoft Excel workbook. You activate the second workbook because you're going to work in it for the following steps. We do a **Repeat** loop of the number of columns. For each column, you set the formula for the sum of the column to the row below the last row of data, and then add some background color. Lastly, you export the file as a PDF. You can see this automation flow in Figure 7.32 and some of its associated activities in Table 7.7.

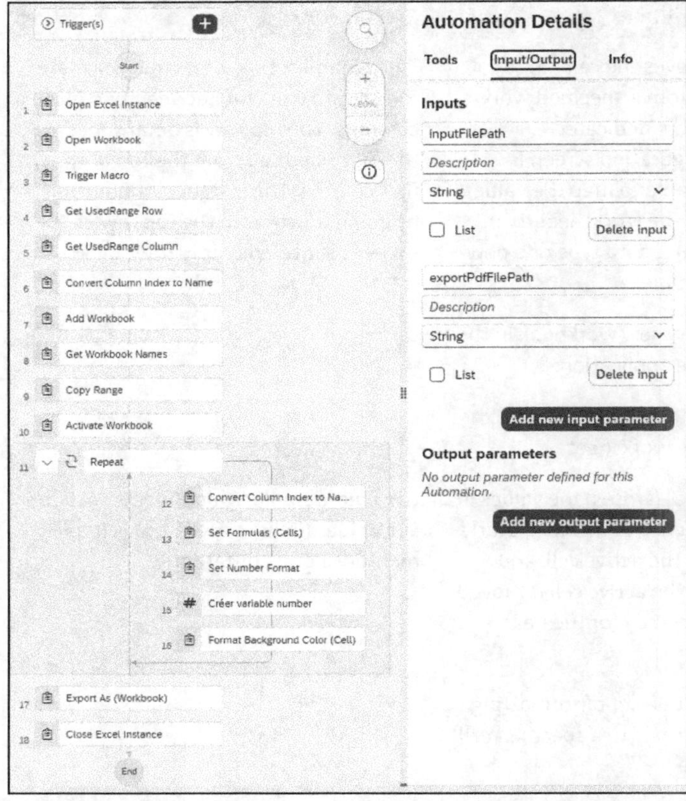

Figure 7.32 Automation for Triggering a Macro, Formatting Microsoft Excel Tables, and Exporting as a PDF

Activity Name	Description	Best Practice
Trigger Macro	Triggers a macro	–
Get Workbook Names	Retrieves the names of the opened workbooks in the current Microsoft Excel instance	–
Copy Range	Copies a specified range into a new workbook/worksheet as specified in the activity parameters	–
Set Formulas (Cells)	Sets (writes) the formulas of a specified cell range in the worksheet	The formulas should be Microsoft Excel formulas (i.e., "=SUM(...)").
Export As (Workbook)	Exports the active workbook as a specified format such as PDF or XPS	–

Table 7.7 Activities Used in Format and Export Automation

Sort Data and Export to CSV

In this automation, you sort the data according to column name and then save the worksheet as a CSV file. Here, you use Microsoft Excel activities to sort your data accordingly. However, you could instead load the data into your automation's context and sort the data using standard automation activities (or with custom code). You can see this automation flow in Figure 7.33 and some of its associated activities in Table 7.8.

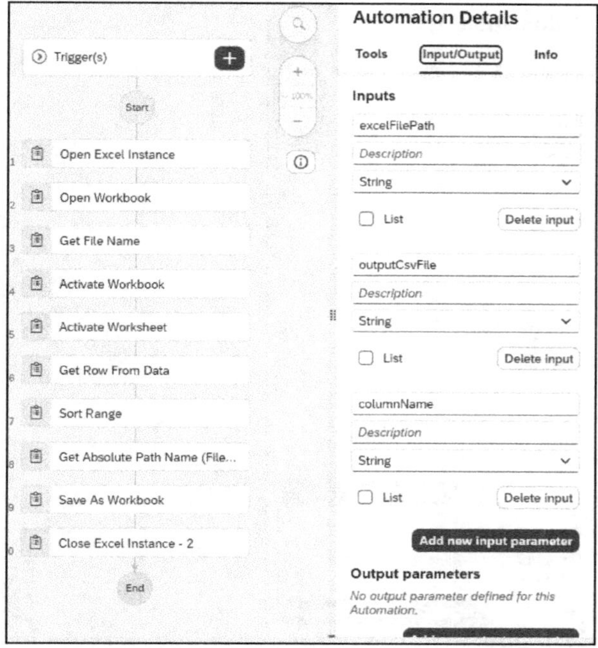

Figure 7.33 Automation for Sorting a Range of Data

Activity Name	Description	Best Practice
Get Row From Data	Retrieves dynamically the last (or first) row index from a group of data using a reference cell	This will be the same result as manually pressing [Ctrl] + [Up] or [Ctrl] + [Down] in the Microsoft Excel worksheet. The active cell of the active worksheet will be used if the given reference cell is empty.
Sort Range	Sorts a range	Sorting behavior will depend on whether the given range is outside or inside a table. Specifically, if the range is inside a table, Microsoft Excel will sort the entire table. An `InvalidArgument` exception is raised if the range provided is across multiple tables or spans outside of a single table.

Table 7.8 Activities Used in Sort Data and Export to CSV Automation

Filter Data and Export to CSV

In this automation, you filter the data to display only rows where content in column A is greater than 100. You then apply another filter to display only rows where content in column B is less than 100. You then copy the filtered ranged to another worksheet and save that worksheet as a CSV file. You can see this automation flow in Figure 7.34 and some of its associated activities in Table 7.9.

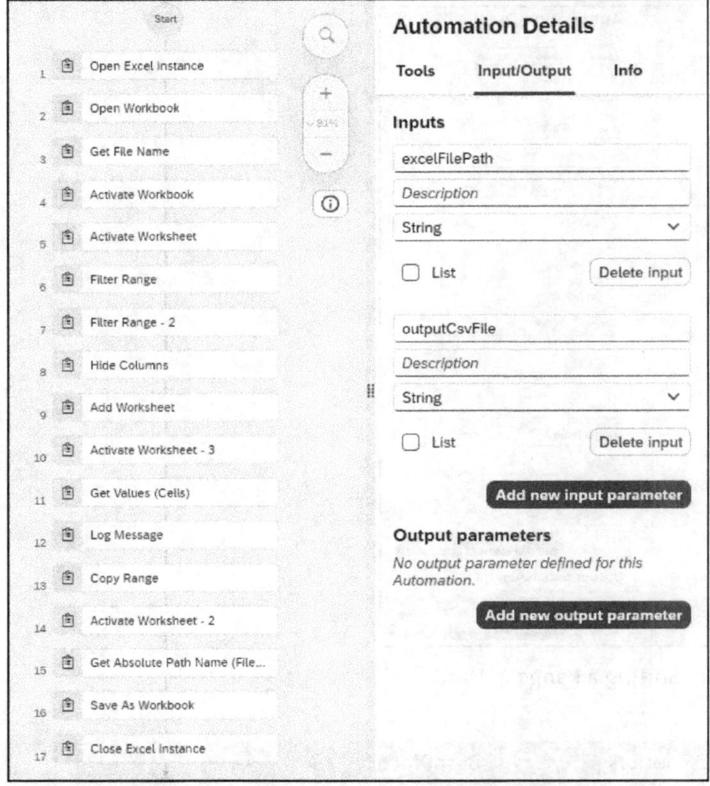

Figure 7.34 Automating for Filtering a Range of Data

Activity Name	Description	Best Practice
Filter Range	Filters a range based on criteria and can be called several times if the filtering must be performed on several columns	Filtering by date is not supported.
Hide Columns	Hides columns corresponding to the reference range	–

Table 7.9 Activities Used in Filter Data and Export to CSV Automation

Activity Name	Description	Best Practice
Add Worksheet	Adds a new worksheet to the active workbook	–
Get Values (Cells)	Reads the values of a specified cell range	The **Get Values (Cells)** activity can be used to retrieve all values as specified by the range, or only retrieve values from a filtered range, which can be toggled by changing the **VisibleOnly** parameter to **True**.

Table 7.9 Activities Used in Filter Data and Export to CSV Automation (Cont.)

7.3.7 Microsoft Outlook Activities

The Outlook SDK is a collection of activities allowing you to create automations using Microsoft Outlook. In this section, we'll describe some use cases that show the power of this SDK.

Send Single Email

In this automation, you create and send a new email. You can see this automation flow in Figure 7.35 and some of its associated activities in Table 7.10.

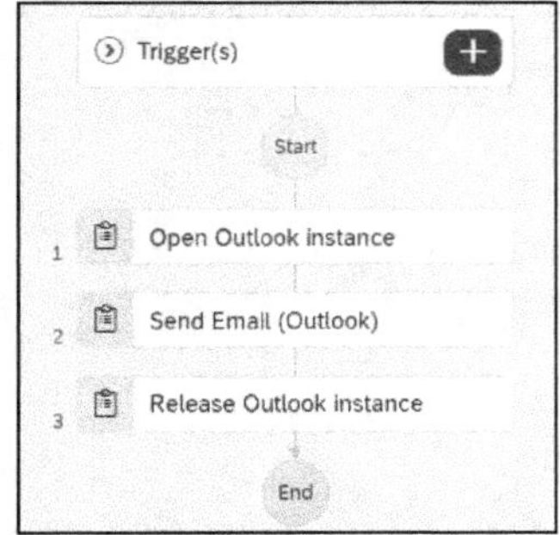

Figure 7.35 Automation for Sending a Single Email

Activity Name	Description	Best Practice
Open Outlook Instance	Opens an instance of Microsoft Outlook	You must always have an open instance of the application running before executing subsequent activities. Make sure that care is taken when managing these instances (as discussed in Section 7.3.5).

Table 7.10 Activities Used in Send Single Email Automation

Activity Name	Description	Best Practice
Send Email (Outlook)	Sends an email	You must provide one or more email item parameters such as : ▪ **To** ▪ **CC** ▪ **BCC** ▪ **Subject** ▪ **Body** ▪ **Attachments**: Must specify a file path
Release Outlook Instance	Releases the instance of Microsoft Outlook without closing the application	Make sure that care is taken when managing these instances. If you do not appropriately release the instance, there may be issues the next time you try to automate on a Microsoft Outlook Instance (as discussed in Section 7.3.5).

Table 7.10 Activities Used in Send Single Email Automation (Cont.)

Search Mails

In this automation, you use search criteria to find specific emails in a mailbox. You then use a **Forever** loop to iterate within the email context to display the information in a popup. You can see this automation flow in Figure 7.36 and some of its associated activities in Table 7.11.

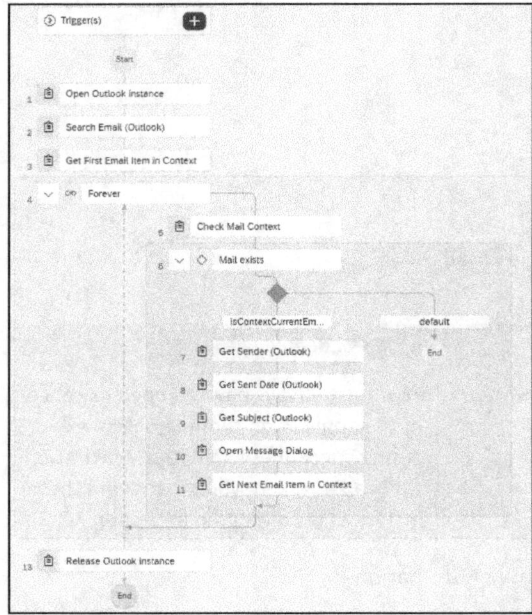

Figure 7.36 Automation for Searching Emails Based on a Set Criteria

Activity Name	Description	Best Practice
Search Email (Outlook)	Searches for a specific email	When using the Outlook SDK, **Email Search Criteria** and **Email Search Criterion** data types are available to be used as parameters for this activity. With these data types, you can search your emails with the following criteria: ▪ **Subject** ▪ **Sender name** ▪ **Recipient name** ▪ **Body content** ▪ **Date Received** ▪ **Date Sent** ▪ **Has Attachment (Boolean)** ▪ **Is Read (Boolean)**
Get First Email Item in Context	Within a context, makes the first email item as the current email item	This activity should follow the **Search Email (Outlook)** activity.
Check Mail Context	Checks if the current context of an email exists	This activity is used to see if you are on an existing email found in the **Search Email (Outlook)** activity. Returns **True** if the context of a current email exists and **False** if it doesn't.
Get Sender (Outlook)	Returns the sender of an email	–
Get Sent Date (Outlook)	Returns the sent date of an email	–
Get Subject (Outlook)	Returns the subject of an email	–
Get Next Email Item in Context	Within a context, makes the next email item as the current email item	–

Table 7.11 Activities Used in Search Mails Automation

Move Emails

In this automation, you use search criteria to find specific emails in a mailbox. You then use a **Forever** loop to iterate within the email context and move emails to another location. You can see this automation flow in Figure 7.37 and some of its associated activities in Table 7.12.

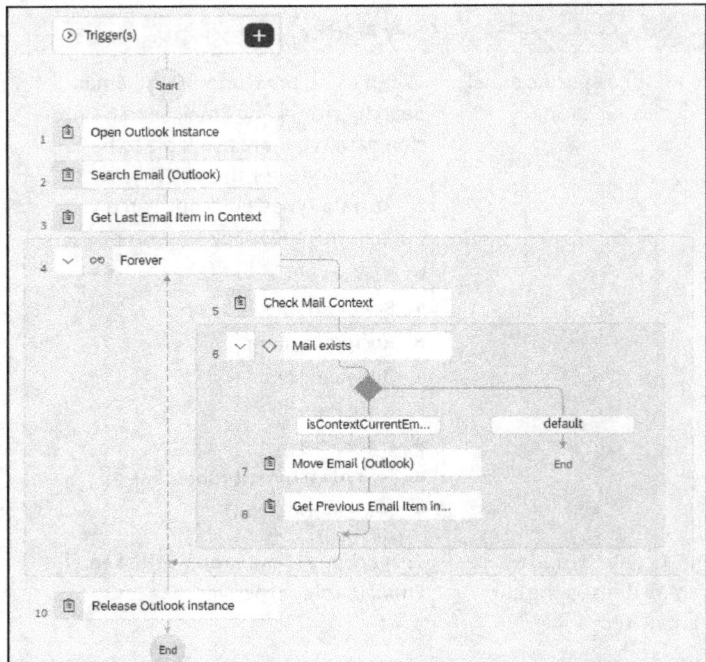

Figure 7.37 Automation for Moving an Email

Activity Name	Description	Best Practice
Get Last Email Item in Context	Within a context, makes the last email item as the current email item	This activity should follow the **Search Email (Outlook)** activity.
Move Email (Outlook)	Moves an email from one location to another	The location you are moving the email to is given as a parameter in the **Move Email (Outlook)** activity, while the location you are moving the email from is defined in the **Email Search Criteria**. To move multiple emails, follow the procedure detailed in the documentation at *http://s-prs.co/v592816*.
Get Previous Email Item in Context	Within a context, makes the previous email item as the current email item	–

Table 7.12 Activities Used in Move Emails Automation

Download Image from Mail

In this automation, you use search criteria to find specific emails in a mailbox. You then use a **Forever** loop to iterate within the email context. For each email found, you parse the content of the email to extract the names of all images. Then, you download the files using the list of names that was generated in the **Get Image Names** activity (**Custom Script**). If several emails have images that have the same name, you should be sure to download them to different locations to not overwrite them. You can see this automation flow in Figure 7.38 and some of its associated activities in Table 7.13.

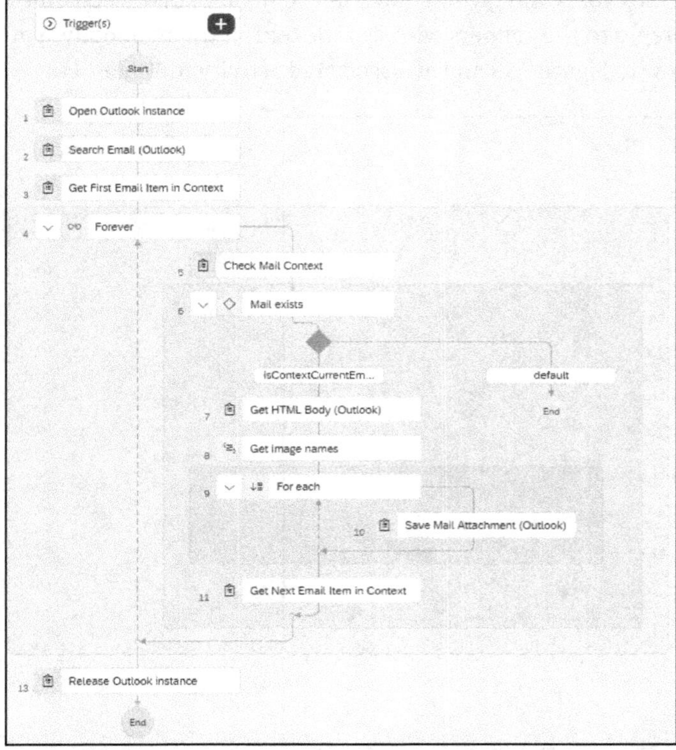

Figure 7.38 Automation to Save Images from an Email

Activity Name	Description	Best Practice
Get HTML Body (Outlook)	Returns the HTML body of an email	–
Get Image names	Finds the image names	This is a custom script that parses the HTML to find the image names.

Table 7.13 Activities Used in Download Image from Mail Automation

Activity Name	Description	Best Practice
Save Mail Attachment (Outlook)	Saves an attachment	A file name with the correct extension must be provided in the **attachmentFileName** field.

Table 7.13 Activities Used in Download Image from Mail Automation (Cont.)

Reply to an Email

In this automation, you'll look for some given emails in the mailbox. For each email found, you automatically reply to the senders with a static text in the mail body. You can see this automation flow in Figure 7.39 and an associated activity in Table 7.14.

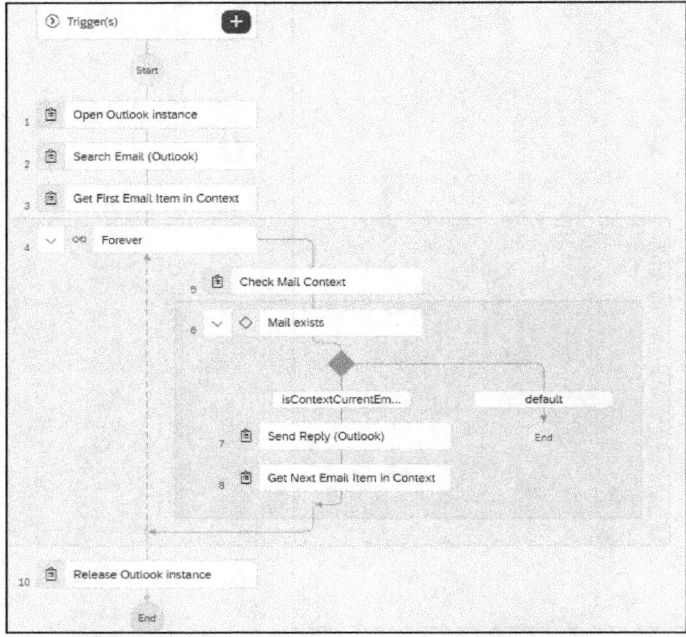

Figure 7.39 Automation for Sending Reply Emails

Activity Name	Description
Send Reply (Outlook)	Sends a reply to the current email in context

Table 7.14 Activity Used in Reply to an Email Automation

Delete Calendar Items

In this automation, you look for calendar items by using criteria set in the **Search Calendar Items** activity and then delete those calendar items. You can see this automation flow in Figure 7.40 and some of its associated activities in Table 7.15.

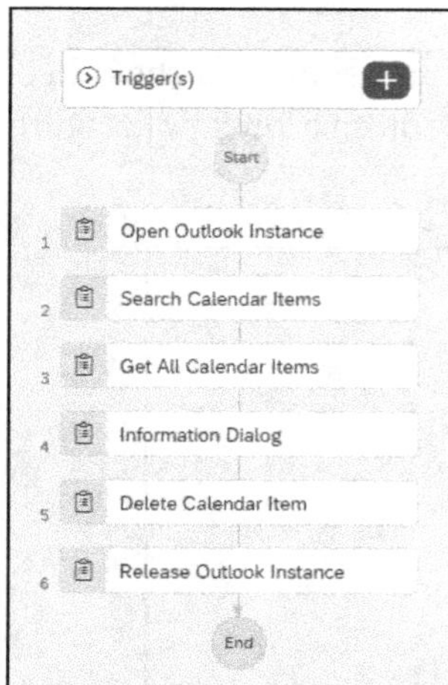

Figure 7.40 Automation for Deleting a Calendar Item

Activity Name	Description	Best Practice
Search Calendar Items	Searches for the calendar items that match the search criteria and fills the context with the result	When using the Outlook SDK, the **Calendar Search Criteria** data type is available to be used as a parameter for this activity. With this data type, you can search your emails with the following criteria: ▪ **Subject** ▪ **Start Datetime** ▪ **End Datetime** Note that the account parameter is optional and can be used to specify which calendar is to be searched. If not provided, the default account will be searched.
Get All Calendar Items	Retrieves all the calendar items from the context	Use this activity to retrieve results from the **Search Calendar Items** activity. In the **propertiesFilter** field, you can filter the items you want to retrieve. Note that the output variable should have the correct value even if it appears shortened on screen.
Delete Calendar Item	Deletes the current calendar item	You should use the **Get First Calendar Item** activity or **Get Last Calendar Item** activity to select a valid calendar item after deletion.

Table 7.15 Activities Used in Delete Calendar Items Automation

Save Attachments

In this automation, you retrieve the list of calendar items and iterate over the collection to save all the attachments if they exist. You can see this automation flow in Figure 7.41 and some of its associated activities in Table 7.16.

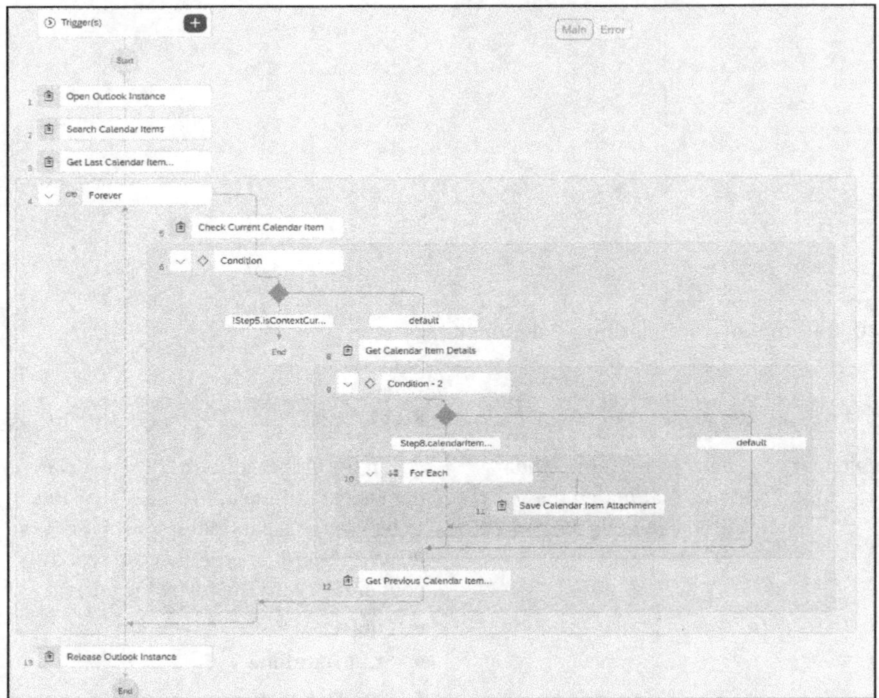

Figure 7.41 Automation for Saving Attachments from Calendar Items

Activity Name	Description	Best Practice
Get Last Calendar Item in Context	Within a context, makes the last calendar item as the current calendar item	This should be used after the **Search Calendar Items** activity.
Check Current Calendar Item	Checks if the current calendar item is defined	This activity is used to see if you are on an existing calendar item found in the **Search Calendar Items** activity. This returns **True** if the context's current calendar item exists; returns **False** if it doesn't.
Get Calendar Item Details	Within the calendar context, retrieves the properties of the current calendar item	The **Get All Calendar Items** activity can be used to get information from multiple items by looping through the results.

Table 7.16 Activities Used in Save Attachments Automation

Activity Name	Description	Best Practice
Save Calendar Item Attachment	Saves the attachment of the current calendar item	A file name with the correct extension must be provided in the **attachmentFile-Name** field.
Get Previous Calendar Item	Within a context, makes the previous calendar item as the current calendar item	–

Table 7.16 Activities Used in Save Attachments Automation (Cont.)

7.3.8 Leveraging Cloud-Based Microsoft Office-Like Applications

SAP Build Process Automation also offers SDKs that allow you to access Microsoft 365 and Google applications. To use these services, you can create external authentication. Blogs, developer tutorials, and documentation are available to create this authentication, so we won't go into significant detail. However, the external authentication (from the SAP Build Process Automation side) can be configured in two different places. First, you can configure the external authentication, as shown in Figure 7.42, from **Control Tower • External Authentication**.

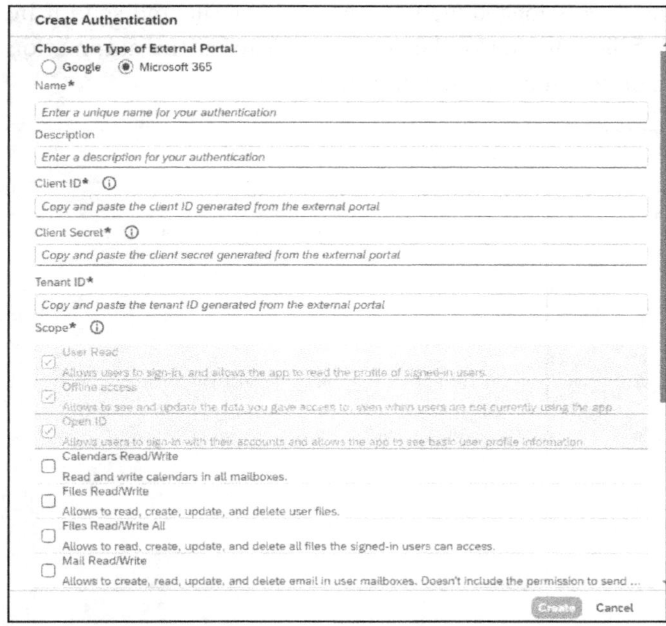

Figure 7.42 Creating Authentication in Control Tower

Once you've made the configurations from the tenant, you can configure the authentication from your locally installed desktop agent (discussed in Section 7.7). This will look something like Figure 7.43.

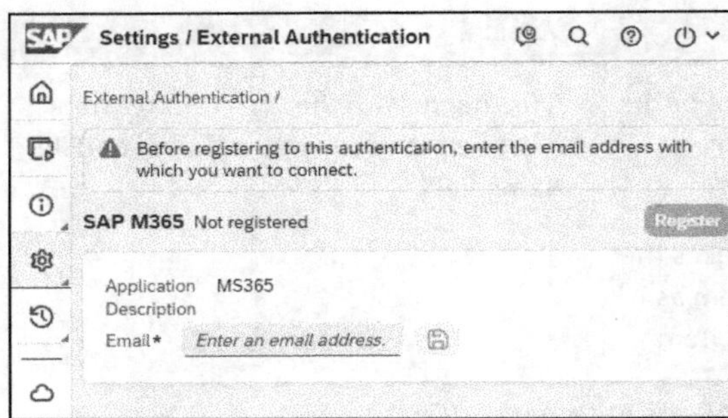

Figure 7.43 External Authentication from the Desktop Agent

7.3.9 Microsoft 365

The Microsoft 365 SDK Cloud package is a collection of activities allowing you to automate Microsoft Office 365 objects, such as shared drives, files, or emails. Let's look at an example to understand how to use this SDK.

In the **Automate Send Email** automation, you automatically connect to Microsoft Office 365 and open the Microsoft Excel document whose URL is used as a parameter to read its content. Once the content is read, you'll log the data in the test console of the automation editor. After the data is logged, you'll download the file and send it by email. Finally, you remove the file from the local machine. You can see this automation flow in Figure 7.44 and some of its associated activities in Table 7.17.

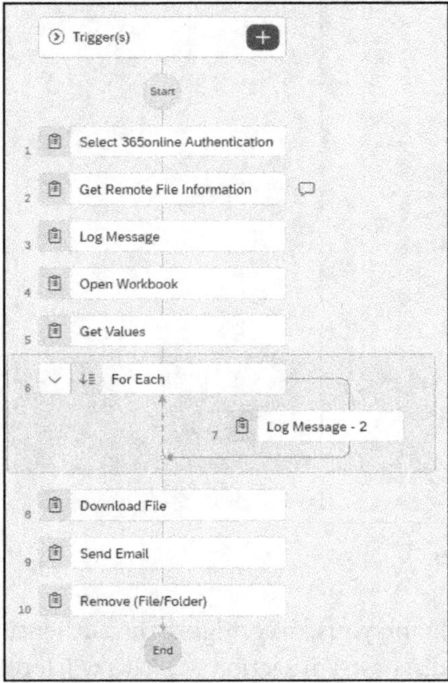

Figure 7.44 Automate Send Mail Automation

Activity Name	Description	Best Practice
Select 365online Authentication	Sets the authentication that will be used for the Microsoft Office 365 online activities	–
Get Remote File Information	Returns useful information about a remote file from its source URL	The source URL provided by Microsoft to open the remote file must be used.
Open Workbook	Opens a workbook if it's not already opened	–
Get Values	Receives the values or formulas from the current worksheet	–
Download File	Downloads a file from Microsoft OneDrive or SharePoint	The output of the **Get Remote File Information** activity will provide the necessary information for this activity. You can use the **driveId** and **fileId** attributes of the output parameter. Leave the **driveId** field empty if the file to download is located on the active user's OneDrive.
Send Mail	Sends an email	–

Table 7.17 Activities Used in Automate Send Mail Automation

Note

To perform many of the activities in this automation flow you must have file read/write permissions on your personal Microsoft OneDrive and the external Microsoft OneDrive or Sharepoint.

7.3.10 Google Applications

The Google SDKs are used in tandem to connect to and work with Google applications, including Google Workspace. Functionality is offered for the following:

- Gmail
- Google Sheets
- Google Slides
- Google Docs
- Google Drive
- Google Document AI
- Google Vision AI

In the following sections, we'll look at some examples to understand how to use these SDKs.

Using Gmail

In the Gmail automation, you complete Google authorization before searching through a mailbox with defined search criteria. If the email is found, you then read and save the email contents, and then reply to it confirming receipt of the email. Finally, you disconnect the Google account. You can see this automation flow in Figure 7.45 and some of its associated activities in Table 7.18.

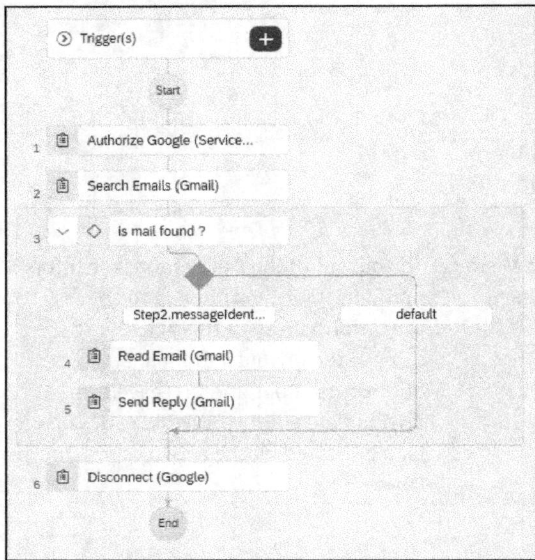

Figure 7.45 Automation Using Gmail

Activity Name	Description	Best Practice
Authorize Google (Service Account)	Gets authorization for a Google account	The service account key path should not be hardcoded and instead should be provided either as a local file system path or an environment variable path. You can also use the **Authorize Google (OAuth Client ID)** activity and the **Select Google Authentication** activity if you've already created the alias in the control tower and desktop agent.
Search Emails (Gmail)	Searches for emails in a Gmail account	–
Read Email (Gmail)	Reads emails in a Gmail account	–

Table 7.18 Activities Used in Gmail Automation

Activity Name	Description	Best Practice
Send Reply (Gmail)	Replies to an email using a Gmail account	Files can be attached from the local file system or the Google Drive.
Disconnect (Google)	Disconnects a Google account	This activity disposes of the authorization object and should be used at the end of an automation flow when connection to the Google account is no longer needed to complete activities.

Table 7.18 Activities Used in Gmail Automation (Cont.)

Using Google Sheets

In this automation, you complete Google authorization before importing a Microsoft Excel file from the local file system and creating a Google spreadsheet. You then rename the spreadsheet and get its details. From here, you copy the first column from the spreadsheet and paste it in a new sheet. You then add a new column of values in the new sheet. Finally, you disconnect from the Google account. You can see this automation flow in Figure 7.46 and some of its associated activities in Table 7.19.

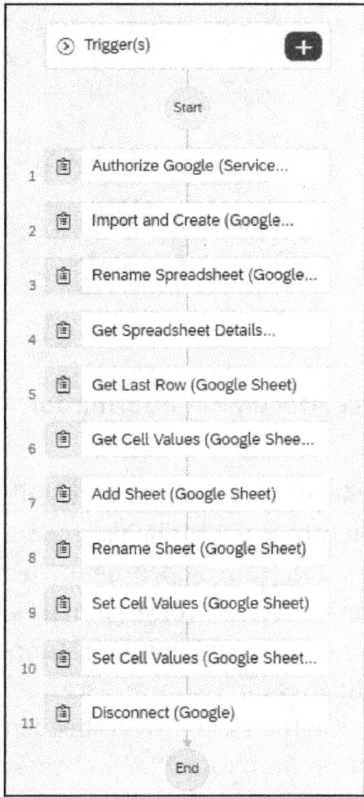

Figure 7.46 Automation Using Google Sheets

Activity Name	Description	Best Practice
Import and Create (Google Sheets)	Imports data and creates a new spreadsheet	This activity can also be used to import data into a new sheet of an existing spreadsheet by using it in conjunction with the **Copy Sheet (Google Sheets)** activity.
Rename Spreadsheet (Google Sheets)	Renames an existing Google spreadsheet	–
Get Spreadsheet Details (Google Sheets)	Retrieves the details of a Google spreadsheet, including the list of the sheets contained in the spreadsheet	–
Get Last Row (Google Sheets)	Gets the last row in Google Sheets	–
Get Cell Values (Google Sheets)	Returns the values of a specified cell range in Google Sheets	–
Add Sheet (Google Sheets)	Adds a sheet to a Google spreadsheet	–
Rename Sheet (Google Sheets)	Renames an existing sheet inside a Google spreadsheet	–
Set Cell Values (Google Sheets)	Sets the values of a specified cell range in Google Sheets	Takes input using "A1" notation.

Table 7.19 Activities Used in Google Sheets Automation

7.3.11 Document Processing with Artificial Intelligence: Document Information Extraction and PDF

Embedded inside your automation editor, you have intelligent document understanding capabilities that can be quickly and easily added to your projects. This helps you use the embedded functionality to extract structured information from a document. This is commonly used to automate the extraction of information from invoices, purchase requisitions, or any data that may be stored in a document. For example, with the functionality we've already looked at in conjunction with this feature, you can scan an inbox and take all emails with a PDF attachment in it. You can then use your embedded document understanding capabilities to extract information from each PDF and enter that information into an online portal. While this is one of the more common use cases for such functionality, this document understanding can be applied to any number of use cases that incorporate uploaded documents. It's an important tool to enhance the power of SAP Build Process Automation.

We'll primarily focus on the Document Information Extraction (DOX) service that comes embedded within your project editor. To use DOX capabilities, follow these steps:

1. In the **Overview** of your project, click **Create • Document Template**.
2. If you already have a template you want to import from another project, you can do that here. Otherwise, choose **Create a New Template**, as shown in Figure 7.47. Click **Next**.

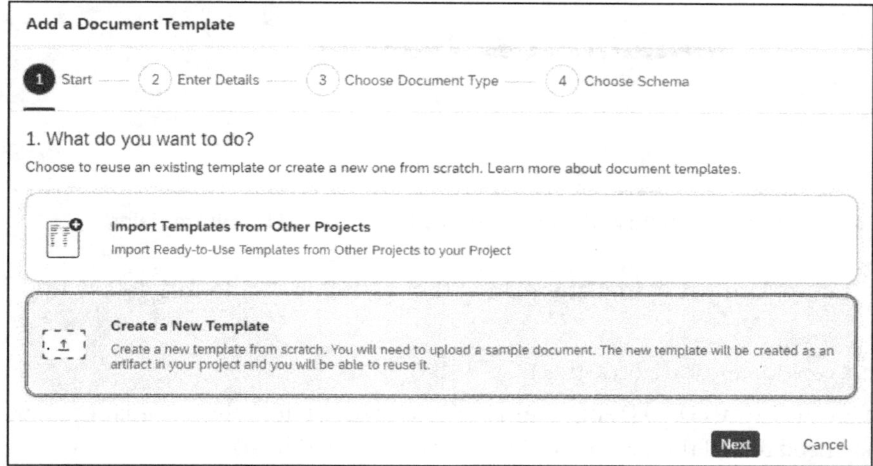

Figure 7.47 Adding a Document Template to Your Project

3. Enter a **Name** and description (optional) for the document template, and upload a sample invoice that you want to use as a template. Click **Next**.
4. Choose the document type. This will help define your schema, which determines which fields you'll pull from the document. You can always choose a custom scheme that you can later define. Click **Next**.
5. Click **New** to create a new schema if one doesn't already exist or choose an existing schema. If creating a new schema, define the header fields and line-item fields. Click **Next**.
6. Check the summary, and click **Add**.
7. You should now see the new document template added to your project. If you click the artifact, you should see something similar to Figure 7.48.

You may also see a new data artifact added to your project that resembles the name of the schema. This data object was created so that when you use this template in an automation, you can neatly extract your data into a data object with clear fields.

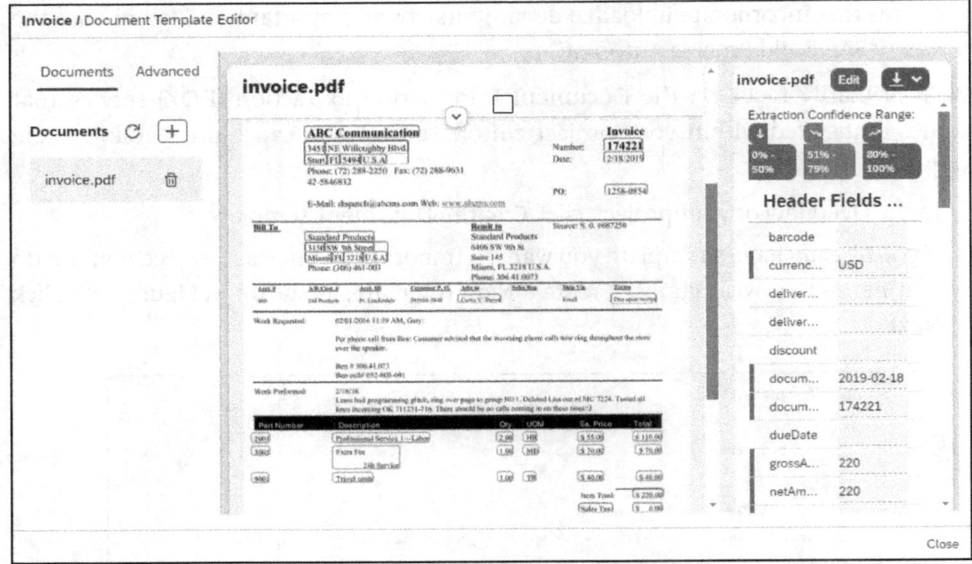

Figure 7.48 Document Template with Identified Header Fields and Line-Item Fields

You now want to use your template in an automation flow. To do this, follow these steps:

1. Navigate to your automation artifact.
2. Search for the **Extract Data (Template)** activity, and drag it into the automation flow. You may need to add the Document Information Extraction SDK to do so.
3. Click the activity, and then click **Add Document Template** on the right-hand panel.
4. Click **Choose a template from the current project**, and then click **Next**. Note that you could have created the template from scratch starting here with the same result.
5. Choose the extraction schema from the dropdown menu. You can then decide what document template to select. However, there's also a **Detect automatically** option. This can be very powerful if you have a few different styles of invoices that you've created templates for. DOX can look at the incoming document, detect which template it's most similar to, and choose the proper template for extraction purpose. This option does work a little slower, but it's extremely powerful for processes that may include different invoice (or other) formats.
6. Click **Next**, and then click **Add** to add the template.
7. There may still be an error in your automation. To fix this, you need to add the document path as an activity parameter. It's common to make this a dynamic variable that you can easily change or to have this refer to a location that you add a file to and then clean up on each iteration.
8. Once you define your document path, you're ready to test your automation using a different invoice (of the same format as your template). If you go ahead and debug

your project, while setting a **Watch** on the output variable of your activity (defaults to **extractedData**) or using a **Log Message** activity to print the variable, you should see the extracted information, saved in a variable, and ready to be used elsewhere in your automation or project (see Figure 7.49).

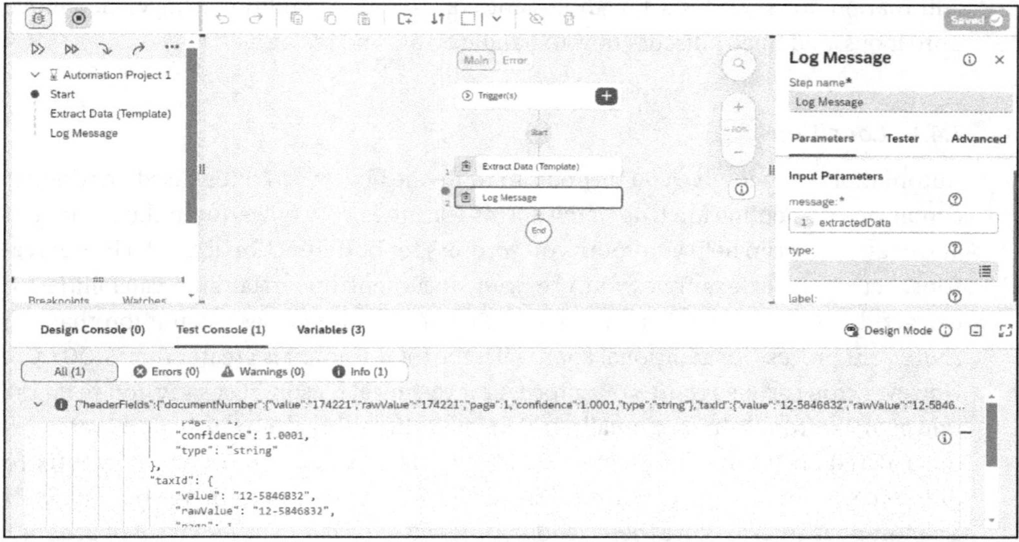

Figure 7.49 Logging Your Variable That Contains the Extracted Data from the PDF Invoice

PDF SDK

The PDF SDK is a collection of activities that allows you to work with PDFs. Typically using the DOX SDK discussed previously is preferable, as it has fewer limitations on what it can extract. The PDF SDK can be powerful when working with PDFs though, especially if you want to quickly find information stored inside the PDF. With the PDF SDK, you can extract data from the PDF and receive it back as a string. One of the most powerful activities for this case is the **Extract Text w/ Reg. Expr.(PDF)** activity. This activity will allow you to use regular expressions to extract specific strings from a page or specific page area (will return the first match found). You can then store this data in a variable to be used elsewhere in the automation or project.

7.4 Controls: Complex Automation Flows and Error Handling

Controls and error management are necessary to create complex functionality from an automation flow. Previously, you learned how to use SDKs and embedded activities to automate a process. But what happens if you want to branch your automation based on a condition or iterate over a list of data such as a Microsoft Excel table? To do this, you can use the controls and error management available to you from within SAP Build

Process Automation. These tools allow you to create complex automation flows and handle errors gracefully to ensure you're automating the process to exactly the requirements you need.

In this section, we'll discuss some of the controls that allow you to create complex automation flows. We'll start by discussing controls with conditional logic, then move onto loops, and finally discuss how to handle errors and create alerts.

7.4.1 Conditions

Automation artifacts give you the tools to route the flow of activities based on whether conditions you define are true or false. This feature is very powerful in the context of automations. It can help you route your process for both functional and technical reasons. Let's say, for example, you use your document understanding capabilities to extract the invoice total from an invoice PDF. You can then set a condition that can route your process for additional approval if the total amount is greater than $100,000. However, you can also route your process for technical reasons. Let's say you're taking information from different emails and logging the information inside different systems. Based on specific information inside the email, you can route the data entry to different systems using a condition. Essentially, for any situation where you can say "If a condition is true, execute these actions, otherwise, execute a different set of actions," you can and likely should use a condition artifact. Conditions are not only binary but also can include many conditional statements to route your automation in multiple ways. To add a condition, follow these steps:

1. In the automation, on the right-hand side panel, go to the **Controls** area, as shown in Figure 7.50. Alternatively, you can search for "Condition" in the **Tools** search bar.

Figure 7.50 Adding a Condition to the Automation

2. Drag and drop the condition into the flow of the automation.
3. Click on the condition to see its parameters on the right-hand panel, as shown in Figure 7.51.

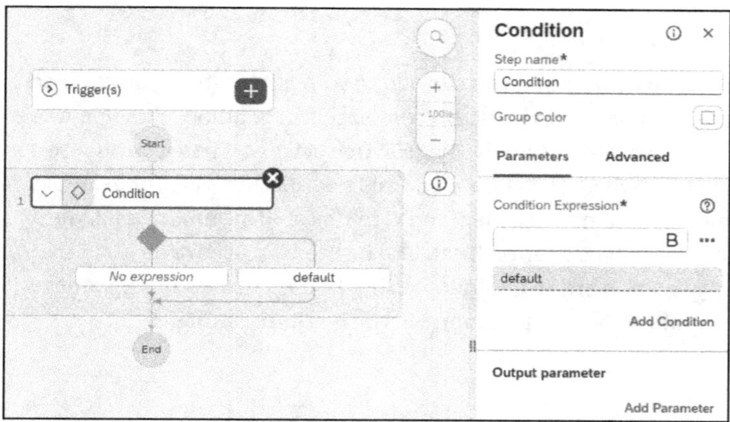

Figure 7.51 Condition Parameters Panel

4. In the **Conditions** side panel, define the conditions to route your automation. When you click on the **Condition Expression** parameter, it will display two values, **True** and **False**. However, this can, and usually should, be changed using the expression editor (access this by clicking the ••• icon, and then the **Edit Formula** option). From within the expression editor, you can create complex IF conditions, as well as test the conditions to make sure they are valid.
5. Click **Add Condition** to add additional conditions.
6. You can also, optionally, decide to define output parameters for each conditional branch. These can be static values or dynamic variables that differ based on which conditional branch the automation follows. This could be useful, for example, if you want to track which conditional branch your automation followed. You could then use this output parameter as the input to another step, such as one where you update a Microsoft Excel file that tracks the automation.
7. Save and test the automation.

Warning

Just because an expression is tested in the expression editor doesn't mean that it will work as expected. It does mean that it's a technically valid expression, but it may not work as intended. To be sure, you must test and debug your automation often to make sure you haven't made a mistake.

Note

The condition control comes with a default condition branch, which the flow of the automation will follow if the stated conditions are not met. This default branch cannot be edited or removed.

Tip

In addition to the **Condition** control that you can add to your automation, you can also route your flow of activities based on another more specific condition: the type of a specified variable. This can be done using the **Type Switch** activity, where you choose a variable and then can check whether its type conforms to a specific type. For example, this may be useful when error handling, where you want to catch an error and handle it in a specific way based on the type of error. It can also be used if you have an input that the user could plausibly enter a string or a number for. In this case, you may want to check what type the variable is, and if it's a number, convert it to a string.

7.4.2 Screen Switch

You may find when automating on the UI of an application, it's possible that more than one screen may load next in your automation. Sometimes, this may occur from seemingly random behavior from the application itself. For example, some applications have different popups for saving, but it may be difficult to ascertain why the application chooses to display one popup over another just some of the time. Other times, you may expect the application to load a different screen based on the user actions, and you may want to proceed differently based on the different screen that arises. To do this, you can use the **Screen Switch** control, which is able to handle multiple possible occurring screens by creating branches of flow for each screen and following that specific branch for whichever screen loads (or is already loaded). To add a **Screen Switch** condition, follow these steps:

1. In the automation, on the right-hand side panel, go to the **Controls** section, as shown in Figure 7.52. Alternatively, you can search for "Screen Switch" in the **Tools** search bar.

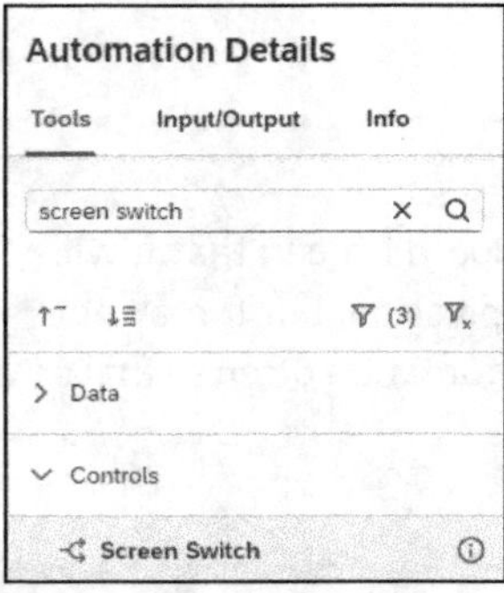

Figure 7.52 Adding a Screen Switch to the Automation

2. Drag and drop the **Screen Switch** into the flow of the automation.
3. Click on the **Screen Switch** to see its parameters in the right-hand panel.

4. The **Screen Switch Expression** fields are displayed in the side panel where you can provide the target screen details. Figure 7.53 shows these screens already chosen. To choose screens, click the ••• icon, and then choose **Edit Target**. A popup will then help you choose which screen to select for the given branch of the automation. You must have at least two branches, each with a screen associated with it.

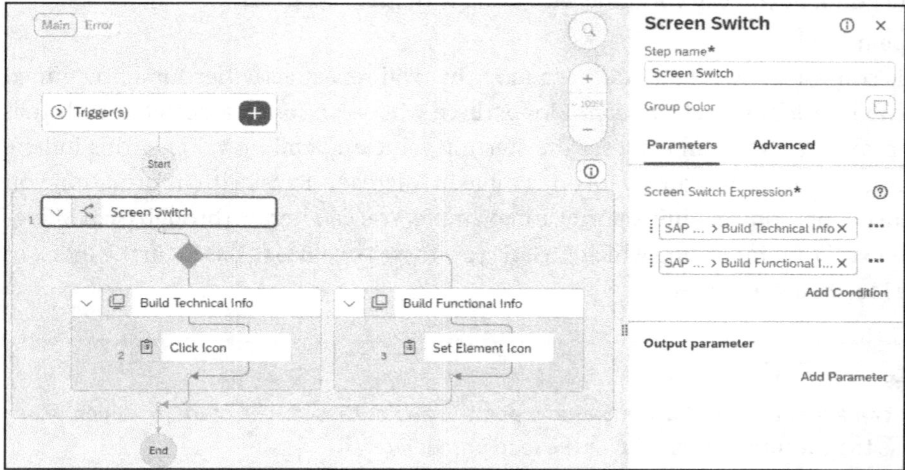

Figure 7.53 Screen Switch Parameters Panel

Note

The **Screen Switch** control will automatically follow the branch of the screen first detected by the system. If no screen is detected after 30 seconds, the control will produce an error, and the automation will fail.

7.4.3 Loops

Automation artifacts also give you the tools to create loops or iterations of activities. The main types of loops that automations support are listed here:

- **For Each**
  This control allows you to iterate through a list of members, executing the defined activities for each member in the list. One example use case for this looping mechanism is if you extract a table from Microsoft Excel and want to iterate through each row of the table and enter the data from that row into a backend system. You can do this with a **For Each** loop.
- **Forever**
  This control allows you to repeat activities until a condition is met. One example use case for this looping mechanism is, as you saw with the Outlook SDK in Section 7.3.7, that you want to iterate through all the emails in your context and stop iterating

when there are no more emails in the context, or when you've found a specific item you're looking for in your emails. As another example, you may want to upload data to a system with only finite processing power or memory. In this case, you can loop through your data entry until you receive a notification saying no further data can be entered. At this point, you can implement a static or dynamic wait to continue data entry, or you can continue with other activities if you want.

- **Repeat**
  This control allows you to define a loop that will repeat activities for a predefined number of loops. Each time you loop, there will be an index associated with that loop number (you can even set the starting value and ending value of this index). This is useful if you want to fill out a table in Microsoft Excel with information you were taking from an online portal, for example. You can iterate through each Microsoft Excel row, indicating which specific row to write values to based on the index of the loop being executed.

Note

You can add a **Loop End** activity at any point in any of these types of loops, which will cause the execution to exit out of the loop immediately.

We'll now discuss how to implement each of these loops.

For Each

To use a **For Each** loop, follow these steps:

1. In the automation, on the right-hand side panel, go to the **Controls** section, as shown in Figure 7.54. Alternatively, you can search for "For Each" in the **Tools** search bar.

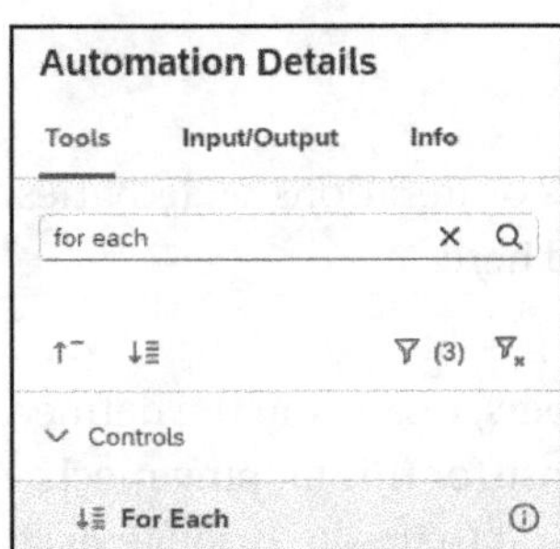

Figure 7.54 Adding a For Each Loop to the Automation

2. Drag and drop the loop control into the flow of the automation.
3. Click on the loop to see its parameters on the right-hand panel, as shown in Figure 7.55.

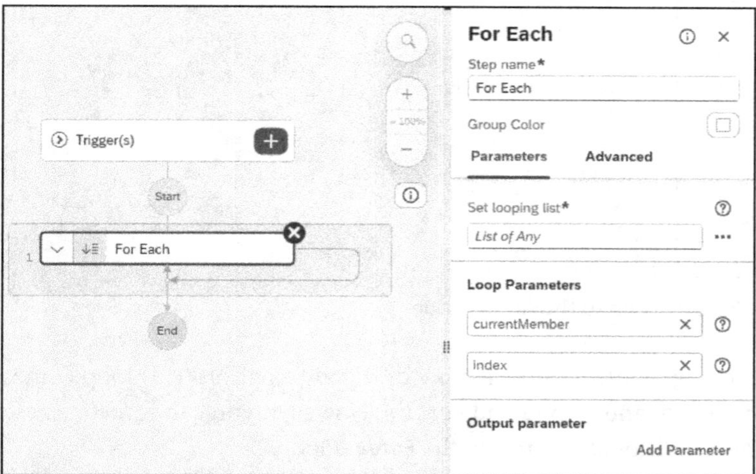

Figure 7.55 For Each Parameters Panel

4. In the loop's side panel, set the looping list that you want the control to loop around. This must be a list of elements. You'll notice there are also two loop parameters, **currentMember** and **index**. The former will refer to the element of the list currently active (i.e., in the third loop, the third element will be active), while the latter refers to the index of the loop you're currently in (this will start at 0, so in the third loop, the index will be 2).
5. Drag and drop other activities into the loop. These activities will be repeated for each iteration of the loop. Make sure to use the **currentMember/index** output variables from the **For Each** loop in the activities if necessary.
6. Save and test the automation.

Tip

You can assign a list of screen elements as the looping list in a **For Each** loop. These elements must be a collection, as discussed in Section 7.2. You can do this either by choosing the proper list from the available choices when clicking in the **List of Any** field under **Set looping list**, or you can click the ••• icon, and then choose **Open the Target Editor** where you can indicate the target selection.

Forever

To use a **Forever** loop, follow these steps:

1. In the automation, on the right-hand side panel, go to the **Controls** section, as shown in Figure 7.56. Alternatively, you can search for "Forever" in the **Tools** search bar.

Figure 7.56 Adding a Forever Loop to the Automation

2. Drag and drop the loop control into the flow of the automation. This loop comes with a **Condition** control and a **Loop End** control. However, you can remove these controls and/or add additional controls in the **Forever** loop.
3. Click on the loop to see its parameters on the right-hand panel, as shown in Figure 7.57.

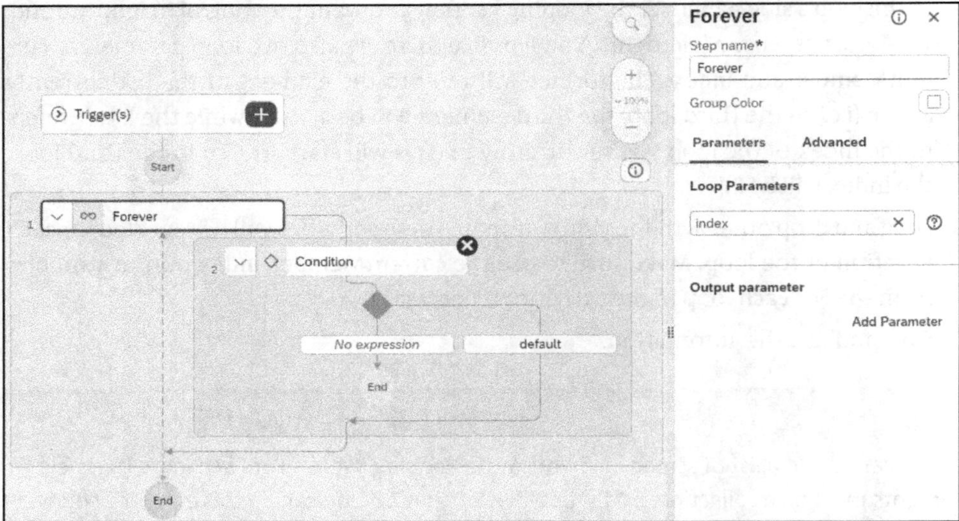

Figure 7.57 Forever Parameters Panel

4. In the loop parameters, only **index** is displayed. This is an integer that starts at 0 and is incremented after each loop. You can use **index** to keep track of which loop you're in, or even use it in an activity (e.g., incrementing the row to push values to a Microsoft Excel file).
5. Set the condition expression as needed. When the condition is met, the **Forever** loop will enter that branch and exit the loop.
6. Place the needed activities into their respective branches of the **Forever** loop.
7. Save and test the automation.

Repeat

To use a **Repeat** loop, follow these steps:

1. In the automation, on the right-hand side panel, go to the **Controls** section, as shown in Figure 7.58. Alternatively, you can search for "repeat" in the **Tools** search bar.

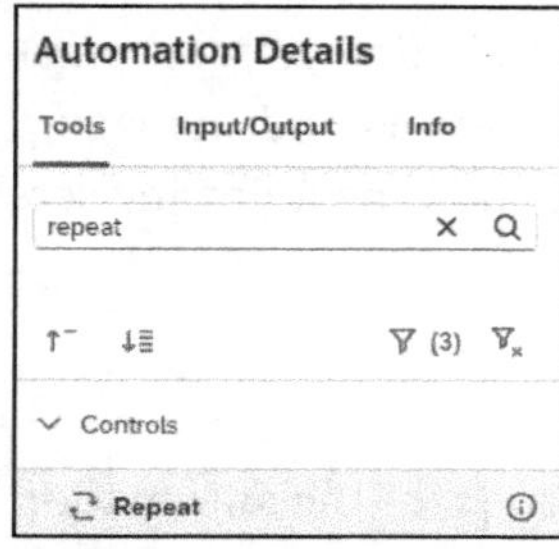

Figure 7.58 Adding a Repeat Loop to the Automation

2. Drag and drop the loop control into the flow of the automation.
3. Click on the loop to see its parameters on the right-hand panel, as shown in Figure 7.59.

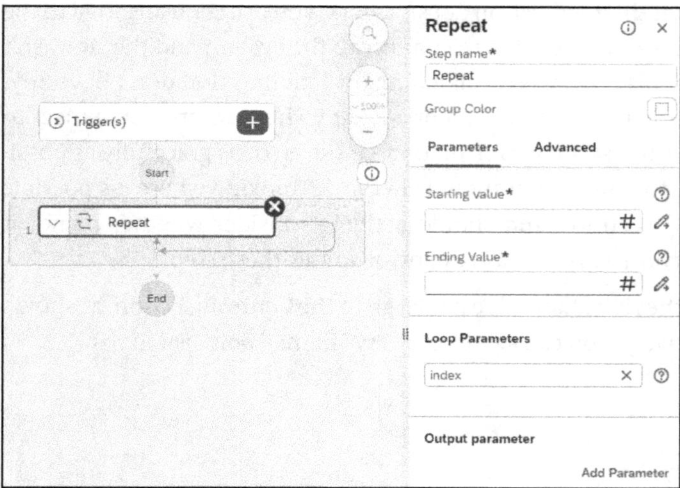

Figure 7.59 Repeat Parameters Panel

4. In the loop parameters, you'll see two values to set: **Starting value** and **Ending Value**. **Starting value** will set the starting number where the iteration begins, and **Ending Value** will set an ending number where the iteration stops. This value will also correspond to **index** and will be incremented after each loop until **index** has reached the **Ending Value**. Similar to the **Forever** loop, you can use **index** during the actual execution of the loop to track which loop you're in or accomplish a specific activity.
5. Save and test the automation.

7.4.4 Error Handling

Over the course of an automation, many errors can be raised during execution. Specifically, many of these errors are predefined in the Core SDK, Excel SDK, Outlook SDK, and SAP GUI for HTML SDK (in addition to business/functional errors raised by the user via the **Stop automation in error** activity). At execution time, errors will automatically fail your automation unless you catch them properly. While designing your automation, you can avoid failed executions by anticipating possible errors and managing them accordingly. You can use some of the techniques described in the following sections to handle these errors.

Try Catch

The **Try** control allows you to catch errors raised on specific steps of the automation and handle these errors using a defined flow of activities on a side branch. This is often helpful if you expect an error to arise during a certain step and would like to handle it in a specific way. This is common when you're connecting to another system, where there may be several errors that can arise during execution. Even if everything is right inside your automation, the system being accessed, for example, may be down when you try to access it. In this case, you can catch the error, and then perhaps try to connect to the system again. Then, if this connection doesn't work, you can send an email to the end user letting them know that you tried to connect to the system and that it wasn't possible. You can also use this type of error handling on activities that don't have anything to do with connecting to a system. Essentially, anywhere where you expect to possibly see an error, you can use a **Try** control to catch the error as gracefully as possible. As you design your automation, you should always be thinking where it's possible that it might fail during execution and should actively consider whether or not it makes sense to add a **Try** control. To add a **Try** control, follow these steps:

1. In the automation, on the right-hand side panel, go to the **Controls** section, as shown in Figure 7.60. Alternatively, you can search for "Try" in the **Tools** search bar.

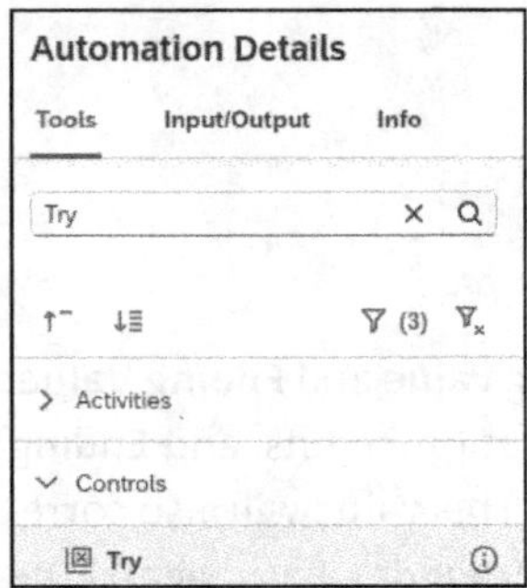

Figure 7.60 Adding a Try Control to the Automation

2. Drag and drop the **Try** control into the flow of the automation.

3. Click on the control to see its parameters on the right-hand panel, as shown in Figure 7.61.

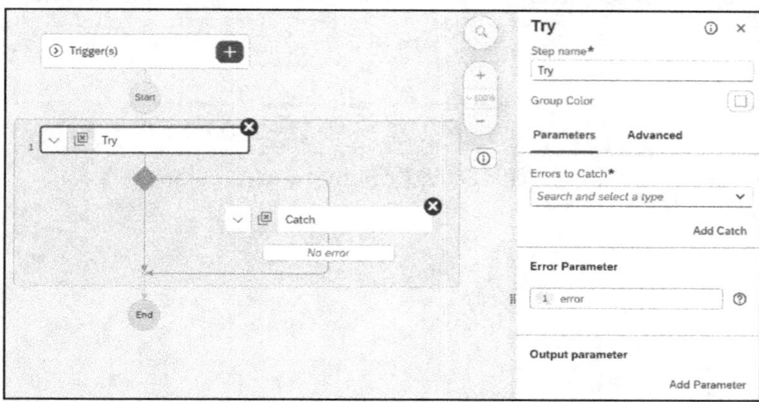

Figure 7.61 Try Control Parameters Panel

4. In the **Try** parameters, you'll see the **Errors to Catch** dropdown. It's best to specify which errors you might be catching to make sure you're handling each possible error exactly as you want to. However, it's also possible to catch a general error if you want by selecting **error**.
5. You can also see that errors related not just to the Core SDK but also added dependencies will show up in the error list. In Figure 7.62 you can see that, when opening a Microsoft Excel worksheet and adding values to its cells, you can catch an **InvalidRangeDefinition** error. If there's a general error that isn't an **InvalidRangeDefinition** error, you'll also catch that. In either case, there's an additional branch added to the automation and you can decide to handle that however you may like.

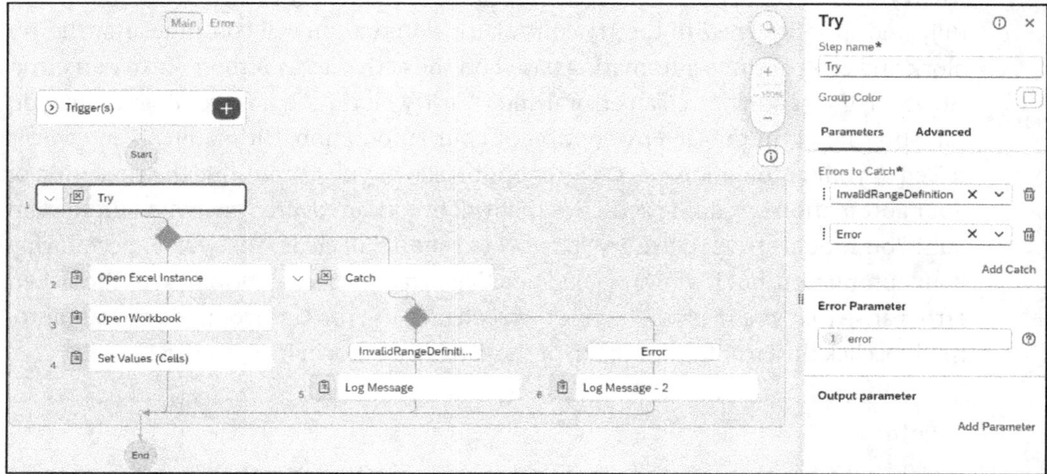

Figure 7.62 Microsoft Excel Specific Try Condition

Tip

You can use the **Log Message** activity in the **Catch** branch to log **error.code** and **error.message**.

Note

You can see the list of errors that a specific activity can throw in its parameters panel, as shown in Figure 7.63.

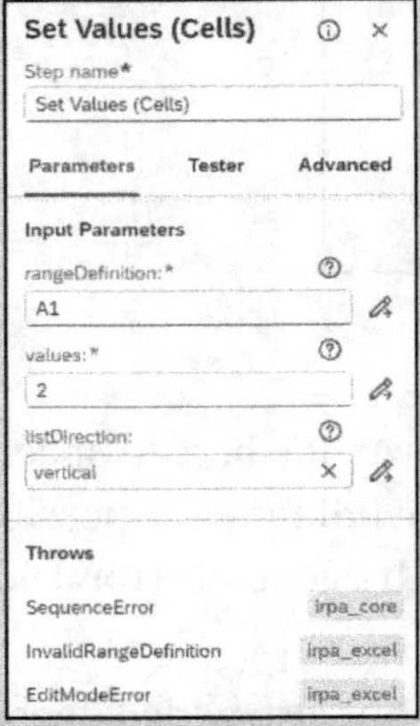

Figure 7.63 Set Values (Cells) Activity with Its Associated Errors in the Throws Section

On Error

We previously discussed handling errors within an automation flow. We discussed that when you believe there will be an error, you should use a **Try** control to catch it gracefully and specifically. With the **Try** control, the automation will continue after the **Try** block and can continue automating based on the activities as if there wasn't an error. However, if you don't catch an error using your **Try** block, and an error does occur, you can still catch it in the **On Error** option of your automation. Ultimately, if an error is raised in your automation, the **On Error** option allows you to continue the execution of your automation by adding activities that will be executed after the main automation fails. You should strive to use **Try** controls to capture all errors that you expect during your automation, but it's always good practice to have a backup flow of activities when errors arise that you may not be able to predict. To use the **On Error** node of your automation, click the **Error** tab at the top of the canvas, as shown in Figure 7.64.

Note

Main automation errors and variables are accessible from the **On Error** node of the automation.

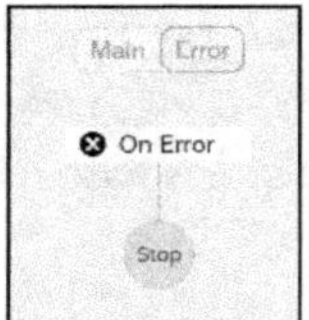

Figure 7.64 Error Tab That Leads You to the On Error Node of Your Automation

7.4.5 Alerts

Alerts allow you to raise business events in your automation and pass along information with them (e.g., **OutOfStock** or **S4LoginFailed** alerts). These alerts can then be subscribed to in environments using alert handlers where we can specify the details of the notification which will be sent via email (recipient, subject, and body of the email with contextual variables available). Once an alert is created, it can be raised from an existing automation. You can use the **Raise Alert** activity from the Core SDK to do this. To create an alert, follow these steps:

1. In your project, navigate to the **Overview** tab and click **Create • Alert**.
2. Enter a **Name**, **Identifier** (prefilled but can be changed), and **Description**.
3. Click **Create**.
4. In the newly opened tab for your alert artifact, as shown in Figure 7.65, you can add parameters in your alert by clicking the **New Parameter** button.

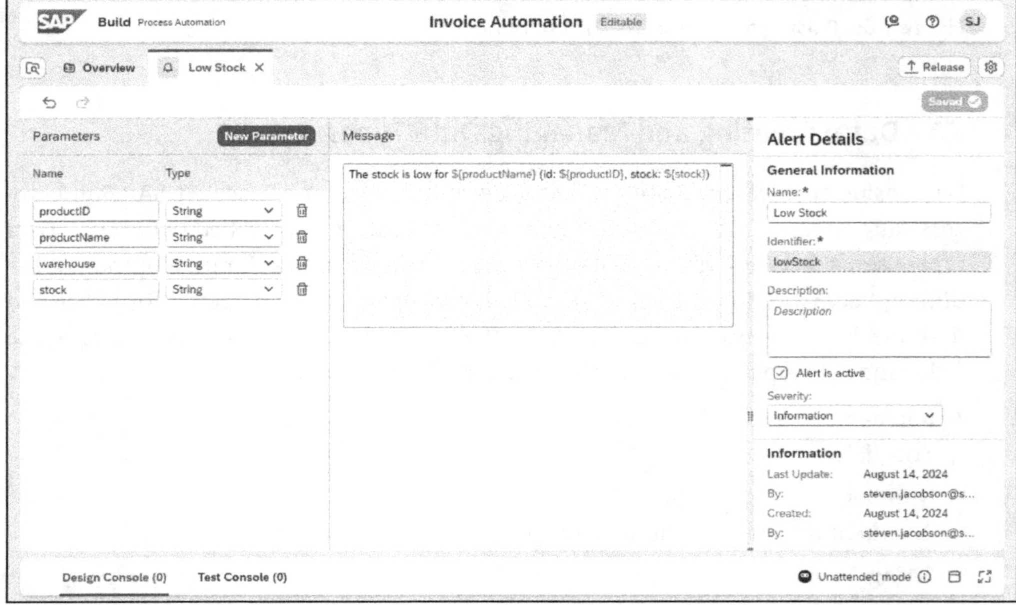

Figure 7.65 Creating an Alert

> **Note**
> Each parameter must have a unique name that is JavaScript compliant and starts with a letter.

5. You can define a parameterized message in the center **Message** field. During execution, the dynamic parameter value will be used.
6. Edit any other necessary settings on the right-side panel, and then save the alert.
7. You can now raise this alert in your automation, as shown in Figure 7.66.

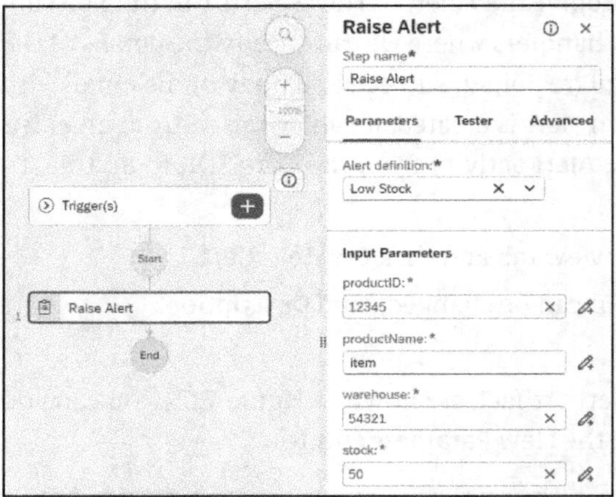

Figure 7.66 Raising an Alert in an Automation

7.5 Data: Creating and Managing Data Types

Data inside business processes can often be complex and multifaceted. SAP Build Process Automation allows you to create your own custom data, called *data types*. Data types are artifacts describing a data structure that can be used as input or output for the different activities in your automation. This concept enables you better formalize and describe data to be used and can facilitate the manipulation and validation of data. The following data types are available to be added to a custom data type:

- **Boolean**
  True/false.
- **Number**
  A valid number (not limited to integers).
- **String**
  A series of characters.
- **Date**
  A date written in the YYYY-MM-DD format (ISO 8601).

- **Time**
  A time written in the HH:mm:ss format (ISO 8601).
- **Date/Time**
  A date and a time written in the YYYY-MM-DDTHH:mm:ss format (ISO 8601).
- **Password**
  A password credential that will be masked when given a value.
- **Data types from other packages**
  Imported data types from other dependencies and packages.

While you can always create these data types manually, some data types may be created automatically when certain SDK packages are imported into the project. You also can import data types by importing a Microsoft Excel file, where the headers of the table will be used as the data type name, and the type of data will be determined by the data in the column. This can be particularly helpful when you plan to work with a Microsoft Excel file containing a large table of information. You can create a data type with dozens of attributes by importing the file.

SAP Build Process Automation also enables you to work with data that only exists over the course of execution with activities, as shown in Figure 7.67.

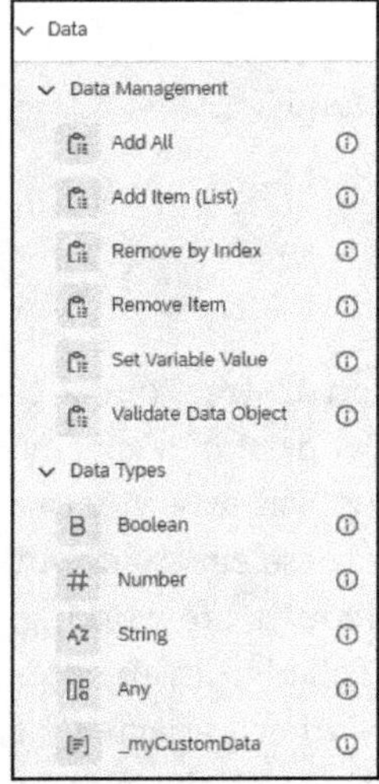

Figure 7.67 Data Management Activities and Types to Be Added to an Automation Flow

Let's say you've loaded a table of data from Microsoft Excel into your process. How would you go about actually using that data to accomplish the different tasks of your process? Using just a few drag-and-drop activities, you could iterate through the list of items, validate specific items, manipulate other items, and concatenate different items before exporting the data to another Microsoft Excel file. Or, you could traverse a legacy web application, pull specific data that you're interested in, format it, and then enter it into a new online portal with only a few activities.

Data will be stored in variables inside an automation artifact. Variables refer to data storage items that have a name, type (Boolean, number, string, or custom data type), and value. They allow you to store data that can be reused throughout the automation, while also allowing you to set and modify their content.

> **Note**
>
> Variables are "type checked" to ensure they are the correct data type defined. However, a variable of type **any** can provide any type of data such as a string, number, Boolean, or custom variable type.

7.6 Automations, Actions, and Processes: Using Existing Artifacts

The previous sections delved into the different options offered to you from SAP Build Process Automation to easily build an automation artifact to automate your process flow. However, frequently when automating end-to-end processes, make sure you're not only automating the process effectively but also designing your flow with more than just function in mind. You should embrace a modular approach by using existing artifacts within your automation flow. Among many benefits, modularization is important for the following reasons:

- Makes the automation flow more readable and easier to understand
- Simplifies debugging
- Encourages fusion development
- Eases maintenance and simplifies future changes

When designing your flow, you should always think where you can abstract out submodules to embrace this approach. Typically, these modules will consist of a specific functionality or theme so that you can subdivide the flow logically. This way, you can split the flow into parts that allow you to design, develop, and test these parts without needing to worry about the complexities of other parts. Once modules are working, you then only need to worry about their inputs and outputs when called from an outer module. This approach also allows you to bring in artifacts from other projects that exist in your SAP Build lobby and use modules from those projects in your current project. Automation artifacts, such as in Figure 7.68, give you the ability to add the following:

- Other automation artifacts
- Process artifacts
- Action artifacts

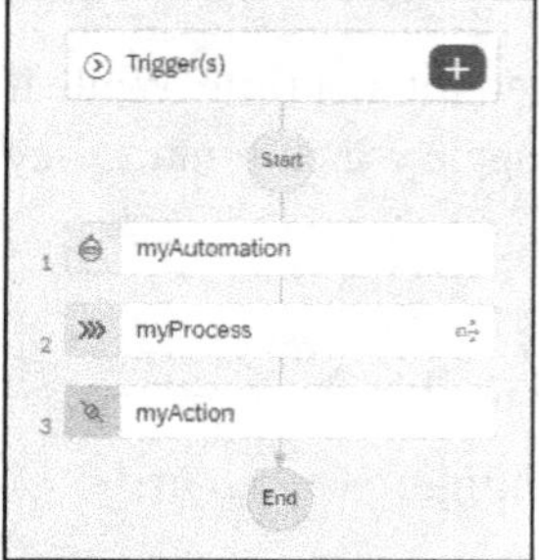

Figure 7.68 Using Existing Artifacts inside an Automation Flow

7.7 Desktop Agent

The desktop agent is a necessary component of automation artifacts that is installed locally on a user's desktop and is required to run automations. While the desktop agent exists as software locally installed, it will connect with your tenant to be able to run and test your projects. This agent will be your local portal to access projects made available from your tenant. It's also helpful from a deployment standpoint, where not all end users may have complete access to the tenant (or understand how to use the tenant, for that matter). With the desktop agent, the details of the tenant are abstracted away from the end user. Instead, the end user now can see all necessary information in an easy to use and consume format.

In this section, we'll discuss how to use the desktop agent to execute and configure the executions of your automations. We'll start by discussing the desktop agent UI and the fundamentals of how to navigate within the agent from the home screen. We'll then discuss the different modes of execution that you can run automations in. Finally, we'll discuss how to collect traces for troubleshooting purposes.

7.7.1 Desktop Agent User Interface

There are multiple pieces of information that you can see even from just the home screen of the agent, as shown in Figure 7.69:

- **Idle**
  Current status of the agent and whether the agent is ready to be used.
- **Agent Mode**
  Current mode that the agent is set to. This can be **Unattended**, **Attended**, **Design**, or **RDP Manager**.
- **Active Project**
  Projects that can actively be consumed by the agent.

- **Selected Tenant**
  Tenant to be activated. The agent can only connect to one tenant at a time. While it can have many tenants stored, the end user will need to toggle which tenant they would like to activate at a given time.
- **Events**
  Number of errors, warnings, and info that the agent has experienced.

There are also several side navigation tabs that allow you to configure the agent:

- **Home**
  Access the overview of your agent. This is where you can see the information in Figure 7.69.
- **Projects**
  Find projects available for execution in **Attended** mode.
- **Info**
  Navigate to **Agent Details** or **Project Details** to see more information.
- **Settings**
  Navigate to **System** (where you can change things such as language and connection type), **Mode** (where you can toggle between **Attended**, **Unattended**, **Design**, and **RDP Manager** modes), or **External Authentication** (where you can inspect your external authentications).
- **History**
  Navigate to **Events** (where you can inspect the **Errors**, **Warnings**, and **Info** events that have occurred with your agent), **Jobs** (where you can inspect the jobs executed inside your agent), or **Reports** (where you can inspect the trace reports for your agent).
- **Tenants**
  This is where you can configure new environments and toggle between the existing environments to activate.

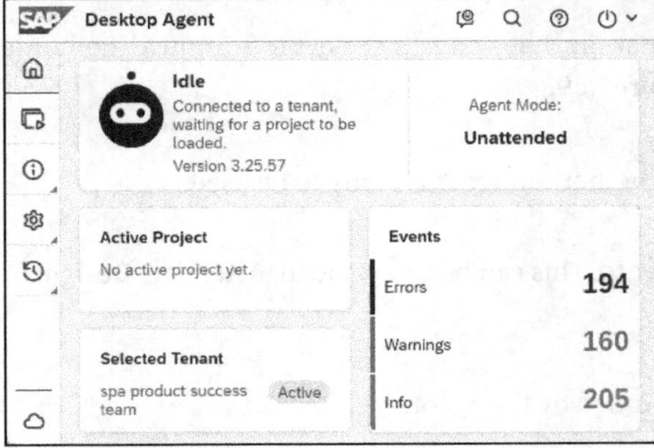

Figure 7.69 Homepage of the Agent

7.7.2 Modes of Desktop Agent

There are a few different modes that you must be aware of when running your automations, which we'll discuss in this section:

- **Attended**
- **Unattended**
- **Design**
- **RDP Manager**

You can only activate one mode at a time for an agent. To do this, open the agent, and then navigate to the **Mode** option inside **Settings** (⚙).

Attended Mode

Attended automations are started manually by a user and can be supervised. In this mode, you must trigger the automation to begin execution. This is useful for automations that can help individuals perform their tasks, but that aren't used on a schedule or as a part of a process artifact. These may be mundane or repetitive tasks that can increase an end user's efficiency and that may be used ad hoc by the user. For example, let's say you want to take all the attendees from a meeting and send out individual emails to each participant providing the notes from the call. This would be an automation best suited for **Attended** mode, as you want to use this functionality for only certain calls and decide to run it when it works for you. To take it a step further, **Attended** mode is also helpful for when the end user wants to interact with the automation during its execution. To take the previous example where you sent out meeting minutes to each individual participant of a meeting, what if you also wanted to send out a personal call to action for each participant? To do this, you could provide a dialogue box that you'll see on each iteration of the automation, where you can provide the individual call to action. The automation can then take this individual information and combine it with the other generalized information before sending it out.

To use an automation in **Attended** mode, you can configure an **Automation Launcher** and add specific automation artifacts to it. This will allow you to customize the different automations you see when looking at your agent. To do this, follow these steps:

1. In your project, navigate to the **Overview** tab and click **Create • Automation Launcher**.
2. Enter a **Name** and **Description** in the popup for the new **Automation Launcher** artifact.
3. On the right-hand panel of the **Automation Launcher** screen, you can see all the automations available in your project. Drag and drop automations into the **Launch Manually from the Agent** section of the **Automation Launcher** page, as shown in Figure 7.70.

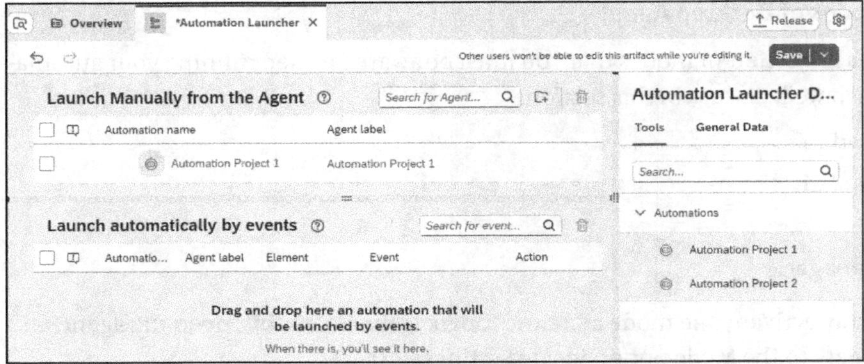

Figure 7.70 Automation Launcher Editor

4. Click on the automation, and edit the label if necessary. This will change the name displayed in the agent but won't change the name of the automation in the main project.
5. Click **Save** to save the **Automation Launcher** artifact.
6. Click **Release** to release the project, and then click **Deploy** to deploy the project.
7. Once your project is deployed, you can configure the **Automation Launcher** for the project in your environment. The **Automation Launcher** will update automatically if the project is redeployed.
8. Open the agent, and navigate to the **Projects** tab.
9. Click **Activate** if the project isn't already active.
10. The package will download, and all available automations as designated when designing the automation launcher will be shown (see Figure 7.71).

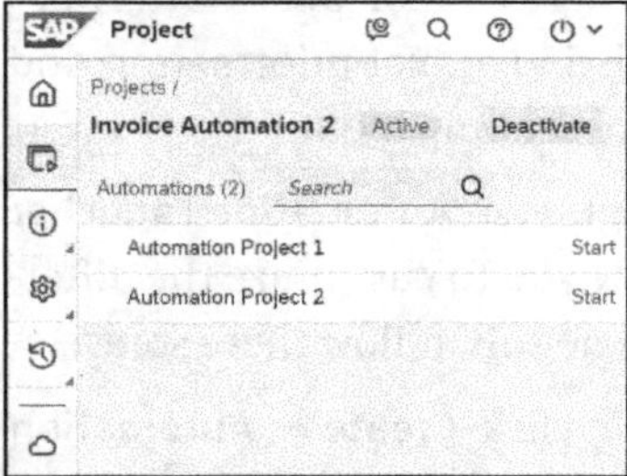

Figure 7.71 Automation Launcher in the Agent

Note

Make sure the agent version listed in the project and the agent version installed locally align. Otherwise, the automation launcher may not appear inside the agent after the project is deployed.

Unattended Mode

In this mode, the agent runs automatically and without human intervention or interaction. Automations running in unattended mode include automations with a scheduled or API trigger, which can be toggled from the **Environments** view in the **Control Tower** screen. **Unattended** mode is also necessary when automations are triggered from within a larger process artifact. Here, as the process will be calling the specific automation artifact, you must ensure that the agent is in **Unattended** mode. If it's not, the tenant won't recognize any available agents to receive the automation, and the process will stall. If you trigger multiple processes that call automations, these automation triggers will queue until at least one agent is changed to **Unattended** mode. When this does happen, the automations will execute in the order in which they were queued.

Design Mode

The agent must be set to **Design** mode during debugging and testing of the project. With newer versions of the agent, the agent will automatically be changed to **Design** mode when you begin testing/debugging your project, as discussed earlier in Section 7.1.

> **Warning**
>
> Make sure to switch the agent mode back to the appropriate mode once finishing testing and subsequently deploying the project. Otherwise, it won't be able to run deployed automations if it remains in **Design** mode.

RDP Manager Mode

The agent can be set to **RDP Manager Mode** to register and manage multiple Windows Remote Desktop Protocol (RDP) sessions. This is helpful if you have installed agents on multiple remote machines where you wish to run unattended automations. During unattended automation, the screen or session may lock, which can cause graphical or user interface actions like mouse clicks or keystrokes to fail. By using **RDP Manager Mode**, you can register and manage these sessions to make sure these automations can execute without issue. The RDP manager will verify every 30 seconds to ensure that the connection is open. If it finds that the connection is no longer open, it will attempt to re-open it twice before failing out. Typically, you'll want to use one agent in **RDP Manager Mode** to manage all your other connections.

7.7.3 Trace Collector

Desktop agent 3 records traces for the agent to monitor the agent and its jobs. (You'll want to make sure that **Support** mode is activated from the **Agents** section of the **Control Tower** page to use traces. Click the ••• icon next to the appropriate agent, and then click **Activate Support Mode.**) Encrypted traces are stored locally and can be pushed to the tenant from the agent, where they can be downloaded and viewed. You can also disable encryption if you would just like to keep the traces locally and not push them to the tenant. If the agent is connected to a tenant, you can upload encrypted traces by choosing ••• and then clicking **Upload Trace** (see Figure 7.72). However, if the agent isn't connected to any tenant, you can choose ••• and then click **Get Traces for Manual Upload** to download the traces and manually upload them to the cloud. You can always look at available traces from inside the **Control Tower** and download them for viewing (see Figure 7.73).

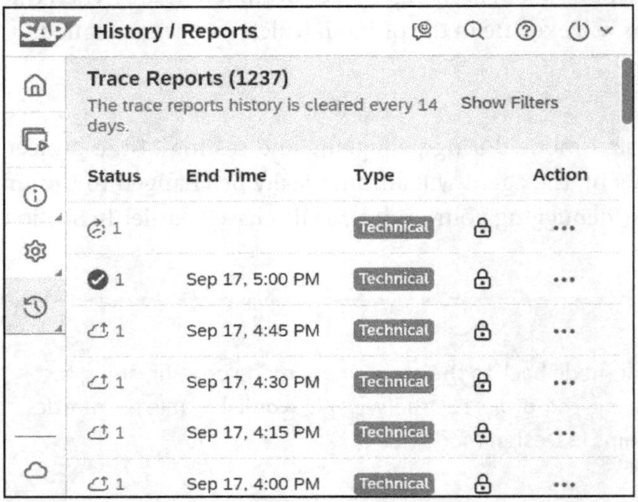

Figure 7.72 Uploading Traces from the Desktop Agent

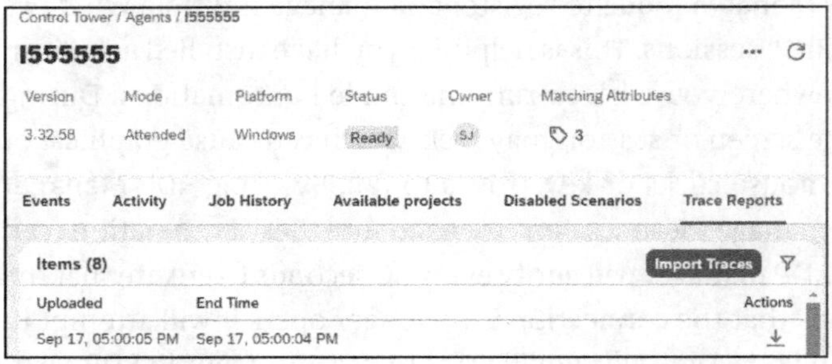

Figure 7.73 Accessing Traces from the Control Tower

7.8 Environments

An *environment* in SAP Build Process Automation is a virtual location inside your tenant where your projects and their related artifacts are deployed to. You can use environments to control how you deploy and execute your projects. When a project is deployed in an environment, users of that environment can access all related artifacts as specified by the security level assigned to the user.

Note

The six different authorization you can give to a user, and their definitions, are listed below:

- **Monitor**
  View environment details and triggers, view and monitor the deployed projects, variables, and other associated resources, and activate traces
- **View logs**
  View detailed execution logs.
- **Deploy**
  Access to all monitor rights, deploy and undeploy projects, deploy all triggers and enable actions excluding **Run Now**
- **Execute**
  Launch forms, processes, and visibility scenarios and has all trigger actions available.
- **Supervise**
  Access the process visibility dashboards and take necessary actions.
- **Administrate**
  Add agents or other resources to the environment, create, edit, share, and delete environments, and has all trigger actions available. This authorization includes both the monitor and deploy permission.

In this section, we'll discuss the concept of environments and learn how to deploy to them. We'll also discuss variables you can create that will pertain only to specific environments and how you can achieve deployment isolation using environments.

7.8.1 Overview

There are two types of environments you can use:

- **Public**
  The public environment is accessible to all users who have access to the SAP Build Tenant. This environment should be used if there is no need to limit user access. This environment also has no restrictions on what artifacts can be deployed to it. If the tenant was available prior to the release of this feature, all existing projects and artifacts will have been deployed here.
- **Shared**
  A shared environment is an environment that has been created by any user inside the tenant. At the time of creation, this environment will only be available to the user that created it. Access, and different authorizations, can then be provided to

other users of the tenant. This allows you to isolate deployment and maintain access, configuration, and execution more cleanly. Thus, you can deploy and run the same project with different configurations and make these different configurations available to different users, depending on line of business (LoB) requirements, execution requirements, security requirements, or any other deployment requirements you may have.

Environments may contain the projects you've deployed to them as well as their related artifacts and configurations necessary for project execution. These artifacts and configurations are defined as:

- **Projects**
  Displays projects deployed to the environment.
- **Triggers**
  Displays the triggers that can start an automation with either a scheduled trigger or an API trigger.
- **Automation launchers**
  Displays the automation launchers in the environments. These automation launchers are created on the project level but configured on the environment level to allow for executing automations in **Attended** mode. The automation launchers are deployed to all agents (as specified when configuring the launcher) inside the environment.
- **Agent management**
  Displays the agents/machines assigned to the environment that can run automations inside the environment.
- **Alert handlers**
  Displays mail alerts for steps in a project.
- **Variables**
  Displays environment variables created in projects for reuse.
- **API keys**
  Displays access to API endpoints and configurations for APIs to monitor artifacts in the environment.

When a project is deployed, all agents declared in the **Agent Management** tab of the environment will be involved in taking deployed projects to be executed. Agent attributes can be used to specify job distribution even further from within an environment.

Note

When deploying a project, all available environments to which the user can deploy will be displayed. When a project version is deployed in a shared environment, if a previous version exists, it's automatically undeployed, and the newer version is deployed in its place. This contrasts with the public environment, where multiple versions can be deployed, and one version is made to be active.

Note

When deploying a project to an environment, you may see any of the following options:

- **Deploy**
  It's the first time you've deployed the project to a specific environment.
- **Upgrade**
  You're deploying a newer version of the project to a specific environment.
- **Redeploy**
  You're deploying the same version of the project to a specific environment.
- **Downgrade**
  You're deploying an older version of the project to a specific environment.

7.8.2 Environment Variables

Environment variables are created on the project layer. To create environment variables, click the ⚙ icon at the top right of your project, and navigate to the **Environment Variables** tab, as shown in Figure 7.74. From there, you'll be able to create variables that will be defined when you deploy your project (see Figure 7.75).

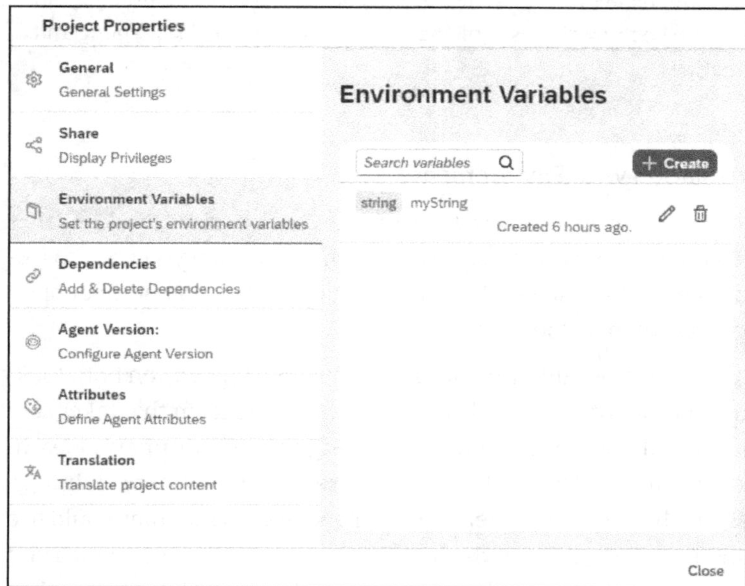

Figure 7.74 Creating an Environment Variable

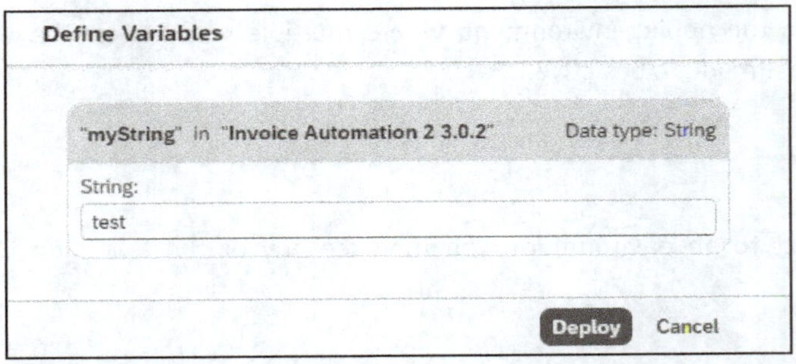

Figure 7.75 Defining the Environment Variable When You Deploy Your Project

This concept of environment variables is powerful when you consider dynamic information that you know may change in the future. You can redeploy your projects with the updated information and allow your automation to not skip a beat. This is also powerful when you consider deployment isolation, where you may want dynamic variables defined based on what environment they are deployed to. Environment variables can be URLs (set as strings), dates, times, destinations, or credentials such as users and passwords.

Warning

The password variable is protected and cannot be seen in the cloud and when it's passed to the desktop agent. However, it's not protected on the client machine, where it can be seen during execution (if designed that way, such as if you paste it into an Excel cell) as well as inside the agent logs. When using a password variable, understand that it may be able to be seen by the user of the session running desktop agent and executing the automation.

7.8.3 Deployment Isolation with Environments

As previously discussed, environments allow you to create containerized locations that hold all of your related projects and artifacts. This can be helpful when you want to isolate your deployment of projects based on LoB, security protocol, or any other deployment strategy an organization may look to put in place.

For example, let's say you have two different groups within an organization both looking to incorporate automation into their daily tasks. You have one team that takes care of invoice management, and another team that is looking to monitor internal learnings and automate emails to personnel who haven't taken the mandatory trainings. These two teams, reasonably, have very different sets of end users. Thus, they could use the environments functionality to dictate which agents/machines, and thus which end users, will have access to the projects in each environment. This would be complete isolation, where each environment would have entirely different projects and artifacts.

However, let's say we have the same set of projects that we want to deploy to two different teams, each with different end users and potentially different environment variables. You could also create this isolation, where each team (and environment) has access to the same project but a different deployed version of that project. During each deployment, you could define different environment variables (e.g., user credentials) that only apply to each specific team (and environment).

Not only can you isolate your environment across different LoBs or teams within an organization, but you can also use this isolation to create environments for development, test/validation, and production. In this case, you can deploy a project to your development environment while you're still testing and iterating during development. Once you begin testing and validating the finished design, you can deploy the project to your test/validation environment. Then, once it has been thoroughly tested here, you can deploy to your production environment. In each environment, you can restrict the users who have access to the environment, ensuring the security protocol is met. Further, you can set environment variables based on the environment you're deploying to, allowing you to dynamically change URLs/destinations to your sandbox or production system based on the appropriate protocol.

Overall, there are several ways you can embrace deployment isolation with environments from within a single tenant. Different organizations have different requirements for their deployment strategy, and there's no definitive way that you must deploy your projects using environments.

7.9 Summary

This chapter discussed the importance of using automation artifacts to automate end-to-end processes. With drag-and-drop simplicity, you now understand how to automate on top of an application's UI, use embedded SDKs to perform activities across a spectrum of SAP and non-SAP applications, manage data, and create complex automation flows with iterations, conditions, and error management.

In the next chapter, we'll discuss decision artifacts, which help you identify, automate, and manage business critical decisions.

Chapter 8
Decision Artifacts

In the previous chapters, we explored the foundational aspects of SAP Build Process Automation to automate workflows and tasks. But what truly elevates process automation from simple task automation to intelligent process execution is the power of decision management. In this chapter, we'll look into the decision capability of SAP Build Process Automation, which is very important in building agile and responsive business processes.

Imagine an e-commerce company processing thousands of orders daily. For each order the company receives, multiple decisions need to be made:

- Is the customer eligible for free shipping?
- Which warehouse is best suited to fulfill the order based on inventory and delivery location?
- What discount can be given to the customer?
- Which other product can be of interest to this customer based on the order history?
- Which payment options should be offered?

Although these decisions seem small and trivial, they impact customer satisfaction, operational efficiency, and, therefore, the overall profit margin of the company.

Decision management is the process of identifying, automating, and managing these business-critical decisions using a business process automation system. It goes beyond hard-coded logic within workflows or applications by empowering organizations to do the following:

- **Centralize business logic**
  Instead of maintaining the same rules in different applications and workflows, a centralized repository for business rules ensures consistency and transparency. With a centralized business rules repository, the organization can ensure that the rules are applied across departments and processes, reducing risk of discrepancies and errors.
- **Empower business users**
  By using user-friendly interfaces that look and behave similar to their favorite spreadsheet application, business experts can directly define and modify business rules without relying heavily on IT.

- **Enable compliance and governance**
  Decision management helps organizations enforce compliance with regulatory requirements, internal policies, and industry standards. By encoding business rules into the process automation system, organizations can ensure that all decisions are in line with legal and regulatory frameworks. For example, a financial institution can automate decision-making processes to ensure adherence to anti-money-laundering regulations.
- **Be agile and efficient**
  The ability to quickly adapt to changing market conditions, new regulations, or updated business strategies is critical for a successful organization. Decision management facilitates this agility by allowing for easy modification and deployment of new rules and policies.

In this chapter, we'll discuss the different SAP solutions for business rules and how to choose the correct solution.

8.1 Selecting the Correct Decision Solution for SAP Scenarios

SAP offers a variety of rule engines to help businesses manage and automate their business rules and processes. Each rule engine is designed for different development environments such as Java, ABAP, SAP HANA, and the cloud, offering flexibility and customization for businesses of all sizes and industries. In this section we'll look at the different solutions from SAP for decisions.

8.1.1 SAP Business Rules Management

SAP Business Rules Management in Java is a rule engine that is part of SAP NetWeaver (Java) 7.1 EhP1 and higher, which allows developers to define and manage business rules in a centralized repository. SAP Business Rules Management is designed to be used in SAP NetWeaver Developer Studio, a Java-based development environment provided by SAP. It provides a user interface (UI) for managing rule projects and defining business rules and allows for easy integration with other SAP NetWeaver components such as SAP Business Process Management. Figure 8.1 shows a decision table maintained in SAP Business Rules Management.

SAP Business Rules Management consists of two main components: *rule composer* and *rule manager*. Rule composer is integrated into SAP NetWeaver Developer Studio as a separate perspective. Rule composer provides interfaces within SAP NetWeaver Developer Studio for business analysts and business rules developers to model complex business rules. On the other hand, rule manager is meant for business managers, administrators, and functional users to modify business rules and effect changes in real time.

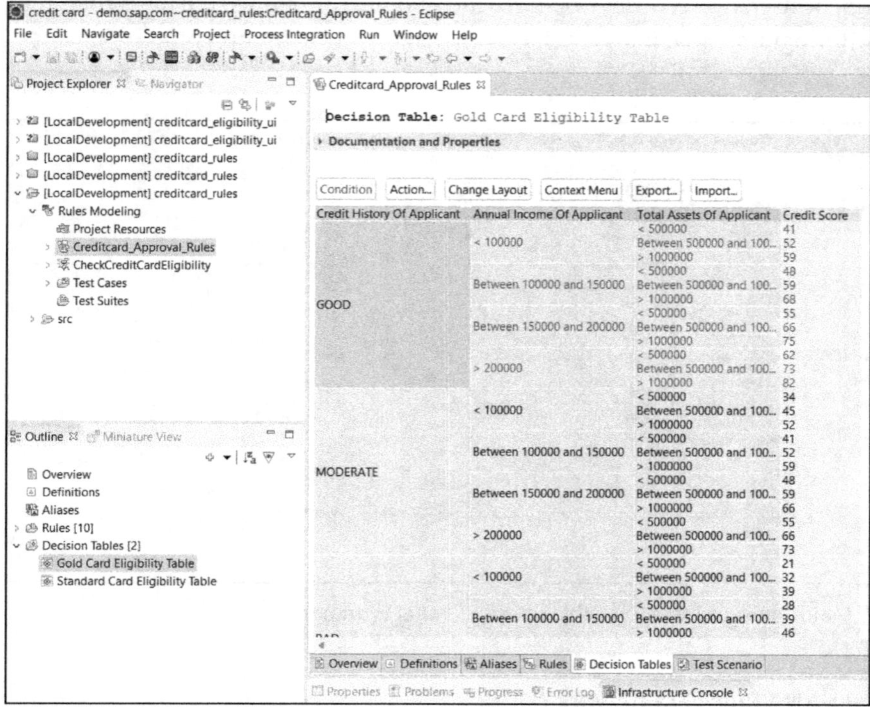

Figure 8.1 Maintaining a Decision Table in SAP Business Rule Management

SAP Business Rules Management is ideal for Java developers who want to define and manage business rules in a unified environment and have the flexibility to integrate with other SAP NetWeaver components. SAP has announced the end of maintenance and extended maintenance timeline of SAP NetWeaver (Java), and this is applicable for SAP Business Rules Management too. SAP Business Process Management and SAP Business Rules Management customers need to look for a migration path: SAP Build Process Automation is the natural choice in SAP Business Technology Platform (SAP BTP).

8.1.2 Business Rule Framework Plus in ABAP

Business Rule Framework plus (BRFplus) is a rule engine provided by SAP for managing business rules in ABAP-based systems such as SAP ERP, SAP Supplier Relationship Management (SAP SRM), SAP S/4HANA, and so on. It allows ABAP developers to define and manage rules using a web-based interface called BRFplus Workbench to define and maintain business rules in a collaborative manner. BRFplus Workbench supports features such as rule modeling, decision table management, function calls, and simulation capabilities. Figure 8.2 shows a decision table being modeled from BRFplus Workbench. Once the business rules are defined and maintained in BRFplus Workbench, they can be integrated with ABAP applications using BRFplus application programming interfaces (APIs) in the form of ABAP *function modules*.

BRFplus will be used by ABAP developers to incorporate the defined business rules into their ABAP programs, making it easier to manage and maintain complex business logic.

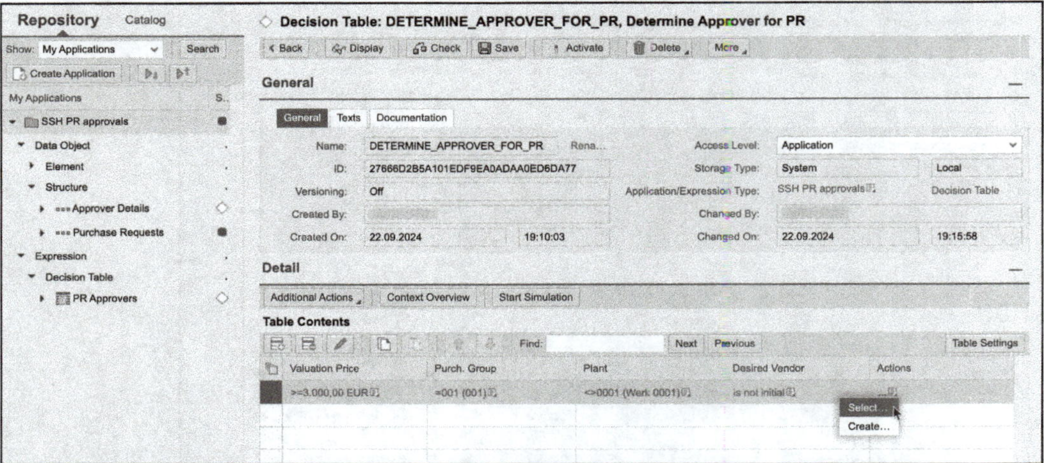

Figure 8.2 Maintaining a Decision Table in the BRFplus Workbench

8.1.3 SAP HANA Rules Framework

The SAP HANA rules framework is a rule engine that allows businesses to automate decision-making processes within their SAP HANA database. The rules are defined on the database artifacts, such as Tables, Views, and so on, within the SAP HANA Studio application (which needs separate installation). On deployment, the rules are converted to SQL procedures in the SAP HANA database. These SQL procedures can then be executed within the SAP HANA database, allowing for real-time decision-making based on the most up-to-date data. Applications that are natively built on SAP HANA such as SAP Business Integrity Screening, SAP Operational Process Intelligence, SAP Customer Relationship Management, SAP Transportation Resource Planning, and so on. They use the SAP HANA rules framework for efficient rule execution on a huge amount of data in the SAP HANA DB. The SAP HANA rules framework is now deprecated and is out of maintenance starting Q1 of 2024. Customers need to carry out a manual migration or remodeling from SAP HANA rules framework to SAP Build Process Automation.

8.1.4 Business Rules and Decisions in the Cloud

When SAP first launched its cloud platform in 2012 (called SAP Cloud Platform Neo at the time), more and more cloud-native applications required a rule engine native to the cloud. It required a rule engine that is easily accessible, easily maintainable, and a cloud-based tooling with robust security measures in place. At the same time, SAP had multiple rule engines and wanted to unify them. Therefore, SAP Cloud Platform Business Rules service was launched in SAP Cloud Platform Neo (previously known as SAP HANA Cloud Platform). This was followed by the business rules capability in SAP Workflow Management and finally the latest and the greatest of all decisions capability in SAP Build Process Automation. Now, let's look at the different cloud rule engines and their current status.

SAP Cloud Platform Business Rules Service in Neo

In the summer of 2017, SAP Cloud Platform Business Rules was launched for Neo trial and factory accounts, marking the initial step in the journey to unify the modeling experience of various rules technologies at SAP. The main objective was to provide customers and partners with access to the top features of different enterprise business rules technologies such as BRFplus, SAP HANA rules framework, and SAP Business Rules Management within a unified framework. This included a simplified SAP Fiori–based rule authoring tool and a central rule repository on one side, and deployment options to different runtimes on the other.

With a rule engine deployment in the cloud, customers were able to use it from any application via the REST APIs, which was much simpler than the existing means of communication: from a workflow deployed in the cloud, routing decisions for system-centric integrations, abstracting business logic for Internet of Things (IoT) applications, and so on. Figure 8.3 shows a decision table modeled in SAP Cloud Platform Business Rules.

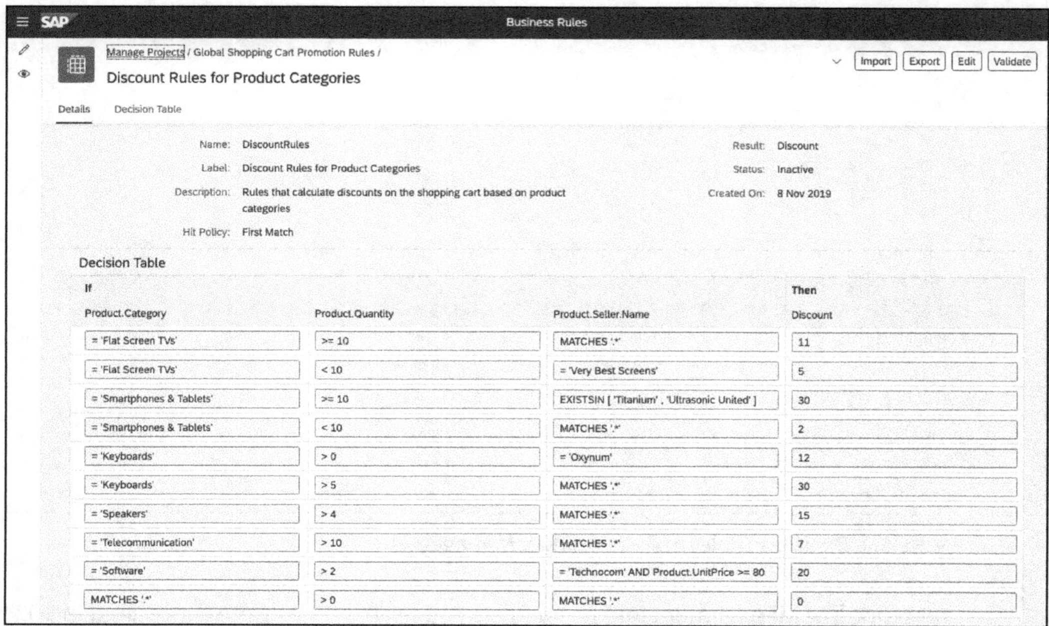

Product.Category	Product.Quantity	Product.Seller.Name	Discount
= 'Flat Screen TVs'	>= 10	MATCHES '.*'	11
= 'Flat Screen TVs'	< 10	= 'Very Best Screens'	5
= 'Smartphones & Tablets'	>= 10	EXISTSIN ['Titanium' , 'Ultrasonic United']	30
= 'Smartphones & Tablets'	< 10	MATCHES '.*'	2
= 'Keyboards'	> 0	= 'Oxynum'	12
= 'Keyboards'	> 5	MATCHES '.*'	30
= 'Speakers'	> 4	MATCHES '.*'	15
= 'Telecommunication'	> 10	MATCHES '.*'	7
= 'Software'	> 2	= 'Technocom' AND Product.UnitPrice >= 80	20
MATCHES '.*'	> 0	MATCHES '.*'	0

Figure 8.3 Decision Table Modeled in SAP Cloud Platform Business Rules

The SAP Cloud Platform Business Rules service is deprecated, and SAP Build Process Automation in SAP BTP is the successor.

Business Rules in SAP Workflow Management in SAP Business Technology Platform

SAP decided to move to the Cloud Foundry platform for its cloud, which offered a more open and flexible platform for SAP to build, deploy, and manage applications. This also aligned with SAP's commitment to openness and interoperability, allowing customers to have more choices in terms of the technologies and cloud providers they can use. Therefore, SAP eventually deprecated the services in the Neo platform for SAP BTP in Cloud Foundry.

SAP also understood that customers had to subscribe to multiple services for the business automation. Therefore, in 2018, it launched a unified service called SAP Workflow Management that included workflow capability, business rules capability, process visibility, and process flexibility capabilities for an end-to-end business process management solution. This integration aimed to streamline business automation and enabled companies to design, automate, and monitor processes in one place, rather than using multiple, disjointed services. Customers could export the rule projects from the Neo platform and import the same into the new service, which reduced the load in moving to the new service. Figure 8.4 shows a decision table in SAP Workflow Management.

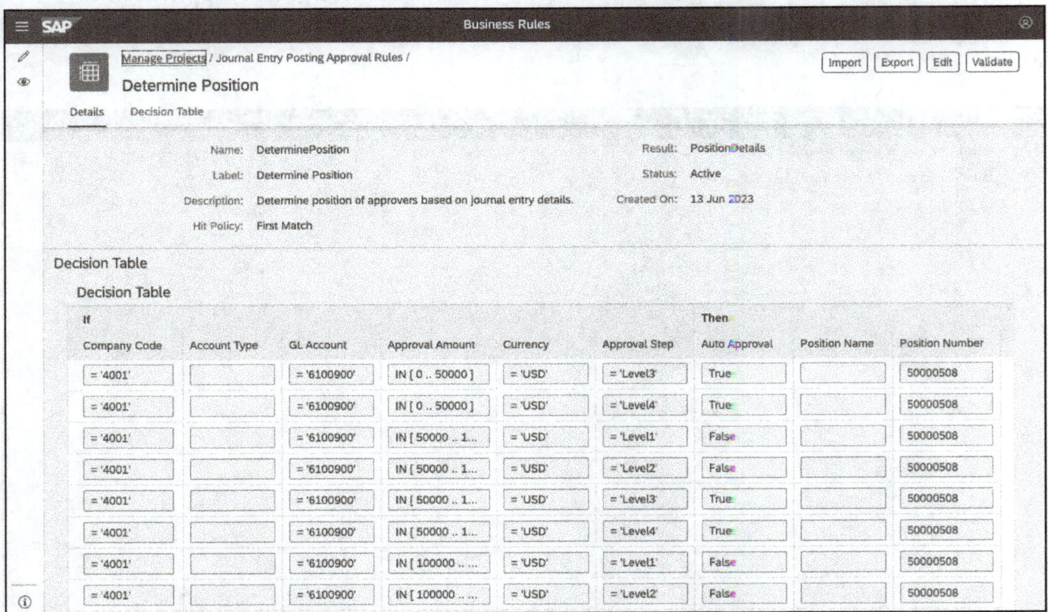

Figure 8.4 Decision Table in SAP Workflow Management

SAP Workflow Management and its business rules capabilities were deprecated in 2023 and is succeeded by the decisions capability within SAP Build Process Automation.

Decisions in SAP Build Process Automation

In 2023, SAP decided to unify the different automation services—SAP Workflow Management and SAP Intelligent Robotic Process Automation (SAP Intelligent RPA)—together to provide a true process automation platform. SAP Build Process Automation combines the capabilities of SAP Workflow Management for workflow and decision automation and SAP Intelligent RPA for task automation. The low-code tooling in SAP Build Process Automation allows nontechnical users such as business analysts, process owners, and so on to create, modify, and automate processes and business rules. This is critical for speeding up process innovation and enables companies to rapidly adapt to changing business conditions.

The *decisions* capability of SAP Build Process Automation is a direct replacement for all business rules in the SAP's cloud environment. It's an enhanced framework with a simpler UI and concepts to create and maintain the rules. It unifies the terms and lifecycle for all artifacts within SAP Build Process Automation such as process, automations, data types, and so on. Figure 8.5 shows a decision table in SAP Build Process Automation along with a *decision diagram* that visualizes the decision.

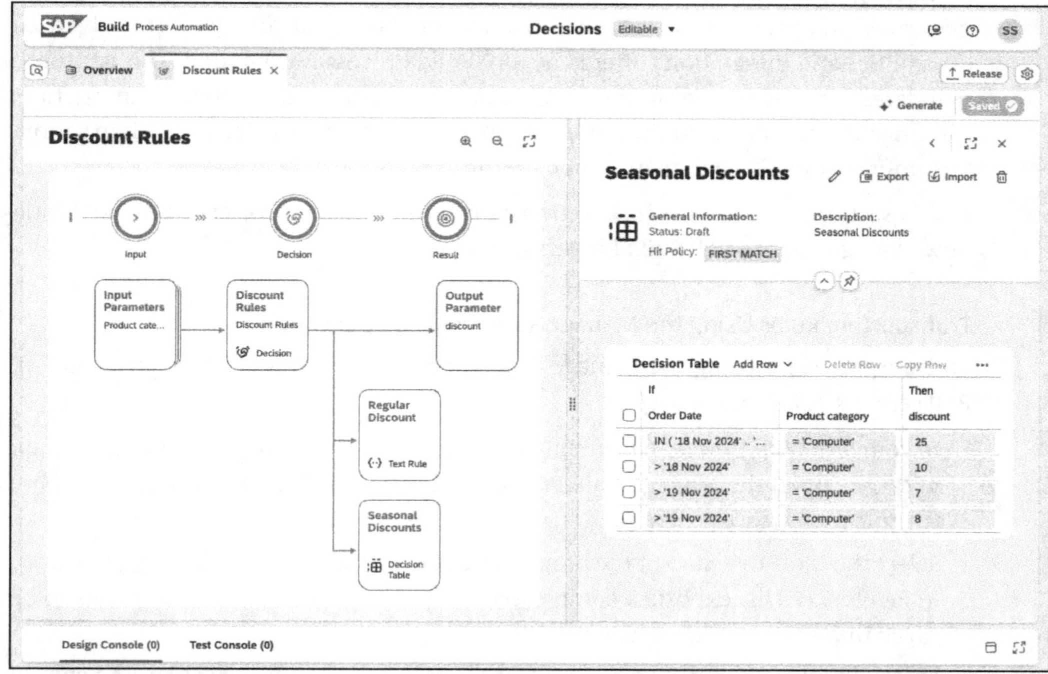

Figure 8.5 Decision Table in SAP Build Process Automation

8.1.5 Transition to SAP Build Process Automation

SAP believes in a future of process automation with SAP Build Process Automation, which includes the decisions or business rules capabilities. SAP is investing in its development to bring even more powerful features and capabilities. While continuing to support the legacy solutions, SAP encourages you to explore the advantages of transitioning to SAP Build Process Automation when the time is right for your organization. Especially when you're using the deprecated SAP BTP services such as SAP Workflow Management or SAP Cloud Platform Business Rules service, it's essential to move to SAP Build Process Automation to benefit from the future-proofed process automation solution.

When migrating from SAP Workflow Management or SAP Cloud Platform Business Rules service to SAP Build Process Automation, organizations have two options: a short-term migration process or a long-term migration process. In the short-term migration process, the goal is to quickly move to SAP Build Process Automation while preserving most of the existing investments in business rules, such as data objects, rule services, rules, and applications that call the rules. This approach is simpler and allows organizations to transition to the new platform without disrupting their existing business processes. The Manage Rule Project application is enabled in SAP Build Process Automation for customers moving from these deprecated SAP BTP services.

In the long-term migration process, the goal is to fully migrate and modernize the business rules and processes within the organization, taking advantage of the advanced capabilities and integrations offered by SAP Build Process Automation. This approach mandates remodeling the *business rules* as *decisions* in SAP Build Process Automation. Although the approach requires more planning and investment, it can result in a more streamlined and efficient environment in the long run.

In this section, we'll discuss how to transport business rules projects from SAP Workflow Management to SAP Build Process Automation.

Transporting Rules Using the Manage Rule Projects Application

To transport rules using the Manage Rule Projects application from deprecated SAP BTP services, follow these steps:

1. In your SAP Workflow Management or SAP Business Rules Management services application, open the Manage Rule Project application. This is the application where the rule projects were created.
2. Select the **DRAFT** version of the project, and choose the download icon ⤓, as shown in Figure 8.6. This exports a compressed file (with extension .zip) that needs to be saved into your local file system.
3. In SAP Build Process Automation, go to the **Monitor** page of the **Lobby**, and choose the **Business Rules** tile under the **Manage** section, as shown in Figure 8.7. This opens a version of the Manage Rule Project available natively within SAP Build Process Automation. The first time you open the application in SAP Build Process Automation, it should be empty. The Manage Rule Project application in SAP Build Process Automation contains all the capabilities except for creating a new rule project. You

can use the decisions capabilities for new requirements and new projects. However, within an existing rule project, you can create new data objects, new rule services, new rule sets, and even new rules.

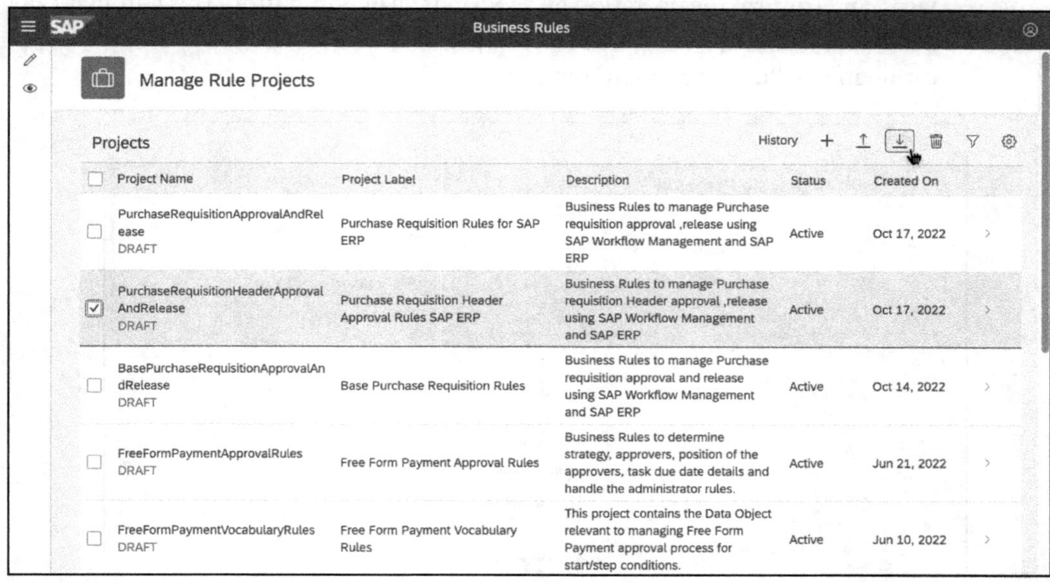

Figure 8.6 Downloading the Business Rule Project to the Local File System

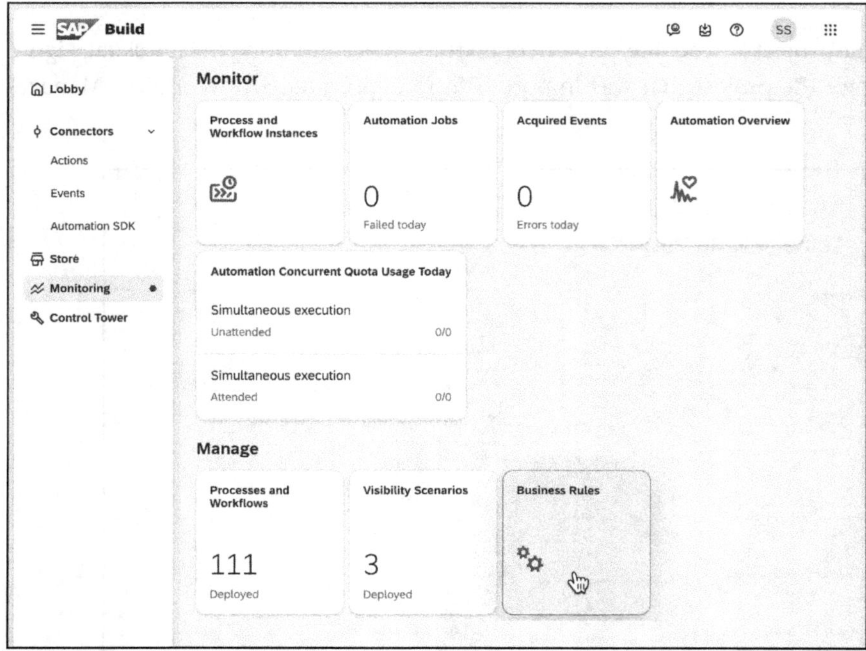

Figure 8.7 Manage Rule Project Application Access from the SAP Build Lobby

4. In the Manage Rule Project application in SAP Build Process Automation, choose the import icon ↑, and select the **Upload Project from File System** option, as shown in Figure 8.8. Select the .zip file that was downloaded from the deprecated SAP BTP services, and confirm the operation by choosing the **Import** button. This automatically adjusts the business rules project and imports it into the Manage Rule Project application in SAP Build Process Automation.

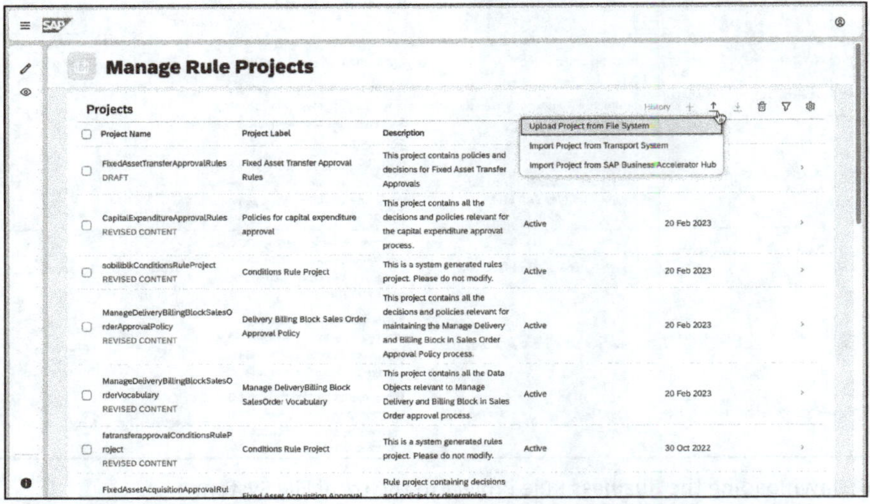

Figure 8.8 Uploading the Business Rule Project into SAP Build Process Automation

5. Once the project is available, you can open the project and work on it as usual. Figure 8.9 shows the project that was imported from deprecated SAP Workflow Management.

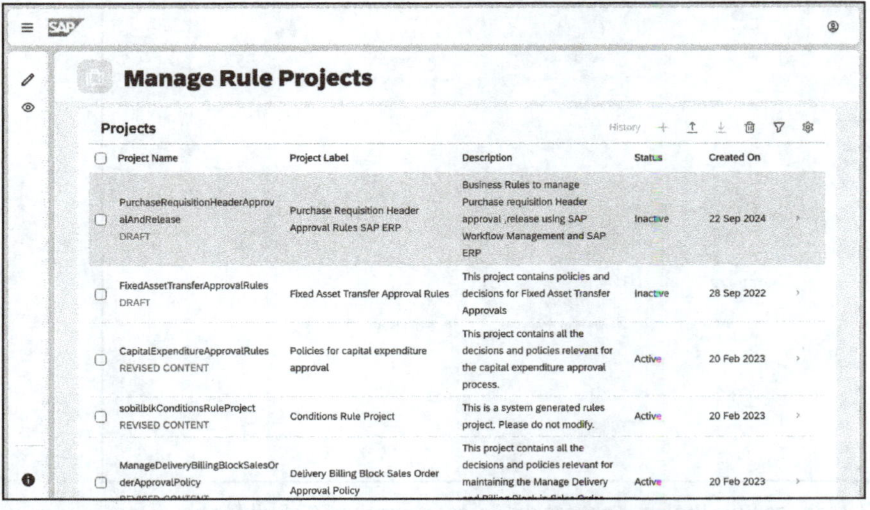

Figure 8.9 Business Rule Project Imported from the Deprecated SAP BTP Service

Business Rules versus Decisions in SAP Build Process Automation

As of today, SAP Build Process Automation contains two different worlds of rules: decisions and business rules. A *decision* is a native artifact of SAP Build Process Automation, which is integrated into its low-code framework. It can be created in the same project as other artifacts such as processes, automations, visibility, and so on, so it can be deployed and transported together with those artifacts.

A *business rule*, on the other hand, is a capability carried forward from SAP Workflow Management and made to work in SAP Build Process Automation for customers migrating to the new service. The framework and lifecycle are unique to it and independent of any other artifacts in SAP Build Process Automation. For example, business rules and decisions have different versioning concepts, different transportation concepts, and so on. The difference is so stark that the repository APIs that are published for business rules won't work for decisions. Decisions can be deployed into environments, but business rules can't.

Although they both help in automating decision logic, they are different artifacts and available in SAP Build Process Automation for different purposes. Table 8.1 compares the different aspects of decisions and business rules.

Feature	Decision	Business Rules
Editor	Decisions are created as part of a process automation project in the lobby.	Business rules are created from the Manage Rule Projects application.
Creation of decision	Creating a decision is simpler, you just define inputs and outputs and create a rule.	Creating a business rule involves creation of the rule, rule set, and rule service, apart from the data objects.
Project	Decisions are created as part of a process automation project, which can contain other artifacts such as processes, automations, visibility scenarios, data types, forms, and so on.	Business rules are created with a rule project, which contains data objects, rule services, rule sets, and rules.
Lifecycle	Decisions follow the SAP Build lifecycle of project release and deployment, along with other artifacts in the project.	Business rules have their own lifecycle and versioning that is applicable only to the rule projects.

Table 8.1 Decisions versus Business Rules in SAP Build Process Automation

Feature	Decision	Business Rules
Transportation	A process automation project can be transported using SAP Cloud Transport Management.	Business rules in SAP Build Process Automation can be transported manually by export and import.
Environments	Decisions will be deployed only in environments.	Business rules aren't deployed in any environment.
API trigger	To execute a decision from an external application, an API trigger should be created.	Because business rules are environment agnostic, they don't need any API triggers.
Deployment supported	Decisions are generally deployed to cloud runtimes only.	Business rules can be deployed into Cloud, ABAP or HANA runtime.
Consumption from process	It's possible to discover available decisions and add them to a process.	Business rules are nondiscoverable in a process. They need to be called as an action using the REST-based execution APIs of business rules.

Table 8.1 Decisions versus Business Rules in SAP Build Process Automation (Cont.)

8.2 Types of Decision Deployment

Decision and business rules have evolved over the years in SAP. With each platform, the business rules technology also evolved with time. We now know that the rule engine is available in ABAP as BRFplus, in SAP HANA as the SAP HANA rules framework, in Java as SAP Business Process Management, and in the cloud as SAP Build Process Automation. With more and more organizations having a hybrid landscape with business solutions spanning on-premises, cloud, SAP, and non-SAP systems, there was an increasing need to have a central business rules management solution where business users can manage rules from one single platform.

Central decision management is therefore one of the important aspects of the decision artifact in the SAP Build Process Automation service. With this functionality, decisions once modeled can be deployed in SAP BTP or SAP HANA, and then consumed in respective applications in these platforms.

In this section, we'll cover the different runtimes where a decision artifact can be deployed.

8.2.1 Cloud

When decisions are modeled and deployed in SAP Build Process Automation, the rules are deployed to the cloud runtime. The deployment involves publishing the decision model from the design phase to a runtime environment from where it can be executed. The most common runtime environment for decisions in SAP Build Process Automation is the cloud runtime. When a business process reaches a point where a decision needs to be executed, the system calls the deployed decisions model in the cloud, which is exposed via REST APIs. Figure 8.10 shows the Determine Authorized Approvers decision artifact within a project named Invoice Approval that has been released and deployed multiple times.

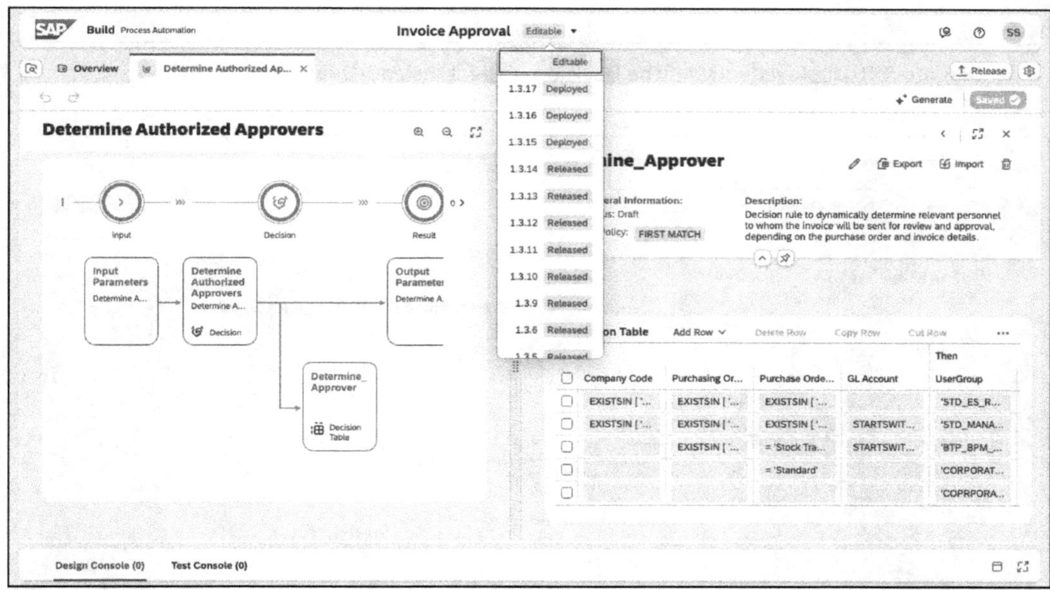

Figure 8.10 Decision Artifact after Being Deployed: Showing All Released and Deployed Versions

To view the content of any deployed decision, you can click on the respective version and then navigate into the decision artifact. The view will open in read-only mode, as shown in Figure 8.11, where you can see the actual rules that got deployed and get information about decision metadata such as the decision ID and version (see Figure 8.12).

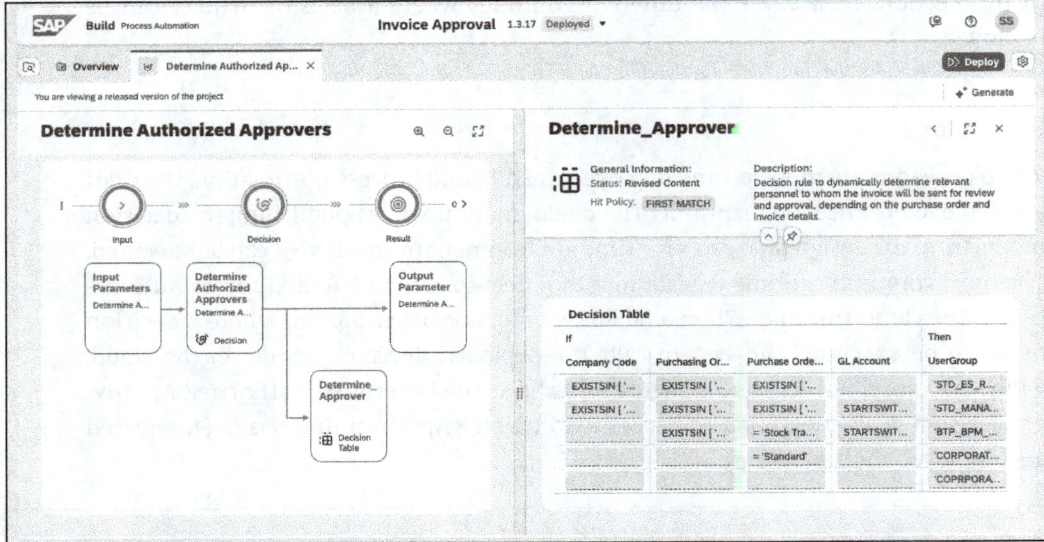

Figure 8.11 Deployed View of the Decision Artifact in Read-Only Mode

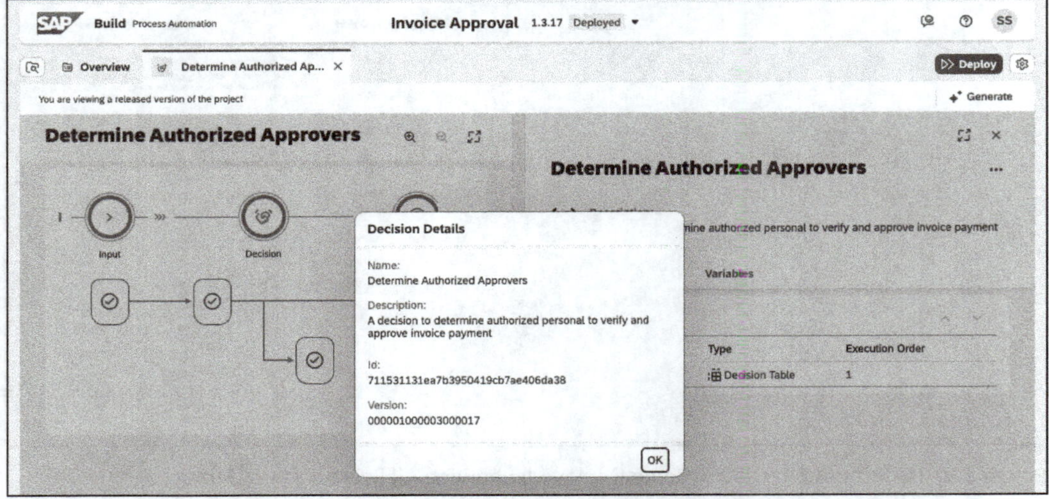

Figure 8.12 Decision Runtime Metadata Information That Can Be Used during API Execution

After deployment, the business rules or decisions can be consumed using a REST-based API. The API to invoke a rule service has been documented in SAP Business Accelerator Hub, as shown in Figure 8.13. The power of a decision is the openness that it brings with the REST-based API. Once deployed, it can be consumed from any application in on-premise, cloud, SAP, or non-SAP systems. Following the lifecycle of the business project, decisions can be versioned any number of times. Each decision version can be deployed in SAP BTP. It's possible to consume any specific version of a decision via the

API. If no version details are provided while triggering the API, then the latest version will be called during runtime execution.

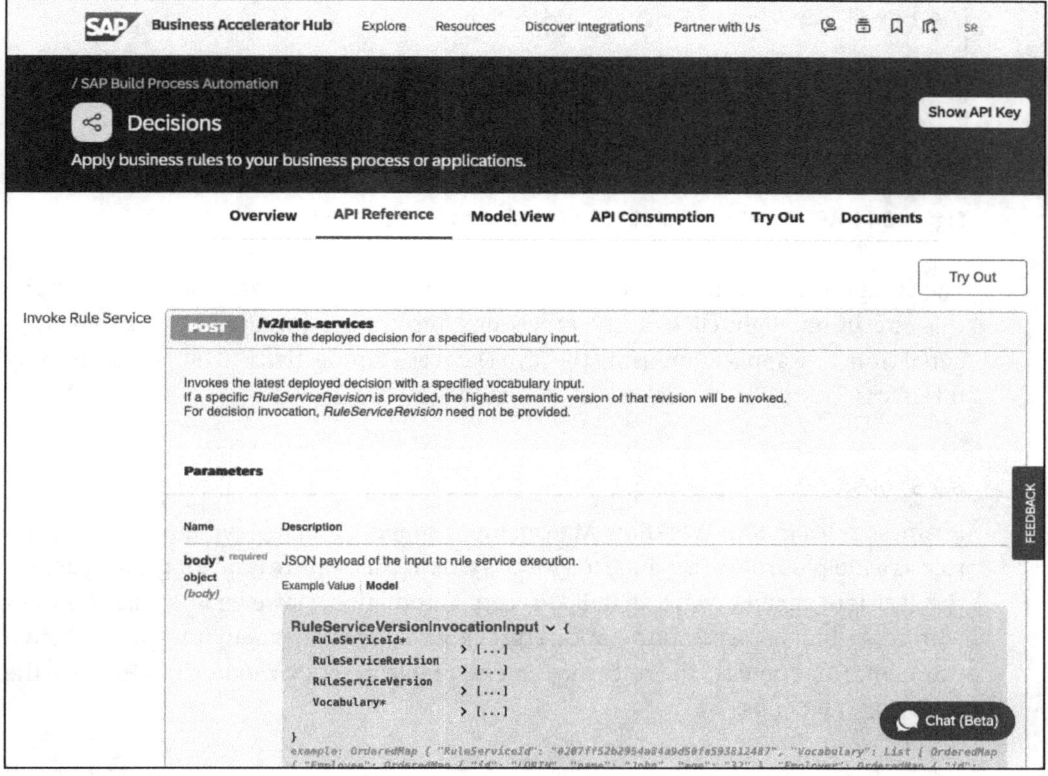

Figure 8.13 API Documentation to Invoke a Decision

8.2.2 SAP HANA

There are various styles of contemplating decisions. Analytical decision-making is one such form where important business decisions are made based on the large set of historical data. SAP HANA is an in-memory data platform supporting both analytical and operational processing, including transactional support—and mixes of these as well—to meet a wide range of requirements. Besides the database capabilities, SAP HANA has many extra features, such as predictive analytics, text search and mining, and so on. SAP HANA is now the database of choice for most of the SAP lines of business (LoB) solutions. To support deployment of decisions into SAP HANA, SAP Build Process Automation supports creation of SQL scripts based on the decisions modeled. Because this isn't a common scenario, the capability can only be accessed via APIs.

Figure 8.14 shows an API called from a REST client to generate an SQL script that can then be run in SAP HANA to create a SQL procedure with defined inputs and outputs.

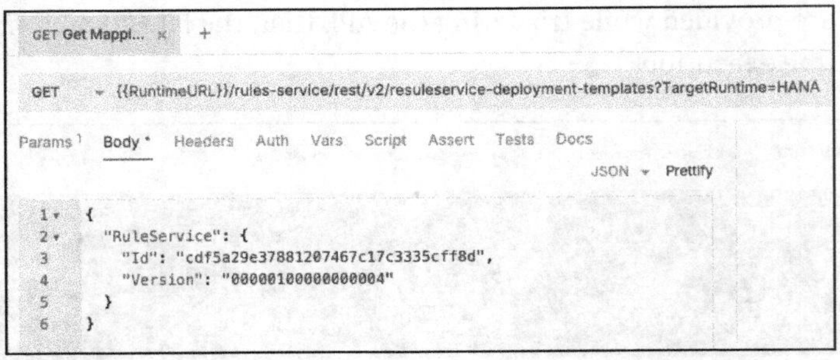

GET Get Mappi... × +

GET ▾ {{RuntimeURL}}/rules-service/rest/v2/resuleservice-deployment-templates?TargetRuntime=HANA

Params[1] Body • Headers Auth Vars Script Assert Tests Docs

JSON ▾ Prettify

```
{
  "RuleService": {
    "Id": "cdf5a29e37881207467c17c3335cff8d",
    "Version": "000001000000000004"
  }
}
```

Figure 8.14 API to Get the Output of a Decision as an SQL Script

The SQL procedure can be executed directly in an SAP HANA database to perform the rule execution. Right-click on the procedure, and choose **Execute**. It can also be triggered from any application using the Java Database Connectivity (JDBC) programming interfaces.

8.2.3 ABAP

Business rules in SAP Workflow Management support the deployment of a business rule as a BRFplus rule in a connected ABAP system. This feature is no longer available in the decisions capability of SAP Build Process Automation. However, the business rules within SAP Build Process Automation still supports the deployment into an ABAP environment. This converts the rules into an executable function module in ABAP like the behavior of BRFplus.

8.3 Decision Models

To work with the decision capability in SAP Build Process Automation, it's important to understand different constructs of a decision artifact. These constructs are the fundamental building blocks used when modeling your decision logic in SAP Build Process Automation. Each construct plays a distinct role in shaping the decision-making process. In this section, we'll discuss the different elements that constitute a decision model.

8.3.1 Decision

In SAP Build Process Automation, a decision artifact represents one or more business rules that are used to automate and streamline decision-making processes within a process or application. It acts as an interface to the different rules and policies that exist for a given purpose.

Consider a scenario involving the approval of a purchase request. There could be multiple rules or policies defined by the business to determine an agent who can approve a purchase request: agent determination based on purchasing organization, material group, or the total net amount, among others. Each of these business rules might need different inputs and can be complex. The decision artifact abstracts this complexity from the applications or processes looking for an approval agent. When an application needs to identify an agent to approve a purchase request, it need not be concerned with the policies defined by the organization. Instead, it interfaces with the decision artifact that handles the execution of the different rules and policies to find the right agent for the given input.

8.3.2 Vocabulary

The foundation of any decision lies in the data that it processes. A set of data is passed as an input to the decision artifact on which one or more rules are applied. The result at the end of the rule execution is returned by the decision artifact as an output. Therefore, the decision artifact requires clear definitions of both input and output parameters, which is also called the *vocabulary* of the decision, during the development or modeling phase.

In SAP Build Process Automation, a *data type* artifact can be created in the process automation project to represent a complex business object. Figure 8.15 shows a **Catalog** business object in SAP Build Process Automation, which is represented as a data type artifact in a **Process Automation** project. It has been created with simple data types such as **String**, **Date**, **Number**, and so on.

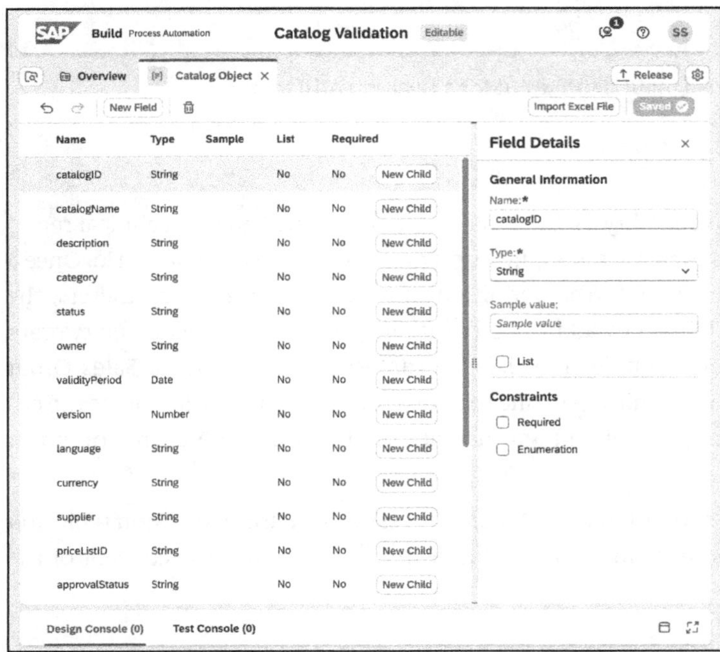

Figure 8.15 Data Types for Representing a Catalog Object in SAP Build Process Automation

Once a data type is created in the project, any artifact, including decisions, can use the data type to represent an object. The input and output parameters of a decision can be defined based on primitive data types such as String, Number, Date, Boolean, and so on, or they can be represented by the data type artifact. Figure 8.16 shows a decision artifact in the **Process Automation** project, where two input parameters are defined: **Catalog** and **Requester Role**. While the **Catalog** is based on a data type artifact, the **Requester Role** is based on a primitive data type **String**.

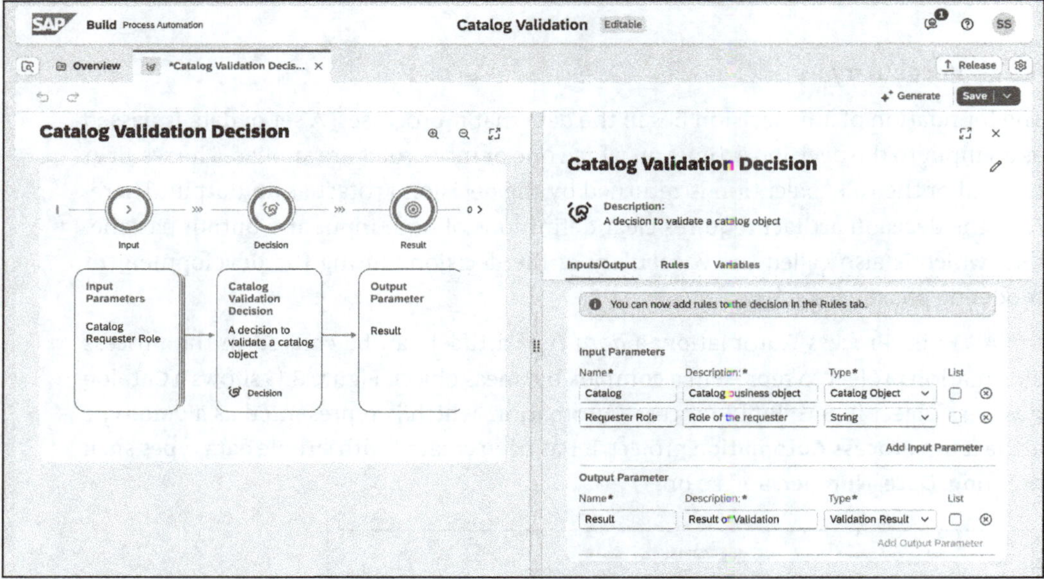

Figure 8.16 A Well-Defined Input and Output for a Decision Artifact

8.3.3 Rules

A rule in the decisions capability of SAP Build Process Automation is a technical representation of a business logic that is applied on a process or a business scenario. Once a decision artifact is created with the definition of input and output parameters, the business logic can be defined by adding rules into it. A decision artifact could contain one or more rules. For example, a decision to determine agents for a Sales Order approval could contain two rules: one rule to determine if the sales order meets the criteria for an automatic approval and another rule to determine the agents only if it needs manual review.

Two types of rules are supported in SAP Build Process Automation: decision table and text rule. We'll discuss these in the following sections, along with the concept of the reusable rule.

Decision Table

A *decision table* is a tool used to map out various conditions and their corresponding actions within a business process. Decision tables allow for the collection of input and output rule expressions in a tabular representation, enabling the configuration of multiple rule conditions and results. This simplifies complex business logic and ensures consistency in decision-making across different scenarios.

Decision tables are designed with a spreadsheet-like interface that makes it easier to understand and use them. This user-friendly interface allows for easier onboarding into the tool, especially for a business user. Therefore, it's a go-to tool for exposing the business logic, which can be directly maintained or modified by a nondeveloper.

Figure 8.17 shows a snapshot of a decision table used for catalog validation. (We'll dive further into navigation and using decisions in Section 8.4.) In every decision, there are two distinct parts—the **If** section and **Then** section. In the **If** section, you define the different conditions that need to be evaluated, which are usually based on the input parameters. These conditions are typically evaluated in the form of "If [condition] is true", for example, Catalog ID = '1001'.

The **Then** section contains the values that will be assigned if the conditions in the **If** section are met. They are usually output parameters to which the values need to be assigned.

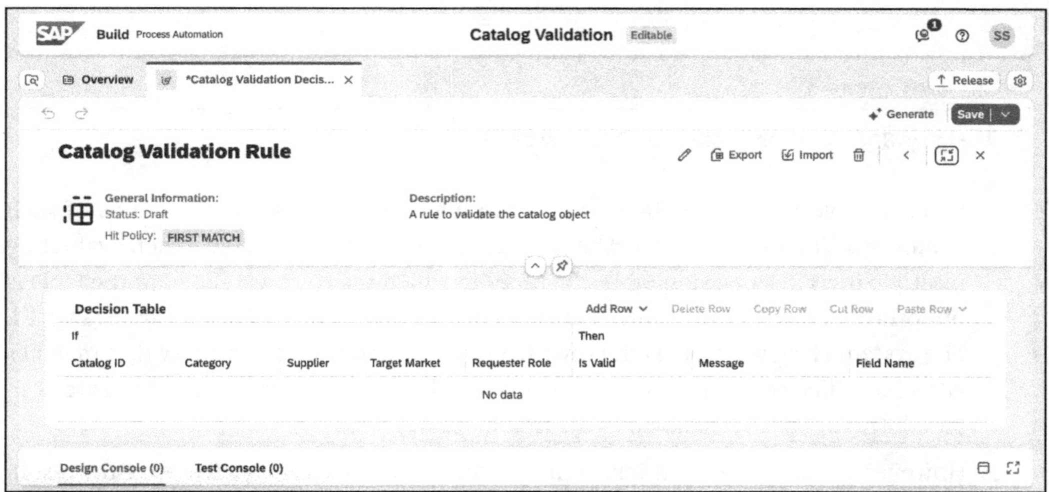

Figure 8.17 Decision Table with If and Then Sections

If there are multiple such conditions that need to be evaluated, then you can add multiple parameters in the If section. When there are multiple conditions in the **If** section, then they are evaluated using the AND logical operator. In a text rule called **Catalog Validation Rule**, shown in Figure 8.18, there are multiple conditions in the **If** section. They are evaluated as shown in Listing 8.1 for the first row.

```
IF
Catalog ID = '1001' AND Category = 'Electronics' AND Supplier = 'Supplier A'
AND Target Market = 'Global' AND Requester Role = 'CatalogAdmin'
THEN
Is Valid = 'TRUE' AND Message = 'Valid Catalog' AND Field Name = ''
```

Listing 8.1 Catalog Validation Rule Condition Example

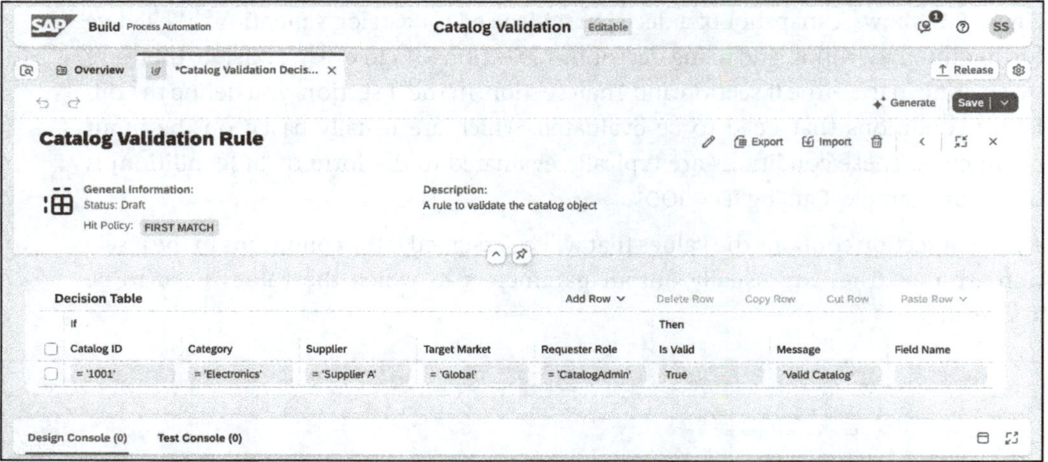

Figure 8.18 Decision Table with One Row of a Rule

In almost every use case, there's a need to maintain multiple combinations of such conditions. Therefore, there will be multiple rows in a decision table. When evaluating multiple rows in a decision table, the order in which the rows are added to the table is important because that dictates the sequence in which the conditions are evaluated. The system always evaluates the rows from top to bottom. The first row that matches all the specified conditions is considered a match or "hit", and no further rows are evaluated. This approach is known as the first hit policy.

However, in some cases, it might be important to evaluate all rows to derive the result. For example, in a validation rule, you would want to evaluate all the rows and identify all possible validation errors in one go rather than stopping the rule evaluation at the first validation error condition, as shown in Figure 8.19. In such situations, the system can be configured to follow the all-match policy. This means that all rows will be evaluated, and rows that match the defined conditions are considered a match. The first hit and all-match policies allow for flexibility in decision-making and can be tailored to suit specific business requirements within SAP Build Process Automation.

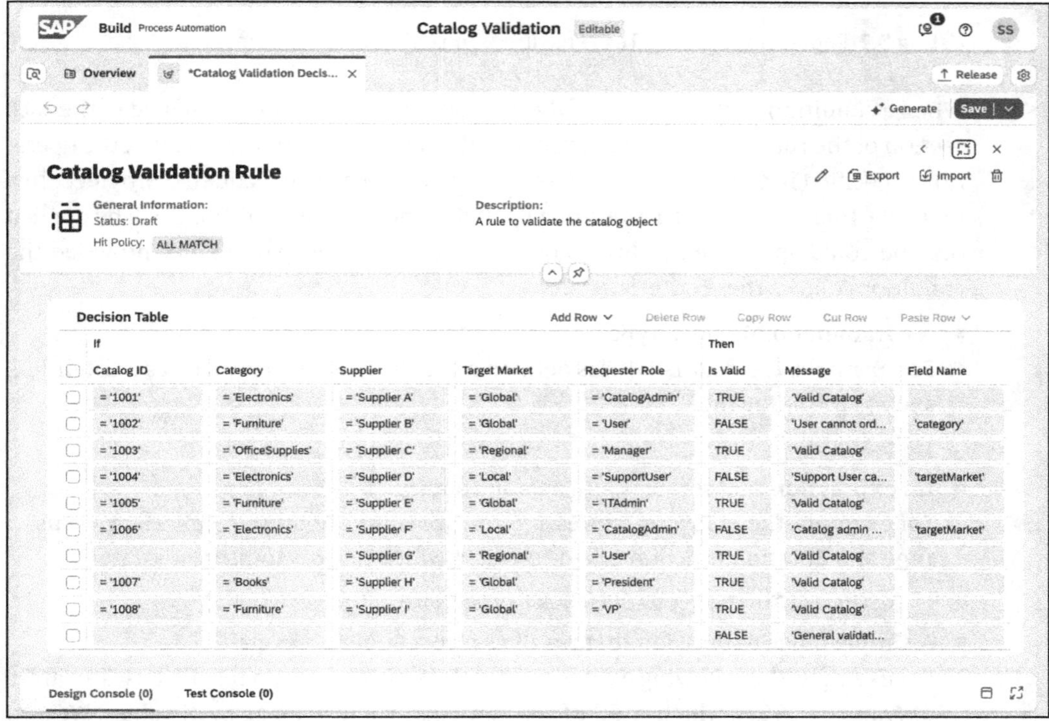

Figure 8.19 Decision Table with Multiple Rows and the ALL MATCH Hit Policy

Text Rule

A text rule in SAP Build Process Automation is a collection of rule expressions written in a simple if-then format. These rule expressions are used to define conditions that must be evaluated as either true or false. It has the structure shown in Listing 8.2.

```
IF
   [Condition]
THEN
   Operations
ELSE IF
   [Condition]
THEN
   Operations
ELSE IF
   [Condition]
THEN
   Operations
```

```
...
ELSE
   Operations
```

Listing 8.2 Representation of a Text Rule in a Simple If-Then-Else Format

When a condition evaluates to true, the corresponding operations specified in the `THEN` section of the rule expression are executed. If a condition evaluates to false, the operations specified in the `ELSE IF` or `ELSE` sections of the rule are executed, if provided. In a text rule, there are no restrictions on which parameters the conditions will be applied on. You could apply the conditions on any of the available parameters, provided the condition follows these criteria:

- **A parameter of Boolean type**
  For example, the field `isValid` is defined as Boolean and can be used as a condition, such as the following:
  `IF (isValid)`
- **An expression** that resolves into a Boolean result
  For example, the following expression resolves into a Boolean result. Here, we use a time and duration function called `TOMORROW`, which returns tomorrow's date:
  `IF (order.totalNetAmount > 100000 AND order.shipDate = TOMORROW())`
- **A simple** range expression **that resolves into a Boolean result**
  For example, the following expression resolves into a Boolean result. It uses an array operator `EXISTSIN`, which checks for a particular value in an array of values. We also use the `IN` operator, which checks for a value in a range of continuous values, such as 7 to 13. These operators return `true` is the value is present. If not, they return `false`:
  `IF (customer.country EXISTSIN['DE', 'IN', 'US', 'FR', 'NL'] AND customer.score NOTIN (7..13))`

Figure 8.20 shows a simple rule for additional validation for a catalog when you open a text rule with the decision artifact. You might be wondering why can't you use a decision table for the same purpose. Why does SAP Build Process Automation support two different types of rules to maintain the same conditions? The main difference between using a decision table and a text rule is the way the rules are represented and structured. While a text rule is written as a text-based expression, a decision table is a tabular representation of rules. Decision tables are often used to manage complex decision-making processes with multiple fixed variables, and they provide a clear and organized way to compare different conditions and results. In contrast, text rules are more flexible and can be written on any parameters—unlike decision tables.

To choose between the two types of rules, ask the following questions:

- **Who will modify the rules?**
  Business users generally find decision tables easier to work with. Decision tables also

support exporting and importing data to/from Microsoft Excel, which the text rules don't support.

- **Do you have a fixed set of manageable parameters for IF conditions?**
  Decision tables generally need a clear definition of which parameters are used in IF, which text rules do not. If you want to give flexibility to the person maintaining the rules to define any conditions of their choice, then go for text rules.
- **How many conditions are needed within your rule?**
  Because it's a tabular representation, a decision table is a more convenient tool when dealing with multiple conditions (rows). In a text rule, for each condition, an `Else If - Then` section needs to be created, which becomes cumbersome for more than 25 conditions.
- **What is the purpose of the rule?**
  If the purpose of the rule is for operations, such as to update or append values to existing input/output parameters or decision variables or to perform a loop operation on an array without assigning any new result, then using text rules is the only option.

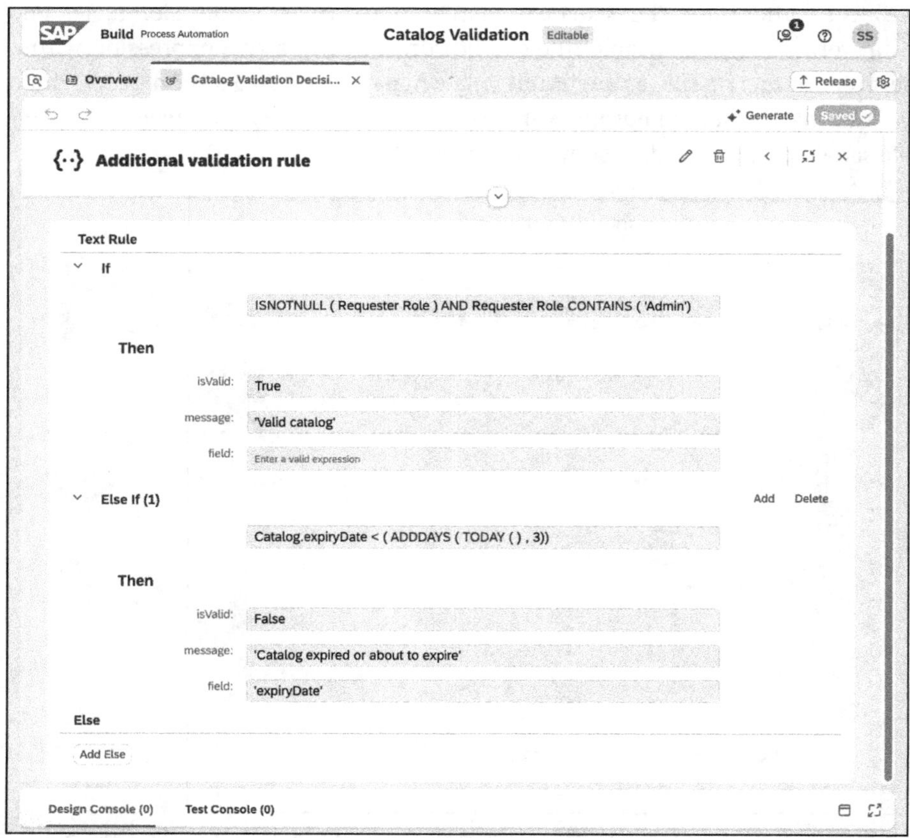

Figure 8.20 Simple Text Rule for Catalog Validation

> **Note**
>
> Decision tables and text rules can't be triggered as standalone objects. The rules can be triggered only in the context of a decision artifact to which one or more rules are assigned. A decision needs at least one rule to be considered a valid decision.

> **Tip**
>
> When there are multiple rules defined within a decision, then the default behavior is to execute them in the order in which they were shown in the **Rules** tab. The **Execution Order** column shows the order in which the rules will be executed. If you want to change the order of execution, then click the more icon ••• in each rule to move them up or down.

Reusable Rules

A decision table or a text rule can be marked as a reusable rule so that they can be accessed from other rules. They can be used in both text rules and decision tables, and the result returned by a reusable rule can be consumed in a rule expression within another rule. To mark a rule as a reusable rule while modeling a decision in SAP Build Process Automation, select **Reusable Rule** from the **Rule Category** dropdown in the **Create Rule** screen of the wizard, as shown in Figure 8.21.

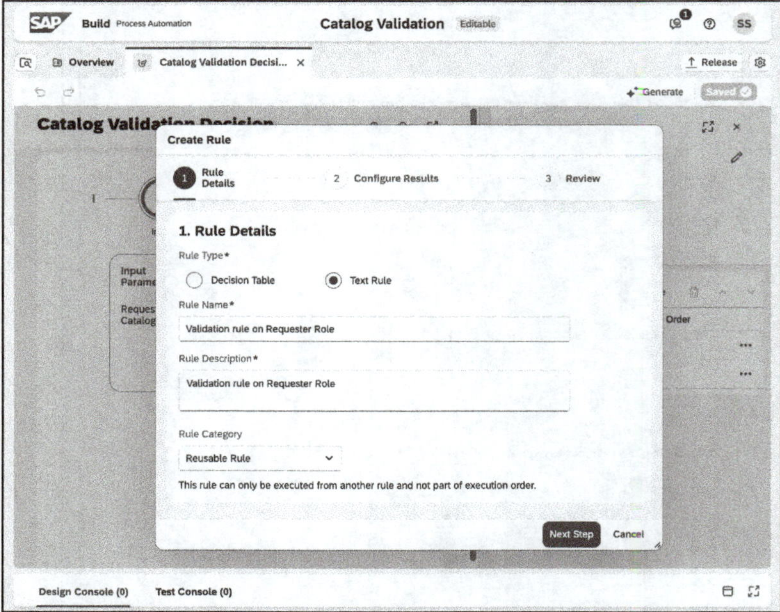

Figure 8.21 Marking a Decision Table or Text Rule as a Reusable Rule by Changing the Rule Category

Once a rule is marked as a reusable rule, it will be listed under the **Reusable Rules** area in the **Rules** tab, as shown in Figure 8.22, making it easy to identify and reuse in other rules within the decision. Because the reusable rules are executed on demand by the **Orchestrated Rules**, the execution order isn't applicable for them.

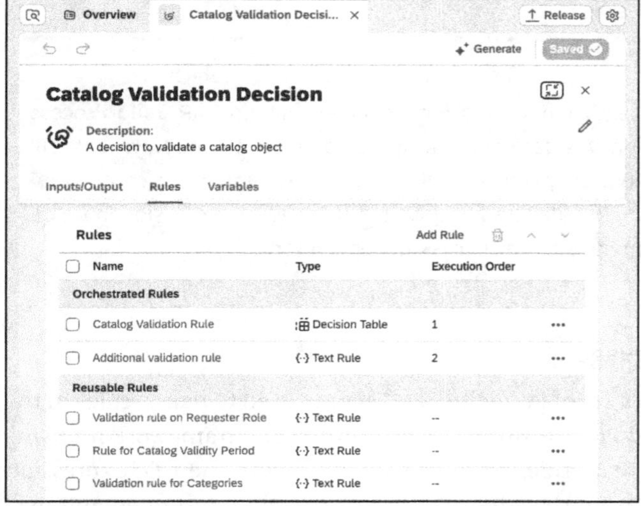

Figure 8.22 Orchestrated Rules and Reusable Rules within a Decision

To use the reusable rule from any other rule, including another reusable rule, simply use them as a function. Figure 8.23 shows three reusable rules used within a text rule when you navigate to **Catalog Validation Rule**.

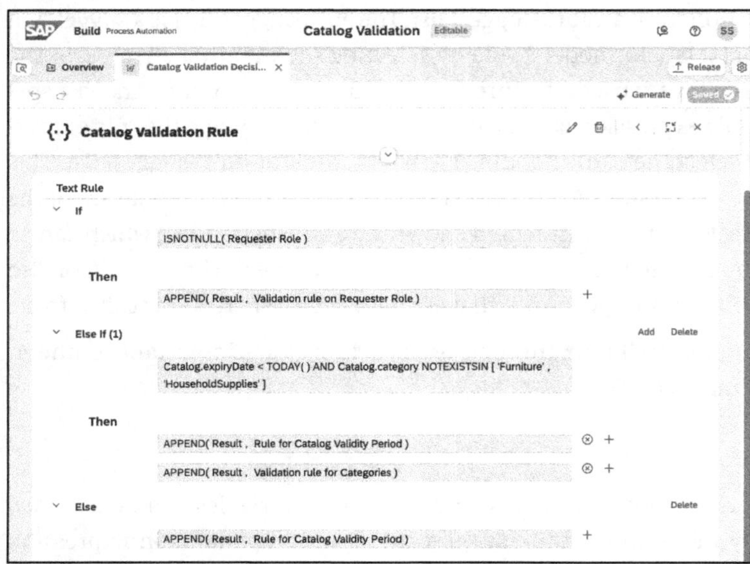

Figure 8.23 Reusable Rules Used within an Orchestrated Rule

Depending on certain conditions defined in the rule, the reusable rules are called as inputs to the **APPEND** function. The result provided by the reusable rules—in this case, one row of catalog validations—is appended into the **Result** output parameter of the decision. The important thing to note here is that the output parameter **Result** is defined as a **List**, so it can accommodate multiple validation results.

> **Warning**
>
> Watch out for cyclic dependency when creating reusable rules in SAP Build Process Automation. If cyclic dependency is detected during deployment, the deployment will fail. An example of cyclic dependency is using Rule1 as an expression in the same Rule1 or using Rule1 as an expression in Rule2 and then using Rule2 as an expression in Rule1. Be aware of this potential issue to ensure successful deployment.

8.3.4 Rule Expression Language

A rule expression language is a set of guidelines and syntax used to define rules in the decision artifacts of SAP Build Process Automation. It provides a framework for using functions and operators to create rule expressions. Adhering to the rule expression language is essential for ensuring that the rules are accurately interpreted and executed by SAP Build Process Automation. This is important as any deviation from the rule expression language guidelines can result in validation errors.

Expression Language 2.0 is based on Decision Model and Notation (DMN) and Friendly Enough Expression Language (FEEL). It's a syntax for defining rule conditions that reduces ambiguity and provides a way to model rule conditions though suggestions and vocabulary selection from an auto-suggest list. The auto-suggest list is a suggestion dropdown menu that lets you model a rule by selecting the required element of the rule expression. This is a very useful feature for a beginner who might not need a special training on the rule expression language syntax. You can also type the rule expression in the field and select the corresponding elements from the auto-suggest list. This mode of defining the expressions is preferred by advanced users who are aware of the rule expression language syntax. Figure 8.24 shows the auto-suggest list, which can be shown on demand by pressing `Ctrl` + `Space` from any of the **If**, **Then**, **Else If**, or **Else** sections and shows only those suggestion that are relevant to where it was called from.

In the following sections, we'll walk through constructs that are important to understand Expressions Language 2.0.

Vocabulary

The input parameters, output parameters, and temporary variables can be accessed from the **Vocabulary** menu in the auto-suggest list. Typical operands in Expression Language 2.0 are as follows: *Catalog.category = 'Electronics'*

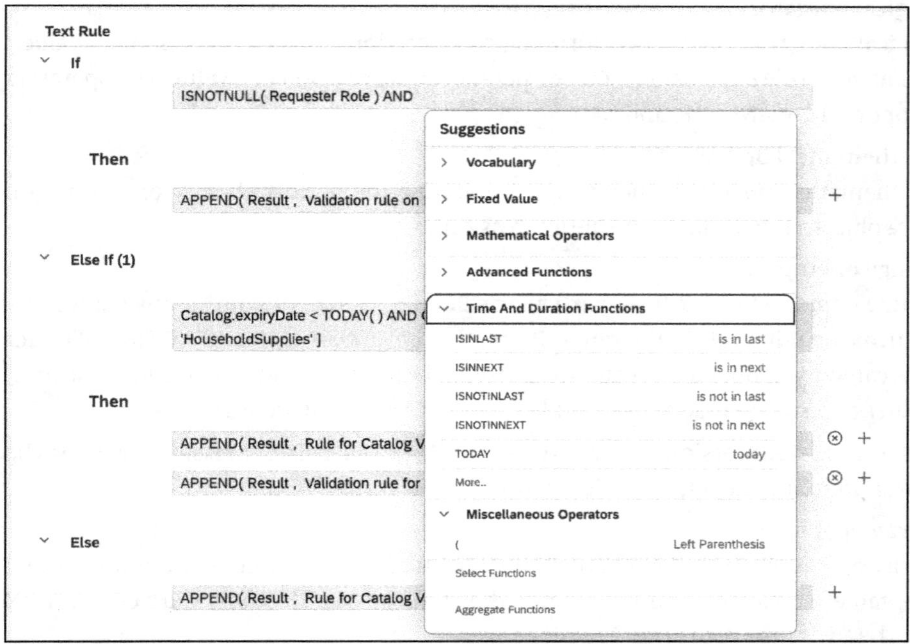

Figure 8.24 Auto-Suggest List Showing Relevant Functions and Vocabulary

Here, **category** is an attribute of the data object **Catalog**. If the **validity** data object is contained within the **Catalog** data object, then attribute **validFrom** can be accessed as **Catalog.validity.validFrom**.

> **Tip**
>
> The data object typically gets complex with multiple levels, making the operands long and unreadable. In a text rule or decision table, you can set human-readable **Labels** to represent them in the **If** section. For example, **Catalog.validity.validFrom** can be represented as **Valid From**.

Operators

As the name suggests, an *operator* is a relational symbol or function used to compare values and define conditions within rules. The operators are listed in the auto-suggest list only if they are relevant for the rule expression. Different kinds of operators are supported in the decisions capability of SAP Build Process Automation:

- **Comparison operators**
  Comparison operators compare fixed values, data objects, or rule expressions and return true or false. Some of the operators are = (is equal to), < (is less than), > (is greater than), != (isn't equal to), and so on.

- **Logical operators**
  Logical operations let you perform logical operations on data objects, fixed values, or between rule expressions. These operations return a Boolean value. The operators supported are AND, OR, and NOT.
- **Mathematical operators**
  Mathematical operators such as +, -, *, and / let you perform binary operations on data objects, fixed values, or rule expressions.
- **Range operators**
  Range operators check for a particular value in a range of continuous values and return true if it's within the range. If not, they return false. IN and NOTIN fall under this category of operators. You can use round brackets—(and)—to exclude the limit values and square brackets—[and]—to include the limit values.

  An example expression is customer.score IN [1..10). This expression includes the lower limit but excludes the higher limit.
- **Array operators**
  Array operators check for a particular value in an array of values and return true if the value is present in the array. If not, they return false. EXISTSIN and NOTEXISTSIN are the two supported array functions.

  An example expression is customer.country NOTEXISTSIN ['DE', 'FR', 'IN'].
- **Functional operators**
  Functional operators perform operations on strings or expressions that return a string value. A lot of functional operators are supported, such as MATCHES, CONTAINS, STARTSWITH, ENDSWITH, and so on.

  An example expression is customer.name MATCHES 'Dr\. Se.* '. In the regular expression 'Dr\. Se.*': "Dr\." matches the string "Dr." exactly. "Se.*" matches any string that starts with "Se" followed by zero or more characters.

Functions

A function is a predefined action or operation that can be performed as part of the expression. As always, functions are listed in the auto-suggest list only if they are relevant for the rule expression. Different kinds of operators are supported in the decisions capability of SAP Build Process Automation:

- **Time and date functions**
  Time and duration functions let you perform operations on date, time, and timestamp values. You can select the date and timestamp values from the date and time picker in the **Fixed Value** section of the auto-suggest menu. Numerous time and date functions are supported in SAP Build Process Automation. Some of those functions are shown in Table 8.2.

Function Name	Syntax	Example
Today	**TODAY()**	Catalog.expiryDate **> TODAY()**
Add days	**ADDDAYS**(<date or timestamp>, <number>)	ADDDAYS(Catalog.expiryDate, 5) **ADDDAYS('2024-12-02', 5)**
Subtract minutes	SUBTRACTMINUTES(<date or time-stamp>, <number>)	**SUBTRACTMINUTES(Employee.log-inTime, 15)**
Hours between	**HOURSBETWEEN**(<date or time-stamp>, <date or timestamp>)	**HOURSBETWEEN(Employee.login-time, Employee.logoutTime)**
Is in last	**ISINLAST**(<date or timestamp >, <key value>)	ISINLAST(startTime, '20 MIN') **ISINLAST(startTime, '2 WK')** **ISINLAST(startTime, '3 HR')**
Month	**MONTH**(<date or timestamp>)	MONTH ('2024-12-02T12:29:07.333Z')
Day of week	**DAYOFWEEK**(date or timestamp) Returns a value between 1 and 7 representing the day of the week with Monday having a value of 1.	**DAYOFWEEK('2024-12-02T12:29:07.333Z')**

Table 8.2 Date and Time Functions

- **Select functions**
  Select functions let you retrieve select values from a table. You can configure the select functions, by choosing the **Select Functions** option in the auto-suggest list, which opens the Configure Select Functions configuration window shown in Figure 8.25 and in Figure 8.26. SAP Build Process Automation supports selecting the number of **TOP** records or using the **SELECT** function to fetch certain attributes from an array based on the conditions.

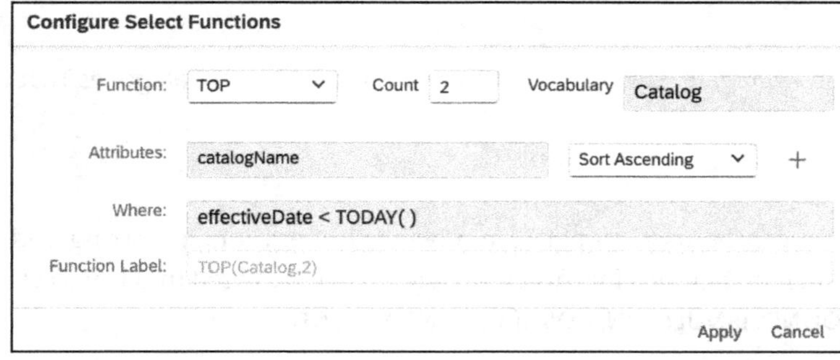

Figure 8.25 Configure Select Functions to Query the TOP 2 Values from an Array

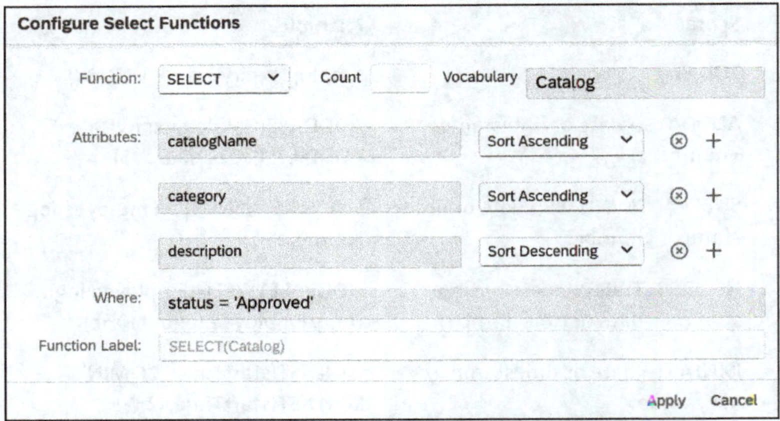

Figure 8.26 Configure Select Functions to SELECT Records from an Array

- **Aggregate functions**
  Aggregate functions let you perform calculations on a set of values and return a single result value or a set of result values. The functions **AVG**, **SUM**, COUNT, **DISTICNT**, **COUNTDISTICNT**, **MIN**, and **MAX** are supported in SAP Build Process Automation. Selecting **Aggregate Functions** in the auto-suggest list shows the **Configure Aggregate Functions** popup window where you can maintain the function based on your needs (see Figure 8.27).

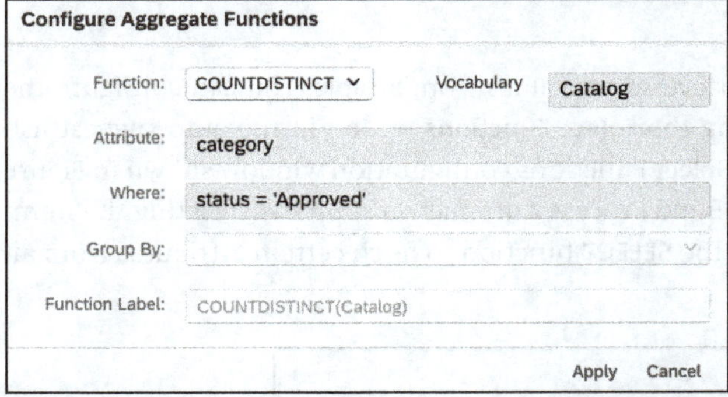

Figure 8.27 Configure an Aggregate Function to Get the Number of Distinct Categories That Are Approved

- **Advanced functions**
  This set of functions can be used to perform advanced operations on string and numerical values. There are quite a few functions under the advanced functions category, including **ROUND**, **ISNULL**, **SIN**, **COS**, and **CONCATENATE**.

Table 8.3 lists the functions that can be used while modeling a text rule only when they are modeled without a resulting data object. Figure 8.28 shows the wizard when you create a new rule, where the **Result Vocabulary** can be changed to **No Result (use execute operations)**.

Syntax	Example	Description
UPDATE(<target>, <source>)	UPDATE(output.catalog.status, input.catalog.status) **UPDATE(output.catalog.status, Reusable rule for validation)**	The attribute status under the data object output.catalog is updated with the attribute status under the data object **input.catalog**. Reusable rules can also be used as the source expression.
APPEND(<target>, <source>)	APPEND (Result, intResult) **APPEND (Result, Reusable rule for validation)**	The data object **Result** of type list/array is appended with the values from **intResult**. The target here must be an array/list, and the source can be a structure, list, or a reusable rule that can return a structure or list.
FOREACH(<vocabulary>)	**FOREACH(catalog)** **FOREACH(My Reusable rule that returns an array)**	This enables the capability for a rule to loop at the catalog and perform the **UPDATE** or **APPEND** functions for each line item.

Table 8.3 Functions That Support Execute Operations

Figure 8.28 Configuring a Text Rule without Resulting Vocabulary

8.3.5 Decision Diagram

Depending on the use case, decisions can get really complex with multiple inputs and rules. A *decision diagram* that is shown when you open a decision artifact can help in visualizing a decision such as a road map and provide the clarity that a developer needs. Figure 8.29 shows a snapshot of a decision diagram that visualizes the different input parameters such as **Requester Role** and **Catalog**, the decision, different rules such as the **Catalog Validation Rule**, and finally the **Output Parameter**. This clear representation makes it easy to understand how each element contributes to the outcome. The diagram is dynamic, and any changes made to the input, rules, and output are immediately reflected.

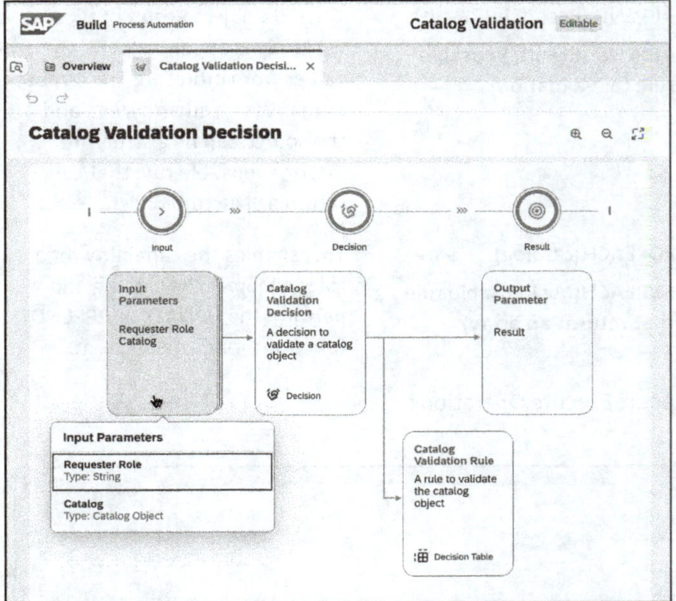

Figure 8.29 Decision Diagram Showing Input Parameters, Output Parameter, and Rules

Decision diagrams are interactive, allowing you to navigate into specific components. Clicking on **Input Parameters**, the decision, the rule (either the decision table or text rule), or the **Output Parameter** will take you directly to the respective editor for that component. This allows for easy navigation and modification of the respective components.

Finally, a decision diagram also gives a visual indicator of the individual component's validation state. Any error in the individual components is reflected in the color of the icon. A decision in error state appears as a red color circle icon (the **Decision** circle shown on the left side of the screen in Figure 8.30). You can get a visual cue from the decision diagram, and click the **Design Console** in the bottom-left corner of the application to know more about the error.

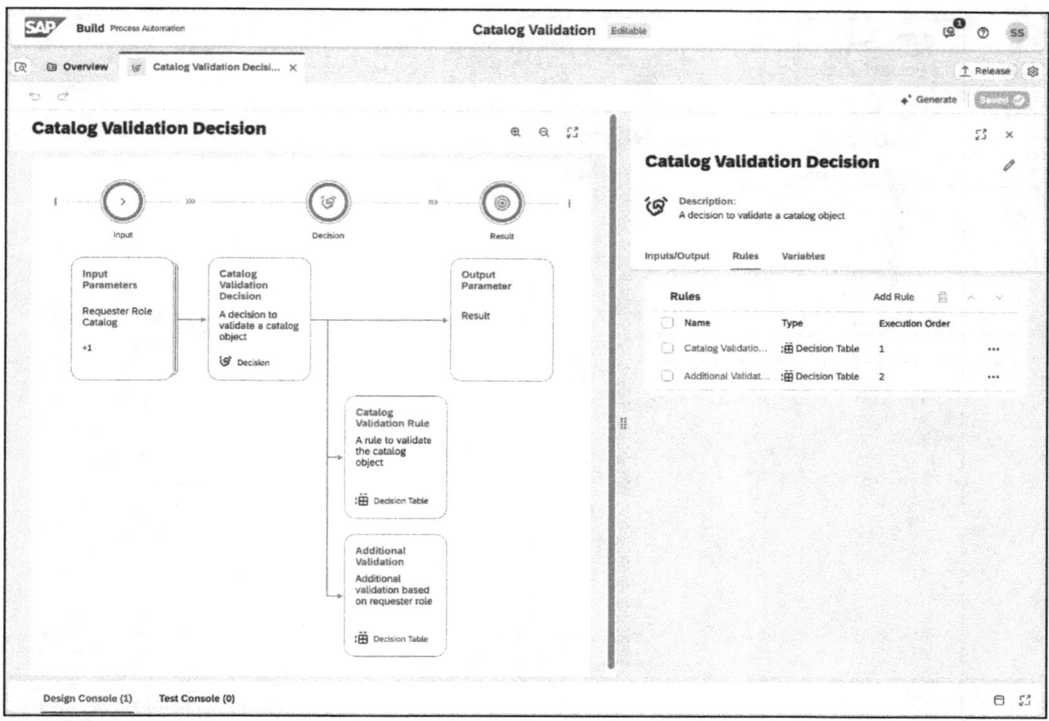

Figure 8.30 Decision in Error State

> **Note**
>
> The rules in a decision diagram are sorted and visualized based on the execution order. Because the reusable rules don't have an execution order, they are hidden in the decision diagram.

8.4 Working with Decisions

In this section, we'll explore the process of working with decisions in SAP Build Process Automation. We'll cover how to create decision tables and text rules within decisions, and how it can be consumed from external applications or within the SAP Build Process Automation processes.

8.4.1 Creating Decision Artifacts

The decision artifact can be created in an existing or new **Process Automation** project from the SAP Build lobby. From a **Process Automation** project, click **Create • Decision**, as shown in Figure 8.31.

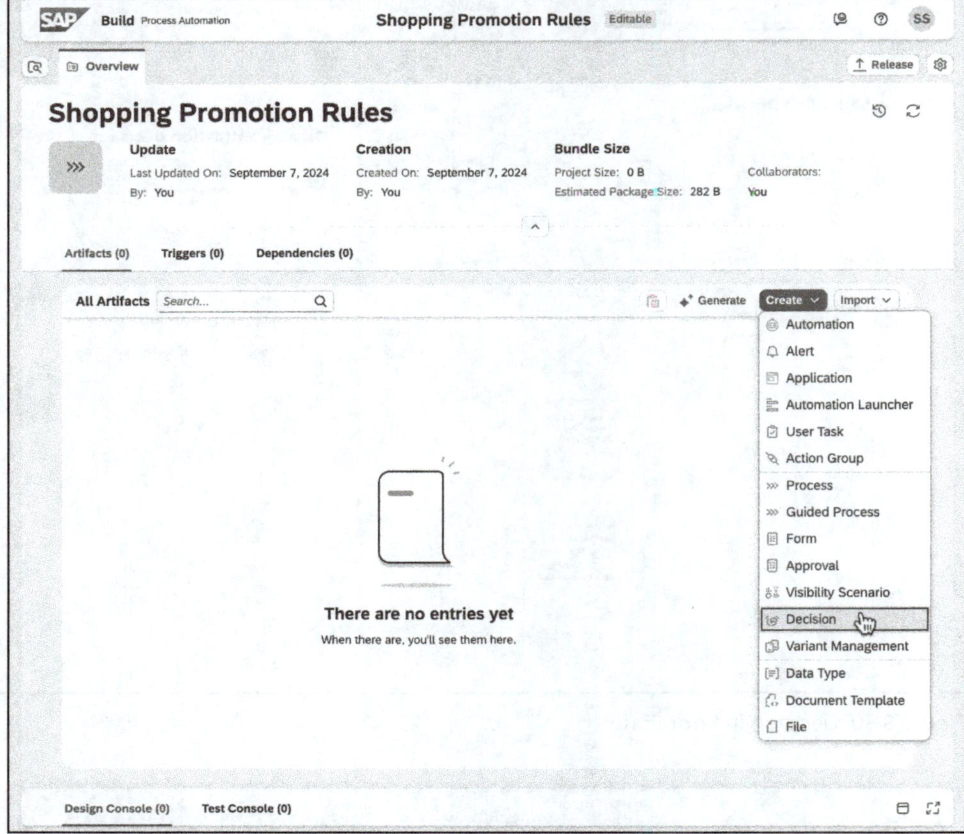

Figure 8.31 Create a Decision in a New Process Automation Project

> **Tip: Decisions-Only Project**
>
> While creating a process automation project in the SAP Build lobby, you must choose between a business process (for processes) and task automation (for RPA bots). If you want to consume decisions from an external application, then it might not make sense to create a process or an RPA automation. In such cases, you can choose either of the project types to create a process automation project. Once a project is created, depending on what you've selected as the project type, a Create Process popup or a Create Automation popup appears. You can simply cancel the popup and proceed with creating a new decision artifact.

In the **Create Decision** dialog, give the **Name** and **Description** of your decision artifact, as shown in Figure 8.32, and click the **Create** button. This action creates a new empty decision artifact in the process automation project.

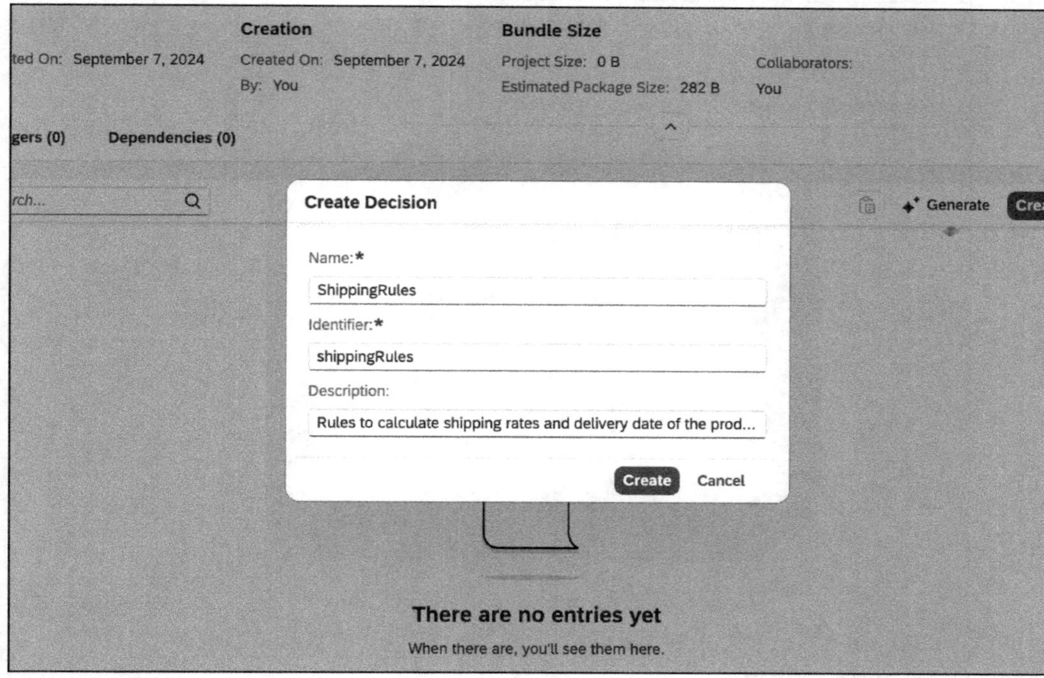

Figure 8.32 Create Decision Popup

8.4.2 Creating Inputs and Outputs

As explained in Section 8.3.2, the decision artifact requires clear definitions of both input and output parameters, which is also called the vocabulary of the decision, during the development or modeling phase. The input and output parameters of a decision can be defined based on primitive data types such as string, number, date, Boolean, and so on. However, for representing more complex data objects, the data type artifact can be used. As you enhance your decision artifact, you can add additional inputs.

> **Warning**
> If you modify the input/output of a decision or modify the data type artifact that is used in decisions, consider the impact it might create on existing decisions.

If your use case needs more complex data objects, then create a data type in the process automation project. Click **Create • Data Type**, as shown in Figure 8.33. Once created, you can add fields in the data type as needed using basic data types such as **String**, **Number**, **Boolean**, **Date**, and so on, or you can use another data type to represent a complex data object. Similarly, create data type artifacts for input and output parameters as needed.

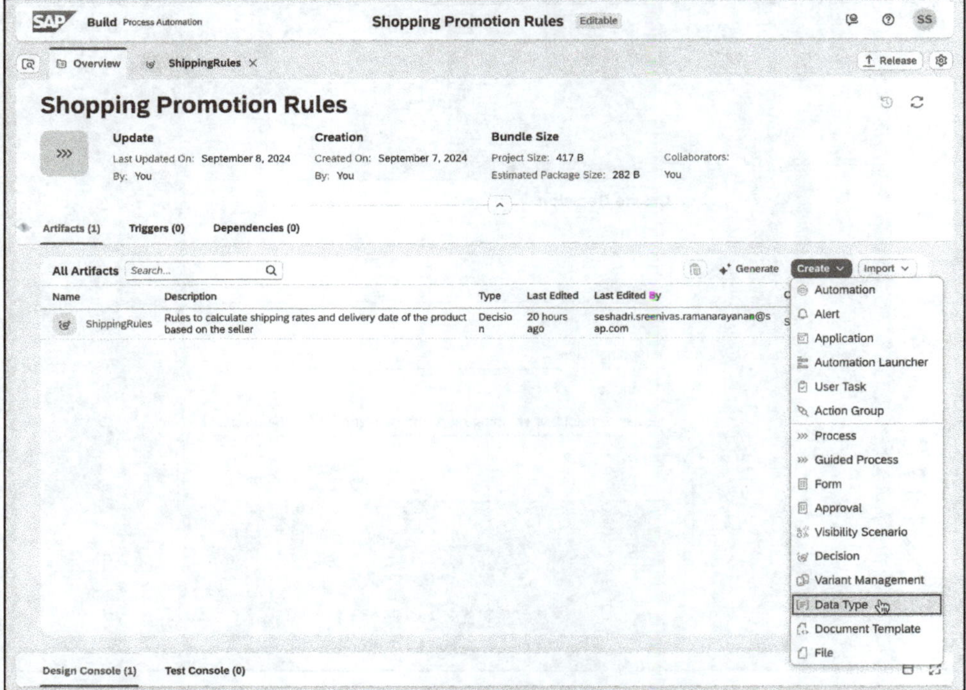

Figure 8.33 Creating a Data Type in the Process Automation Project

Once the data type artifacts are created, open the decision artifact to set up the input and output parameters. Click the Add Input Parameter button to define one input parameter. Provide the **Name**, **Description**, and **Type** of the input parameter. Repeat the steps to add all input parameters that are necessary for the decision. Similarly, to add an output parameter, click the Add Output Parameter button to define exactly one output parameter. Once an output parameter is added, the **Add Output Parameter** button is disabled because decisions in SAP Build Process Automation support exactly one output parameter.

Tip: Define Multiple Outputs for a Decision Artifact

A decision can have exactly one output parameter. To define multiple parameters as output of a decision, create a data type with as many parameters as needed, and use that data type to define the single output parameter. You can also set the output as a **List** if you expect multiple records will be part of the decision output.

Figure 8.34 shows a decision artifact with **Product Input** as input parameter and **Delivery Rates Output** as output parameter, and both are based on a data type artifact available within the project. It's possible to mark an input parameter and the output parameter as a **List** provided they aren't defined by basic data types such as number, string, and so on. It means that you can pass an array of data to the decisions, and the decision can return an array of data.

Note: Input and Output Parameter Names

While naming the input and output parameters and while adding new fields to a data type, keep in mind that these names will be part of the API of the decision. For the decision shown in Figure 8.34, the API interface will be defined with `Product_Input` as the input to execute the decision. On execution, the decision will return a parameter called `Delivery_Rates_Output` where the results of the rule execution will be available.

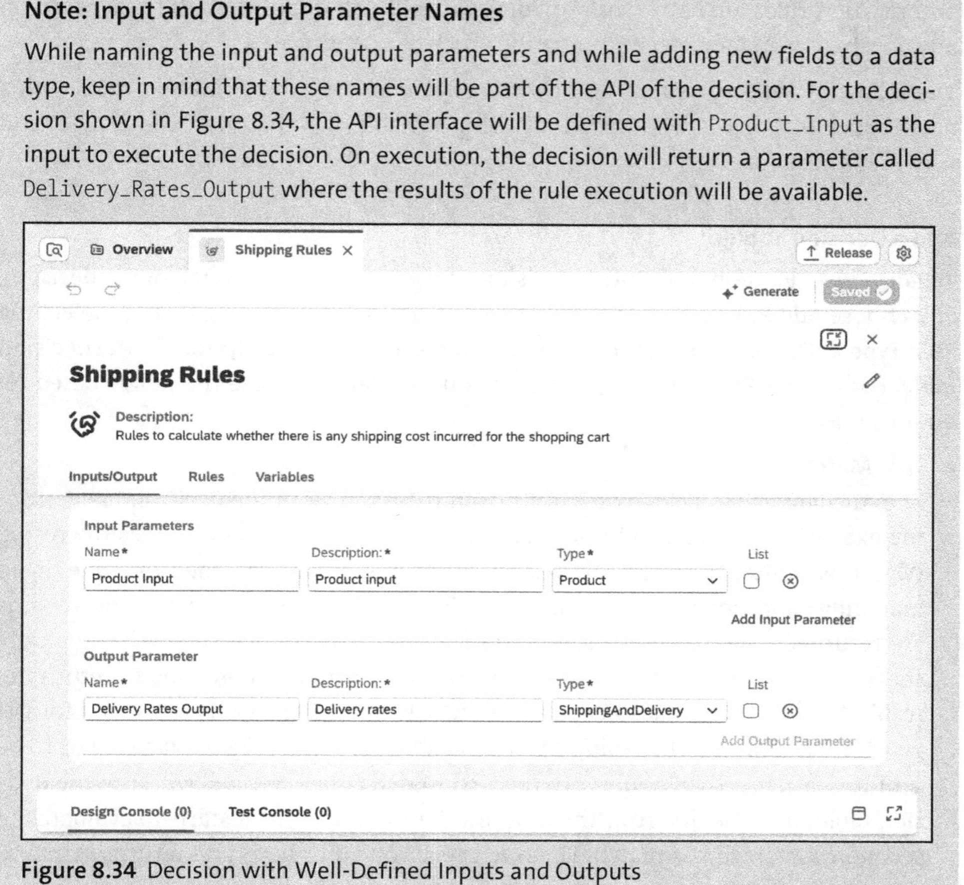

Figure 8.34 Decision with Well-Defined Inputs and Outputs

Warning

If you're changing the input and output parameter names, then the decision interface will also change, and the calling application needs to adapt the changes. Therefore, avoid changing the input and output parameter names once the decision is consumed.

8.4.3 Creating Rule Artifacts

The next step in setting up a decision in SAP Build Process Automation is to add a decision table or text rule to your decision artifact. Both decision tables and text rules are equally capable, but there are subtle differences between the two artifacts.

Decision tables are easier for business users to modify and support exporting/importing from Microsoft Excel. Though they need clear definition of parameters for IF conditions, they are more convenient when you want to maintain multiple matching criteria. Text rules are better for providing flexibility in defining conditions based on any vocabulary and for specific operations such as updating or appending values.

The following sections will cover how to create decision tables and text rules with a decision artifact. It will also introduce using variables within a decision artifact and why that's needed.

Add a Decision Table

To add a new decision table into a decision artifact, go to the **Rules** tab of the decision, and choose **Add Rule**. A **Create Rule** dialog appears, as shown in Figure 8.35. Select the **Rule Type** as **Decision Table**. Provide a **Rule Name** and **Rule Description** for the decision table. For the **Hit Policy**, there are two options available. Choose the policy based on your use case:

- **First Match**
  The rule engine scans the conditions from the first row of the decision table, stops the execution when a condition matches, and returns that condition as the result. With this policy, there will always be either one record returned or none, if no matching conditions were found. There will never be a case where multiple records are returned. The execution time for a rule with **First Match** selected is typically shorter, as the rule doesn't have to traverse the entire decision table to return the result. But of course, it also depends on which condition row has matched for the given input. Consider the shipping rules in Table 8.4. If a 28 kg package needs to be sent to Germany, then a rule with the **First Match** policy selected will scan the decision table from the first row. The first row doesn't match, so it starts to evaluate the second row, which is a match. The rule execution stops here, and returns standard shipping at €30.

Destination	Weight (kg)	Shipping Method	Cost (€)
Germany	<=5	Standard shipping	10
Germany	>5	Standard shipping	30
Germany	>20	Freight shipping	75

Table 8.4 Example Shipping Rule

- **All Match**
  The rule scans the entire decision table and returns all the matching records. With this policy, there can be 0..n number of records returned by the decision table depending on how many conditions match. For the same decision table represented

in Table 8.4, if the **Hit Policy** is **All Match**, then the result will be different. The first row doesn't match, so it starts to evaluate the second row, which is a match. It will continue the evaluation and proceed to the third row, which is also a match. Therefore, the rule execution returns two matching records: standard shipping at €30 and freight shipping at €75.

Tip

When using a decision table with a **First Match** hit policy, the conditions or rows must be carefully ordered because the evaluation stops as soon as a match is found. Organize the table with the specific conditions at the top and the generic conditions at the bottom.

For the **Rule Category**, use either **Orchestrated Rule** or **Reusable Rules** depending on your use case. We've discussed the rule category in detail in the "Reusable Rules" section under Section 8.3.3. Choose **Next Step**.

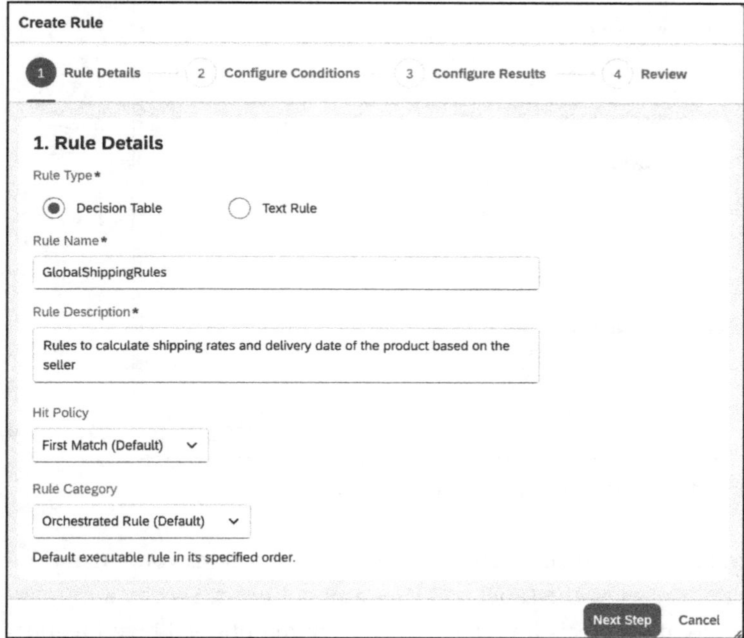

Figure 8.35 Creating a Decision Table

Tip

Mark an output parameter as **List** when using the **All Match** hit policy in a decision table. This allows you to fetch all records that meet the specified conditions and return them as an array of multiple values in the result.

Configure Conditions is the next step (see Figure 8.36) where you can add the attributes that will be available in a decision table to configure the conditions from the available list of vocabulary. You can also define condition attributes based on expressions using the different functions. For example, a condition attribute can be defined as `CONCAT(salesOrder.org, salesOrder.division)` or `AVG(salesOrder.items.totalNetAmount)`.

Make sure that the attributes you've chosen will have values while the decision table is executed. The input parameters will usually have the values provided by the caller (process or external application). The other attributes in the vocabulary, including the output parameter, can be filled by previously executed rules within the same decision.

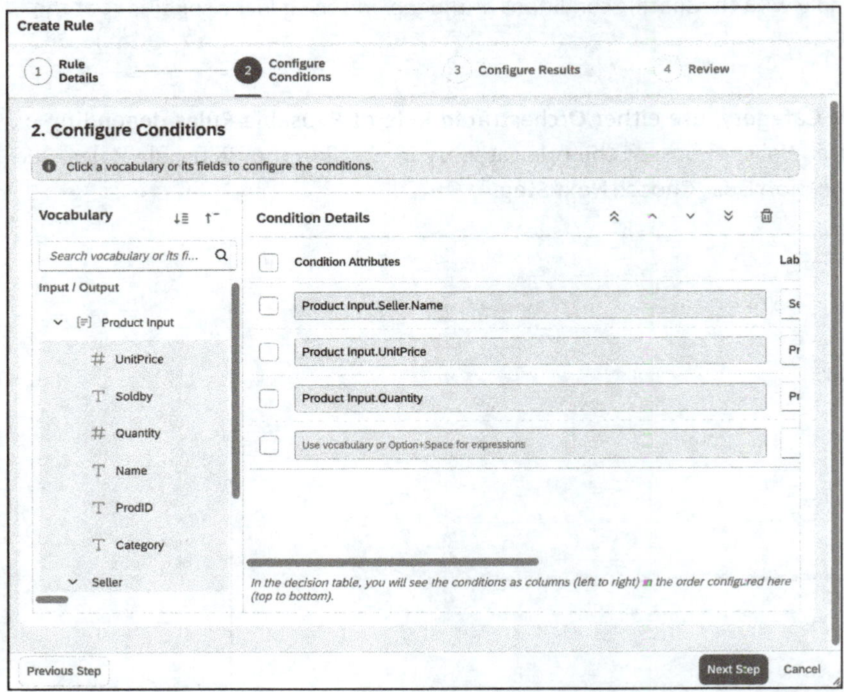

Figure 8.36 Defining Condition Attributes of a Decision Table

Once a **Condition Attribute** is added, there are two other configurations you can optionally do. Scroll to the right to view the **Label** and **Operator** columns. Specify a human-readable **Label**, especially if the decision table will later be maintained by a business user. Otherwise, the default label is the name of the condition attribute such as **Product Input.Seller.Name**. It particularly makes sense to add **Label** when using expressions to define the condition attributes.

The next configuration that you can enable, depending on your use case, is to provide an **Operator** for the conditions. Figure 8.37 shows the different operators you can maintain for each condition. If an operator is maintained here, for example, <=, then the same operator is applicable for all the records in the decision table for that **Condition**

Attribute. It makes things simple for maintenance and readable for the business users, but at the same time, it removes the flexibility to maintain different operators for different rows.

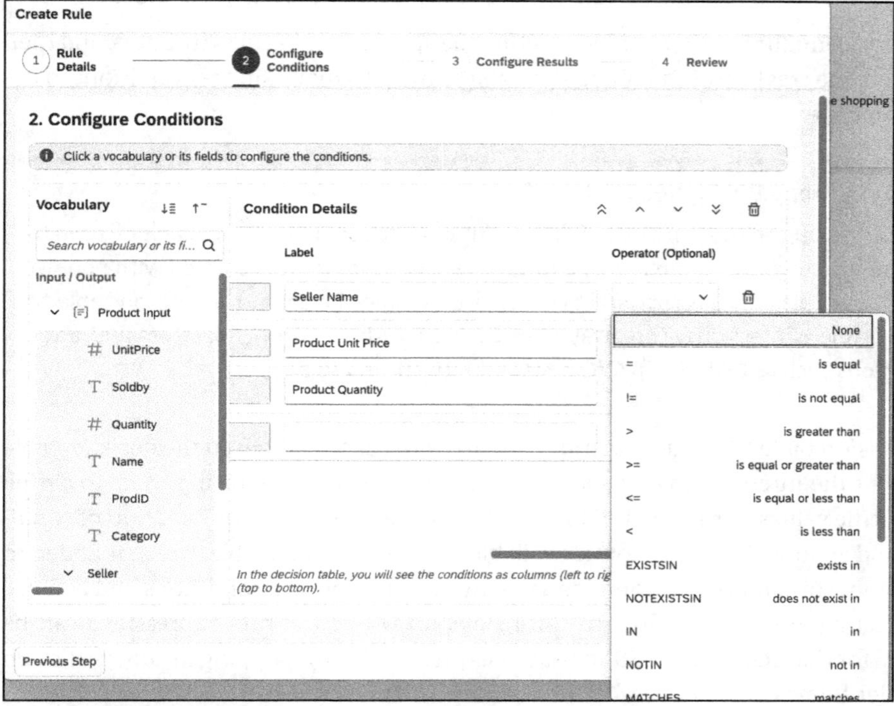

Figure 8.37 Defining Label and Operator for the Condition Attributes

Figure 8.38 shows a decision table with one of the condition attributes with the Product Quantity operator. Because the other condition attributes don't have a default operator, the end user can choose any operator of their choice to maintain the conditions. The conditions for the Product Unit Price column uses < for some rows and > for other rows. The Product Quantity column doesn't have that flexibility, but it's easier to maintain. Depending on the use case, you can enable a default operator for the condition attribute.

Decision Table — Add Row ∨ | Delete Row | Copy Row | Cut Row | Paste Row ∨

If			Then	
Seller Name	Product Unit Price	Product Quantity >=	ShippingRate	ShippingDuration
= 'Ultrasonic United'	> 0	0	100	'Same-Day Delivery'
= 'Technocom'	< 500	0	25	'5 to 12 Days'
MATCHES '.*'	< 350	10	250	'5 to 10 Days'

Figure 8.38 Decision Table with Operator >= for the Condition Attribute Product Quantity

Because the rule engine evaluates the decision table from left to right and top to bottom, the order of the columns in a decision table plays a vital part in the performance of a decision table. Move the condition attributes up or down using the arrow icons ⇈ ^ ⌄ ⇊ so the columns that filter out most of the data should come first.

Similarly, configure the **Results** by selecting the appropriate **Result Attributes**, and then complete the creation of a new decision table. Now, let's maintain the conditions in the decision table.

Note: Defining the Input as List

Although you can pass an array of data as input to decisions, this data can't be used to define the condition attributes of a rule. Once a decision table is created with non-list data as conditions, the arrays/list can be used within the cells of a decision table to model rule expressions. The arrays can be used with the array operators and array functions such as **EXISTSIN()**, **AVG()**, **SELECT()**, **TOP()**, and so on.

The decision table UI includes a grid structure with cells. The cells in the decision table represent the intersection of a specific condition and action, allowing users to define the specific values or rule expressions that will be used for each combination of conditions and actions. An empty row is available in a new decision table, which is added to a decision. This empty row contains as many cells as there are conditions and actions. On selecting any of the cells, an editor pops up where the rule expressions can be entered for that cell. Figure 8.39 shows the expression editor popping up after a click on the **Seller Name** column.

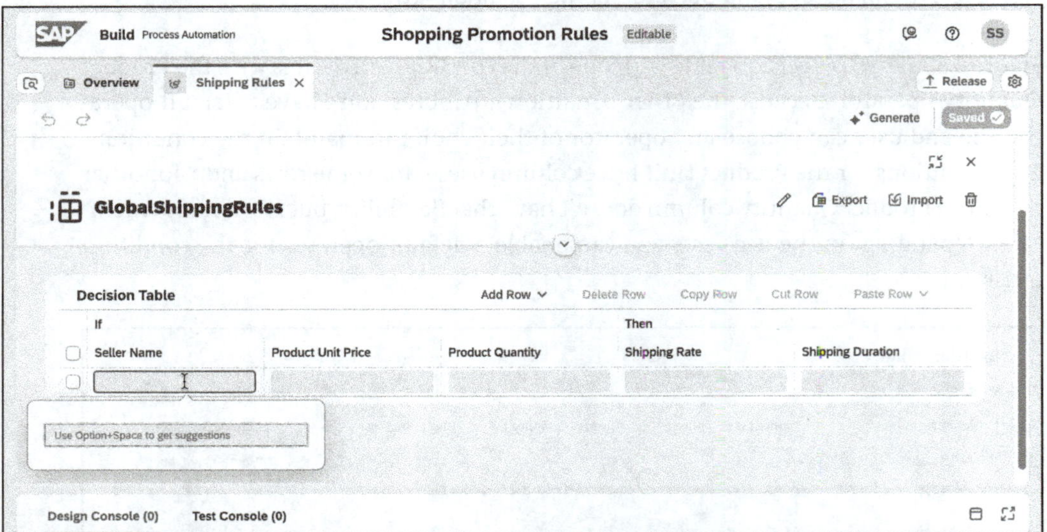

Figure 8.39 Editing a Cell in the Decision Table

Pressing Ctrl+Space opens the auto-suggest list to help you enter the rule expression. Because an operator isn't included in the **Seller Name** condition column, the rule expression should start with an operator that is relevant for the condition column, as shown in Figure 8.40. Click outside the expression editor to confirm the changes.

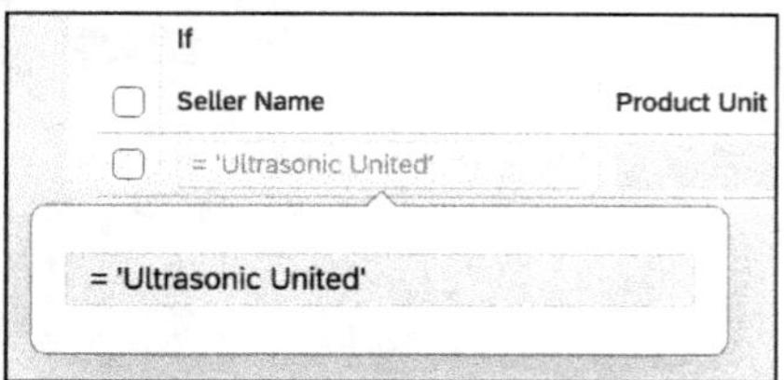

Figure 8.40 Maintaining a Rule Expression within the Condition Cell in a Decision Table

If there are any errors while maintaining the rule expression, then the **Design Console** tab on the screen shows the appropriate error message after you click the **Save** button, as in Figure 8.41.

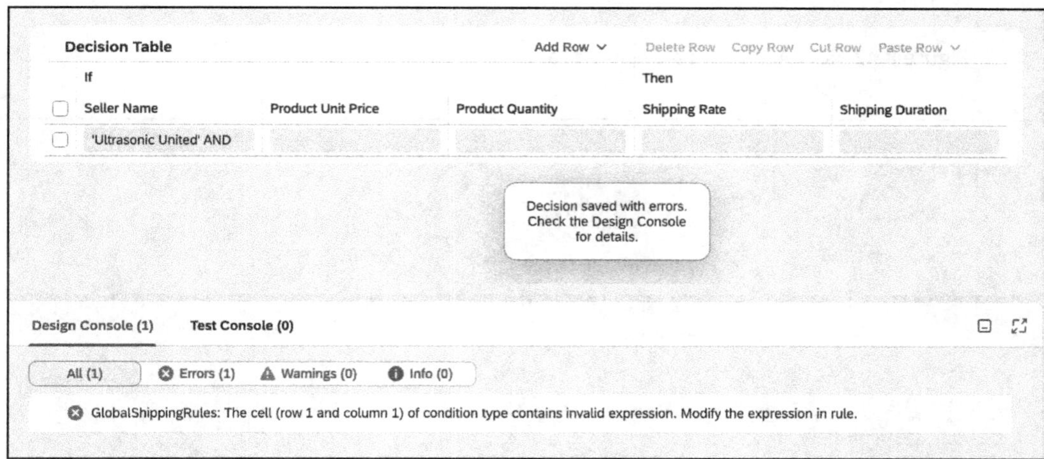

Figure 8.41 Validation Error Message after Saving the Decision Artifact

Once all the cells have been maintained with appropriate rule expression, the next step is to add further rows. Select the row, and choose **Add Row • Insert First** or **Insert After**, as shown in Figure 8.42, depending on where you want to create a new row. Similarly, a row can be copied, cut, or deleted by choosing the appropriate buttons. A row that was copied or cut can be pasted before or after the selected row. Choose the **Save** button to confirm and save the changes made.

Another way of editing the decision table is using your favorite spreadsheet application. The content of a decision table can be exported in *.xlsx* format. The spreadsheet file can be used to maintain the different rules in the form of rows and saved in the local file system. Once the spreadsheet is complete, a user with the `ProcessAutomation-Developer` role can import the data within the spreadsheet to update the content of the decision table, as shown in Figure 8.43.

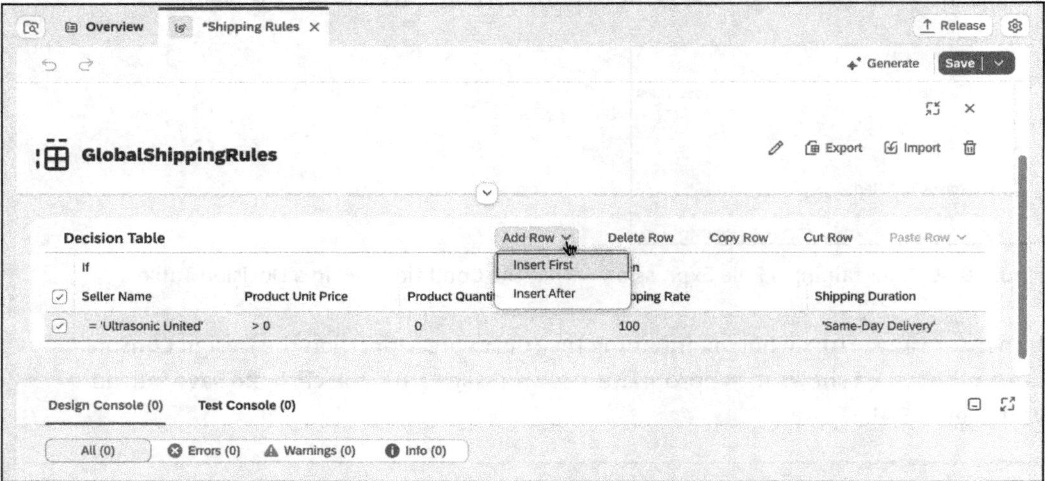

Figure 8.42 Row Actions in a Decision Table

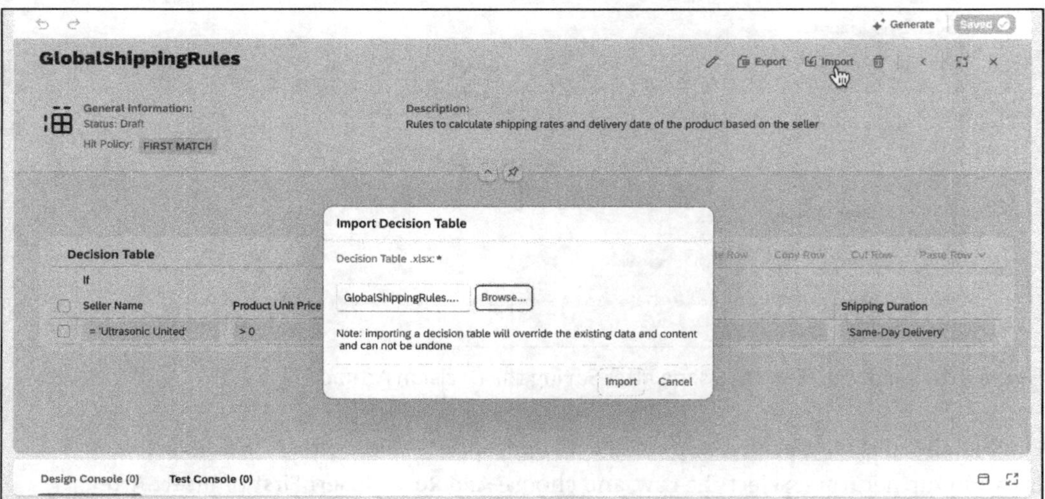

Figure 8.43 Import Data from Spreadsheet into Decision Table

There are a few things to be noted while working with the spreadsheet to maintain decision table data:

- Exporting to a spreadsheet isn't possible if the decision table contains any errors as represented in the **Design Console**.

- The exported spreadsheet contains certain technical details such as project name, project ID, rule name, rule ID, condition columns, and so on. This data won't be modified by the user.
- The import operation relies on data consistency; any discrepancy between the technical details in the spreadsheet and the decision table will result in failure. It means that you can't import a spreadsheet of a different decision table even if the name matches. However, if the same project is transported to a different tenant, then the IDs of the artifacts remain the same, and the import will work across the tenants.
- No new columns will be added or removed from the spreadsheet; otherwise, the import operation will fail.
- New rows can be added, but be sure to consider the format of the data in the cell. The cell value usually starts with a single quotation mark ' to denote an escape character which is ignored during import. For example, the condition MATCHES '.*' will be maintained in the spreadsheet as 'MATCHES '.*' or > 10 will be maintained in the spreadsheet as ' > 10.

It's possible to change the definition of a decision table, such as changing the condition columns, changing the result columns, and changing the **Label** or **Operator**, after it has been created. To do so, click the edit icon, which brings up the same wizard that was used during the creation. Any changes made in that wizard will be reflected as soon as the wizard is completed with the Finish button.

Warning

If a condition column of an existing decision table is removed, the data maintained in that column will also be removed. When you add the condition column back or add a completely new condition column, the decision table will have empty cells for the column. Unless it's filled, the condition column won't be used for evaluation.

For ease of maintaining values in the decision table, SAP Build Process Automation supports enumeration-based value helps for those fields with a fixed set of values. For example, days of the week, gender, clothing sizes, and blood types have a fixed set of values, and these can be set as enumeration in the data type. For example, Figure 8.44 shows the **ShippingDuration** field with a fixed set of values used as **Enumeration**.

When such a data type is used to define the condition or result columns, it's possible to choose one of the values from the enumerations while maintaining the cell. To use the value help for such a cell, press Ctrl+Space to open the auto-suggest list in the condition editor. When choosing a **Fixed Value** option in the auto-suggest list, the value help icon can be used to open the value help dialog. In cells for result columns, the value help icon appears directly in the condition editor, as shown in Figure 8.45.

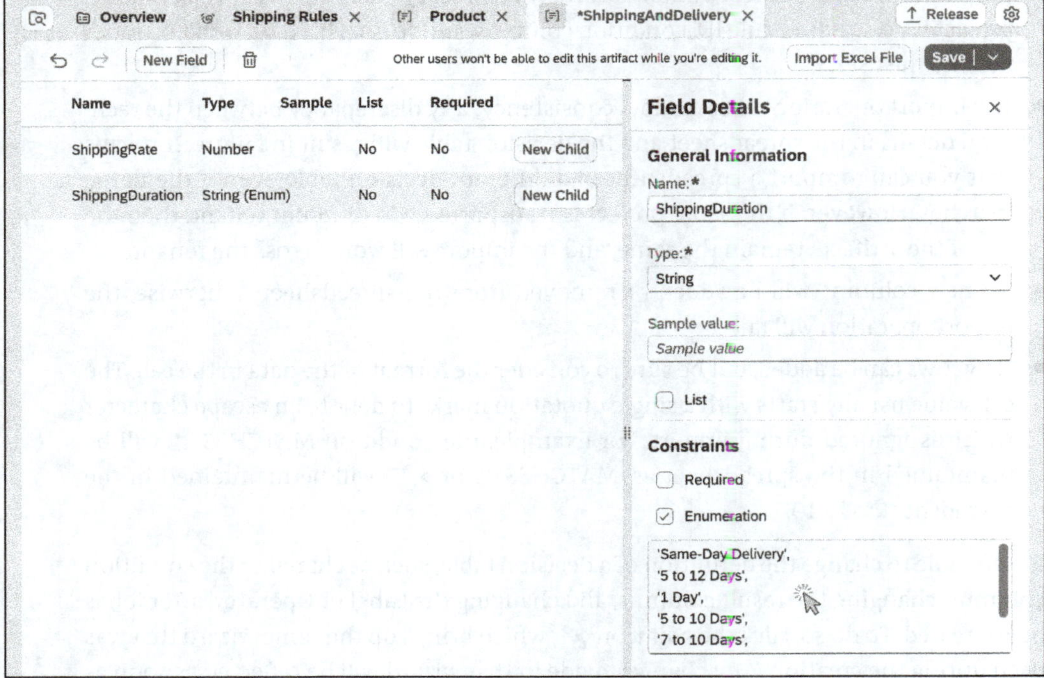

Figure 8.44 Enumeration Setup in the Data Type with Comma-Separated Values

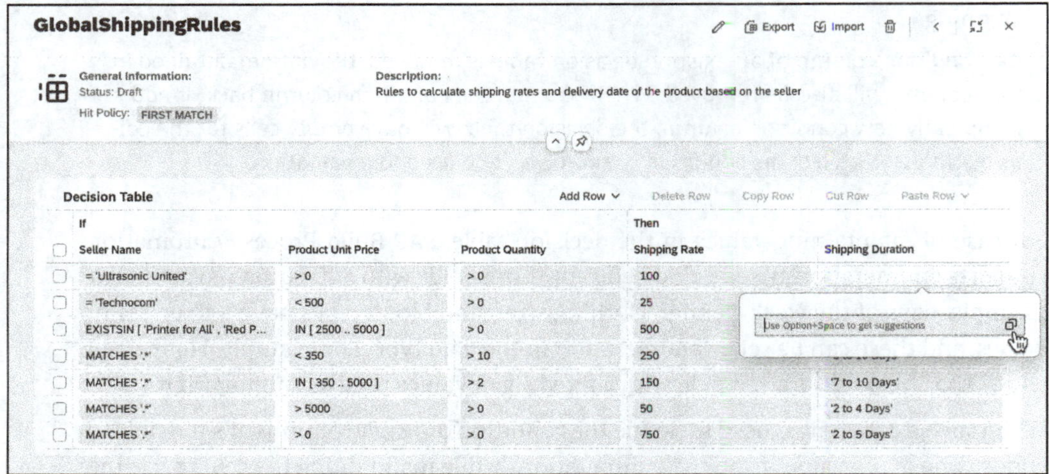

Figure 8.45 Value Help Icon in the Condition Editor for Cells in the Result Column

Add a Text Rule

A text rule in SAP Build Process Automation is a collection of rule expressions written in a simple if-then format. To add a new text rule to a decision artifact, go to the **Rules** tab from the decision artifact, and choose **Add Rule**. In the **Create Rule** dialog, select the

Rule Type as **Text Rule** (see Figure 8.46). Provide a **Rule Name** and **Rule Description** for the text rule. Choose the **Rule Category** as **Orchestrated Rule (Default)**.

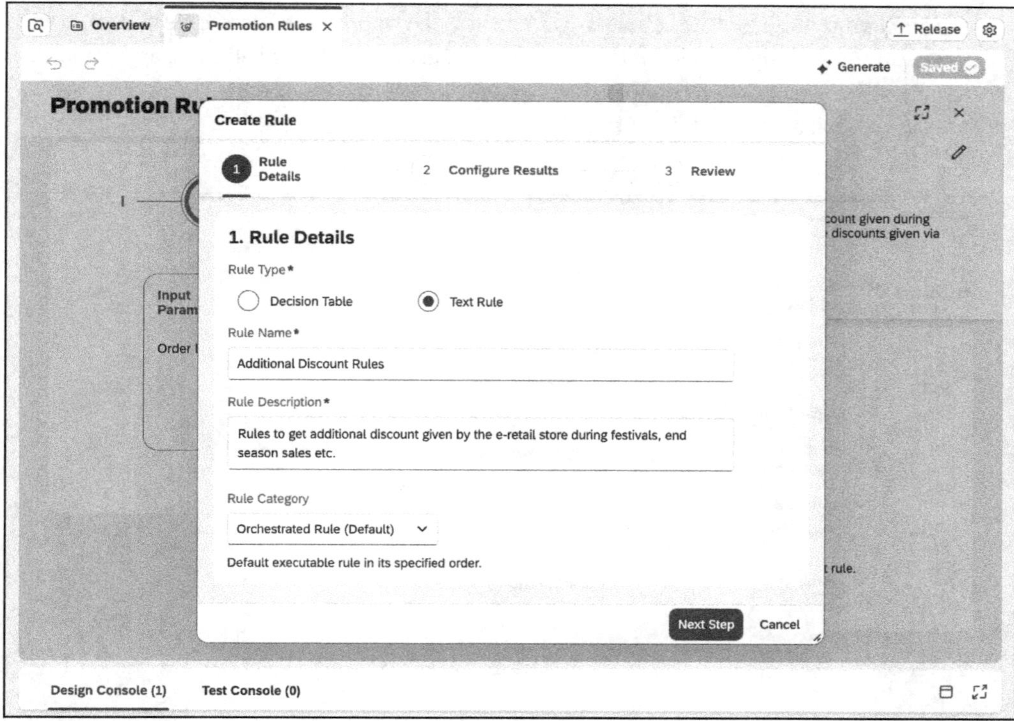

Figure 8.46 Create a Text Rule with Rule Category as Orchestrated Rule

> **Note**
>
> A decision table or a text rule can be marked as a reusable rule so that they can be used from other rules. Once they are marked as reusable rules, these rules can be used in text rules as well as in decision tables. The result returned by a reusable rule can be consumed in a rule expression within those text rules and decision tables. Refer to Section 8.3.3 for more details on reusable rules.

The main difference in the creation of a text rule when compared to creation of a decision table is that there is no condition vocabulary needed for a text rule. The text rule can work with any available vocabulary. You need to define only the result while creating a new text rule, as shown in Figure 8.47. Move the **Result Attributes** up or down using the arrow icons ⌃⌃ ⌃ ⌄ ⌄⌄ as needed. The order of the **Result Attributes** doesn't affect the performance of the text rule in any manner, but you have the option of choosing the order in which the **Result Attributes** are shown to the user. Use the delete icon 🗑 to remove any result attribute of a text rule. Proceed to the **Review** step, and click the **Create** button to complete the creation of a text rule.

Note

By default, the result data object is shown for the selection of **Result Attributes**. You can change the result vocabulary by choosing the appropriate data objects from the **Result Vocabulary** dropdown UI element. The **Result Vocabulary** also has an option to choose **No Result**, which enables you to create a text rule for execute operations. We'll cover that in a later section of this chapter.

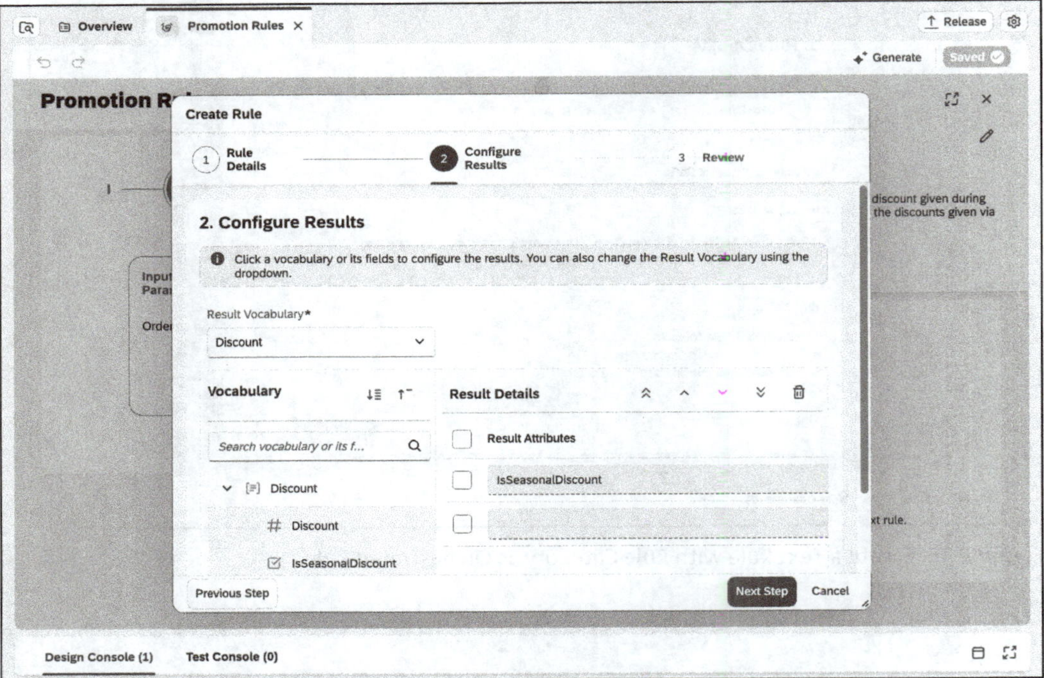

Figure 8.47 Configure Result Attributes of a Text Rule

Once you click the **Confirm** button in the **Review** step, an empty text rule is created based on the **Result Attributes** selected, and it's added to the decision. A text rule contains three main sections—**If**, **Else If**, and **Else**. When a text rule is created, only the **If** section contains (an empty) definition, and the **Else If** and **Else** sections aren't added by default, as shown in Figure 8.48. You can use the Add Else If and Add Else buttons to create the respective sections. Each of these sections has two further sections: one where you can enter an expression that should be evaluated and a **Then** section where the results are assigned with values.

Use the [Ctrl]+[Space] keys to open the auto-suggest list to enter a valid expression that will be evaluated and assign the values for the result fields as well. In Figure 8.49, an expression checks if the order date lies within a date range of **20 Dec 2024** and **3 Jan 2025**. If this condition evaluates to be **true**, then it gives a seasonal discount of 5%.

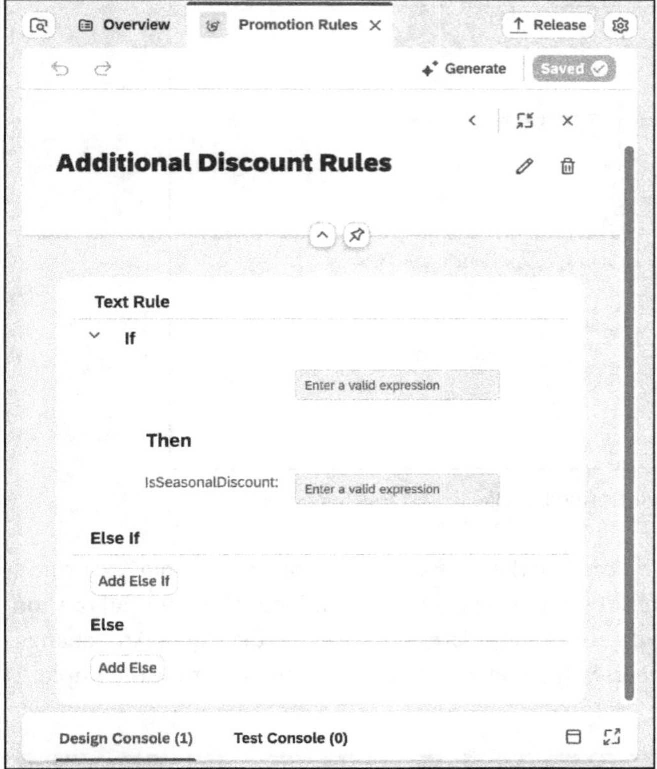

Figure 8.48 Empty Text Rule with If, Else If, and Else Sections

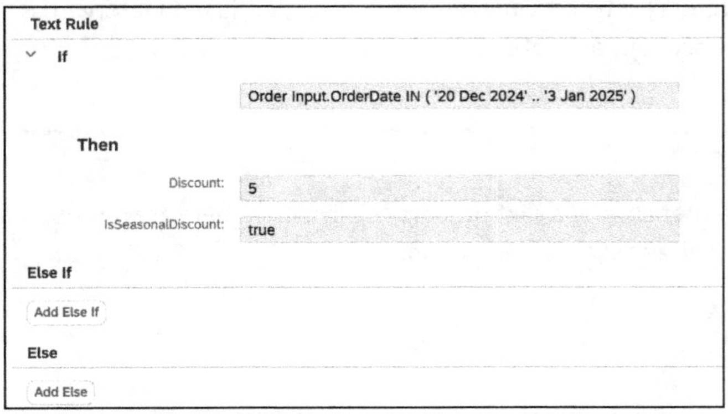

Text Rule
If
Order Input.OrderDate IN ('20 Dec 2024' .. '3 Jan 2025')
Then
Discount: 5
IsSeasonalDiscount: true
Else If
Add Else If
Else
Add Else

Figure 8.49 Expression That Will Be Evaluated in the IF Condition of a Text Rule

Depending on the use case, use the **Add Else If** and **Add Else** buttons to create the respective sections and add expressions to evaluate and assign the result values. In Figure 8.50, the **Else If** section and **Else** sections are added to define some special seasonal discounts only when the products are ordered during certain promotional periods.

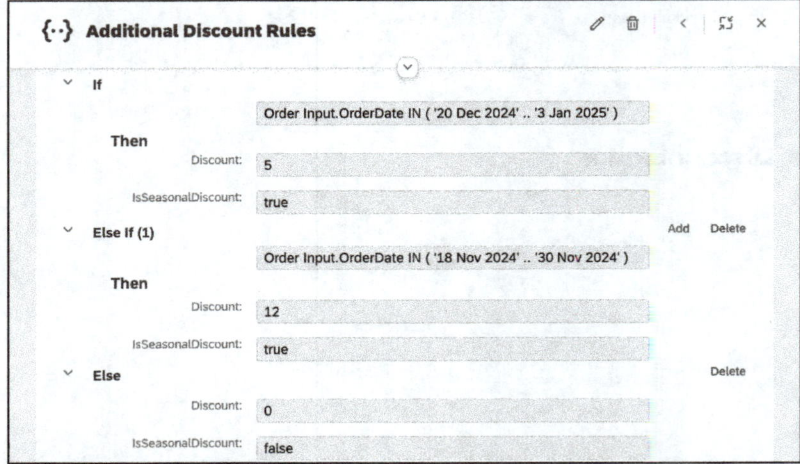

Figure 8.50 Text Rule with Well-Defined If, Else If, and Else Sections

Once a text rule is created, you can use the icon to edit the definition of text rule at any given point. This launches the same **Create Rule** wizard, but this time, all the configurations from the text rule are prepopulated for you to edit them. Any changes made in the wizard that appears will be reflected as soon as the wizard was completed with the **Finish** button.

Note

There is always one If section and one Else section in a text rule. Depending on the use case, you can add one or more **Else If** sections. While the **If** section is mandatory in a text rule, the Else If and **Else** sections are optional.

Tip

For easy readability and maintenance of a text rule, especially when there are multiple **Else If** sections, you can expand or collapse the section.

Add Variables to Decisions

Apart from the input and output vocabulary that are available within a decision, depending on the use case, you might want to have a temporary variable that can be used within the decision. For example, for a given product and order date, there might be multiple discounts that are possible: loyalty discounts, seasonal discounts, voucher code, volume discounts, first-time purchase discounts, and so on. Each of these discounts can be modeled either as a text rule or decision table. Figure 8.51 shows the decision artifact **PromotionRules** with multiple text rules. Each of these text rules could use a temporary variable as the result parameter.

Tip

When there are multiple rules within a decision of the Orchestrated Rules rule category, it's important to define an execution order. The execution order determines the sequence in which the rules are evaluated, and it can significantly impact the outcome of the decision. Rule1 rejects travel request high-risk countries, and Rule2 approves requests below $500. Because the output of Rule2 can overwrite the output of Rule1 in an orchestrated rule, the sequence in which they are executed is important. If you want to prevent travel to a high-risk country, then that rule should be the last rule in the decision and vice versa. You can adjust the order by selecting a rule and using the up ^ and down ∨ buttons, as shown in Figure 8.51, or by selecting the more icon ••• and choosing **Move Up** or **Move Down**.

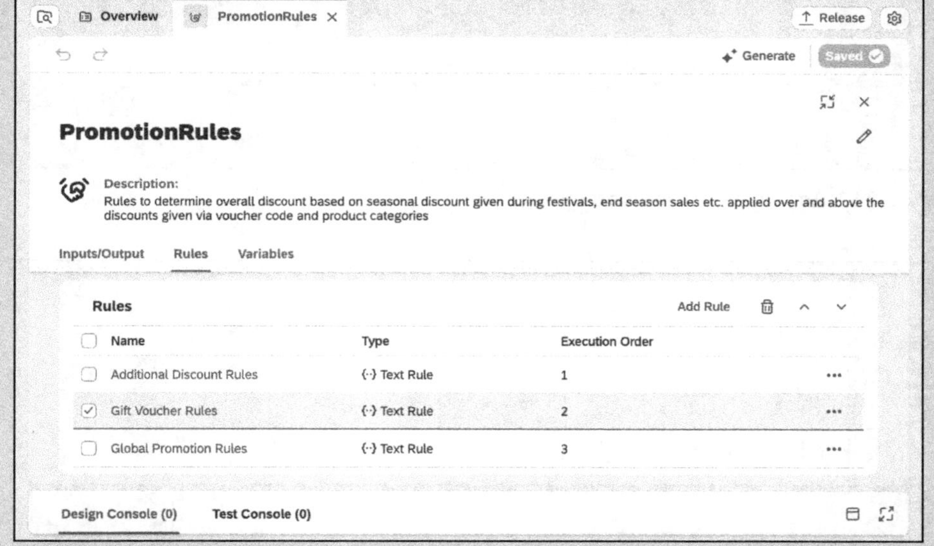

Figure 8.51 Multiple Orchestrated Rules in a Decision with a Specific Execution Order

A final rule can then use these variables to apply a discount based on these variables and return the final discount. To create a variable within a decision, go to the **Variables** tab, and choose the **Add Variable** button. Enter the **Name** and **Description**, and choose the data **Type** of the variable, as shown in Figure 8.52.

Once a variable is created, it's available as vocabulary. It can then be used to define the condition or result of a text rule or decision table. Figure 8.53 shows the **Edit Rule** wizard that appears when you edit the definition of the text rule using the ✎ icon. The variables can also be used in the auto-suggest list when you press Ctrl+Space in a text rule, as shown in Figure 8.54.

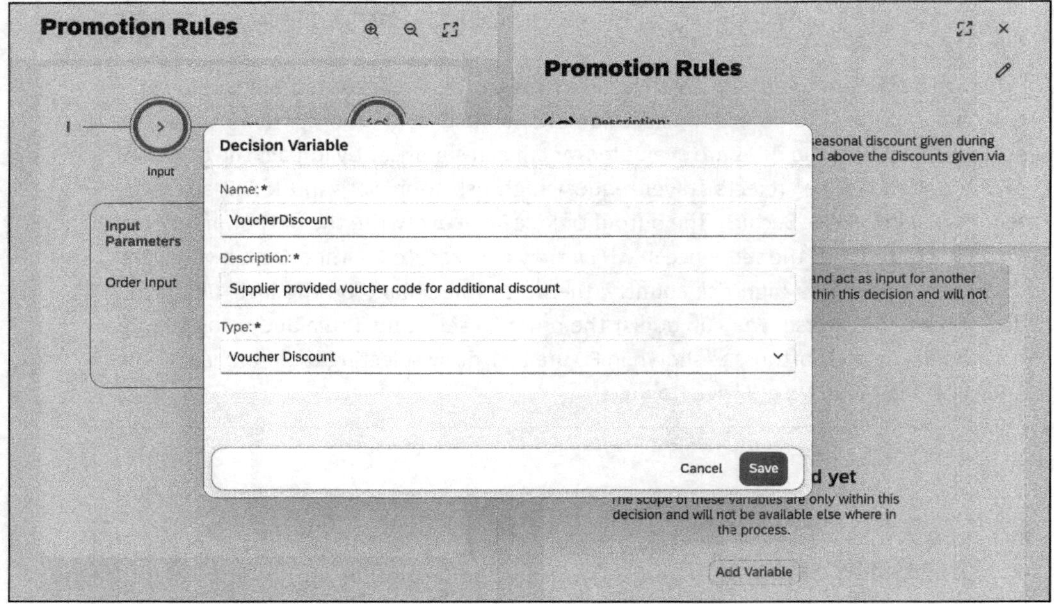

Figure 8.52 Adding a Decision Variable

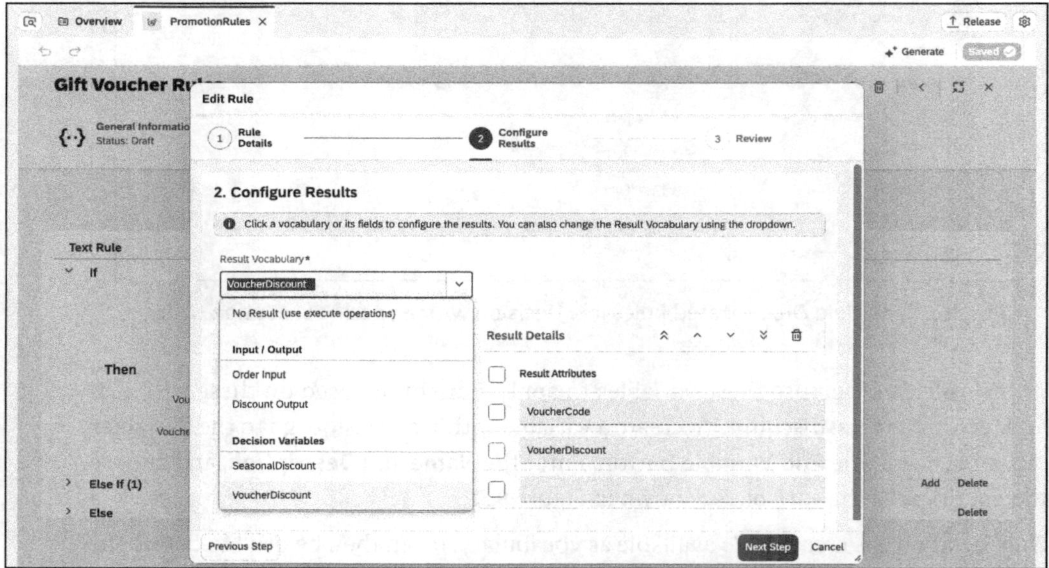

Figure 8.53 Decision Variables Available as the Result Vocabulary of a Text Rule

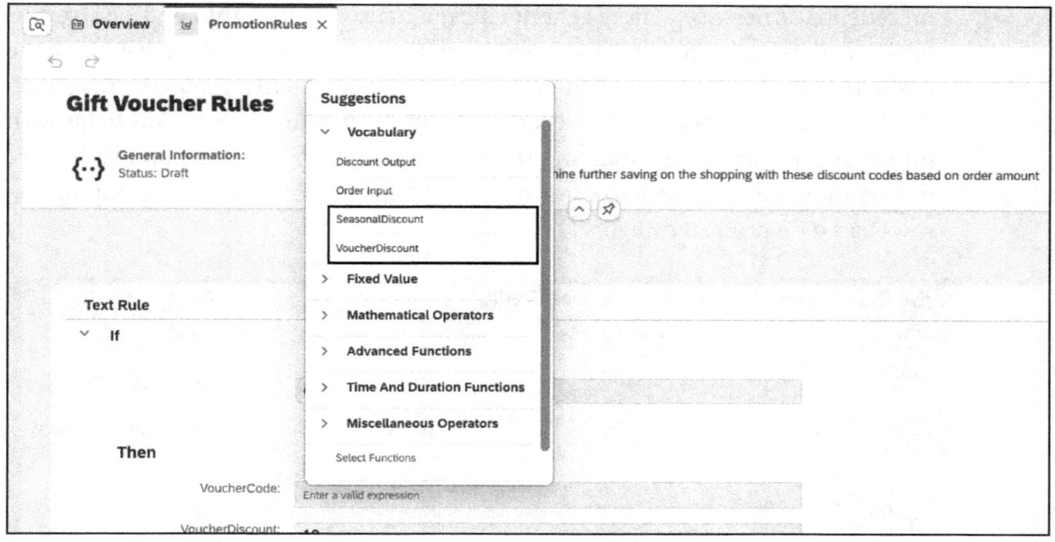

Figure 8.54 Decision Variables Available in the Auto-Suggest List

Modeling Text Rule for Execute Operations

If you don't assign a result vocabulary to a text rule in SAP Build Process Automation, as shown earlier in Figure 8.28, you can perform execute operations on existing input/output parameters or variables within a decision. This may involve updating or appending values to these parameters or variables, and you can also include multiple execute operations within a single text rule. For example, you can use **FOREACHITEM** to perform a loop operation on **List** type parameters. The other execute operations supported in SAP Build Process Automation are **APPEND** and **UPDATE**, which works on a **List** type parameter. Refer to Table 8.3 for more details on the functions that support execute operations, syntax, and examples.

> **Note**
>
> If you have any validation errors while modeling, they will appear in the **Design Console** at the bottom panel of the editor.

8.4.4 Best Practices for Modeling Decision Tables and Text Rules

Creating performant decision tables and text rules in SAP Build Process Automation involves a combination of good design practices and careful data management. Here are some best practices you can follow:

- **Order of the rows**
  The rule engine processes the rules with a decision table from top to bottom. Therefore, in a decision table, place more specific rules higher up to ensure that they take

precedence. Otherwise, a more generic rule might be satisfying the condition, resulting in an unexpected result. This is relevant with the First Match hit policy where the evaluation stops as soon as a matching row is found. Similarly, group the rows with similar conditions together for easier readability and maintenance. This helps with future updates and understanding the table logic. Table 8.5 shows an example of having more generic rules toward the end of the decision table and grouping the rows based on product category.

Total Order Amount (EUR)	Remaining Credit (EUR)	Product Category	Action
< 1500	> 5000	MATCHES '.*'	Auto Approve
IN [2000 .. 5000]	< 2000	Standard Products	Manager Approval Required
<= 2000	< 1000		Reject
IN [5000 .. 10000]	IN [2000 .. 5000]	High-Value	Manager Approval Required
> 10000	< 5000		Reject
<= Customer Credit Limit	< 100	MATCHES '.*'	Manager Approval Required
> 0	> 0	MATCHES '.*'	Reject

Table 8.5 Decision Table with Specific Rule in the Top and Grouped Based on Product Category

- **Order of columns**
  The rule engine processes each row in a decision table from left to right. Therefore, order your columns from the most specific conditions to the least specific. Columns that filter out most of the data should come first. This can speed up the evaluation process as the engine can quickly eliminate nonmatching rows. Place columns that represent primary or mandatory conditions before those that are optional or less critical. In the decision table represented in Table 8.5, if the Product Category was placed as the first column, then multiple inputs would pass that condition requiring the second column to be evaluated. The same applies for text rules as well, where you need to have specific conditions evaluated first.
- **Remove redundancy**
  Ensure that conditions and actions aren't redundantly duplicated across multiple rows. This not only maintains cleaner tables but also enhances performance.
- **Use wildcards smartly**
  If you have conditions that apply to a broad range of inputs, then you might use

wildcards MATCHES '.*' or leave the cell empty. However, overuse can degrade performance, so apply judiciously. Prefer MATCHES '.*' over leaving the cell empty and prefer STARTSWITH or ENDSWITH over MATCHES '.*' for simple patterns and not regular expressions.

- **Use categorical conditions over numerical conditions**
  Categorical comparisons (e.g., using enumerations) are generally faster than numerical range comparisons. For example, prefer Category IN ('Type1', 'Type2') over Value > 10 AND Value < 20 or Value IN [100 .. 500).
- **Reduce complexity**
  If your decision table or text rule is exceedingly large or complex, consider splitting it into multiple, smaller rules that handle different segments of the logic. This can be more efficient and easier to manage. This can be more efficient and easier to manage. Complex logical expressions involving many AND, OR, and nested conditions can slow down rule evaluation.
- **Reduce size of rules**
  For a decision table, repeated values should be modeled in the left columns. This reduces the size of a decision table by optimizing the model (merging the columns) and increases the execution speed. This is very important, especially for huge decision tables.
- **Use advanced functions**
  Use advanced features such as reusable rules and variables to reduce redundancy of the rule conditions and to also reduce the overall decision footprint.

8.4.5 Best Practices for Consuming Decisions from a Process

A decision can be easily executed from a process by adding a decision activity and choosing the appropriate decision artifact available in the project. The catch here is that the decision artifact should be available within the project where the process is created. It's fine to create a decision and a process that calls that decision in the same project. But there are multiple reasons why decisions and processes should be modeled in separate projects:

- **Independent change management**
  Decisions often change more frequently than the process. Separating them allows updates to decision logic without needing to redeploy or modify the entire process. This makes maintaining and iterating rules more efficient.
- **Version control**
  If decisions and processes are in the same project, any updates to the decision logic will only be reflected in new process instances created after redeployment of the updated project version. This leads to version dependence, meaning that ongoing process instances will continue using the old decision logic until they complete,

resulting in inconsistent outcomes if the decision logic changes frequently. By contrast, separating decisions into different projects allows immediate application of updates across both new and existing process instances, without requiring the process project to be redeployed. This decoupling enhances flexibility and consistency.

- **Improved flexibility**
  Decoupling decisions from workflows allows for greater flexibility. The same decision can be reused across multiple processes or projects, making it easier to scale automation and ensure consistency across different processes.
- **Better design**
  The choice of modeling decisions and processes in separate project promotes better design by separating the business logic (decisions) from the process logic (workflow). This makes both aspects easier to manage, test, and maintain.
- **Better permission control**
  If decisions and processes are in separate projects, permission control becomes easier and more granular. You can assign project-level authority to control who can access or modify decisions versus who can manage the processes. For example, you can restrict decision logic access to specific business users or administrators responsible for decision-making, while keeping process logic accessible only to workflow designers or developers. This segregation of duties ensures that only authorized personnel can make changes to the decision models, improving security, accountability, and governance.
- **Faster iteration and deployment**
  Changes to decision logic can be rapidly implemented and deployed without needing to redeploy the entire process, enabling faster time-to-value and more agile updates.
- **Enhanced testing and validation**
  Because decisions are separated, they can be tested in isolation, ensuring that logic updates work as expected without affecting the overarching process.

But how do we call a decision that isn't available in the same project as the process? The answer lies in project dependency. Here are the steps to make that happen:

1. Publish the deployed version of the project containing decisions as the library. From the **Lobby**, select the project, which opens the project details section on the right. Go to the **Versions** tab, select the ••• icon, choose **Publish to Library**, and confirm the action by selecting the **Confirm** button. The version is marked as **Published to Library**, as shown in Figure 8.55. Once a project version has been published to the library, it will be discoverable for consumption from any other process automation project.
2. In the project where processes are created, go to the **Project Properties** section, and choose **Add a Business Process Project Dependency**, as shown in Figure 8.56.

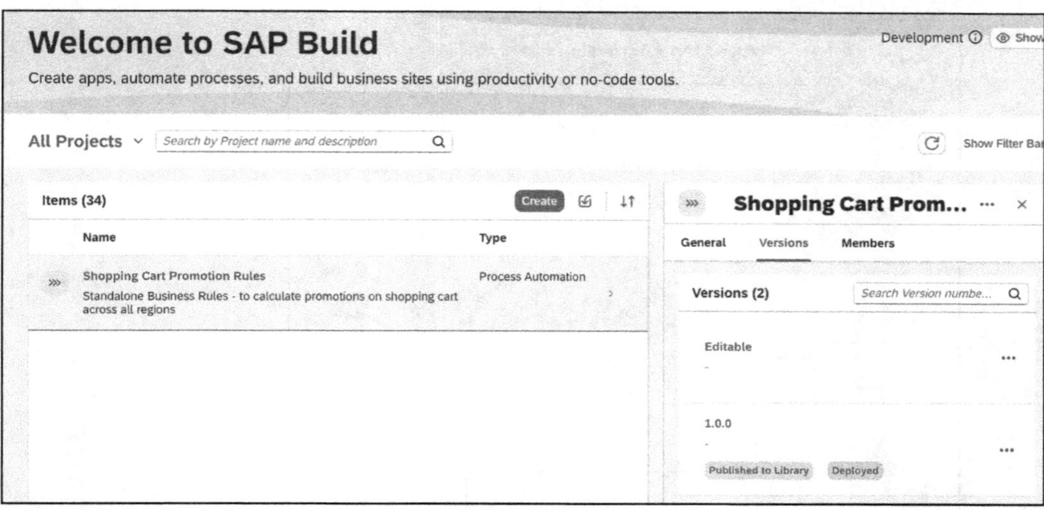

Figure 8.55 Project Version Published to the Library

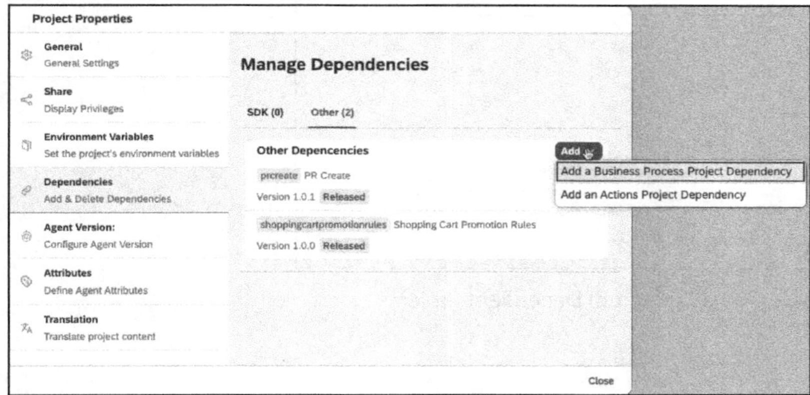

Figure 8.56 Add a Project Dependency to the Project Containing Decisions Artifacts

3. In Figure 8.56, a dependency to the **Shopping Cart Promotion Rules** project was added. This makes all artifacts in the **Shopping Cart Promotion Rules** project, including decisions, data types, and so on, will be available for consumption in the current project.
4. In the process builder, when you try to add a decision activity, you'll be able to see the decisions from the dependent project. In Figure 8.57, you can see the decisions available within the current project (**Determine Approver**) and decisions available from the **Shopping Cart Promotion Rules** project (**Promotion Rules** and **Shipping-Rules**). You can also see the project that contains these decisions. It's useful especially when you have multiple project dependencies. You can now map the inputs and outputs as needed, and then work with it such as a decision artifact available within the project.

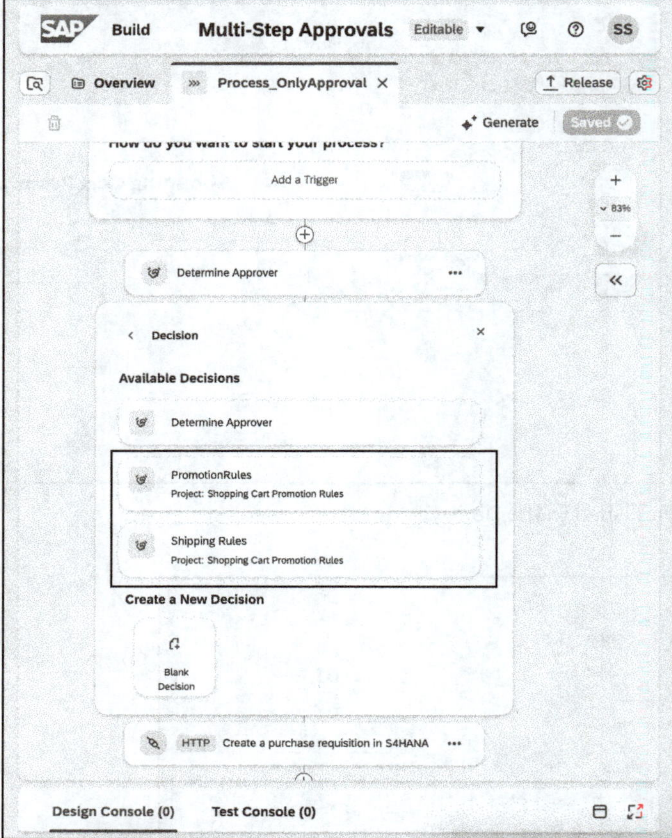

Figure 8.57 Decisions Available from Dependent Projects

8.5 Summary

In this chapter, we explored the decisions capability of SAP Build Process Automation and its pivotal role in automating and managing business logic. We began by defining decision management, emphasizing its importance in centralizing and automating decision-making across processes. Next, we examined various SAP decision solutions, guiding users on selecting the appropriate rule engine. We discussed different deployment options—cloud runtime, SAP HANA runtime, and ABAP runtime. We broke down and understood the building blocks of decision models, including vocabulary, decision tables, text rules, and decision diagrams. Finally, we covered best practices for creating decisions, including how to define inputs, outputs, decision tables, and text-based rules. This framework enables robust, scalable process automation.

Now that you've understood how various artifacts in SAP Build Process Automation are created, we'll look at how they can be deployed and executed in the next chapter. We'll also briefly introduce to the concept of business and technical monitoring.

Chapter 9
Business Project Lifecycle Management

Developing robust and scalable automation solutions requires more than just technical expertise. It demands a commitment to structure, clarity, and maintainability from the very first line of code. This chapter explores the lifecycle management within SAP Build Process Automation, ensuring your projects are not only successful today but also remain adaptable, manageable, and scalable for years to come.

Business project lifecycle management comprises the entire spectrum of activities involved in the creation, release, deploy, and maintenance of a process automation project. It involves managing various artifacts such as processes, workflows, decisions, automations, process visibility scenarios, data types, and so on, all of which play a crucial role in the successful implementation of process automation within an organization.

Throughout this chapter, we'll explore the key stages of the project lifecycle, including project creation, configuration, deployment, execution, and monitoring. We'll also discuss the best practices for managing each stage that can significantly impact the overall project outcomes.

9.1 Lifecycle Phases of a Business Process Project

In this section, we'll explore the various stages that a business process project goes through from creation to deployment and execution. We'll cover the key phases, including creating and configuring a project, releasing and deploying a project, and finally executing and processing the tasks. Understanding these phases is important for efficiently managing a business process project and ensuring its smooth implementation.

9.1.1 Create and Configure

In SAP Build Process Automation, setting up a project involves creating a business process project that integrates various process-focused steps, referred to as artifacts. Artifacts are important components that allow you to configure and manage the tasks necessary for your process to run successfully. They can vary from simple steps to more complex building blocks, and their specific configuration will depend on the process you're designing.

In the following sections, we'll discuss the steps in creating and configuring a project. You'll understand what exactly should be included in a project. Then, you'll learn how to manage the project dependencies and project members such as assigning roles and privileges to developers and administrators. Next, we'll explore the process of enabling translation for the project and important project actions such as copying, importing, and exporting projects from one tenant to the other.

Define a Project

If you're wondering what exactly a project should be and what artifacts should be created within a project, then let's consider an example with a scenario in the procurement department. While creating a project in SAP Build Process Automation for procurement, follow these best practices:

- **Define clear objectives.**
  Set specific and relevant goals that should be achieved with the project. Determine the exact part of the procurement process you want to automate, such as purchase requisitions, purchase order approvals, vendor management, and so on. Based on the business requirements, prioritize the most impactful areas first, such as automation high-volume tasks or those with significant time savings. For a procurement project, the objective might be to automate the approval of purchase orders under $12,000, aiming to reduce manual approval time from seven days to one hour while ensuring compliance with company policies.
- **Use a modular design.**
  Structure your project into distinct reusable artifacts so that each can handle a specific function within the overall process. A modular design of your project and artifacts enables flexibility, maintainability, and scalability. To achieve that, break down your project into smaller manageable processes such as approval workflow, approver determination process, automation to create purchase order, purchase order data type, and so on. Ensure that these modules have well-defined interfaces (inputs and outputs) and can function independently so that they can be used in different projects if needed.
- **Design for reusability.**
  Design artifacts and components in a way that they can be used across multiple projects or with the same project. Use parameters to make artifacts adaptable to different contexts. For instance, a process could accept parameters, such as the approver, allowing it to be reused in a multistep approval process. Identify common processes, application programming interfaces (APIs), or business rules that multiple projects might share and centralize them. For example, a vendor verification process can be used in both procurement and accounts payable automation projects. Another

example for reusable artifacts could be the delegation of authority decisions, business rules, or even data types. This approach saves time to develop features, reduces errors, and ensures consistence in the automation processes. Another example of reusable artifacts is the software developer kits (SDKs) that are predelivered in SAP Build Process Automation such as the Excel SDK, PDF SDK, Google Vision AI SDK, Google Authorization SDK, Outlook SDK, Microsoft 365 Cloud SDK, SAP GUI for HTML SDK, SAP Ariba Solutions SDK, and so on, which can be used in the creation of automation bots.

Manage Project Dependencies

Based on the best practices to define a project, you can then identify the granularity of a process or even a project. You can create data types, decisions, processes, and forms that can be reused in more than one project and publish a released version of the project as libraries, as shown in Figure 9.1.

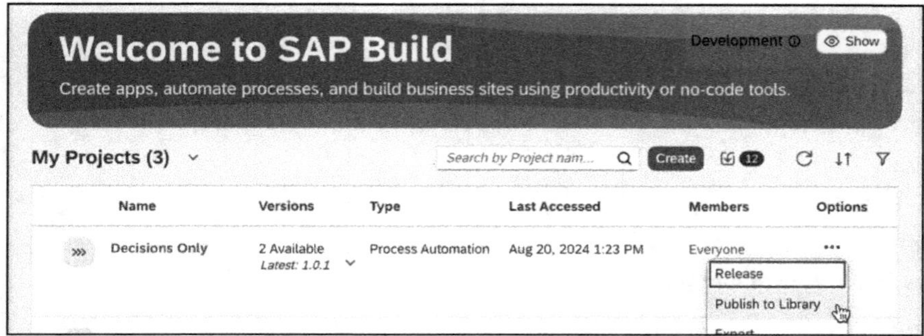

Figure 9.1 Publish a Released Version of a Project to Library

Once a project has been published to the library, it can be used from any other project in the same subaccount by adding a project dependency. A *project dependency* in SAP Build Process Automation refers to linking your project with artifacts from other sources, such as SDK packages or different process automation projects, event projects, or action projects. This allows you to easily select and use these preexisting elements in your processes, automations, and decisions.

To add a process automation project or SDK project as a dependency, open the project settings from the project settings icon ⚙. Then, click **Add • Add a Business Process Project Dependency**, as shown in Figure 9.2, to add a project dependency.

Once the project dependency has been added, the artifacts from the dependencies can be consumed in your project from the respective editors. For example, when you try to add a subprocess in the process editor, all the processes from the current project as well as processes from the dependent project are listed for your selection. Figure 9.3 shows how a process (**Process_OnlyApproval**) from a dependent project (**Multi-Step Approvals**) is shown to the user for selection.

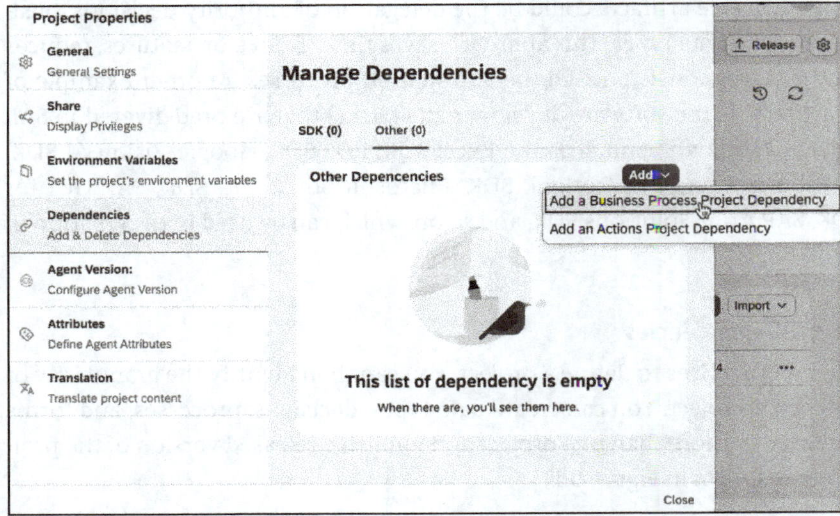

Figure 9.2 Adding a Project Dependency to a Business Process Project or an Action Project

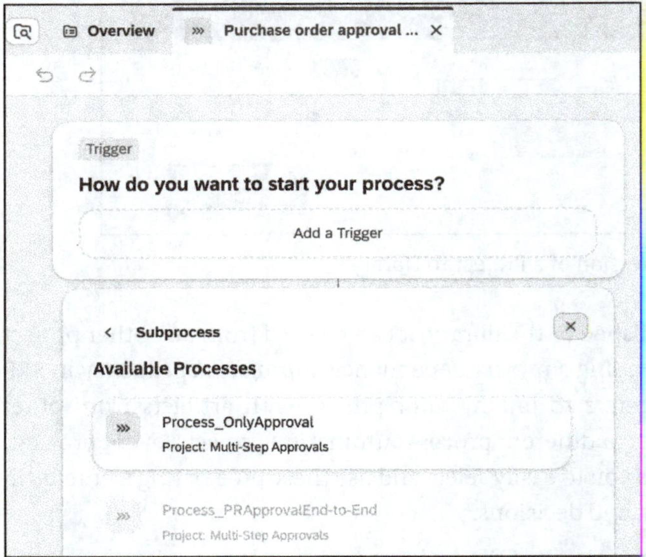

Figure 9.3 Consuming a Process from a Dependent Project

Note

You can't add a process with a form-based trigger as a subprocess. Those processes will be disabled for selection in the subprocess dialog.

Figure 9.4 shows how a data type from a dependent project can be consumed in the **Configure Process Inputs** dialog by choosing the appropriate data type as **Type**. The

TYP_Approvers and **TYP_Comments** data types are available from the **multistepapprovals** dependent project. Similarly, other artifacts such as decisions, automations, and forms are available in the respective editors.

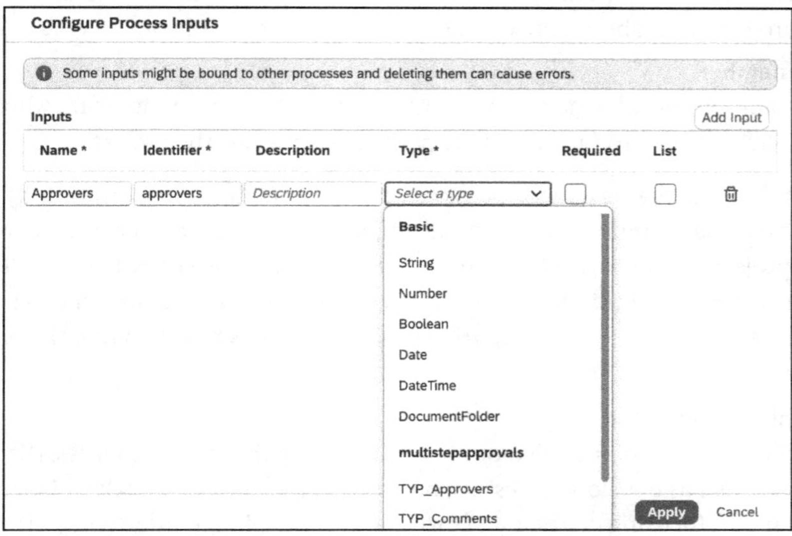

Figure 9.4 Consuming Data Types from a Dependent Project

Manage Project Members

In SAP Build Process Automation, you can access the application development lobby only if your user is assigned with either the **ProcessAutomationDeveloper** or **ProcessAutomationAdmin** role collections. In the lobby, it's possible to configure different members for a project, assigning them roles based on their responsibilities and the tasks that they need to perform. There are three main authorizations available for a project in SAP Build Process Automation:

- **Viewer**
  Users with the viewer authorization can view, export, and deploy the project. While they have full read access, they can't modify the project's existing content.
- **Developer**
  In addition to all viewing permissions, this grants the ability to make changes to the project, such as modifying existing configurations, and then releasing and publishing the project. However, they can't control permission of other users.
- **Administrator**
  This gives you the highest level of access, letting you manage all aspects of the project and team. You can control permissions, add or remove members, and do everything a viewer or developer can do. As the project creator, you'll have this by default

Apart from this, the project can also be toggled with the following **General Access** permissions:

- **Everyone with access**
  Allows anyone with the `ProcessAutomationDeveloper` role to access the project. Though it's not recommended to provide everyone access to a project, there might be cases where this is needed. If the team size is small or for experimental projects or educational projects, it might not make sense to enforce access control.
- **Only project members**
  This is the default option when a new project is created. Only those members who have been explicitly added to the project are permitted to access the project.

When a user with either the developer or the admin role creates a new project, by default only that user has complete access to the project. No other developers can view or manage the project and its artifacts. It's easier to make the project available to all developers in the account, but this increases the risk of unauthorized changes and various other potential security issues. Configuring project members based on their roles has several advantages:

- **Efficient collaboration**
  Clear role definitions streamline collaboration by outlining the responsibilities. Different personas such as developers, testers, business users, and other stakeholders can work in parallel, focusing on their tasks without overlapping or interfering with each other's work.
- **Security and access control**
  Assigning specific roles ensures that each member has access that is necessary for their responsibility. This minimizes the risk of unauthorized changes and can even protect sensitive project data.
- **Compliance**
  Proper role configuration supports compliance with industry regulations and internal company policies. It also simplifies the audits by providing clear records of who made what changes to the project artifacts and when.

To manage members of a project in the lobby, click the options icon ••• of the project, and select **Manage Members**. An overview of all the members currently assigned to the project is shown in a list along with their authorizations. In the **Type** field, choose either **Member** for granting privileges to one or more individual users by their email address (see Figure 9.5) or **User Group** for granting privileges to a group of users (see Figure 9.6). Then, choose the appropriate authorization for the user(s) or user group(s), and save the changes.

Tip

If you had maintained a destination with the name **Identity_Authentication_Connectivity_IDS** to connect to the identity directory of Identity Authentication, then you can choose the email address and user group from the available list using the value help option. Otherwise, the value help won't appear, allowing you to only enter the values for email address and user group, which can be an error-prone process.

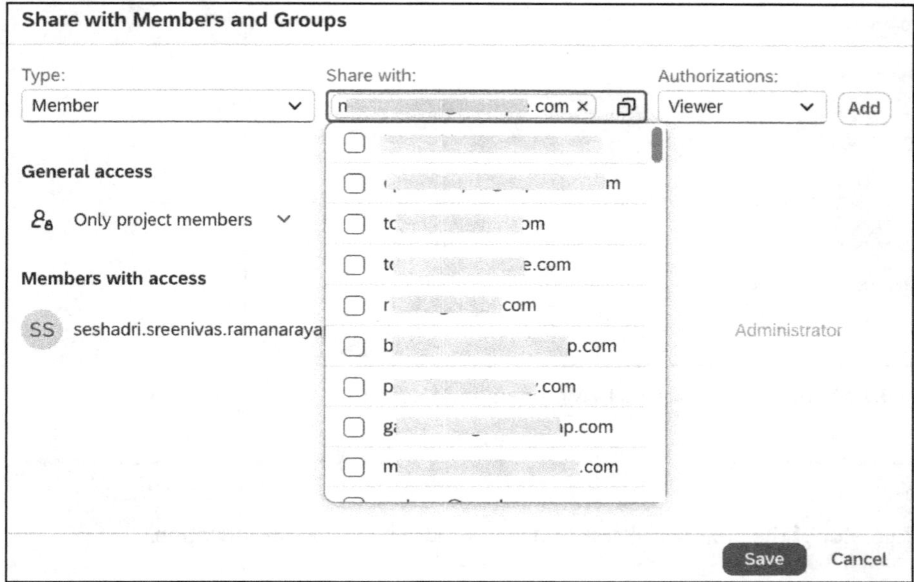

Figure 9.5 Grant Project Privilege to One or More Individual Members

Figure 9.6 Granting Project Privileges to One or More User Groups

Figure 9.7 shows how the project authorization will look when more than one user and user group have been granted privileges.

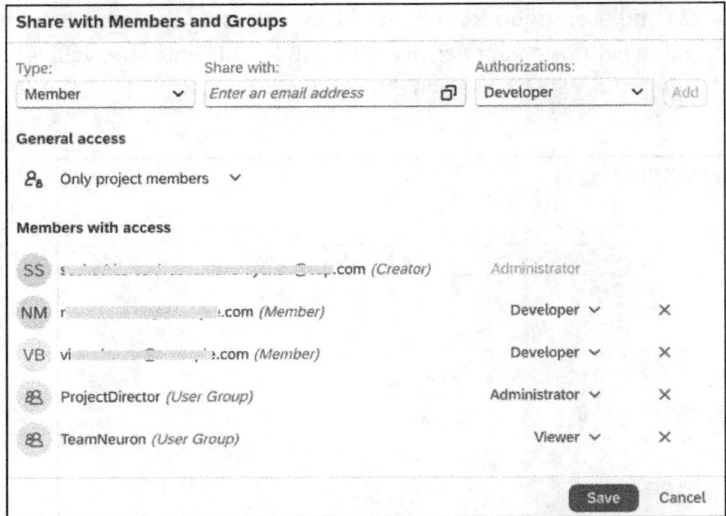

Figure 9.7 Sharing Project Access to Members and User Groups

> **Tip**
> The creator of the project will always have the administrator authorization.

Translation

Before deploying a process automation project to the production environment, it's important to verify whether the project includes any texts that require translation. As a best practice, establish a standard development language, for example, English, for all project and artifact creation. This means that all texts, including artifact names, descriptions, field labels, task titles, and so on, should be maintained in this development language. However, end users interacting with the process, such as accessing start forms or tasks from My Inbox or SAP Task Center, will need to see these texts in their preferred language. Making the texts within applications available in different languages makes it more accessible to a wider range of users within your company. Although SAP Build Process Automation doesn't auto-translate the texts into different languages, it supports the usage of translated texts for end user–relevant content such as data type name and description, process inputs and outputs, process name and description, approval task subject, task outcomes, and so on.

To export the translation-relevant texts, follow these steps:

1. Open the project settings from the project settings icon ⚙, and choose **Translation**.
2. Choose **Export File** to download a translation properties file. The file contains a list of keys and the content to be translated.

In the snapshot of the translation file shown in Listing 9.1, you can see different texts available for translation such as the activity name, approval task subject, artifact name, and so on. Create one copy of the file per target language, and name it according to the ISO 639-1 language code format. For the file containing German translations, use *translation_de.properties*; for the file containing English translations, use *translation_en.properties*; for the file containing French translations, use *translation_fr.properties*; and so on. Translate the content that is to the right of the = sign.

```
#XTXT
process_
OnlyApproval.c.a90a9c79c35641ca984d37affce76101.data.artifact.model.contents.de
cision_determineApprover_1.name=Determine Approver
#XTXT
process_
OnlyApproval.c.a90a9c79c35641ca984d37affce76101.data.artifact.model.contents.fo
rm_form_Approval_1.subject=Approve Purchase Request -
${context.startEvent.purchaserequest.PR_Name}
#XTXT
TYP_PurchaseRequest.md.name=TYP_PurchaseRequest
#XTXT
process_OnlyApproval.h.inputs.title=inputs
```

Listing 9.1 Snapshot of the Translation Properties File Named translation.properties

Warning

Make sure that you only change the content during translation and that the key remains unchanged.

To import the translation-relevant texts, follow these steps:

1. Open the project settings from the project settings icon ⚙, and choose **Translation**.
2. Choose **Import** to upload a translation properties file.
3. Browse and choose the files from the local file system.
4. If you've followed the file naming convention per the ISO-639-1 language code format, then the language should be automatically detected. If not, you can manually select or change the language of the file that is being uploaded.
5. Click **Import**.

Figure 9.8 shows the dialog to export, import, and manage the translation file. The translation-relevant texts are applied to the project artifacts after deployment and are used in the end user–facing applications such as My Inbox, SAP Task Center, and so on.

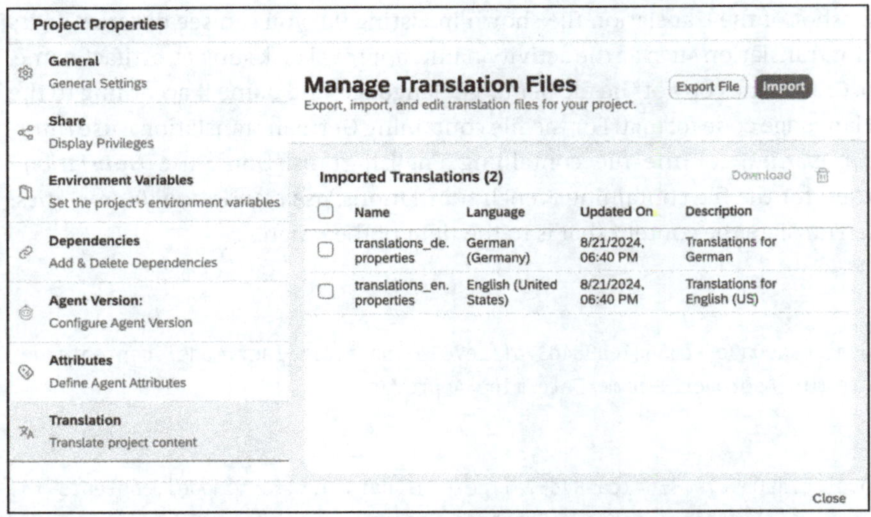

Figure 9.8 Export and Import Translation-Relevant Texts for the Project

Copy, Import, and Export Project

You can also create a new project by copying from an existing project in the lobby or import a version of a project from another system. Copying a project provides a starting point for new process automation projects while preserving the original project's integrity. Importing a project from another system enables the reuse or sharing of process automation assets across different tenants or landscapes.

To create a new project as a copy of an existing project, follow these steps:

1. Open the existing project in the lobby.
2. Go to **Options** •••, and select **Save as New Project**.
3. Select the version to copy if multiple versions exist for the project.
4. Update the **Project Name** and **Description**.
5. Choose **Save as New**.

To export a project to your local system, follow these steps:

1. Open the existing project in the lobby.
2. Go to **Options** •••, and select **Export**.
3. In the **Export Project** dialog, select the version that you want to export.
4. Choose **Export**.
5. This generates a file with extension *.mtar*. Save the file to the local system.

To import a project from a local file system, follow these steps:

1. Navigate to the lobby.
2. Choose the import icon .

3. In the **Import** dialog, choose **Browse Files** to search for the **.zip* or **.mtar* file in your local system, or choose projects from the list for import, if available.
4. The file will be uploaded to the lobby.
5. Select the uploaded file and choose **Import**. On successful import, the project will be available in the lobby.

> **Note**
>
> The maximum file size of the project *.mtar* file that can be imported is 400 MB.

> **Tip**
>
> If you export the project and import it in the same tenant, then the import might fail due to a naming conflict. SAP Build Process Automation requires a unique name for the project. Change the name of the existing project in the lobby before importing the project *.mtar* file.

Best Practices for Process Automation Projects

Now that you have a good understanding of the different project settings in the creation and configuration phase of a process automation project, here are some of the best practices that will help you create efficient, maintainable, and scalable process automation projects:

- **Granularity of the project and artifacts**
  As explained in the "Define a Project" subsection earlier, consider having a modular design for the project with reusability in mind. A well-defined project structure can be more efficient, flexible, and scalable, ultimately saving time and reducing errors in the process.
- **Data types**
  Data types play a crucial role in SAP Build Process Automation by providing a structured and consistent way to define and handle data within your automation processes. Because the data types for common business objects such as purchase orders or purchase requests don't change, consider creating them in a separate project so that they can be reused across different procurement-specific projects. When you add a dependency to action projects, you also inherit data types of the project that can be used in your project.
- **Usage of decisions**
  Using a decision or business rule in a process automation project helps to involve subject matter experts to define the decision logic. If the same logic needs to be applied in more than one place in the process, including decisions will reduce errors and improve consistency.

- **Naming conventions**

  In a process automation project, you can create as many as 15 different artifacts such as automations, processes, decisions, data types, forms, and so on. As your project grows, managing the different artifacts within the project could be a challenge.

  Ensure than the names accurately and clearly reflect the purpose and function of the artifact such as process, decision, automation, data type, and so on. This will help users easily understand the purpose and functionality of the element.

  Use standardized naming conventions across all processes, decisions, automations, and data types to ensure consistency and clarity for all users. This could include using prefixes or suffixes to indicate the type of element, such as "PROC_" for processes, "DEC_" for decisions, "DT_" for data types, and so on.

  Document the naming conventions used for processes, decisions, automations, and data types to provide a reference for all users. This documentation should be easily accessible and regularly updated as new elements are added.

  Finally, incorporate user suggestions and insights to continually improve the naming practices that suit your organization.

Tip

When there are numerous artifacts within your process automation project, finding an artifact that a user is looking for can be difficult. In such cases, there are a few ways to easily find artifacts. If you know the name or part of the name, then you can search for the artifact in the search bar. If you've maintained a prefix for the artifact type (e.g., DT_ for data type, PROC_ for process) then you can search for the prefix to filter the artifact type. You can also click the explorer icon to view the different artifacts based on the artifact type.

9.1.2 Release and Deploy

In this section, we'll explore the steps involved in releasing and deploying a project. We'll begin by discussing the project lifecycle and the meaning of editable, released, and deployed states. Then, we'll learn how to release a version of a project and deploy it to different environments. Finally, we'll learn about restoring a version of the project, publishing it to a library, and transporting it between development, quality, and production tenants.

Project Lifecycle

Project lifecycle status in SAP Build Process Automation is a structured framework that guides the various stages a project can go through, from creation to deployment, and potentially, to retirement. By using lifecycle statuses, you can effectively manage different versions of your projects. There are three lifecycle stages of a process automation project in SAP Build Process Automation:

- **Editable**
  This is the default status assigned to a project upon creation. An editable project is basically in its draft phase. This status allows for modifications and serves as a sandbox for developers to iterate on the project.
- **Released**
  The stage is assigned when a project is generated or imported, signaling that the project is out of the draft phase and potentially ready for broader consumption. Released projects are read-only but can be released again to create new released versions or to be deleted. Moving a project to **Released** sets the stage for deployment.
- **Deployed**
  This is the status that signifies a project can be used within a productive environment or that the artifacts in the project can be executed. A deployed project is neither editable nor deletable to prevent changes. However, you can undeploy a project to reset its status back to **Released**.

When you open a project from the lobby, the **Editable** version of the project is shown to you. By clicking the **Editable** dropdown list, you can access the different versions of the project, if they exist. Figure 9.9 shows the three different lifecycle stages of a process automation project.

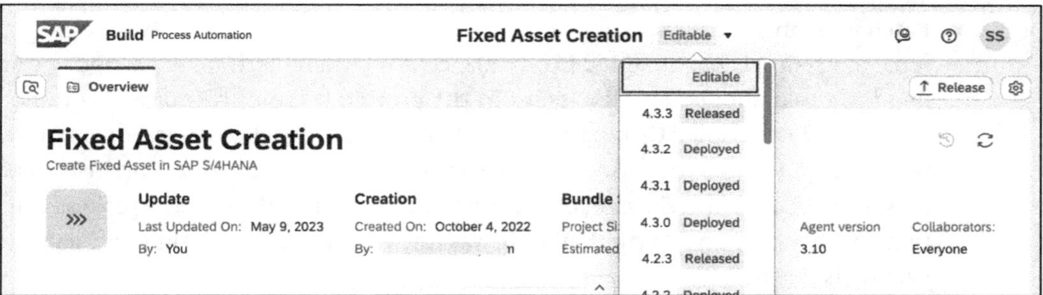

Figure 9.9 Project Lifecycle Stages

Undeployment in SAP Build Process Automation has specific behaviors depending on the contained artifacts. For projects with automations, application captures, data types, decisions, project launcher, or user tasks, the version and related artifacts such as triggers and variables will be undeployed, possibly canceling ongoing jobs. However, completed or canceled jobs remain accessible.

Warning

For projects with forms, processes, or visibility scenarios, administrative rights are necessary, and *all* project versions will be undeployed, removing workflow definitions and therefore ending running instances, deleting pending approval forms, and deleting completed workflow instances. Undeployment is enabled for nonprocess projects even in tenants marked as production (**Control Tower • Tenant Details**), but undeployment is disabled if the project includes a process to prevent any loss of data. Don't undeploy a project containing processes until you check that it's no longer needed, that it's no longer running, and that the corresponding processing data is kept elsewhere. Refer to SAP Note 3245676 for more details.

Releasing a Version

Versioning in SAP Build Process Automation is the process of tracking and managing changes to a project throughout its lifecycle. This includes keeping a record of each iteration of the project, as well as the ability to revert to previous versions if necessary. It enables teams to track the history of a project, understand who made certain changes, and maintain a clear audit trail. This is important in quality control and being compliant with regulations, as well as for troubleshooting and identifying issues that may arise during the building process.

SAP Build Process Automation uses the concept of *semantic versioning* in the format of x.y.z (major.minor.patch). This scheme uses version numbers, and their changes convey the extent of change to the underlying process automation project from one version to the other:

- **Patch version**
  A patch version is incremented (from 1.0.0 to 1.0.1) when there are small updates or bug fixes that address specific issues in the project. It typically indicates a minor change that is designed to improve the stability of the project artifacts. For example, when fixing a bug in the form where it was displaying sales organization instead of sales order ID, when the user task title is enhanced with fields from the process context, or when additional error handling was implemented in an automation, a patch version is released.
- **Minor version**
  A minor version is incremented (from 1.0.0 to 1.1.0) when new features are added in a backwards-compatible manner. For example, when new fields are made available in a form, when additional condition columns are enabled in a decision table, or when new email notifications were sent to approvers, a minor version can be released.
- **Major version**
  A major version is incremented (from 1.0.0 to 2.0.0) when there are significant changes that aren't backwards compatible. For example, when using new decisions for delegation of authority, when the process trigger is changed from event-based to API-based, or when a process was modularized to use variant template, a major version will be released.

Note

When a project is released for the first time ever, the version is always marked as 1.0.0, and this behavior can't be changed.

When the project is deemed ready for versioning, you can click the **Release** button to create a new version, which brings up the **Release Project** dialog. You must choose the extent of change that has gone into the project artifacts since the last version, based on which the version number is automatically determined by SAP Build Process Automation. Figure 9.10 shows a patch version being released for the project.

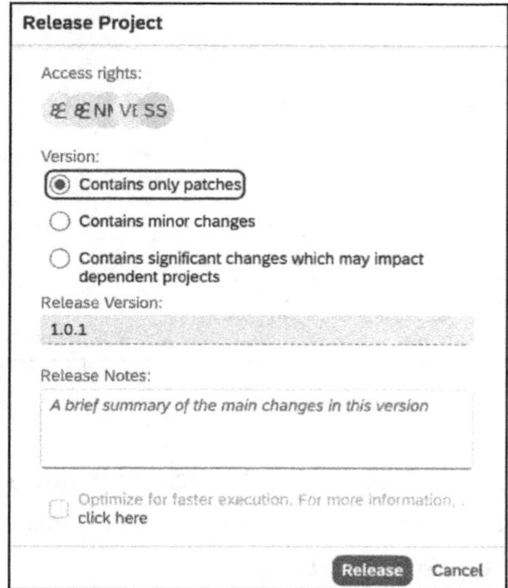

Figure 9.10 Releasing a Patch Version for a Process Automation Project

For projects with automations that automate the screens in SAP WebGUI, a dependency to the SAP WebGUI SDK is automatically added. For such projects, choose the **Optimize for faster execution** option while releasing a version.

Note

Don't use the **Optimize for faster execution** option for debugging purposes because the actions are executed in batch processing, and the data in the tester window might not be valid. Therefore, we recommend that you only select this feature once your automation is finalized.

Deploying a Project

The artifacts within a process automation project become executable only when the project is deployed, and only released versions of a project can be deployed. Deploying a project is simple—from a released version of a project, click the **Deploy** button, as shown in Figure 9.11.

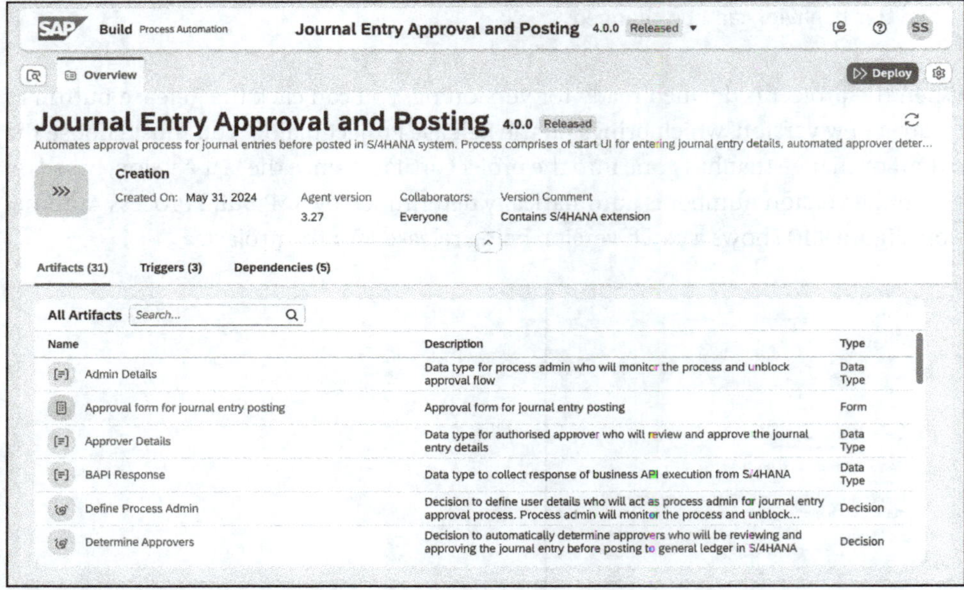

Figure 9.11 Deploy Released Version of a Project

However, you need to understand that a project can be deployed and executed inside a functional area called an *environment*. We talk more about environments in Chapter 10, but here we'll discuss what an environment is and why is it needed.

You can think of an environment as a shell or a container within SAP Build Process Automation where the different artifacts (e.g., processes, automations, decisions, etc.) are deployed and executed from. In SAP Build Process Automation, there are two key types of environments: public and shared. The *public environment* is the standard space everyone shares, governed by security permissions. *Shared environments* are a bit different. They can be set up by users with the `ProcessAutomationAdmin` role and can be shared with specific users or groups who have the `ProcessAutomationDeveloper` role. You could create multiple shared environments in SAP Build Process Automation and deploy the same project into different environments. The environments can have its own settings and configurations, and you can even choose which user(s) can be part of an environment. This allows for more tailored project deployments, with the flexibility of different settings and variables, and better control over who can execute processes, automations, decisions, or access a specific form.

Imagine a situation where you want to transport your content from the development/quality tenant to the production tenant. Once you transport and deploy a project into production, then the project and its artifacts are basically live. However, you might want to validate the artifacts in the production tenant before making it actually live. If there were two environments in the production tenant, called **Pre_Production** and **Production**, each with their own configurations, then you could deploy the project into the **Pre_Production** environment and execute the artifacts. Figure 9.12 shows such a selection of environment when you click the **Deploy** button in the project. A few business users could be given access to the **Pre_Production** environment to test the project. If the test was successful, then you can deploy the same project version into the **Production** environment. This will make sure that the business isn't affected due to bugs or issues that are discoverable only in the production landscape.

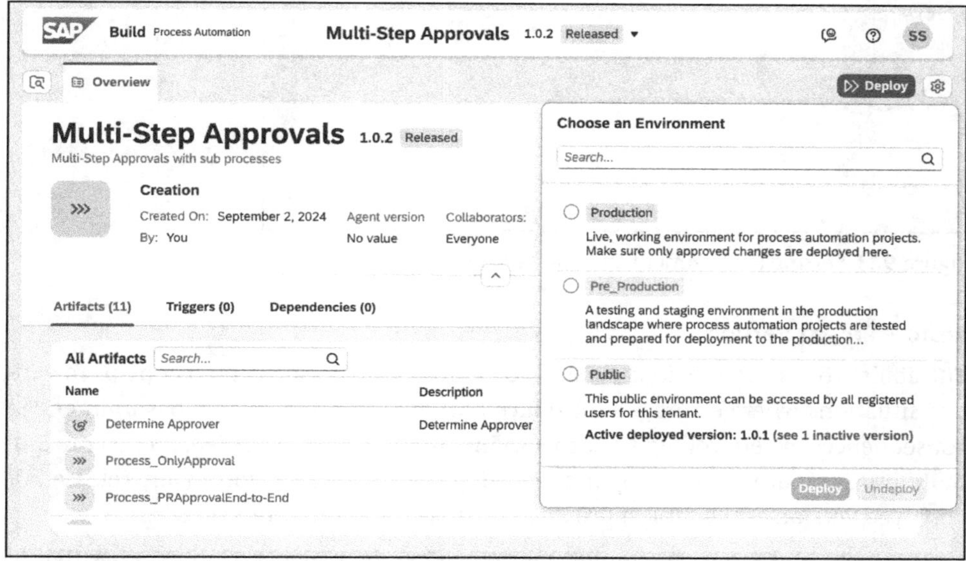

Figure 9.12 Selection of Environments during Deployment

Selecting the environment enables the **Deploy** button in the **Choose an Environment** dialog. Click the **Deploy** button in the dialog, and proceed to the next step of **Define Variables**. If any of the artifacts used an environment variable, then those variables need to be assigned values during deployment of the project, as shown in Figure 9.13. This will allow the artifacts to use the values of the variables automatically when they are executed. Environment variables can be reused across different projects within an environment. Therefore, during deployment, you can either set a new value or use existing values available in the environment. Examples of environment variables include destinations, dates, times, users, and so on.

> **Note**
>
> You can't deploy individual artifacts within a process automation project. Rather, the entire project will be deployed. However, you can disable certain artifacts (e.g., an automation) before releasing a version and enable it back in the subsequent versions if needed.

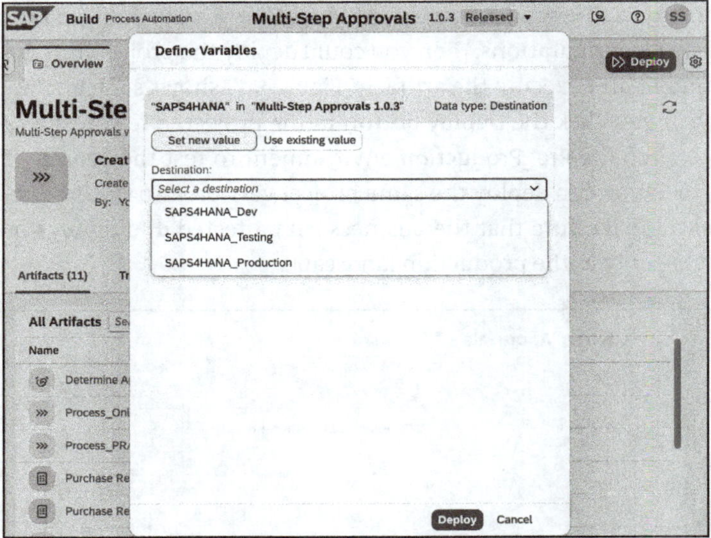

Figure 9.13 Setting Environment Variables during Deployment

Restore Project Version

The ability to restore previous versions of a process automation project helps in various situations where errors are introduced, where the latest version has unintended consequences, where rolling back to specific functionality is needed, and so on. SAP Build Process Automation supports reverting the editable version to any released or deployed version of the project provided the following conditions satisfy:

- The major and minor version of the released or deployed version matches that of the latest built package.
- None of the artifacts in the editable version are currently being edited.

If the previous condition satisfies, then from the editable version of the project, click the **Restore a version** icon, as shown in Figure 9.14. Select the version to restore, and click **Confirm**.

Figure 9.14 Restore a Version of the Project

Publish a Project

Publishing a project to the SAP Build Process Automation library will make a version of the project discoverable and reusable from a different project. Publishing a project is an optional step, but you would want to publish a project into the SAP Build Process Automation library at least for the following use cases:

- You have some reusable artifacts such as data types, processes, automations, decisions, and so on in the project that are consumable from a different project.
- You have an artifact in your project that you want to trigger from an application created in SAP Build Apps.
- You have an action project that needs to be consumed from a process automation project.

To publish a project from the lobby, select the more icon •••, and choose the **Publish to Library** option, as shown in Figure 9.15. You can then choose the specific version that needs to be published and click the **Publish** button.

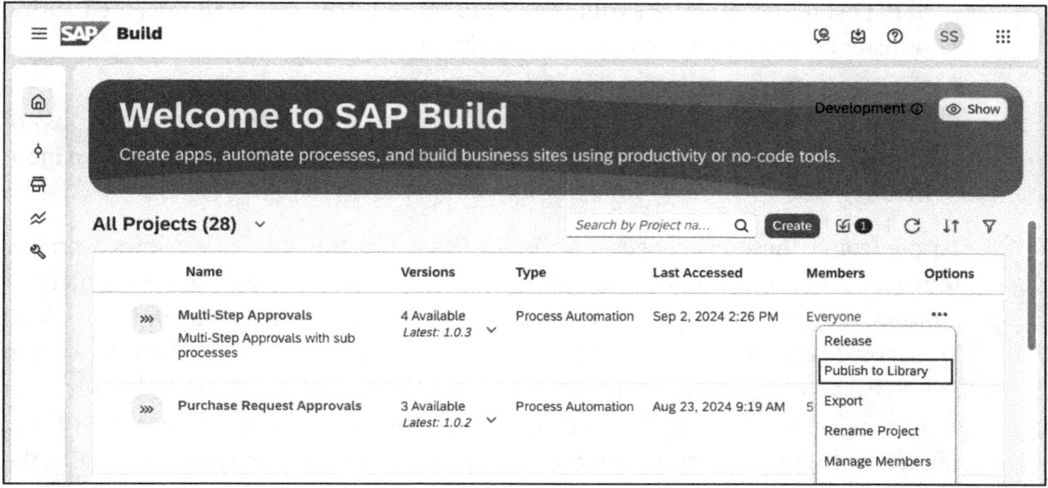

Figure 9.15 Publishing a Process Automation Project to the Library

> **Tip**
>
> The options to enter **Line of Business** and **Product** are relevant only when you want to publish the project into SAP Store for consumption by all SAP Build Process Automation customers and partners. Otherwise, you can leave it empty.

Transport a Project

Transporting a process automation project across systems is a crucial step in the development lifecycle, especially when working on complex enterprise applications. It allows the project to be transferred between different tenants, such as development, test or quality, and production, ensuring that the project and its artifacts are tested before releasing them into production. This results in a more reliable and efficient development process.

The easiest way to transport a process automation project from one system to another is to export the project into your local file system and import the same into the target system. However, such a file-based transport of projects isn't recommended in an enterprise setup for a variety of reasons. Some of the most important reasons are as follows:

- **Error-prone process**
  Because it's a manual process, human errors during the export, transfer, or import can lead to project failures, data loss, or misconfigurations that could have significant operational impacts.
- **Security risks**
  Exporting and importing files increases the risk of unintentional data breaches, loss, or exposure. Sensitive data could be compromised if the files are intercepted during transfer or accessed by unauthorized individuals.
- **Data integrity**
  The process of exporting and importing files can lead to data corruption, loss of metadata, and integrity issues, especially if files are manually handled or even modified.

To overcome these challenges, SAP Build has a functionality that allows seamless movement of the process automation project to be transported from one tenant to another. The process setup for transporting projects takes advantage of SAP Cloud Transport Management and SAP Content Agent service on SAP BTP. These services in SAP BTP enable developers to have more control over their processes and ensure that any changes made to applications are thoroughly tested and vetted before being released into the production environment. Here are the high-level steps to enable the transport capability in SAP Build:

1. **Set up SAP Cloud Transport Management.**
   The SAP Cloud Transport Management tenant needs to be set up in advance to manage the transport process between source and target tenants. This requires specific configuration to subscribe to the SAP Cloud Transport Management service, set up role collections, and configure the landscape to create transport destination, nodes, and routes. It's best set up on a special subaccount for single tenant applications/services.
2. **Set up the SAP Content Agent service in the source tenant.**
   In the source tenant, for example, in the development tenant of SAP Build Process Automation, create a service instance of the SAP Content Agent service to work with SAP Cloud Transport Management. Create a destination in the SAP BTP cockpit with

the name "TransportManagementService" using the OAuth client details of the SAP Cloud Transport Management service instance. Once these are in place, you'll need to get the credentials from the service key of the SAP Content Agent service instance and create a new destination in the SAP Cloud Transport Management subaccount representing the source tenant (development).

3. **Set up the SAP Content Agent service in the target tenant.**
   Like the source tenant, the target tenant (e.g., production) also requires an instance of the SAP Build Process Automation and an SAP Content Agent service instance. Once these are in place, you'll need to get the credentials from the service key of the SAP Content Agent service instance and create a new destination in the SAP Cloud Transport Management subaccount representing the target tenant (production).

Once the SAP Content Agent service and SAP Cloud Transport Management service have been set up, you can select a specific version of the project in the development tenant, which is tested successfully, and choose **Transport**, as shown in Figure 9.16. In the transport dialog that opens, enter the description for initiating the transport in the **Transport Description** field, and choose **Transport**. This will make the project available in the SAP Cloud Transport Management system, and a user with a specific role (SAP Cloud Transport Management import queue operator) needs to explicitly choose the target node to import the transport request into. Once the operator has placed the transport request in the import queue, the process automation project is now ready for import into the target tenant.

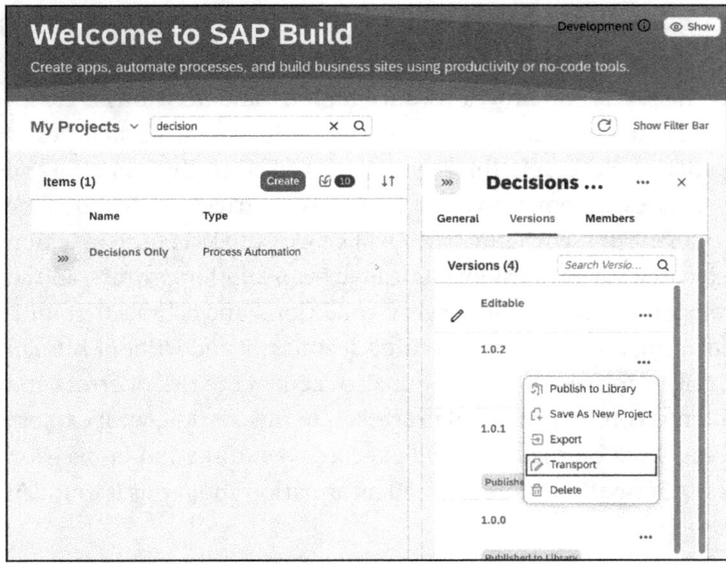

Figure 9.16 Transporting a Project Version

In the target tenant (e.g., production), navigate to the lobby, and select the import icon. In the **Import** dialog, you can click the **Browse Files** button to upload the project file.

Once the file is uploaded, it appears in the table, but it's not yet imported into the lobby. To import the project into the lobby, select the project in the table, and click **Import**, as shown in Figure 9.17.

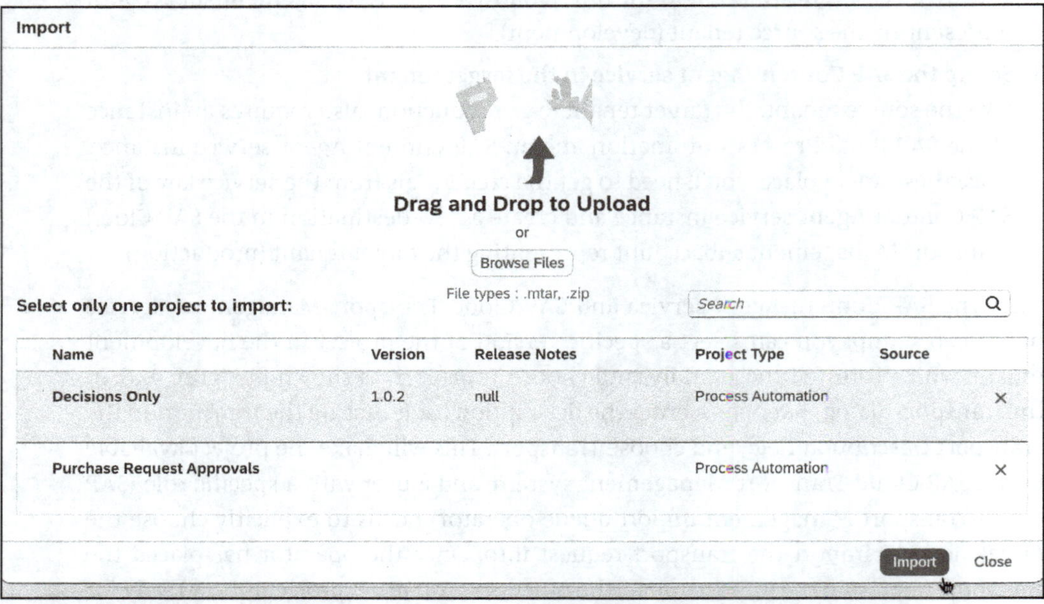

Figure 9.17 Importing a Project from the SAP Cloud Transport Management Queue

9.1.3 Execute and Process Tasks

Understanding how to effectively run processes, automations, and decisions is essential for business users, developers, and administrators to fully use the capabilities of SAP Build Process Automation. Success in this area depends on carefully designing and configuring how the processes, automations, and decisions are triggered. It should be clear during the development phase whether they will be executed via human action from an application, executed via business events, called from another process, and so on. Based on that, developers can set up the triggers, conditions, and actions that initiates their processes, ensuring that tasks are executed seamlessly and without human intervention. This not only improves efficiency but also reduces the risk of errors and inconsistencies that often accompany manual processes. In this section, we'll explore how to execute a process, work on the tasks, and execute an automation or decision within SAP Build Process Automation. More detailed information about this is found in Chapter 10.

During the development phase or at least in the initial stages, you wouldn't have the entire application to initiate the process, automation, or decision. So, starting an automation or process with sample data or mock data can help simulate interactions and validate the workflow logic before the full application is available. Different artifacts in

SAP Build Process Automation support different ways to be called, especially during the development process. For example, a developer can test run an automation from the automation editor by clicking the test icon ▷, as shown in Figure 9.18. This feature enables developers to execute the automation's functionality and adjust during the development phase.

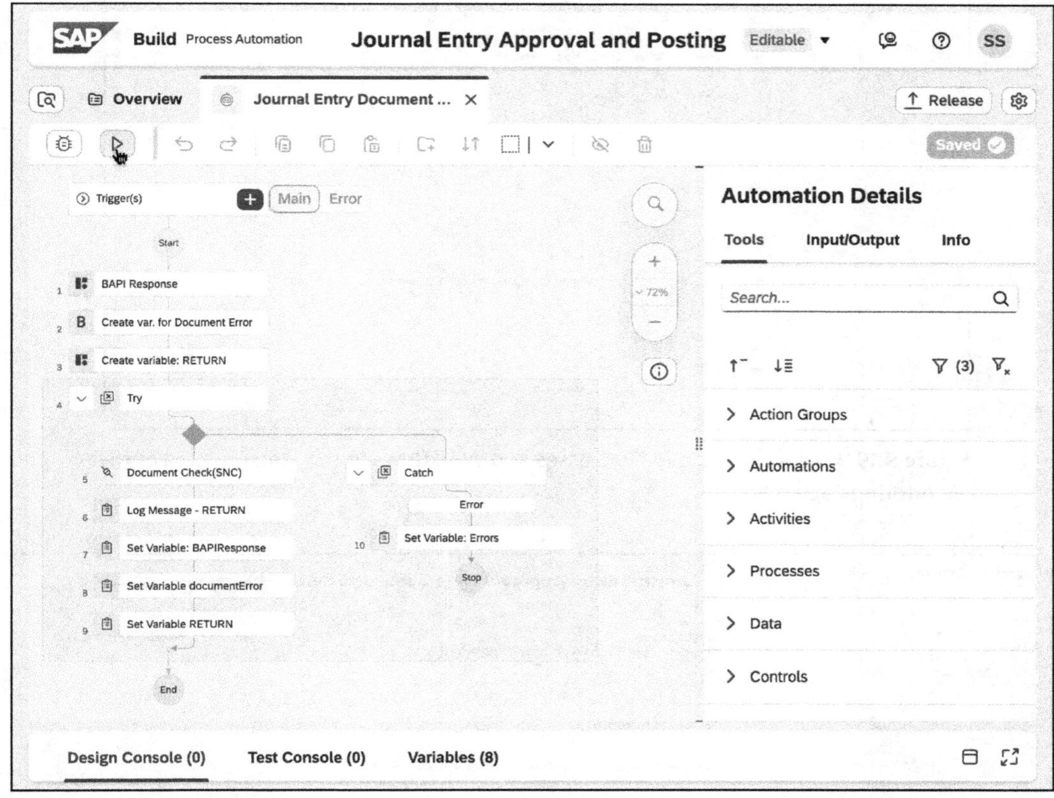

Figure 9.18 Test Automation during the Development Phase

Similarly, administrators and developers can use the **Processes and Workflows** application in the **Monitoring** page to access all deployed processes and workflows in the tenant, as shown in Figure 9.19.

In the Processes and Workflows application, you can initiate an instance of a specific process or workflow as well as provide a sample context or mock data during the process initiation, as shown in Figure 9.20. Search for the process that you want to initiate, click the **Start New Instance** button, provide the sample or mock JavaScript Object Notation (JSON) data that will be sent to the process instance, and finally click the **Start New Instance and Close** button. This allows developers to execute the process as modeled, test, and make corrections where needed during the development phase.

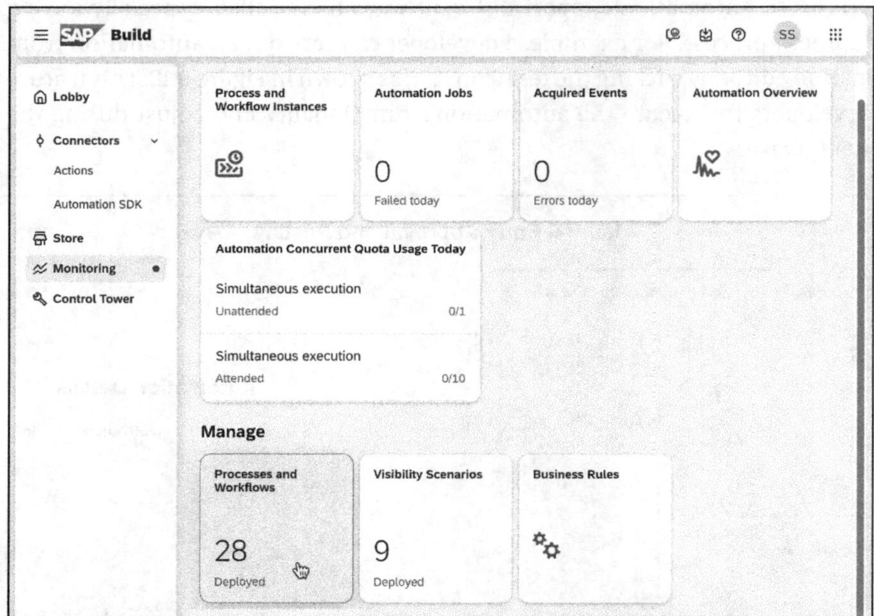

Figure 9.19 Access All Deployed Processes and Workflows in the Manage Area of the Monitoring Page

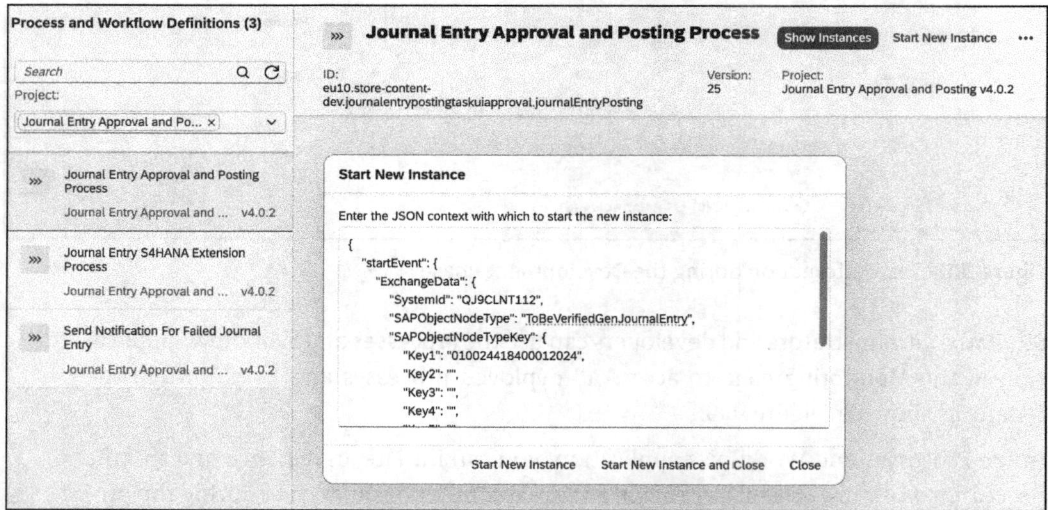

Figure 9.20 Starting a New Process Instance with Sample or Mock Data

To access the tasks that are created as part of the process, click the My Inbox icon in the lobby, which opens a developer-version of the My Inbox application. Here, developers can access the tasks that are assigned to their usernames, as shown in Figure 9.21.

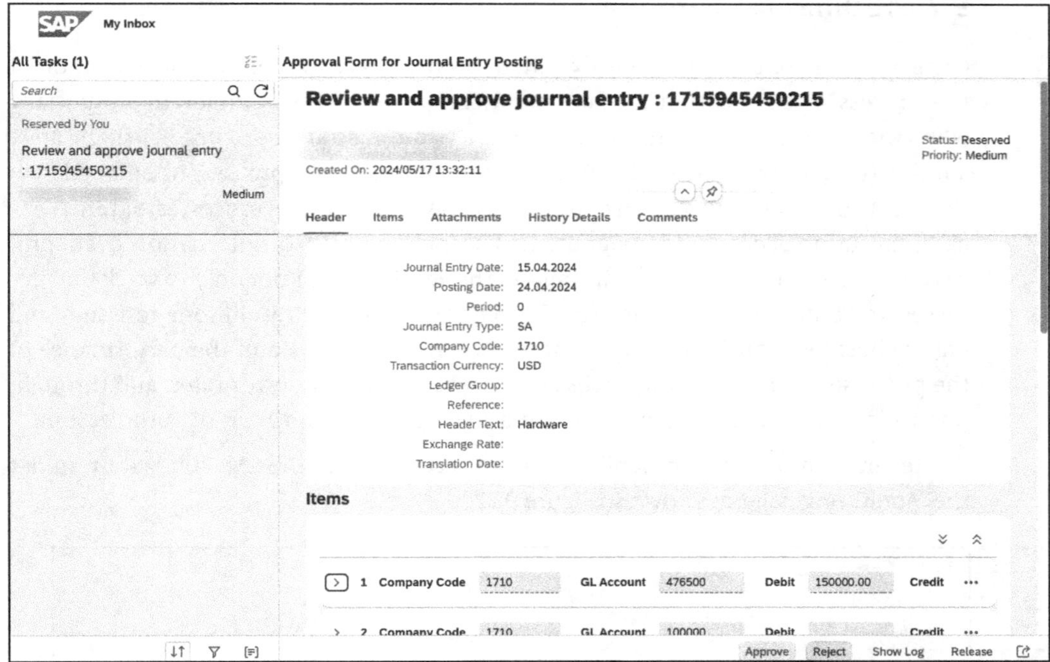

Figure 9.21 Developer Version of the My Inbox Application Accessed from the Lobby

> **Note**
>
> The developer version of the My Inbox application displays the same tasks and information as the SAP Fiori My Inbox app that is predelivered with SAP Build Process Automation. Accessing the developer version of My Inbox doesn't need SAP Build Work Zone configurations and doesn't come with SAP Build Work Zone–only capabilities such as substitutions, in-app notifications, SAP Mobile Start, and so on.

It's important to note that these methods are typical ways of starting a process or an automation during the development phase. Depending on how and where the artifacts are consumed from, the triggers for each of the artifacts can be defined during the development phase. For example, if a process will be triggered automatically based on a business event such as a sales order being blocked for delivery in SAP S/4HANA, then the process will be modeled to start with an event trigger. If an application will call a decision defined in SAP Build Process Automation, then an API trigger will be created for the decision. If an automation will be executed every 30 minutes, then a schedule trigger will be created for the automation. In Chapter 10, we'll cover in detail how a process or automation can be started in a typical process automation use case.

9.2 Technical Monitoring

Once a process or automation is executed, it's necessary to monitor the execution of the process or automation for successful completion. This is imperative especially when the process encounters errors. Users with administrative privileges, usually an IT administrator or process administrator, can use monitoring applications embedded in the lobby to track the execution of processes, the context of the processes, external API calls, execution of decisions, assignment of user tasks, health of automation, event processing, and so on. This monitoring allows the administrator not only to track the processes and automation but also to take actions to provide resolutions to issues and ensure business continuity. Administrators can also understand the performance of the processes and automations jobs such as execution time, error rates, and throughput, which can help the organization identify bottlenecks and areas of optimization.

The technical monitoring capabilities can be accessed from the SAP Build lobby, under the **Monitoring** page, as shown in Figure 9.22.

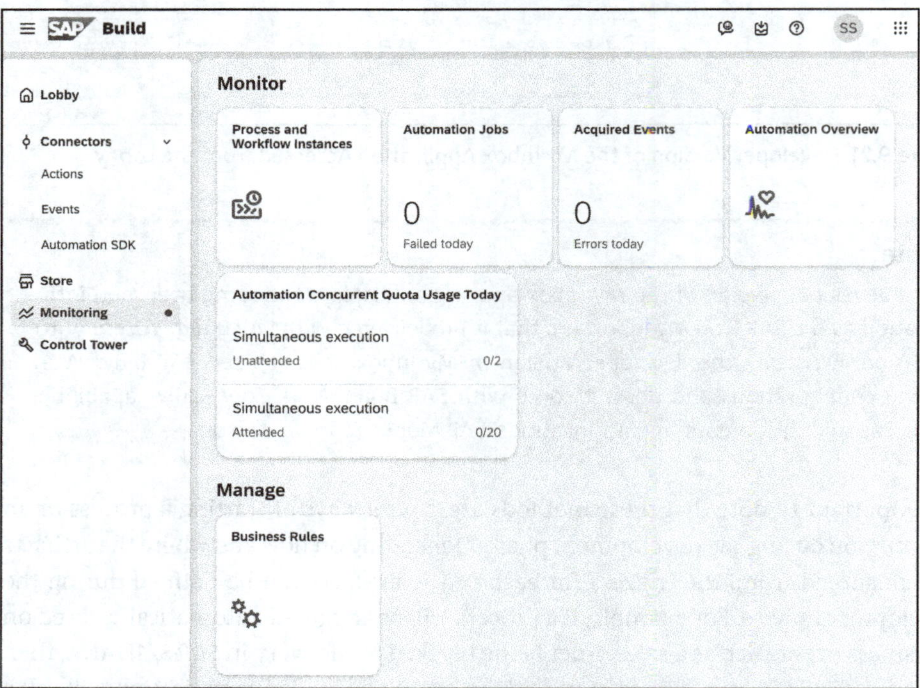

Figure 9.22 Monitoring Page in the SAP Build Lobby

There are two areas in the page for monitoring—**Monitor** and **Manage**. The apps in the **Monitor** area focus on knowing what is happening to the processes, workflows, automations, and events such as viewing the process instances, context data of the process, logs of an automation job, visibility events, errors, and business events. The apps in the **Manage** area enable the administrator to perform actions such as viewing deployed processes, starting a process instance, processing the visibility event data, downloading the workflow model, and so on.

> **Note**
>
> Technical monitoring provides a lot of capabilities that can change the outcome of the process. For example, the users can terminate a process, view the context using APIs, and even change the context using APIs. Therefore, it's behind role collection `ProcessAutomationAdmin`, and only those users who have been explicitly assigned this role collection can access the technical monitoring capabilities.

Technical monitoring is explained in detail in Chapter 11. However, in the following sections, we'll introduce you to the different technical monitoring capabilities in SAP Build Process Automation.

9.2.1 Monitoring Processes and Workflows

Clicking the **Processes and Workflows** tile in the **Monitoring** page in the **Manage** area takes you to an application to view all the deployed processes and workflow definitions in that SAP BTP tenant. Figure 9.23 shows a snapshot of the application to manage the process and workflow definitions. Here you can search for the process or workflow based on its name or ID, filter by project, start a new instance, and show existing instances.

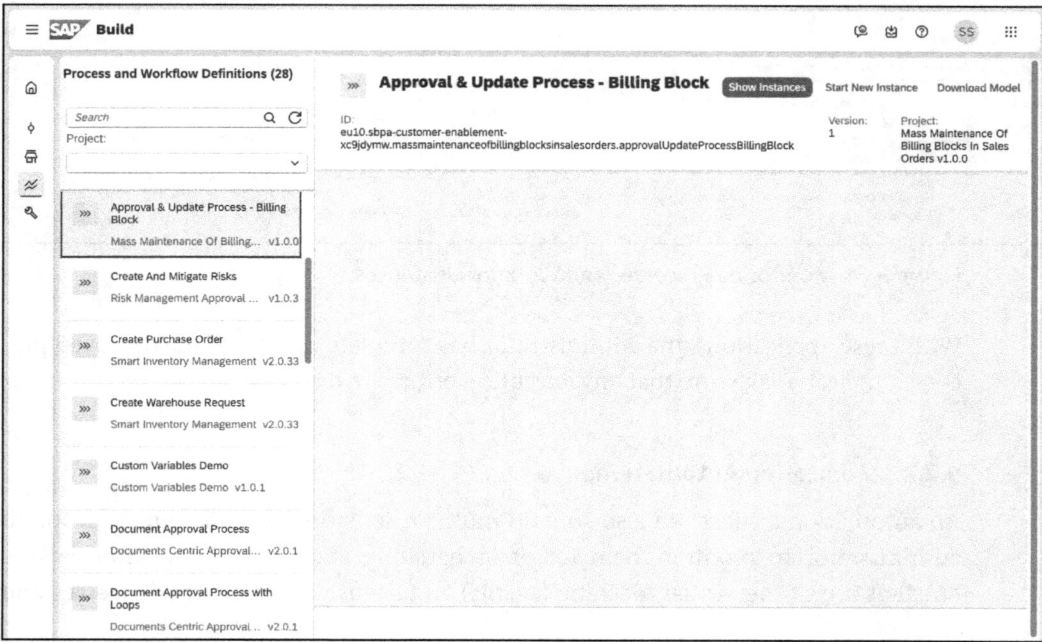

Figure 9.23 Managing the Processes and Workflows

Clicking the **Process and Workflow Instances** in the **Monitoring** page in the **Monitor** area enables you to monitor the instances of processes as well as workflows in that SAP BTP tenant. You can also navigate to the application by clicking the **Show Instances** button at the top of Figure 9.23. This takes you to an application, as shown in Figure 9.24. By default, a status filter is applied so that only **In Progress** instances are shown. But you can reset that filter by clicking the ∇_{x} button to see the completed or canceled instances as well. This application provides all the information about the running instance such as who started the instance, when it was started, ID and version of the process or workflow that was used to start the instance, environment, and instance ID, along with the event logs, error logs, and context of the process or workflow. The administrator can retry the process in case of errors, put the process on hold, or even cancel the process if needed.

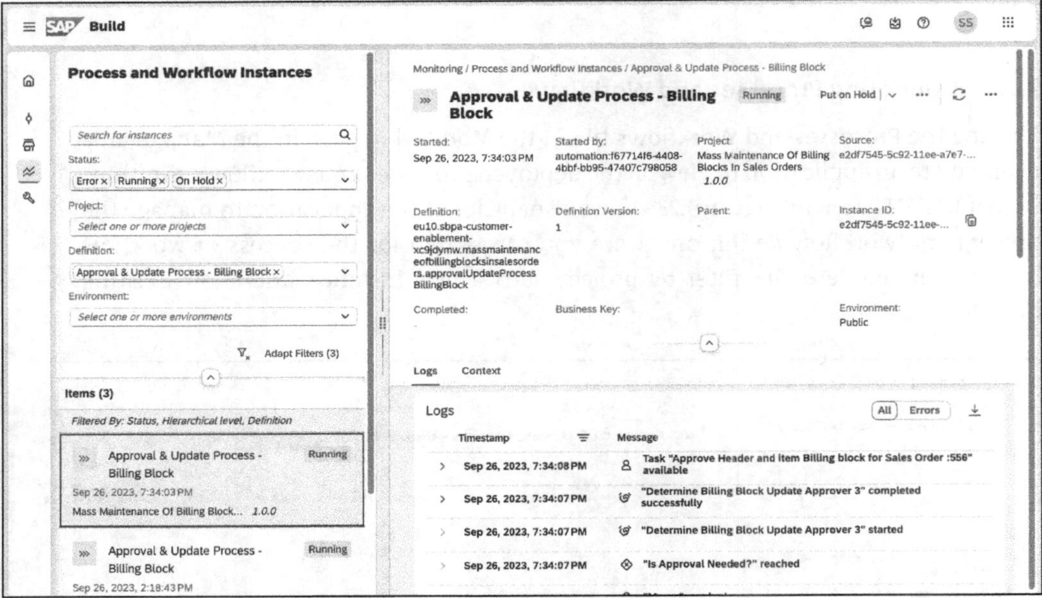

Figure 9.24 Monitoring Processes and Workflow Instances

With these applications, the administrator has complete control of the deployed processes and can make sure that any execution errors are resolved.

9.2.2 Monitoring Automations

An automation artifact can also contain multiple activities, and it's important for an administrator to monitor them for error handling and at the same time ensure efficient use of the virtual machine (agents) and licenses. Clicking the **Processes and**

Workflows tile in the **Monitoring** page in the **Manage** section takes you to an application to view all the deployed processes and workflow definitions in that SAP BTP tenant.

The **Automation Jobs** application in SAP Build Process Automation allows users to monitor and manage all automation jobs in their projects. It provides an overview of individual jobs, including their status, trigger, and duration. Users can filter job information by date range, automation, status, project, and machine. They can also cancel a job with the **Ready** status, refresh the table, and retry any failed, expired, or canceled jobs under certain conditions.

A table displays columns with information about automation jobs, such as their name, status, project and version, environment, machine, trigger, last updated time, duration, and traces availability. Users can sort data in the **Last Updated** column and export or view traces if available for a specific job. When a specific job is selected, the details of the automation job are shown (see Figure 9.25). Here, users can see the different completed or in-progress steps of an automation run, including the input context.

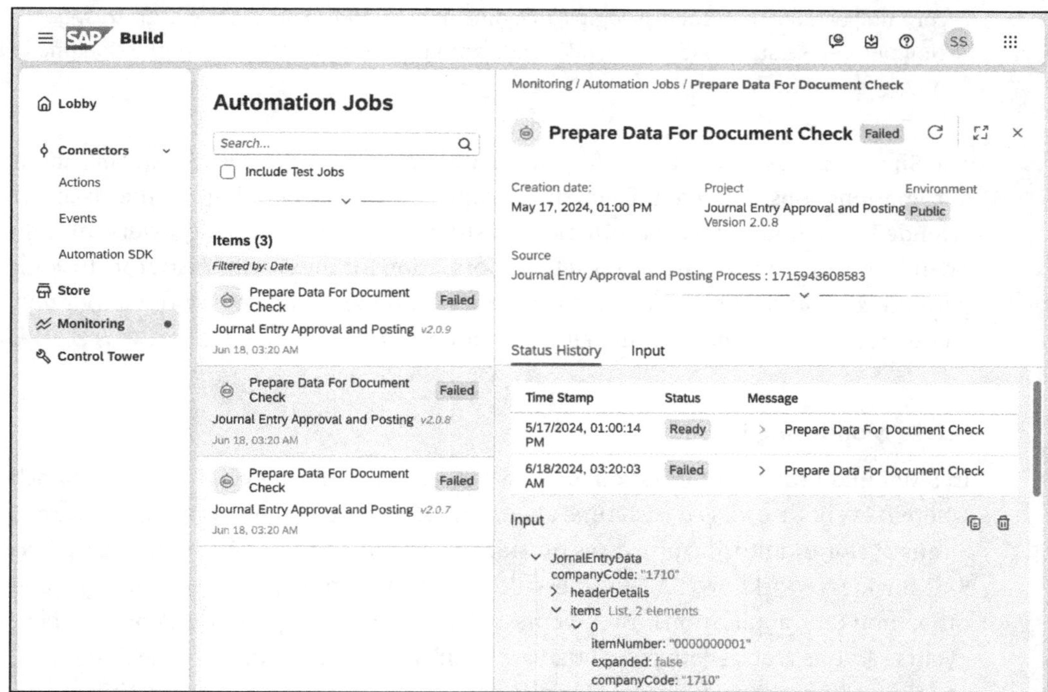

Figure 9.25 Monitoring Automation Jobs

The **Automation Overview** in SAP Build Process Automation provides a centralized dashboard to monitor the health and performance of your automations. It shows real-time and historical data about the following:

- **Job Status**
  Track the number of ongoing jobs by status—**Ready**, **Running**, **Pending**, or **Waiting**.
- **Execution History**
  View the number of finished jobs by status—**Successful**, **Canceled**, **Failed**, or **Expired**.
- **Agents Status**
  Tracks the number of automation agents and their status—**Starting**, **Idle**, **Running**, **Busy**, or **Disconnected**.

Users also have the option to view general event logs by clicking the **Logs** button, filter the logs by required criteria, and download the logs if needed. Users can view logged alerts and download a comma-separated values (CSV) file containing data records for review and analysis. The **Data Export** feature allows downloading various types of records, including historical job data, distribution records, and business activity data.

Warning

While downloading the business activity data, users should exercise caution as it could contain personal data such as logins, machine names, and so on. It's the users' responsibility to make sure that the downloaded content is used per the data protection and privacy regulations.

The Simultaneous Quota Usage for Automation application shows the details about automations jobs that have been run simultaneously. Depending on the usage of attended and unattended automation, customers are billed usually every month. Therefore, this is a very important set of information for the IT administrators to compare the simultaneous automation jobs to the number permitted by the quota purchased for attended and unattended automation runs.

9.2.3 Monitoring Events

In SAP Build Process Automation, process visibility adopts an event-driven approach wherein events are logged each time an activity starts and completes within a process. Similarly, for monitoring processes outside of SAP Build Process Automation, such as SAP Business Workflow, SAP Business Process Management, Qualtrics surveys, and other non-SAP applications, the events should be sent to process visibility. These events serve as crucial data points that are eventually used to populate the dashboard, enabling comprehensive process monitoring. However, inaccurate or missing events can critically impact the accuracy of business key performance indicators (KPIs), leading to a flawed analysis. This flawed analysis can misguide business users who rely on precise data to track process inefficiencies or bottlenecks.

For tracking and analyzing visibility and business events, administrators can use the **Event Acquisition** application in the **Manage** area of the **Monitoring** page. Given the significant amount of event data generated, users must apply explicit filters based on parameters such as process definition ID, process instance ID, and timestamps to isolate relevant events. Based on the filter criteria, the application lists all relevant events, as shown in Figure 9.26. If there were any errors during the event correlation run, those are shown in the **Visibility Event Errors** tab. Though administrators can't fix the errors directly in this application, they can share their findings with the developers of the process visibility scenario to fix them.

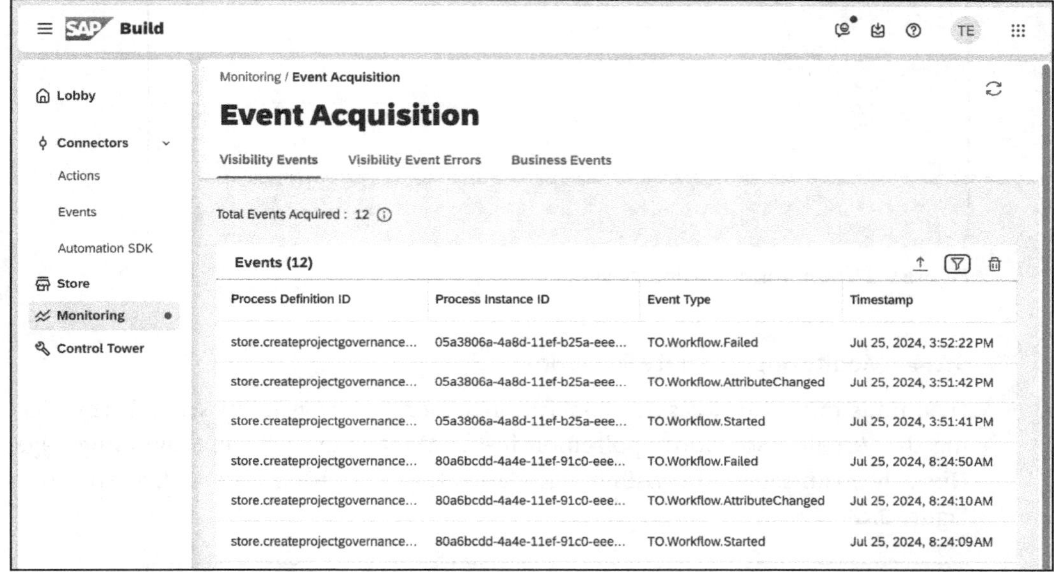

Figure 9.26 Monitoring Events from Process Visibility

A process in SAP Build Process Automation can be configured to start on a business event that occurred in SAP S/4HANA. In such a case, there should be an SAP Event Mesh or SAP Advanced Event Mesh setup that receives business events from SAP S/4HANA or other systems. The **Business Events** tab in the Event Acquisition application records all events that are received in SAP Event Mesh or SAP Integration Suite, advanced event mesh, which is configured to start at least one process in SAP Build Process Automation. Figure 9.27 shows a list of sales order–related events observed in SAP S/4HANA. This can help administrators identify issues when a process isn't initiated or when a business event is raised in the respective system.

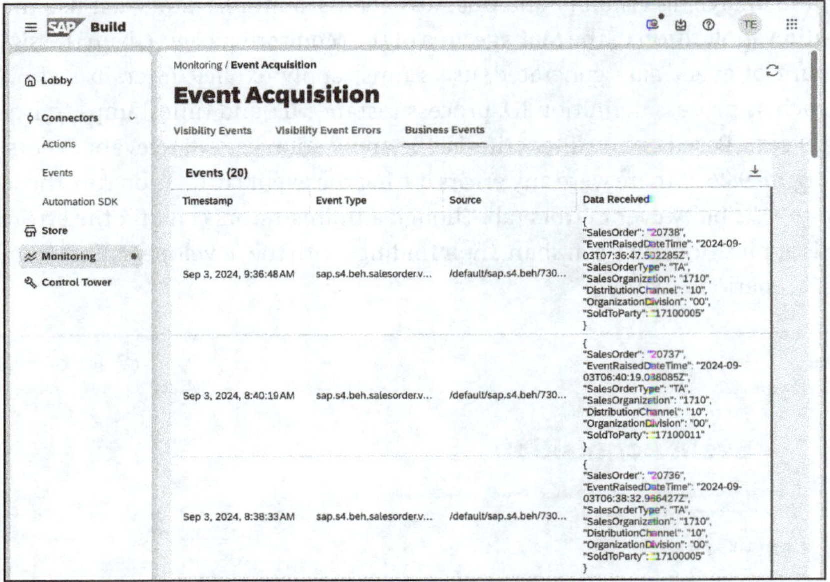

Figure 9.27 Monitoring Business Events

9.2.4 Monitoring Visibility Scenarios

IT administrators can access the application to monitor the visibility scenarios by clicking the **Visibility Scenarios** application in the **Manage** area of the **Monitoring** page. Here, they can see all the visibility scenarios that have been activated, as shown in Figure 9.28.

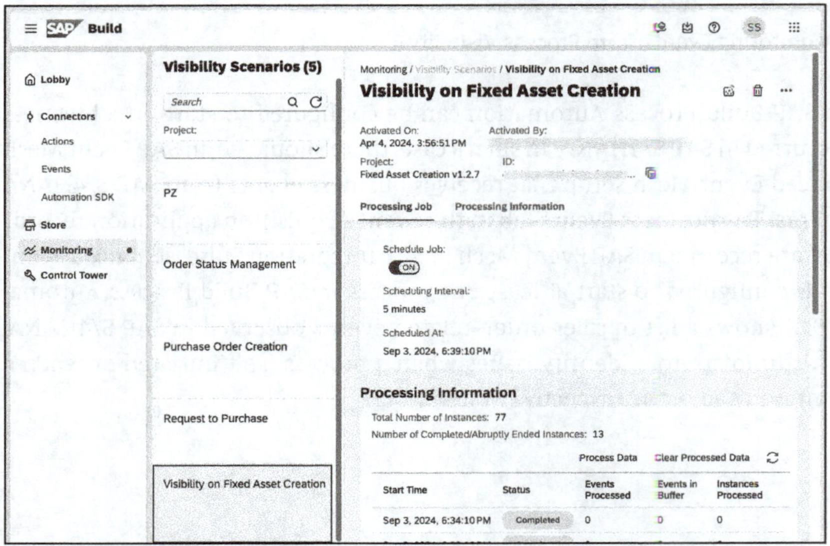

Figure 9.28 Monitoring Visibility Scenarios

Consider a business scenario such as purchase request approval that's broken into workflows with multiple steps—level 1 approval, level 2 approval, and so on. Each step generates events that indicate whether the step has started or completed. You might have several purchase request approvals happening in parallel, each at a different level of approval, generating a varying number of events. Gathering and processing these events accurately is crucial for maintaining an up-to-date dashboard that reflects the process situation in real-time for the business. This is taken care of by a job that runs every five minutes to process and correlated the visibility events. However, not everything runs smoothly all the time. There can be disruptions—network glitches, system outages, or hiccups in execution. In such cases, events might not be recorded properly or could even be delayed. This is where administrators have the responsibility to rectify the data. They can delete the correlated data logs by clicking the **Clear Processed Data** button and initiate reprocessing of the visibility event data by clicking **Process Data** to reinstate the dashboard's accuracy once the underlying issues are resolved.

As the business processes continue day in and day out, massive amounts of data can accumulate over a period of time. Because the data storage is metered in SAP Build Process Automation, having a large amount of historical data will incur costs. Administrators can archive older data from their applications to conserve resources and manage expenses. Later, if needed for audit or analysis, this archived data can be retrieved.

9.3 Business Monitoring

In SAP Build Process Automation, business monitoring refers to the ability of business users to track business processes in real time. SAP Build Process Automation enables business monitoring through its process visibility capability, which provides a near real-time dashboard for business users. With the *process visibility dashboard*, end users can gain one view of the process no matter where it runs (SAP, non-SAP, cloud, or on-premise applications) and understand both present and past process performance. They can also drill down into individual instances or requests of the process for detailed analysis. One of the key benefits of this functionality is its proactive nature—it enables companies to identify and address any issues as and when they arise through actionable insights on a running process instance or request. This allows companies to proactively identify and address any issues or inefficiencies and make data-driven decisions to enhance their processes.

For example, you can configure a visibility scenario for a purchase order approval process to ensure that the orders are approved on time. You can configure various aspects such as phases, states, statuses, and actions to know when a purchase order is in a critical state and to take corrective actions to resolve an undesired status. You can configure process performance indicators such as the total value of purchase orders where approvals are overdue, approved purchase orders by purchasing organization, and so on. You can use the process visibility dashboard to track the approval requests, take corrective actions on the critical orders, and understand the performance of the process using the configured process performance indicators. The best part is that the visualization of KPIs in the process visibility dashboard is automatically generated based on configurations in the visibility scenario.

When a visibility scenario is created for gaining visibility on one or more processes in SAP Build Process Automation, it automatically understands two things:

- The various events that are observed in the processes
- The context of the process when each event is observed

Based on this information, developers can easily define the KPIs that the business user wants to monitor in near real-time and can configure how these KPIs will be visualized, such as in bar charts, line charts, donut charts, and so on. Once the project containing the visibility scenario is deployed, a process visibility dashboard is generated to show the KPIs. The link to the dashboard is available in the deployed version of the project, or it can also be configured for access from SAP Build Work Zone. When the process is executed, the system automatically observes the events raised and processes them to show in the dashboard. There is no other configuration required to enable monitoring a process in SAP Build Process Automation.

Figure 9.29 shows a process visibility dashboard accessed from SAP Build Work Zone, which is created for an accounts payable process. (We'll cover process visibility dashboards in more detail in Chapter 11, Section 11.3.)

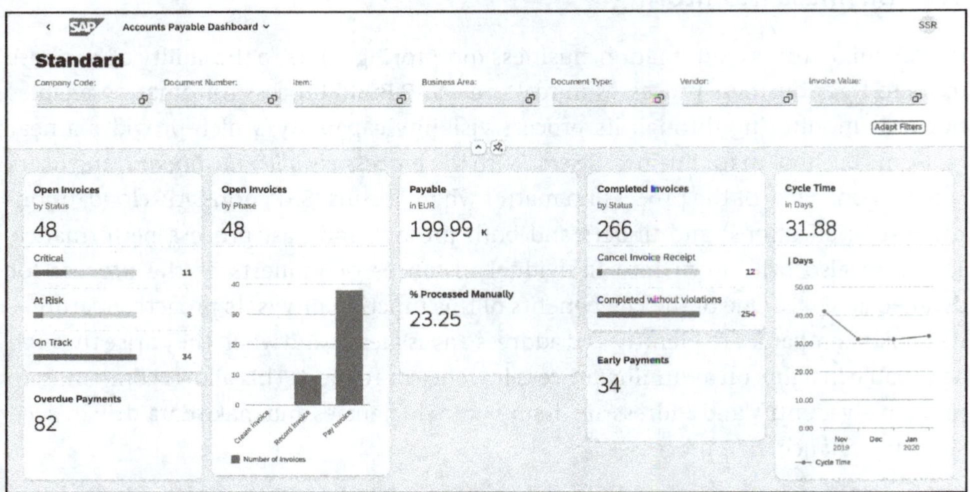

Figure 9.29 Process Visibility Dashboard for Monitoring the Accounts Payable Scenario

It has been configured to show the open invoices in different states based on how critical they are with respect to the desired completing time, the different phases of the process, and where the invoices are, and it also tracks various business KPIs such as cycle time, number of early payments, percentage of manually process invoices, and so on.

The end users, for example, process owners, who are responsible for the KPIs will want to make sure the invoices within specific business areas are worked on based on priority. They can apply filters that are available based on the context of the scenario to filter the entire dashboard to show KPIs for the specific business area, as shown in Figure 9.30.

Figure 9.30 Process Visibility Dashboard Filtered Based on Context Values

If a dashboard is created to track purchase order approvals, the process context will be based on the purchase order business object, so the filters will also be specific to the purchase order business object.

Each of the KPI tiles in the process visibility dashboard can be clicked to navigated into the details of the instances (e.g., invoices or purchase orders) that contributed to the KPI value. Figure 9.31 shows how users can see the list of invoices that are open and in critical state by clicking the **Critical** status under the **Open Invoices** KPI. This automatically applies the filter **Critical** for the attribute **Status** and **Open** for the attribute **State**. Users can also navigate into a specific instance (e.g., invoice number 1000102 or purchase order number 900000018) to see more details of completed steps or phases, current step or phase, process path taken until now, status, cycle time, and any available actions.

> **Note**
> Detailed information on how such a dashboard can be configured and the authorization concepts involved is given in Chapter 11.

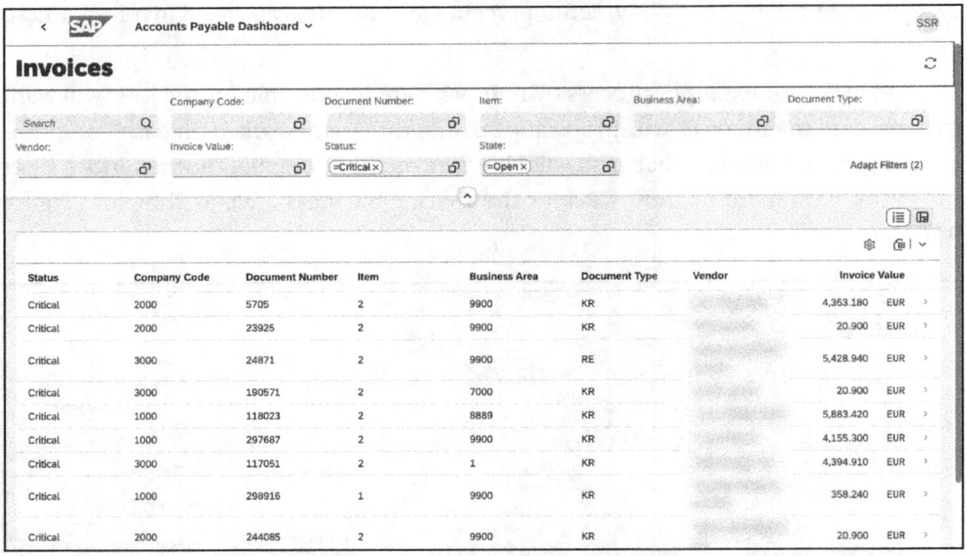

Figure 9.31 List of Invoices Contributing to the Critical Invoices KPI

9.4 Summary

Business process lifecycle management involves the entire lifecycle of a process automation project in SAP Build Process Automation. The lifecycle begins right from identifying the granularity of a project, creation of a project, the different artifacts within it (e.g., processes, automations, decisions, visibility scenarios, etc.), deployment, and continuous monitoring of the deployed artifacts. It also includes the status management, releasing versions or publishing a project into the library for enabling reuse. Successfully leveraging SAP Build Process Automation requires a clear understanding of how to design, configure, and trigger processes, automations, and decisions. Whether they are executed through human action, business events, or another process, SAP Build Process Automation ensures seamless execution, increased efficiency, and reduced risk of errors and inconsistencies. Once the processes are in the execution phase, administrators with the right authorization can use embedded monitoring applications to track processes, automations, and events to identify errors; take actions for resolution; and assess performance for optimization. Finally, you learned that SAP Build Process Automation offers business monitoring through the process visibility capability, allowing end users to track processes in real time to proactively address issues for enhanced efficiency.

In the next chapter, we'll look at how processes, automations, and decisions can be executed from an application either through an API, an event, or UI application depending on the use cases. In addition, you'll learn about the different consumption options and how to configure the artifacts to enable such use.

Chapter 10
Executing Processes, Automations, and Decisions

In today's fast-paced business environment, the ability to execute processes with efficiency and accuracy is crucial for achieving success. This chapter covers the execution of processes, automations, and decisions in SAP Build Process Automation, exploring the purpose of runtime environments and application programming interfaces (APIs). By the end of this chapter, you'll have a comprehensive understanding of how these concepts within SAP Build Process Automation can be applied for efficient and effective execution of their business processes.

Having gained a comprehensive understanding of how to build processes, decisions, and automations in SAP Build Process Automation, the next step is to understand how to execute these artifacts. There are multiple ways to execute each of the artifacts within SAP Build Process Automation such as calling a process from a business add-in (BAdI) in SAP S/4HANA, calling a decision from a non-SAP application, or calling an automation from a process. It's important to have a clear understanding of when and why to use each approach. By understanding the various execution strategies available within SAP Build Process Automation, users can choose the approach that best suits their specific needs and requirements.

In this chapter, you'll learn the significance of runtime environments—what are they and how to use them for your specific use cases. We'll also explore SAP Build Process Automation APIs and how to use them before we cover the different ways to executing process and decisions.

10.1 Understanding Runtime Environments

When you want to deploy a released version of a process automation project, you can choose which environment it should be deployed into. But what are environments, and why should you deploy process automation projects into environments? It's important to understand the concept of environments and the reasons for deploying process automation projects into them. We'll walk through the deployment behavior in public environments, the project lifecycle of an environment, and configurations with an environment in this section.

10.1.1 Overview of Environments

When you click the **Deploy** button in a project, you're presented with a dialog to choose an environment into which the project should be deployed, as shown in Figure 10.1.

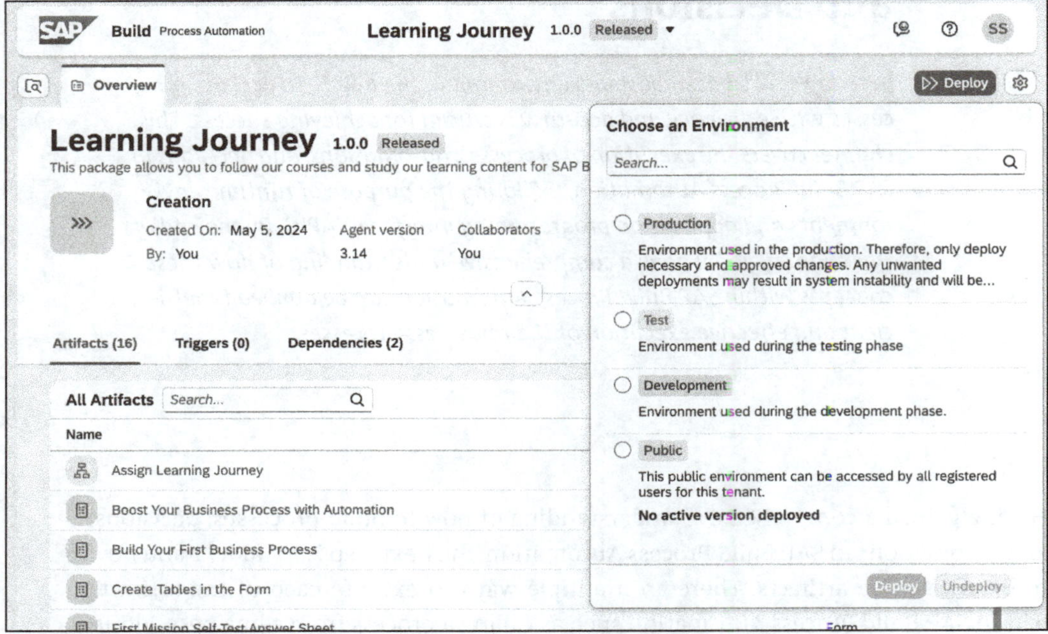

Figure 10.1 Environment Selection during Deployment

Each version of a process automation project can be deployed and run from a functional area called an *environment*. It can be thought of as containers or contexts within which processes, decisions, or automations are executed. An environment provides a segregated space with its own set of configurations, resources, and permissions, allowing teams to work on multiple projects simultaneously without interference. It enables a distinct deployment or runtime environment, such as development, testing, or production, where the artifacts are deployed and executed. Environments can be accessed and maintained in the **Control Tower** page of the SAP Build lobby, as shown in Figure 10.2.

There are two types of environments:

- **Public environment**
  Public environments are the default environments available in SAP Build Process Automation. It's a shared space where the users can deploy and execute processes, decisions, and automations. The projects deployed in the public environment are available to all users based on their assigned security permission.
- **Shared environment**
  Users with the `ProcessAutomationAdmin` role can create a shared environment, which

can then be shared with specific users or groups within the organization having the `ProcessAutomationDeveloper` role. By establishing multiple shared environments within the SAP Build Process Automation tenant, you can segregate the deployment of various projects or project versions. This enables you to deploy and execute the same project with distinct configurations using environment-specific variables. Additionally, you can restrict automation tasks to specific users and machines, enhancing control over the bot execution.

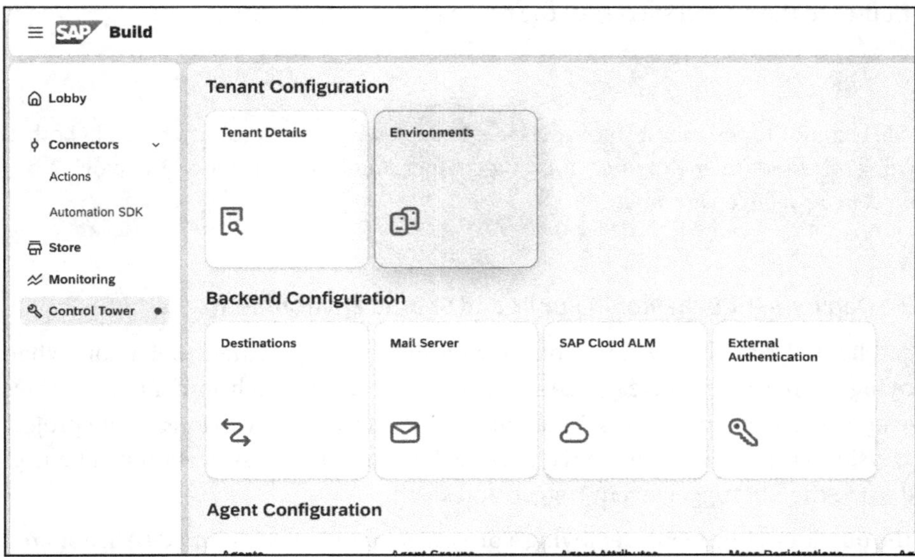

Figure 10.2 Maintaining Environments from the Control Tower Page

To understand the value of environments, let's take look at an example of process automation projects for HR and procurement. Process automation projects for HR focus on employee onboarding, offboarding, leave management, and so on, while the procurement-specific projects focus on processes for purchase requisitions, purchase orders, supplier management, invoice processing, and so on. By deploying the HR-related projects and procurement-related projects in their respective environments, it's possible to achieve access control—access to the HR environment will be restricted to HR personnel and authorized users responsible for managing HR processes. With the help of environments, such role-based access control can be implemented to ensure that only authorized individuals can initiate, monitor, and execute the HR-related workflows and automations or procurement-related workflows and automations.

The same project version can be deployed into different environments of the same tenant, for example, *development*, *test* and *production* environments. Because environments provide a dedicated space for each stage of the deployment lifecycle, it allows for isolation and segregation of development, testing, and production activities. This separation ensures that the changes made in one environment don't impact the stability or integrity of the others. For example, the developers can experiment freely in the development environment without affecting the reliability of the production environment.

In essence, environments are essential for not only controlling access to processes but also for maintaining the integrity and stability of different stages within the deployment lifecycle. They provide a structured approach to managing and executing process automation projects, ensuring that each environment serves its specific purpose and contributes to the overall success of the project.

Tip

While technically possible, it's not advisable to deploy the same project version in both public and standard environments. Doing so undermines the purpose of deploying a project in a shared environment.

10.1.2 Deployment Behavior in Public and Shared Environments

The public and shared environments have a distinct deployment behavior. When deploying a project version (e.g., version 1.0.0) in the public environment, it becomes active and accessible to all users. When deploying subsequent versions of the project (e.g., version 1.0.1), all versions remain active, but only the latest version can be triggered using the API trigger, event trigger, and so on.

Like the public environment, deploying a project version (e.g., version 1.0.0) in a shared environment makes it active within the specified environment. However, when deploying subsequent project versions (e.g., 1.0.1), only the latest version is active, while all other versions are deactivated. If there are running instances of lower versions of the project, then they are terminated during deployment of higher version.

Tip

Though the shared environment provides flexibility in access control and data isolation, organizations should carefully consider their deployment strategies and choose the most suitable environment configuration based on their specific requirements and operational constraints.

10.1.3 Project Lifecycle with Environments

You can deploy a released version of a project into one or more shared environments. When you try to deploy the *same version* of a project to the same environment, then it's considered a **Redeploy**, as shown in Figure 10.3.

When you try to deploy a *higher version* of a project to the same environment, then it's considered an **Upgrade**, and when you try to deploy a *lower version* of a project to the same environment, then it's considered a **Downgrade**, as shown in Figure 10.4.

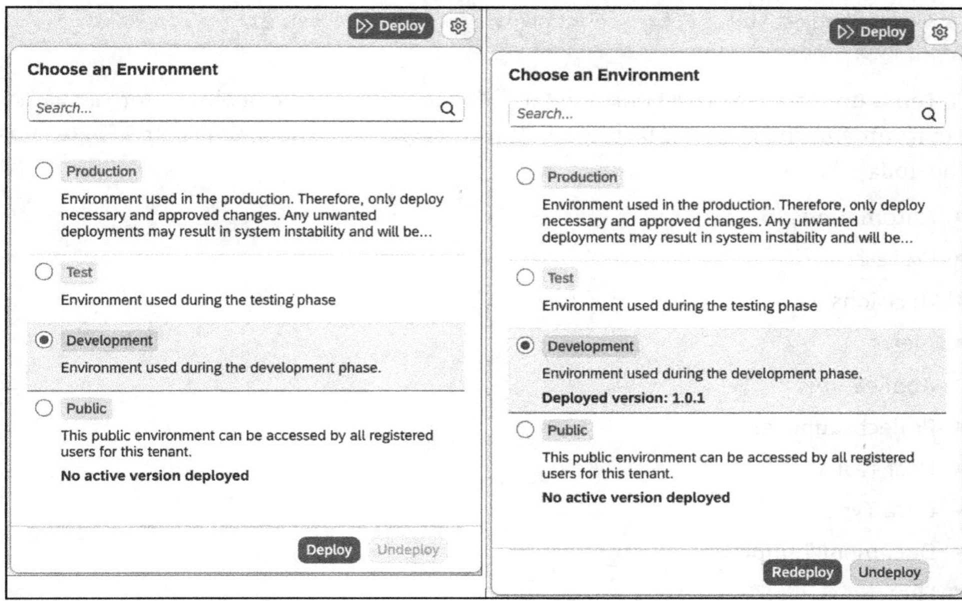

Figure 10.3 Deploy and Redeploy of a Project Version

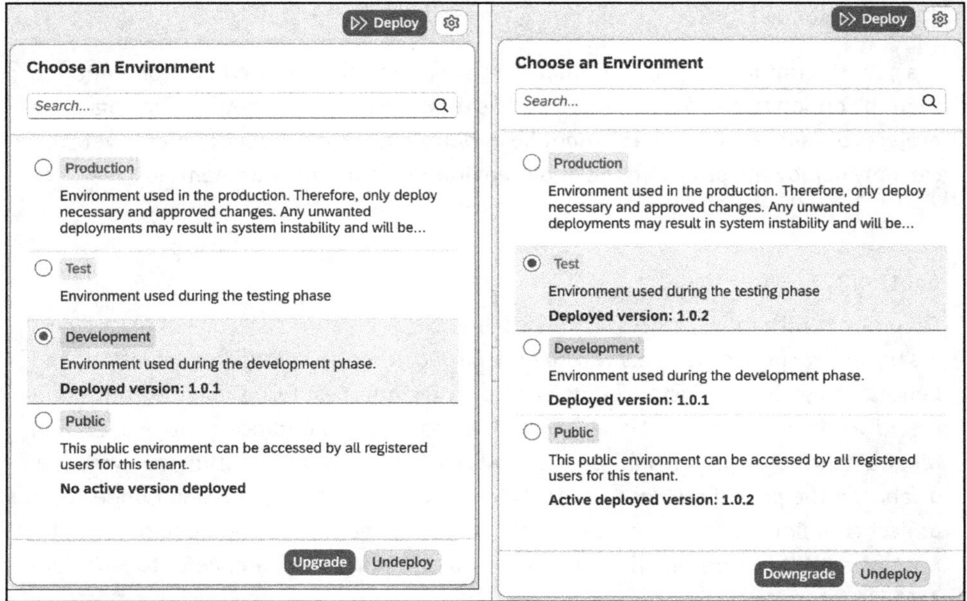

Figure 10.4 Upgrade and Downgrade of a Project Version

Note

In a public environment, you can upgrade or downgrade to deploy more than one version of a project, but only one version is the **Active** version. The **Active** status is shown with the project status in the project overview, next to the name of the project.

At the time of writing this book, the following artifacts can be deployed into a shared environment, and more artifacts will be added over time (refer to SAP Note 3394899 for the updated list):

- **Automations**
- **Processes**
- **Decisions**
- **Alerts**
- **Applications**
- **Project Launcher**
- **User Tasks**
- **Data Type**
- **Document Template**
- **File**
- **Environment Variables**

Note

If a project contains any artifact that isn't supported for a shared environment yet, then the option to deploy the project will be disabled with the following message: **The project contains artifacts that cannot be deployed into shared environments yet. You can only deploy the project in the public environment. (ID: environment_id).**

Caution

For projects with forms, processes, or visibility scenarios, administrative rights are necessary, and *all* project versions will be undeployed, removing workflow definitions and hence ending running instances, deleting pending approval forms, and deleting completed workflow instances. Undeployment is enabled for nonprocess projects even in tenants marked as production (**Control Tower • Tenant Details**), but undeployment is disabled if the project includes a process to prevent any loss of data. Don't undeploy a project containing processes until you check that it's no longer needed and no longer running, and the corresponding processing data is kept elsewhere. Refer to SAP Note 3245676 for more details.

10.1.4 Configurations within an Environment

Each environment contains distinct configurations that define the conditions for running automation projects. The following configurations can be accessed from the **Environments** tile available in the **Control Tower** page:

- **Projects**
  By accessing the projects within an environment, it enables you to manage, collaborate on, and monitor the projects that have been deployed. Upon entering the environment, you'll be presented with a comprehensive list of all your deployed projects, providing a centralized and convenient overview to easily locate and select the specific project you want to work on.
- **Unattended Triggers**
  Manage the scheduled triggers and API triggers in your environment to automatically initiate processes, automations, or decisions. This allows for efficient and timely execution of tasks without the need for manual intervention, streamlining operations and improving productivity within your system.
- **Attended Triggers**
  You can manage the triggers that initiates an automation in your environment, allowing you to manually start automated processes. This gives you the ability to initiate tasks at your discretion, adding flexibility to your task automation needs.
- **Agent Management**
  In an environment, you can allocate agents to various projects. This includes assigning individual agents or agent groups to specific automation tasks within a project. It's also possible for an agent to be assigned to numerous environments, including the public environment.
- **Alert Handlers**
  In the context of task automations in SAP Build Process Automation, alert handlers enable the sending of emails for alerts generated within a project. Alerts are used to define important business events and, once created, can be triggered within existing task automations.
- **Variables**
  To view the variables created in the projects deployed in your environments, you can access the **Variables** tab of the relevant environment. From there, you can see a list of the variables created, their values, and additional details such as project and version information, as well as any related variables.
- **API Keys**
  In SAP Build Process Automation, you can generate an API key for a specific environment to access the public endpoints. Once created, you can view the complete API key, copy it once, and store it in a safe place for future use. Note that once the key is added to your environment, you won't be able to copy or paste it again.

- **Visibility Scenarios**
  As an administrator, you can monitor and control the visibility scenarios that have been deployed into a particular environment. This includes the capability to search for specific visibility scenarios, access and review event processing logs, and navigate to the visibility dashboard for further analysis and management.
- **Processes and Workflows**
  As an administrator, you can manage the processes and workflows that are deployed into a particular environment. This includes the capability to search for specific processes, monitor and handle active instances, and initiate new process instances as needed.

10.2 SAP Build Process Automation Application Programming Interfaces

The APIs for SAP Build Process Automation are essential as they allow external applications to interact with and use the functionalities of the service. There are various reasons why an organization would be using the APIs of SAP Build Process Automation:

- **Integrate with external systems**
  By exposing APIs, SAP Build Process Automation can easily communicate and exchange data with external systems. This integration allows for a more efficient and streamlined workflow, enabling organizations to automate processes based on events or interactions from external systems. For example, an organization using a customer relationship management (CRM) system can leverage the APIs of SAP Build Process Automation to trigger processes or automations based on customer interactions or events within the CRM system. This integration helps in streamlining the overall business process and ensuring that data and workflows are synchronized across different systems. APIs play a crucial role in enabling this level of integration and automation within SAP Build Process Automation.
- **Create customized user interfaces (UIs)**
  Exposing APIs allows UI developers to create custom UIs that can seamlessly integrate with SAP Build Process Automation. By using inbox, workflows, business rules, automation and process visibility APIs, UI developers can design tailored interfaces that streamline and automate various processes within the SAP and non-SAP systems. For example, a custom inbox application can be built to display tasks from multiple source systems, including SAP Build Process Automation. This custom UI not only provides users with a centralized view of their tasks but also allows them to initiate and manage processes, automations, and decisions directly from other applications they use, such as a company's intranet portal or a mobile app. This level of customization and integration enhances the user experience (UX) and empowers business users to efficiently navigate and manage their workflows from a UI they are

familiar with. The use of custom UIs made possible through APIs contributes to increased productivity, improved decision-making, and a more seamless UX for business users interacting with SAP Build Process Automation.

- **Extend functionality**
  APIs provide organizations with the means to extend the core functionality of SAP Build Process Automation. By leveraging SAP Build Process Automation APIs, developers can tap into the system to access and manipulate data and processes in ways that go beyond the standard capabilities. For example, developers can use APIs to retrieve and analyze historical data associated with specific sales orders, allowing them to make more informed decisions when setting approval deadlines. Another example is the use of APIs to a create custom process monitoring application that offers functionalities not found in the standard application, such as updating process context in real time. This flexibility in extending SAP Build Process Automation through APIs empowers developers to tailor the software to their organization's unique needs and processes.

10.2.1 Prerequisites to Access the Application Programming Interfaces

To call an SAP Build Process Automation API, the SAP BTP subaccount must have a service instance for SAP Build Process Automation and a subscription to the standard plan. A subscription to the standard plan enables access to the capabilities of SAP Build Process Automation via the application development lobby, while a service instance on a standard plan helps you access the resources of SAP Build Process Automation via APIs.

As shown in Figure 10.5, follow these steps to identify the base URL and the credentials to call any SAP Build Process Automation API:

❶ In the SAP BTP cockpit, navigate to **Instances and Subscription** under the **Services** section.

❷ Select the service instance for SAP Build Process Automation. If you had used a booster to set up your subaccount, then there would be a service instance **sap_processautomation**.

> **Tip**
>
> You can apply a filter to only see the subscription and instances of SAP Build Process Automation. If there are no service instances for SAP Build Process Automation, then click the **Create** button to create one.

❸ If there are no service bindings available for the SAP Build Process Automation service instance, then click the **Create** button to create one. In the popup, provide a binding name and create a service binding.

❹ If there is at least one service binding available for the SAP Build Process Automation service instance, then the **View Credentials** button is enabled. Click the **View Credentials** button.

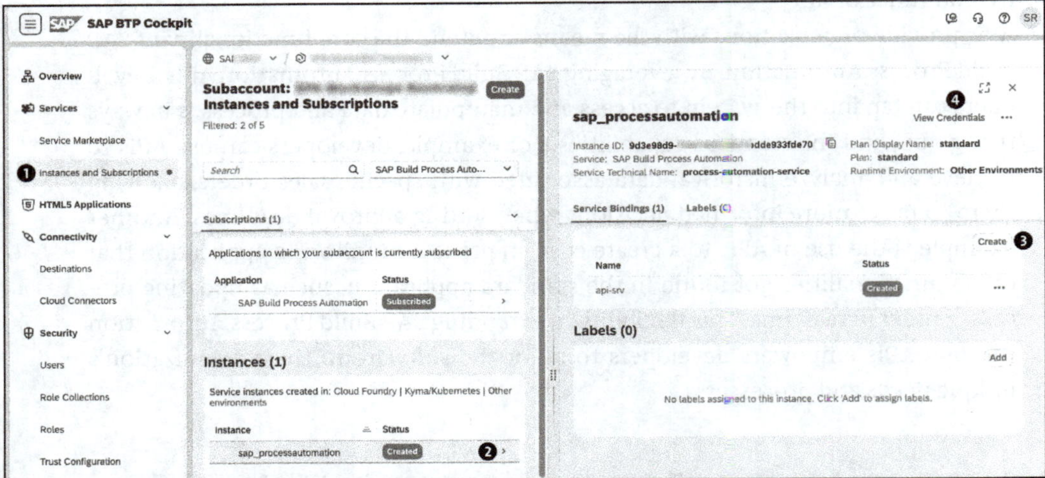

Figure 10.5 Viewing the Credentials of the SAP Build Process Automation Service Instance

The credentials will be shown in a popup (see Figure 10.6). You can view them either in JavaScript Object Notation (JSON) format or in a form as shown here.

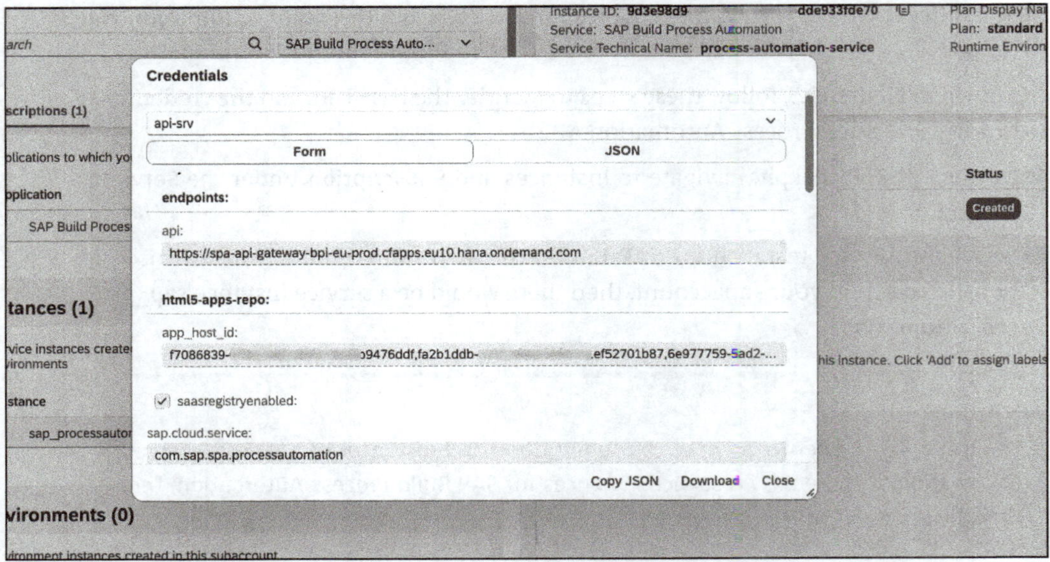

Figure 10.6 Credentials of SAP Build Process Automation

The **Credentials** popup displays a lot of properties. To call the APIs of SAP Build Process Automation, the following properties are needed:

- **api** under the **endpoints** section is the base URL to access any SAP Build Process Automation resource via APIs.
- **url** under the **uaa** section (user account and authentication [UAA]) is used by applications such as SAP Build Process Automation to authenticate users and obtain access tokens for accessing the secured resources within SAP BTP.
- **clientid** under the **uaa** section is a public identifier for an application such as SAP Build Process Automation, which is registered with the UAA service. It will be used to uniquely identify the application when requesting access tokens.
- **clientsecret** under the **uaa** section is a private key or password associated with a client ID. It's used to authenticate the application when it requests access tokens from the UAA service.

Warning

Using the UAA URL, client ID, and client secret, the application resources can be accessed by any external applications via APIs. Therefore, this information should be kept confidential and not shared publicly.

10.2.2 Understanding SAP Build Process Automation Application Programming Interfaces

In this section, we'll provide an overview of the APIs in SAP Build Process Automation, explore how to access the API documentation from SAP Business Accelerator Hub, and understand the differences between using REST and OData APIs in SAP Build Process Automation.

Overview of APIs in SAP Build Process Automation

All APIs of SAP Build Process Automations are documented as part of SAP Business Accelerator Hub (*https://api.sap.com/*). As shown in Figure 10.7, you can find the API package from SAP Build Process Automation by entering "SAP Build Process Automation" in the search box to search for all content related to SAP Build Process Automation, including APIs. From the search results, scroll to select the **SAP Build Process Automation** API package.

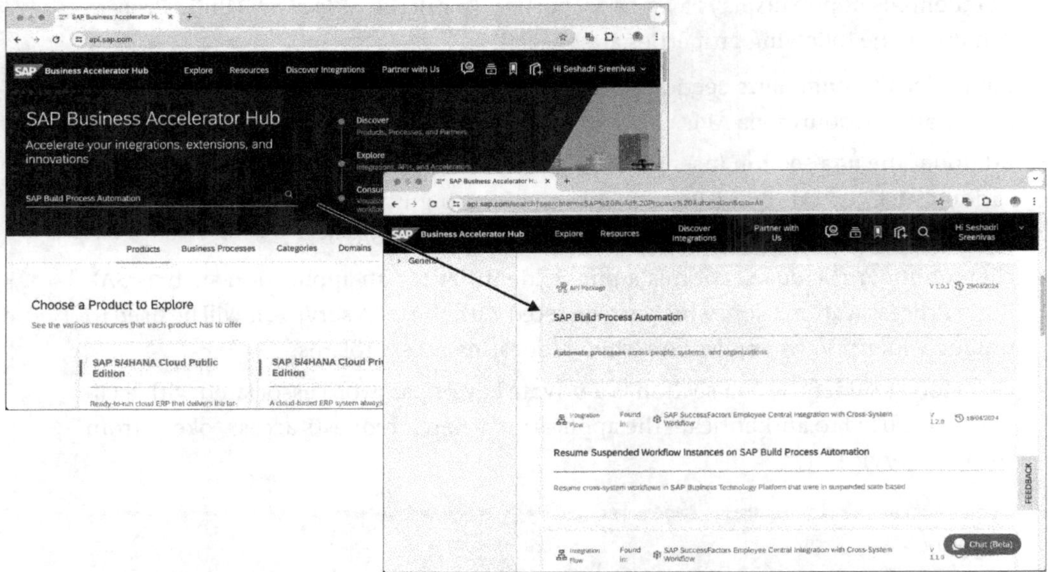

Figure 10.7 Search for SAP Build Process Automation API in SAP Business Accelerator Hub

Tip

SAP Business Accelerator Hub allows users to explore and test a variety of content to help in integration, extension, and digital transformation. The platform offers discovery of content for integration and extension needs; exploration of various products, categories, and business processes; and the ability to test APIs in both sandbox and configured systems.

SAP Build Process Automation exposes two kinds of API to address different use cases—OData APIs and REST APIs:

- **OData APIs**
  The OData APIs play a crucial role in providing access to user task related data following the Task Consumption Model (TCM). These APIs are important for applications such as My Inbox and SAP Task Center, as they enable the display of tasks, as well as the creation and management of substitution rules. For example, when a process participant logs into My Inbox or SAP Task Center, the application can present tasks that are specifically relevant to the logged-in user. This functionality is made possible through the OData APIs of SAP Build Process Automation. Developers can leverage these OData APIs to create customized inbox applications designed to align with the

unique needs and requirements of their business. By using the OData APIs, developers have the capability to tailor the application to suit their specific workflows and processes. This provides a great deal of flexibility and adaptability, ensuring that the application aligns with the organization's existing systems and practices. Figure 10.8 shows the OData API package.

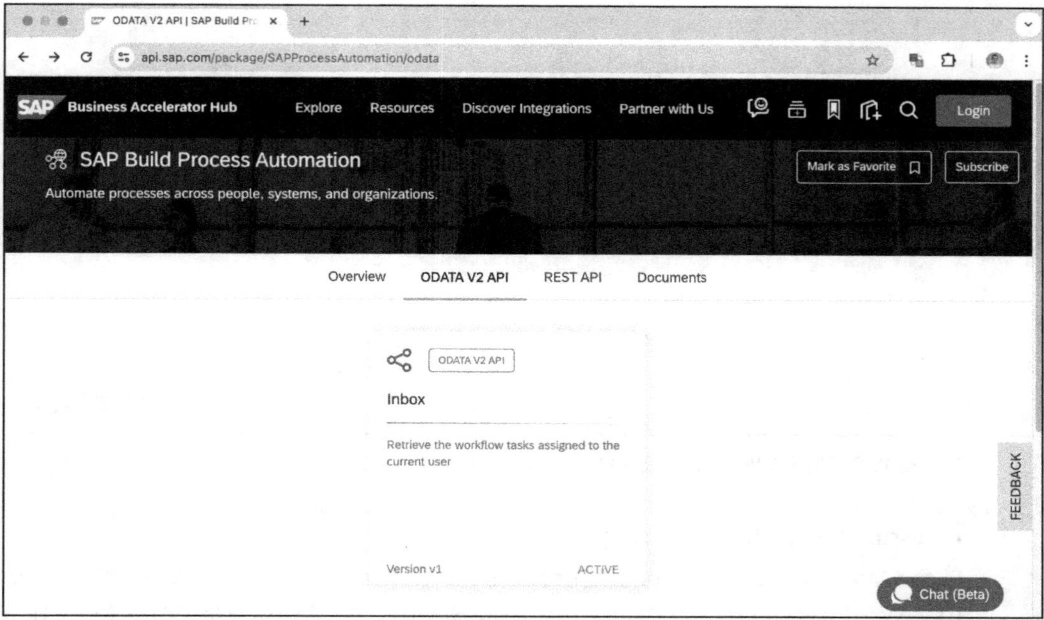

Figure 10.8 OData API to Retrieve Workflow Tasks Assigned to the Current User

- **REST APIs**
  REST APIs provide a way to access and manipulate various functionalities within SAP Build Process Automation. These APIs allow users to perform a wide range of tasks related to processes, automations, decisions, and visibility scenarios. Depending on the roles and authorization levels, users can initiate new processes, view and manage process instances, and access details such as process context and execution logs. They also can manage the process instances by deleting, canceling, or suspending them as needed. For user tasks within a process, REST APIs enable users to list and inspect details, including the task context and processor. Users can also complete a user task instance or send message events to a specific process. In addition, the APIs allow for the invocation of decisions and the initiation of new task automations. In addition, users can access all available API triggers for an automation, as well as push events to visibility scenarios, and obtain details about the visibility scenario phase, path, and step. Figure 10.9 shows the REST API package.

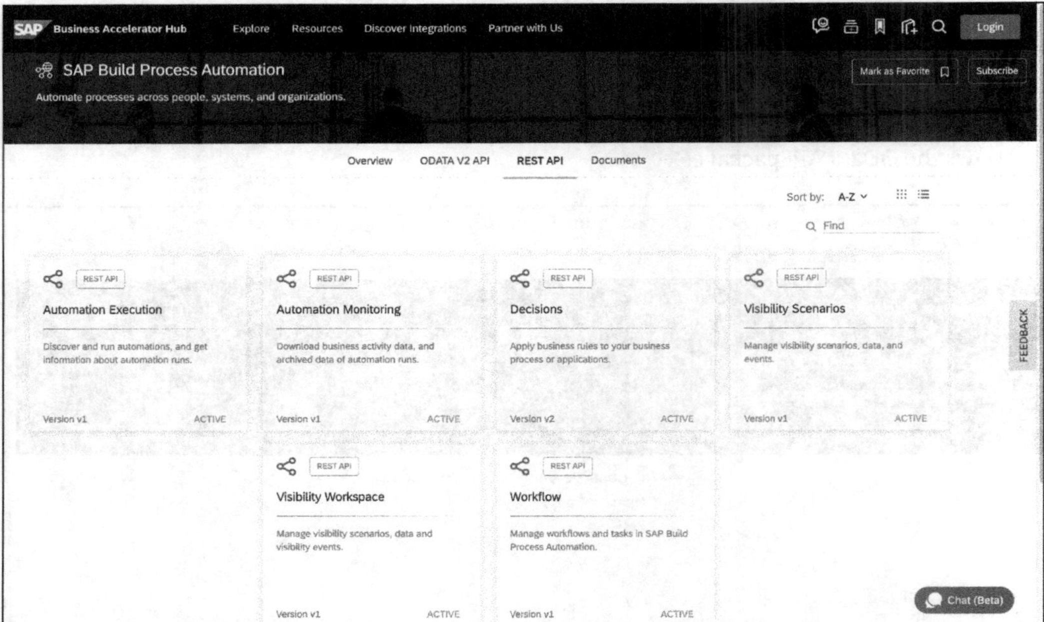

Figure 10.9 REST APIs of SAP Build Process Automation

Accessing the API Documentation

In the API package of SAP Build Process Automation, you can click the **REST API** section to see various resources that can be accessed, such as Workflow, Decisions, Automation Execution, and Visibility Workspace, and so on. To access a specific resource, for example Workflow, click on the corresponding tile to navigate further. Figure 10.10 shows the overview page of **Workflow** resource with the following details:

❶ **API Specification**
The API specification can be downloaded as JSON or XML. This is needed if you want to call the APIs from an *action project* of SAP Build Process Automation.

❷ **Authentication Methods**
The different supported by SAP Build Process Automation to authenticate the user for accessing the specific resource. SAP Build Process Automation APIs are protected with OAuth2.0 and supports the following authentication methods:

- OAuth 2.0 Access Code Flow: This is a standard authorization flow used by applications to access resources on behalf of users. It's suitable for applications running on SAP BTP or in trusted environments where the client secret can be protected. It's ideal for scenarios where the user interaction is required, or the resource access is based on the user's roles and authorization. In SAP Build Process Automation, this flow can be achieved with the authentication type OAuth2SAMLBearerAssertion, OAuth2Password, and OAuth2UserTokenExchange.

- OAuth 2.0 Application Flow: This technical user flow is also known as a client credentials flow. It's primarily used for server-to-server communication, where the client application is considered trusted, and there is no need for user involvement.

❸ **Configuration Details**

This section provides a base URL to access the specific resource. For example, to access the workflow-specific APIs, the **SANDBOX URL** is *https://hostname/public/workflow/rest*, where */public/workflow/rest* is the base path to access the workflow resource. Figure 10.11 shows the configuration details of the workflow API.

To derive the fully qualified URL to access the workflow APIs, replace the *hostname* with *api endpoint* that was identified earlier in Section 10.2.1. Therefore, the API would look like the following for EU10: *https://spa-api-gateway-bpi-eu-prod.cfapps.eu10.hana.ondemand.com/public/workflow/rest*.

10

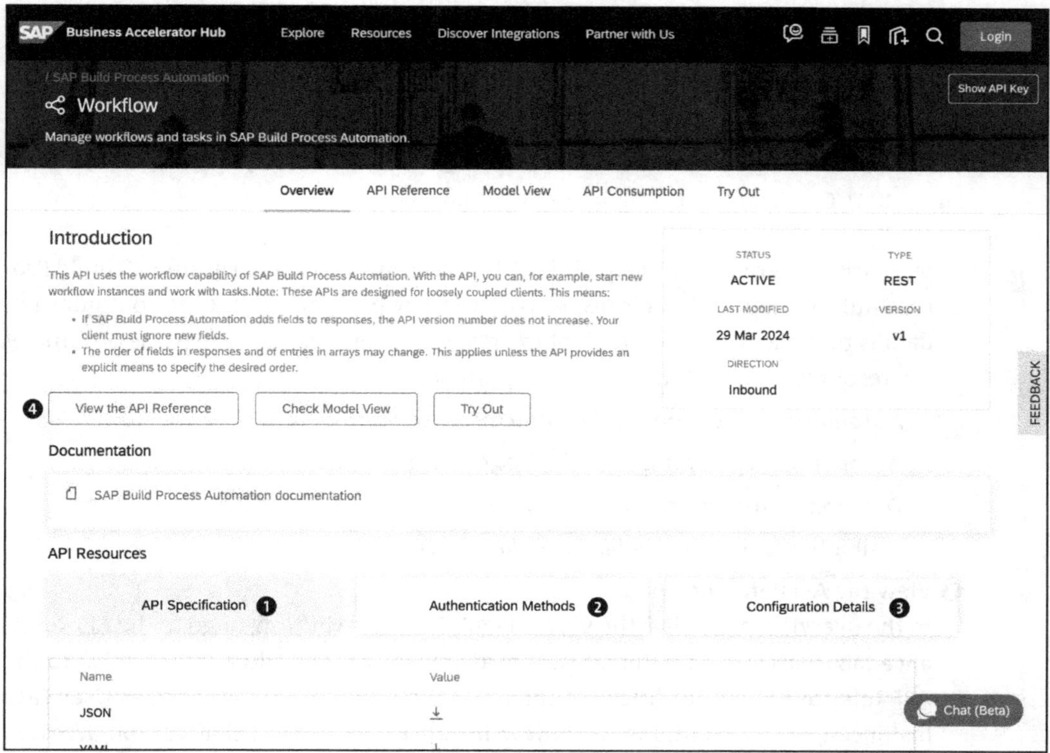

Figure 10.10 Overview of the Workflow Resource under SAP Build Process Automation

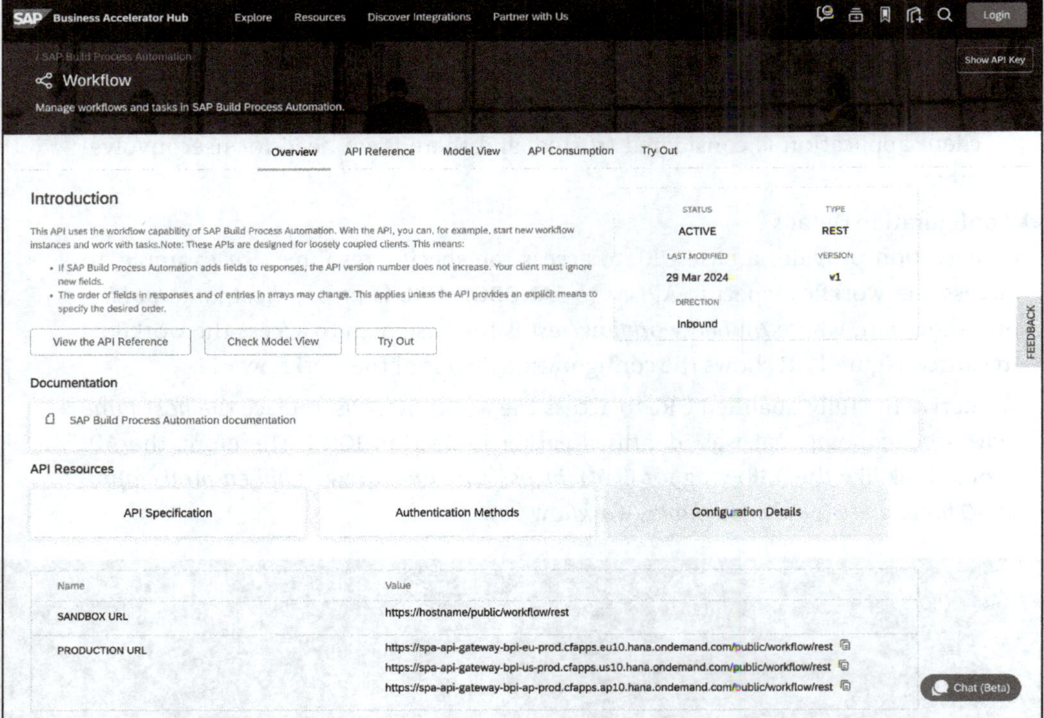

Figure 10.11 Configuration Details of the Workflow API

Similarly, to derive the fully qualified API URL of any resources within SAP Build Process Automation, visit their respective overview page, and check the configuration details to identify the base path of the resource. Here are the base paths of some of the resources in SAP Build Process Automation:

- **Automation Execution**: */public/irpa/runtime/v1*
- **Automation Monitoring**: */public/irpa/monitoring/v1*
- **Decisions runtime**: */public/rule/runtime/rest*
- **Visibility Workspace**: */public/visibility/runtime*

❹ **View the API Reference**

In the overview page, click the **View the API Reference** button, or go to the **API Reference** tab to access the different APIs available for the workflow resource. Go to the **API Refence** section for information about the different entities such as **User Task Instances**, **Task Definitions**, **Workflow Instances**, and so on that can be accessed within the workflow resource. Figure 10.12 shows the **API Reference** section where you can access the information about the resource, entities, and their methods. On selecting a specific entity, all the methods the entity supports are listed. A detailed documentation of that API, including the inputs, outputs, possible errors, and so on, can be accessed by selecting the individual methods such as **GET**, **PATCH**, and so on. Each method has path and query parameters that determine the URL of that API. For example, to retrieve a task instance by ID, the path is */v1/task-instances/{taskInstanceID}*, and the method is *GET*. By combining the URL that was derived in the **Configuration Details** section together with the path, the URL of the API can be derived as the following: *https://spa-api-gateway-bpi-eu-prod.cfapps.eu10.hana.ondemand.com/public/workflow/rest/v1/task-instances/{taskInstanceID}*.

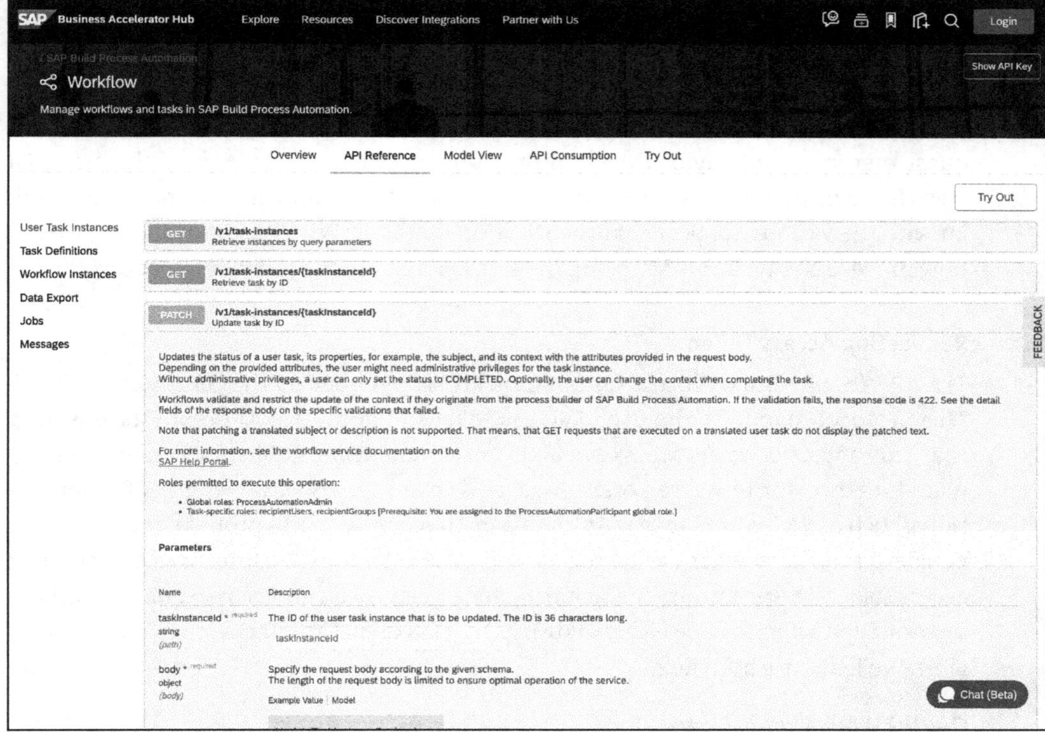

Figure 10.12 Accessing the Methods of Entities within a Resource

Which APIs to use in SAP Build Process Automation: OData vs. REST APIs

OData APIs of SAP Build Process Automation are used for accessing user task–related data and creating customized inbox applications where sorting and filtering are necessary. REST APIs of SAP Build Process Automation are used for accessing and manipulating various functionalities within SAP Build Process Automation, including process management, executing a decision, starting an automation, and so on. Depending on the use cases, the respective APIs can be used.

Note

It's important to note the difference between REST APIs and OData APIs. While both are commonly used for accessing and manipulating data, OData is a protocol built on top of the HTTP standard that allows for the creation and consumption of query-able and interoperable RESTful APIs. On the other hand, REST APIs are a more general concept that refers to a set of architectural principles for designing networked applications, and they can be implemented using various protocols, including HTTP.

10.2.3 Using SAP Build Process Automation Application Programming Interfaces

When it comes to invoking an API, it's crucial to understand the underlying structure and the expected outcome. This includes understanding the specific endpoints, request methods, and payload data. In previous sections, you've learned the essential steps for identifying the API, deriving the fully qualified URL of an API, and being familiar with the various supported authentication methods. With this information, we can now proceed to invoke an API using the steps laid out in the following sections.

Requesting Access Token

The APIs for SAP Build Process Automation are safeguarded by the OAuth 2.0 protocol. This protocol mandates the use of access tokens to authorize API requests, ensuring that only approved applications can access user data. This establishes a secure and regulated method for resource access. An access token serves as a credential for applications, such as SAP Build Process Automation, to securely access protected resources on behalf of a specified user or technical entity. The UAA APIs must be used to obtain an access token. The procedure for acquiring the access token varies based on the authentication type, which can be either OAuth 2.0 Access Code Flow or OAuth 2.0 Application Flow. We'll discuss both next.

OAuth2.0 Application Flow

The OAuth2.0 application flow or client credentials flow will be used to access the APIs directly without any user involvement. It's ideal for automated processes or machine-to-machine communication. Let's look at how to request an access token for the OAuth2.0 Application Flow. Figure 10.13 shows how the OAuth2.0 Application Flow API endpoint can be executed in a REST client:

❶ Set the method as POST.

❷ Use the UAA URL obtained from Section 10.2.1, and form the URL: *{uaa.url}/oauth/token?grant_type=client_credentials*.

❸ Choose **Basic Auth**, with the UAA client ID as **Username**, and the UAA client secret as **Password**.

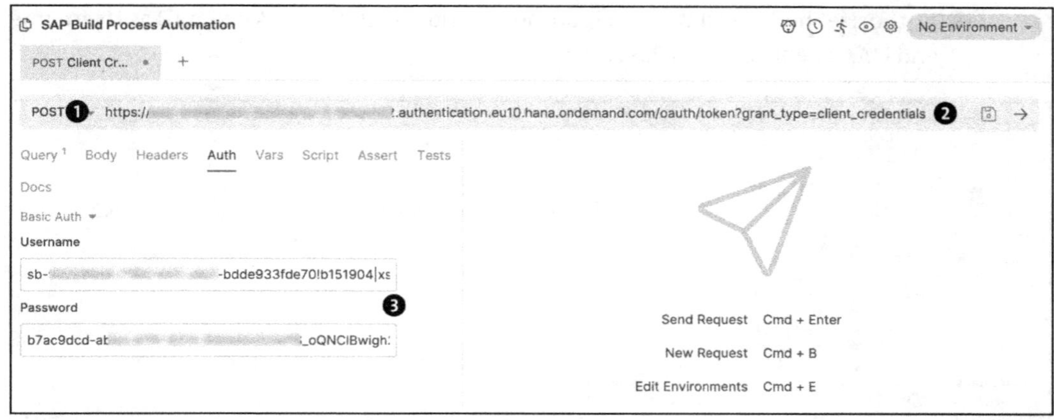

Figure 10.13 Request Access Token for the OAuth2.0 Application Flow

On executing it, the API should be successfully returning the access token in the following JSON format. The property `access_token` can be used to authenticate the API calls of SAP Build Process Automation, as shown in Listing 10.1.

```
{
  "access_token": "xxxxxxxxxxx",
  "token_type": "bearer",
  "expires_in": 43199,
  "scope": "uaa.resource",
  "jti": "xxyxxxx"
}
```

Listing 10.1 Sample Response Body When Successfully Calling the API for the Grant Type Client Credentials

OAuth2.0 Access Code Flow

The OAuth2.0 access code flow will be used to access the APIs on behalf of an actual user. It provides a more secure and granular approach to accessing protected resources such as process instances. Let's look at how to request an access token for OAuth2.0 Access Code Flow with password grant. Figure 10.14 shows how the OAuth2.0 Access Code Flow API endpoint can be executed in a REST client:

❶ Set the method as `POST`.

❷ Use the UAA URL obtained from Section 10.2.1 and form the URL: *{uaa.url}/oauth/token?grant_type=password*.

❸ Set the **Body** type as **Form URL Encoded**, and add two key-value pairs:

- **username**: Username of the user who has the authorization to access the resource
- **password**: Password of the user

❹ Authorize the API call with basic authentication, with the UAA client ID as **Username** and UAA client secret as **Password**.

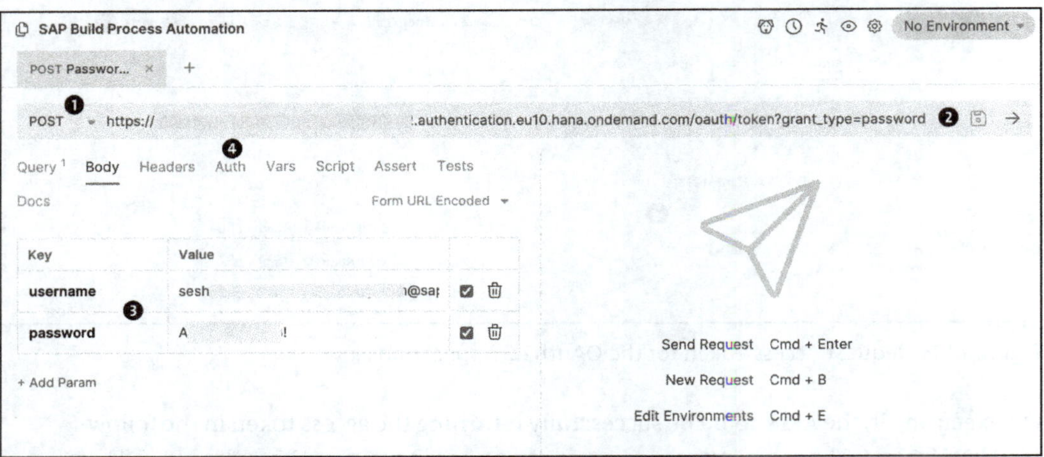

Figure 10.14 Request Access Token for OAuth2.0 Access Code Flow with Password Grant

On executing it, the API should be successfully returning the access token in the following JSON format. The property `access_token` can be used to authenticate the API calls of SAP Build Process Automation. The `scope` property gives the exact privileges that the user has, as shown in Listing 10.2.

```
{
  "access_token": "xxxxxx",
  "token_type": "bearer",
  "id_token": "yyyy",
  "refresh_token": "xxxyyy-r",
  "expires_in": 43199,
  "scope": "xsuaa!b120249.ReadFormDesignTimeArtifacts
xsuaa!b120249.ReadDeployedForms xsuaa!b120249.SPA_ADVANCED_USER openid
xsuaa!b120249.ExecuteFormsValueHelp xsuaa!b120249.UploadFormsDocuments
uaa.user",
  "jti": "yyxxyy"
}
```

Listing 10.2 Sample Response Body When Successfully Calling the API for the Grant Type Password

Access the API Endpoint

As an example, we'll demonstrate calling a Workflow API to retrieve all running workflow/process instances. It's important to note that the same approach can be used for calling any API within SAP Build Process Automation, such as executing a decision, initiating an automation, or starting a process.

To retrieve all running workflow instances, we need to access the API under the workflow resource. We've understood from Section 10.2.2, under "Accessing the API Documentation," that the fully qualified URL to access the workflow resource is the following: *https://spa-api-gateway-bpi-host.hana.ondemand.com/public/workflow/rest*.

To access the workflow instances entity, the entity specific path */v1/workflow-instances* needs to be appended to the preceding URL. Therefore, the API endpoint to access the workflow instances would be the following: *https://spa-api-gateway-bpi-host.hana.ondemand.com/public/workflow/rest/v1/workflow-instances*.

Figure 10.15 shows how a workflow API can be executed from the REST client:

1. Set the method as GET.
2. Use the derived URL to get all workflow instances.
3. Set the Authentication as **Bearer Token**.
4. Paste the access token obtained from the previous section, "Requesting Access Token."

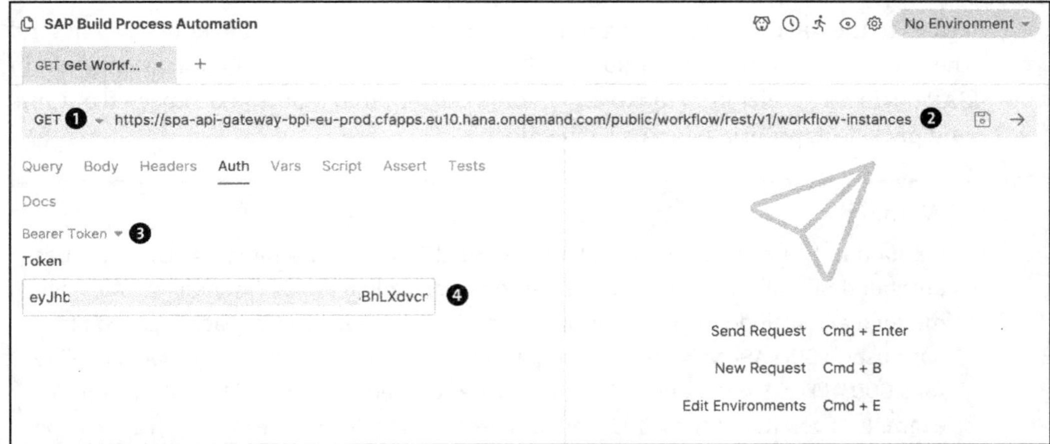

Figure 10.15 Execute the API to Get the Workflow Instances

On executing it, the API should be successfully returning all workflow instances in the following JSON format. You can refer to the API documentation to add further query parameters such as `id`, `definitionId`, `startBy`, `subject`, `businessKey`, and so on to find the exact workflow instance. Listing 10.3 shows a typical response while calling the GET Workflow Instances API.

```
[
  {
    "id": "ea7fd2b9-0459-11ee-95b4-eeee0a9a6198",
    "definitionId": "eu10.fixedassetcreation.createFixedAsset",
    "definitionVersion": "12",
    "subject": "Create Fixed Asset for Company 1010 in Class General",
```

```
    "status": "RUNNING",
    "businessKey": "",
    "parentInstanceId": null,
    "rootInstanceId": "ea7fd2b9-0459-11ee-95b4-eeee0a9a6198",
    "applicationScope": "own",
    "projectId": " eu10.fixedassetcreation",
    "projectVersion": "4.3.2",
    "environmentId": "default",
    "startedBy": "user@domain.com",
    "startedAt": "2024-05-13T11:04:32.791Z",
    "completedAt": null
  }
]
```

Listing 10.3 Typical Output of the API to Get a Workflow Instance

The POST, PUT, or PATCH APIs typically require a JSON body to be sent in the request. The request body allows transmission of complex data structure such as objects and arrays to the API endpoint. For example, to start an automation, the requester will send the invocation context and inputs via `POST` method. Check the API documentation in SAP Business Accelerator Hub to understand the required inputs and expected outputs and provide appropriate inputs to the API.

Warning

Inbound API calls (application calling SAP Build Process Automation APIs) as well as outbound API calls (SAP Build Process Automation calling any SAP or non-SAP APIs) are metered and include the price calculation of the service usage. For each licensed Standard user, 1,000 API calls are allowed per month while each Advanced user is eligible for 5,000 API calls per month. The API calls are counted at the SAP BTP subaccount. For example, if you have licensed 10 standard users and 1 advanced user, then there can be up to 15,000 API calls per month irrespective of if it was called from a single user or multiple users. Anything more than 15,000 API calls will incur additional charges. In API-driven use cases or with many process instances, additional licensing may be needed for API calls.

10.3 Running a Process, an Automation, and a Decision

Whether you're a business user, developer, or administrator, understanding the nuances of running a process, an automation, and a decision is essential for maximizing the value of SAP Build Process Automation. The key to successful automation lies in the design and configuration of processes and task automation within SAP Build Process Automation. Organizations can define the triggers, conditions, and actions that make up their automated processes, ensuring that tasks are executed seamlessly and without human intervention. This not only improves efficiency but also reduces the risk of errors and inconsistencies that often accompany manual processes. In this section, we'll explore the practical aspects with examples of running a process, an automation, and a decision within SAP Build Process Automation.

10.3.1 Starting a Process and Automation as a Developer or Administrator

In SAP Build Process Automation, developers or administrators have different ways of starting a process or an automation during the development phase. These methods allow for testing and troubleshooting before the automation or process is deployed in a production environment.

Automation developers can use the ▷ button in the automation editor to test run the automation. Figure 10.16 shows how an automation developer can trigger a test run. 10

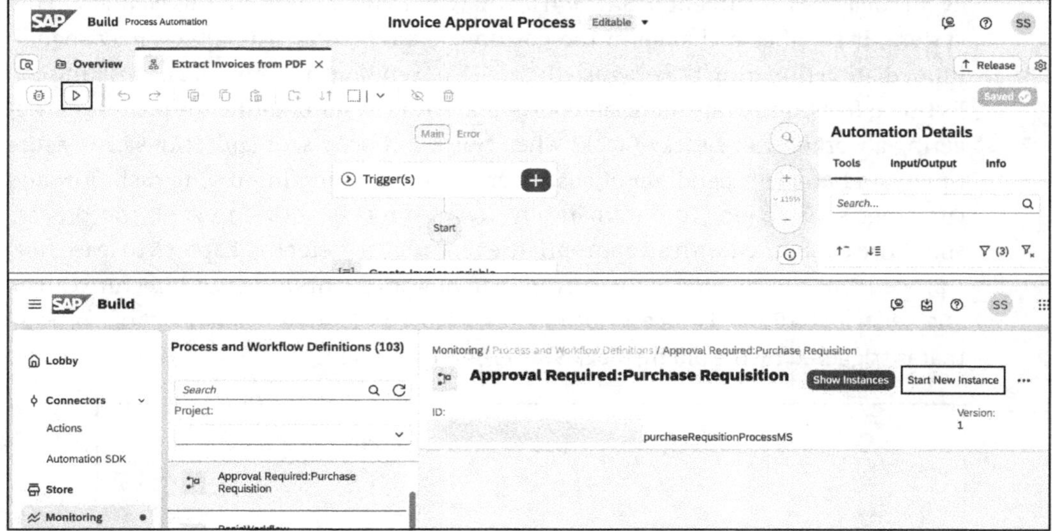

Figure 10.16 Testing an Automation as a Developer and Starting a Process as an Administrator

This allows developers to verify that the automation is functioning as expected before integrating it into the larger process. By clicking the test button, the developer can see how the automation will behave in the system and can make any necessary adjustments or corrections before finalizing the automation. Similarly, workflow administrators or developers can use the **Process and Workflow Definition** application available in the **Monitoring** section to start an instance of a deployed process by providing a start context. This allows administrators to initiate a process and monitor its progress, ensuring that it's functioning as intended and resolving any issues that may arise during the development phase.

It's important to note that these methods are typical ways of starting a process or an automation during the development phase. In subsequent sections, we'll discover how a process or automation can be started in a typical process automation use case.

> **Note**
>
> Process with any triggers (API, event, or form) can be initiated from the Process and Workflow Definition application. It's the responsibility of the administrator to provide an appropriate start context depending on the trigger; otherwise, the process instance might end up in an error state as the process might expect a different input.

10.3.2 Event-Based Triggers

A process can be modeled to start automatically based on an event observed in another system, such as SAP S/4HANA. Let's take an example of a **Billing Block Status Changed** event for a sales document in SAP S/4HANA. When a billing block status changes in the system, it signifies a change in the customer's payment status, which may require immediate action from the organization, such as sending out reminders, initiating collection efforts, updating financial reports, and so on. By automating the process of triggering a workflow or series of tasks when this event occurs, organizations can ensure timely and efficient handling of customer payment issues, improving cash flow and customer satisfaction. To automatically start a process with an event, the process should be configured with an appropriate event trigger. Refer to Chapter 5 to learn how a process can be triggered with an event and how an event-driven architecture can be enabled with SAP Build Process Automation. Figure 10.17 shows a snapshot of a process that is triggered by the **Billing Block Status Changed** event.

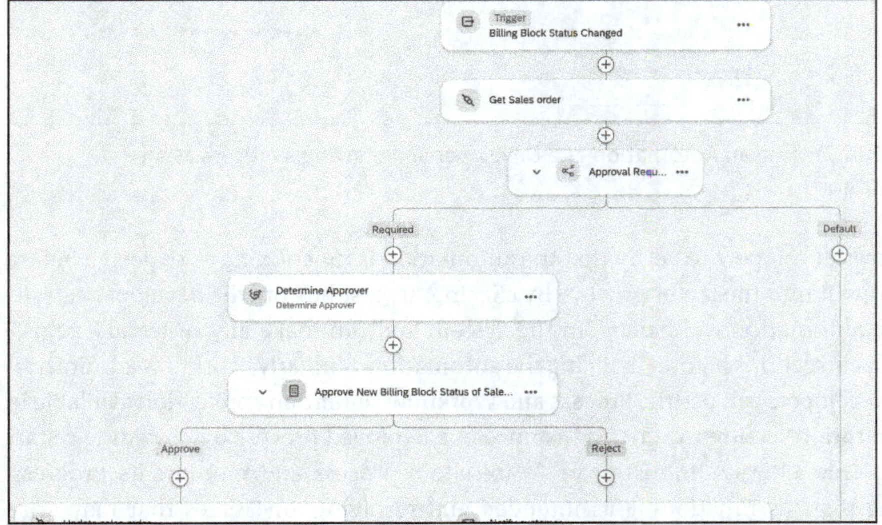

Figure 10.17 Process with an Event Trigger

Tip

Only a process can be initiated with an event as trigger. If you want to start an automation or decision from an event, then create a process with an event trigger that can call an automation or decision as required.

10.3.3 Application Programming Interface Triggers

An API trigger creates a dedicated endpoint that allows you to initiate the execution of a process, automation, or decision. They allow for seamless integration and communication between different systems or applications with SAP Build Process Automation. When compared to triggering a process, automation, or decision via an event, the API-based triggers allow for more precise control over when and how a process, automation, or decision is triggered. When a sales order is changed in SAP S/4HANA, there's usually one event triggered (Sales Order Changed); however, there are multiple user exits or BAdIs when a sales order is saved. A decision can be triggered from a BAdI to validate a sales order before saving, an automation can be triggered from a BAdI on modification of tax data in the header, and so on. This level of control is essential for complex business processes and allows for greater flexibility in determining the appropriate conditions under which a process, decision, or automation will be initiated. Process, automation, and decision artifacts in SAP Build Automation support execution by dedicated endpoints.

In the following section, we'll understand how to create API triggers to execute processes, decisions, and automations. We'll explore the process of defining API triggers and identifying the necessary inputs that need to be sent to the API trigger.

API Triggers for Processes

To start a process based by calling an API, the process should be configured with an appropriate API trigger as the start. By configuring mandatory and nonmandatory inputs for the process, the input to the API can also be adjusted. Refer to Chapter 5 to understand how to define an API trigger for a process and configure its input.

API Triggers for Automations

There are a couple of ways to enable API triggers for automation in SAP Build Process Automation. Like defining a trigger for a process, in the automation editor, you can define that the automation will be started with an API trigger. Or you can define an explicit API trigger for any deployed version of automations. You can do that by accessing the environments from the **Control Tower** page. Navigate to the **Unattended Triggers** tab of the environment, and click the **Add Trigger** button. From the list of deployed projects, choose the project where the required automation is available, and proceed to create an API trigger.

API Triggers for Decisions

The API triggers for decision artifacts are created from the **Unattended Triggers** tab available in the specific environment (e.g., **Public**) found in the **Control Tower** page. Like creating an API trigger for automation, click the **Add Trigger** button, choose the project where the decision is available, and proceed to create an API trigger for the decision.

Figure 10.18 shows how an API trigger can be created for decisions and automation.

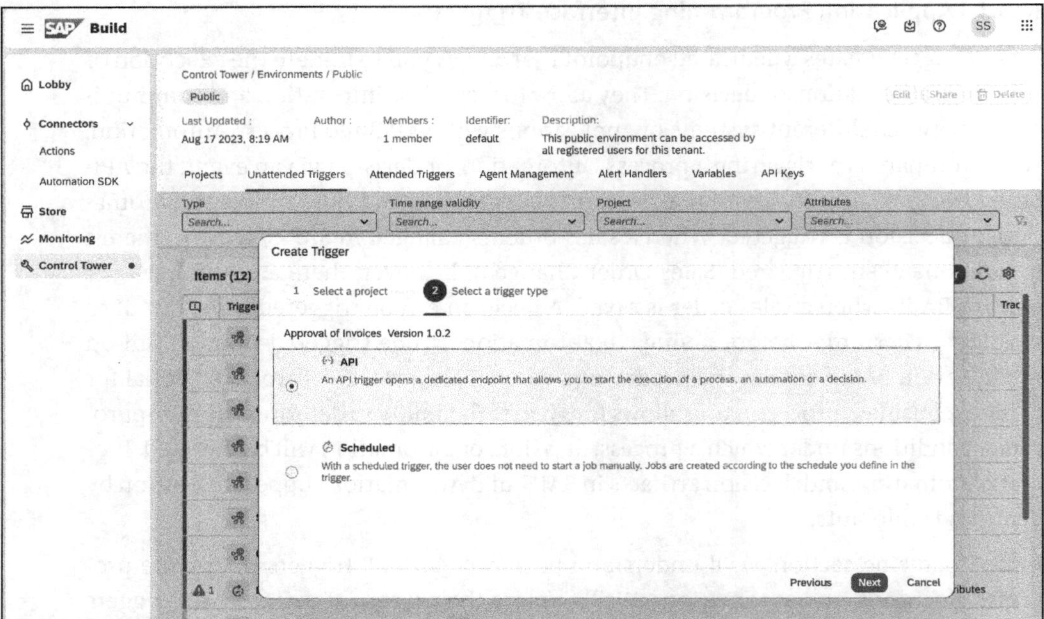

Figure 10.18 Creating an API Trigger for Automation and Decision

Identify the API and Inputs

Irrespective of how the API trigger was created and what the API trigger will execute, the endpoint for each API trigger and their corresponding input payload can be accessed from the environment. From the **Unattended Triggers** tab under the specific environment, search for the API trigger. In the **Actions** column, click the overflow button, and select **View**. Figure 10.19 shows how an API-based unattended trigger can be viewed.

This displays the endpoint that executes the selected artifact, the payload that is expected as an input, the sample output schema, and any other details used to configure the API trigger. The REST endpoint can be called from any system or application, including an SAPUI5 app deployed in SAP BTP, Cloud Foundry environment. The endpoint will be authenticated with the OAuth2.0 protocol as explained in Section 10.2.3. Figure 10.20 shows how the details about the trigger can be viewed.

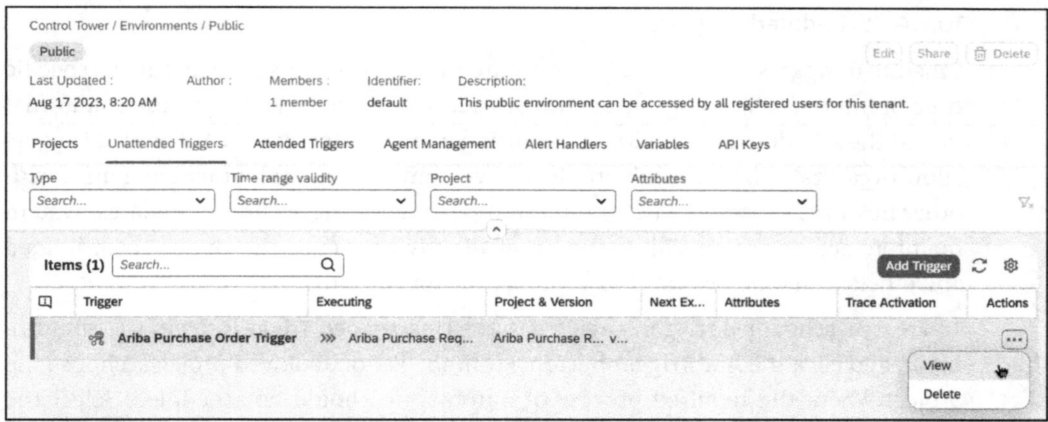

Figure 10.19 View the API-Based Unattended Trigger

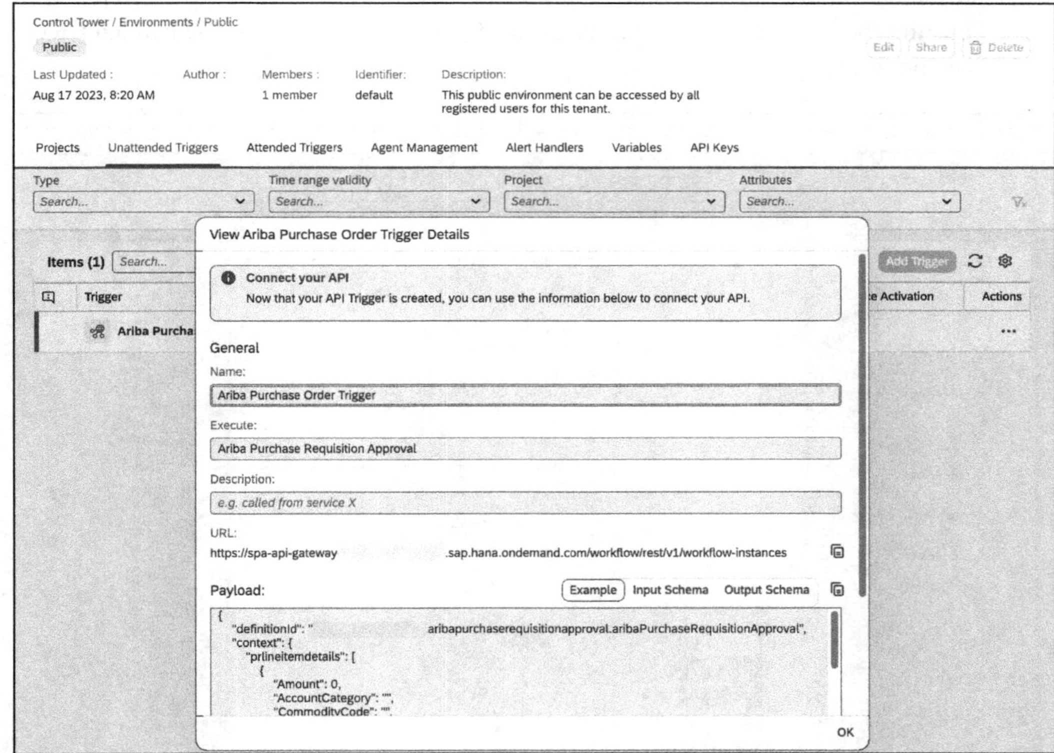

Figure 10.20 View the API URL and Payload Required to Start the Process, Automation, or Decision

10.3.4 Scheduled Triggers

Scheduled triggers in SAP Build Process Automation are events set to occur at a specific time or interval. These triggers are used to automatically initiate a process or automation at the scheduled time, without requiring manual intervention. Scheduled triggers allow organizations to coordinate their development and deployment activities with other business processes, such as end-of-month reporting, customer updates, system maintenance, and so on. This ensures that all activities are carried out in a coordinated and efficient manner, without disrupting other operations.

To create a scheduled trigger, navigate to the **Unattended Triggers** tab of the environment, and click the **Add Trigger** button. From the list of deployed projects, choose the project where the required process or automation should be scheduled. Select the **Scheduled** option for trigger type, and proceed to configure how frequently the process or automation should be run, as well as the inputs that might be needed. For scheduling an automation, you need to additionally provide distribution restriction policy of picking the automation agent to be used. Figure 10.21 shows how a scheduled trigger can be setup.

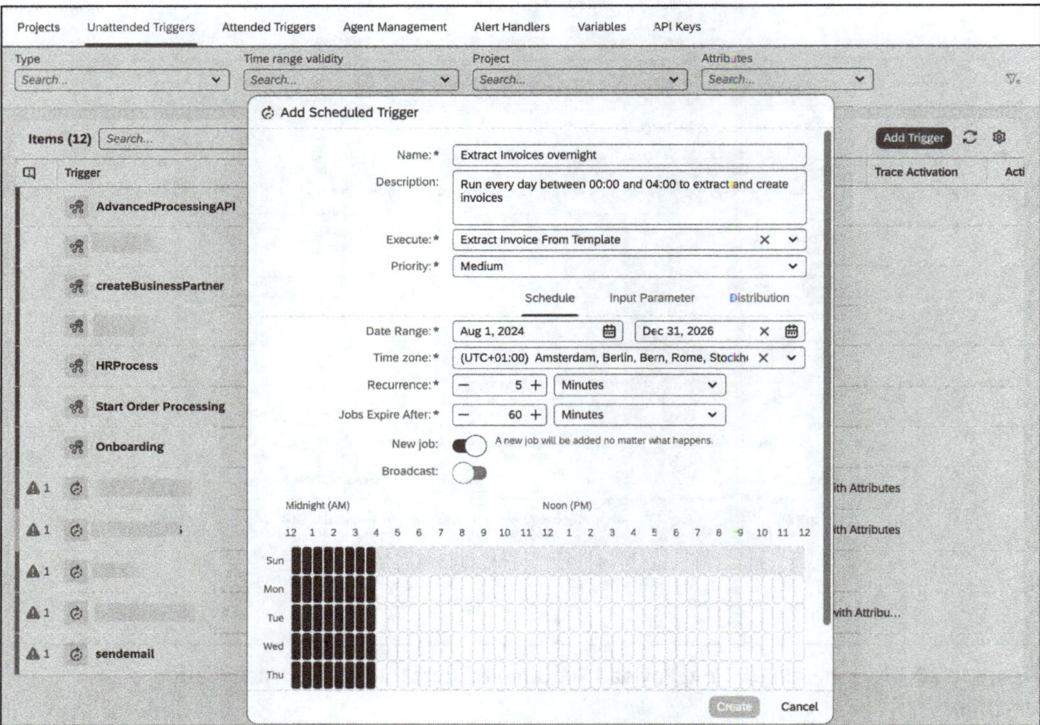

Figure 10.21 Setting Up Scheduled Trigger for Processes and Automations

After providing an appropriate **Name** and **Description** of the scheduled trigger, choose the specific automation or process that needs to be executed in the **Execute** field. Choose an appropriate **Priority** that determines which trigger should be executed first if multiple triggers are eligible at the same time. In the **Schedule** section, specify the **Date Range** and **Time Zone** when the trigger should be active. Then, you can configure the **Recurrence** field and **Jobs Expire After** to define how frequently the job should be executed and how long the system should wait before marking the job with status **Expired**. Disable the **New Job** toggle if the system shouldn't create a new job when a job with a similar **Input Parameter** is running. If you enable the **Broadcast** toggle, then the automation job would run in multiple agents, and therefore the **New Job** toggle will be disabled. In the **Weekly Calendar** section, you can choose at which days of a week and time the job should be enabled. If the automation or process requires any inputs, then go to the **Input Parameters** tab and configure them. Finally, click the **Create** button to create the scheduled trigger for an automation or process.

Note

Scheduled triggers aren't supported for decisions because the decisions are typically evaluated based on specific triggers or events, rather than on a regular schedule. Instead, decisions should be designed to be triggered based on specific events or conditions from a process, automations, or an external system, and the system should be configured to react to the outcome of the decisions.

10.3.5 Form Triggers

In SAP Build Process Automation, only a process can be initiated from a UI form-based trigger. The *form* artifact of SAP Build Process Automation provides a user-friendly and intuitive interface for end users to initiate processes. Refer to Chapter 5 to understand how a process can be initiated with a *start form*. Once the process automation project is deployed, **Form Link** and **Launchpad Configuration Parameter** will be available within the deployed process. **Form Link** provides a direct URL to access the form to start the process. If the form needs to be available as a tile in SAP Build Work Zone, then you can use the predelivered Process Trigger application and configure the URI parameter of the application as the **Launchpad Configuration Parameter**.

10.3.6 Execute from Another Artifact

Within SAP Build Process Automation, a process can call another process or workflow, call an automation, and call a decision using the respective activities. On the other hand, it's important to note that an automation can't trigger a process or decision directly. If there is a need for this functionality, it can be achieved by creating an action project that uses APIs to initiate a process or execute a decision and then consume that action from the automation. However, a decision can't trigger an automation or a process.

Table 10.1 illustrates the list of artifacts in SAP Build Process Automation and the different triggers that they support.

Artifact Type	Event Trigger	API Trigger	Schedule Trigger	Form Trigger	Execute from Another Artifact
Process	Yes	Yes	Yes	Yes	No
Automation	No	Yes	Yes	No	Yes (from process)
Decision	No	Yes	No	No	Yes (from process)

Table 10.1 Artifacts and Supported Triggers in SAP Build Process Automation

10.4 Summary

In this chapter, we covered the important components of executing processes, automations, and decisions in SAP Build Process Automation. We started by explaining the concept of runtime environments, which provide segregated spaces for deploying and running different versions of a project. Next, we discussed the SAP Build Process Automation APIs, including the technical prerequisites for accessing the APIs, how to access them, and the documentation available from SAP Business Accelerator Hub. You also learned how to execute the APIs and the supported authentication methods. Finally, we explored the various methods for running processes, automations, and decisions, including event-based triggers, API triggers, form-based execution, scheduled triggers, and calling from other artifacts. We now have a good understanding of when to use each method to ensure efficient and effective execution of processes, automations, and decisions in SAP Build Process Automation.

In the next chapter, we'll explore the benefits of monitoring and the two different types of monitoring: technical monitoring and business monitoring.

Chapter 11
Monitoring

In this chapter, we'll explore both technical monitoring and business monitoring of SAP Build Process Automation artifacts such as processes, decisions, and automations. Monitoring is important in business process management to ensure that the process and its related artifacts are operating proficiently within the designated standards and key performance indicators (KPIs).

Monitoring is an integral part of business process management and provides insight into end-to-end process execution, which can be used to analyze process outcomes. Business processes are core to organizations' success, and it becomes important that these processes run smoothly for healthy and profitable business transactions. A process is composed of many steps, and continuous monitoring gives a complete overview of how the process is executing, helps to identify the errors before they occur, finds bottlenecks and opportunities for improvements, and thus keeps the business running. The key is to detect the errors before they can cause any critical damage and disrupt the business performance.

Using the monitoring feature of SAP Build Process Automation, you can do both technical and business monitoring to track the processes in real time, identify the root cause of failures, collect execution data that can help in optimizing the process, and improve process efficiency. Monitoring eventually leads to continuous improvement strategies to streamline and enhance the processes over the years.

In this chapter, you'll learn about business and technical monitoring in SAP Build Process Automation, the benefits to be gained, and how to use these different monitoring techniques to enable business users and administrators to improve process efficiencies and identify bottlenecks early to enhance product outcomes and customer satisfaction.

11.1 Benefits of Monitoring

Every business needs a robust monitoring capability in process automation and management. Besides the standard benefits of improving process efficiency, continuous optimization, tracking errors and inconsistencies, cost savings, better risk management, and enhancing customer experience, there are a few more that you get specifically from the monitoring feature in SAP Build Process Automation:

- **Centralized access**
  You can access monitoring applications for different artifacts, such as workflow and automation, directly from the SAP Build Process Automation lobby. With centralized access and a unified experience, administrators or authorized individuals can efficiently monitor the processes from one workbench without navigating through multiple applications and tools. This not only saves time but also enhances user interaction.
- **Governance enabled**
  Governance is a key and differentiating aspect of monitoring in SAP Build Process Automation. By introducing governance in the monitoring, you can control the access of these applications to designated personas in your organization. Process data contains sensitive and confidential information about the company's core business processes such as procurement, finance, supply chain, master data, and so on, so securing this information becomes extremely important. Imagine if such information was leaked to unauthorized personal and exploited to derive a company's outcomes, performance, and KPIs. Having governance as a foundation ensures reliable, resilient, and robust process monitoring.
- **Real-time monitoring dashboard**
  With rapid digitalization transformations and changing market demands, real-time monitoring is a must in analyzing trends and getting actionable insights to adapt the processes based on market requirements. The intuitive visual dashboard with graphical charts can be used to efficiently track the process in real time, detect irregularities, and take quick actions to improve the process. It helps business process owners get a clear and transparent view of the processes to make informed decisions and derive optimization strategies.

11.2 Types of Monitoring

Monitoring caters to two personas, business process owners and business process administrators. Having diverse monitoring capabilities, SAP Build Process Automation offers both technical monitoring and functional/business process monitoring, which will both be discussed in this section. The objective of these different types of monitoring is to provide a holistic approach to all monitoring needs under one platform to track milestones, analyze process performance, predict cycle times, and address issues before that affect the business.

11.2.1 Business or Functional Monitoring

Process owners can monitor the business processes and manage the performance in real time using the dashboard with visual representation. Using these graphical tiles built on processed data, the visibility application enables users to analyze processes, track business KPIs, derive data-driven insights, and take corrective actions to quickly resolve issues. While technical monitoring is for administrators from the IT department to manage process operations to ensure that processes are completed successfully and keep up the daily business, business monitoring is more functional in nature where process owners can examine the key business KPIs, analyze trends, identify anomalies, and derive opportunities for process improvements.

Taking the same purchase requisition example as used earlier, a business user can use the visibility dashboard to monitor how purchase requisition processes are running overall in the company or region, determine the cycle time for process completion, track KPIs (e.g., new purchase requisitions raised today or purchase requisitions trends in each quarter), or view which step in the process is taking time for execution.

Business monitoring is at a level higher than technical monitoring and provides a holistic view of the overall performance of the business where the process may span different landscapes and systems. It's therefore key to manage customer demands and adapt to changing market trends with insights to process improvements. Figure 11.1 shows an example of business monitoring using the process visibility dashboard.

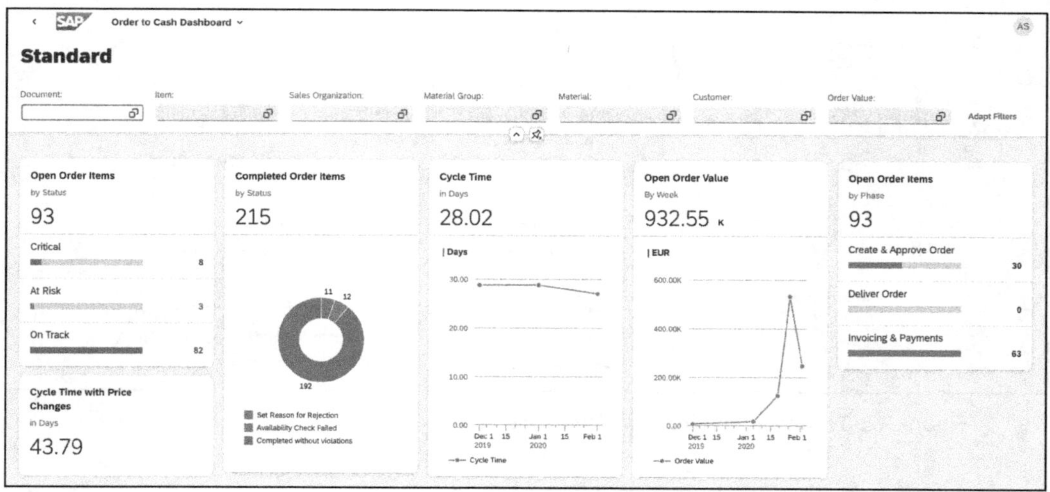

Figure 11.1 Business Monitoring Using the Process Visibility Dashboard

11.2.2 Technical Monitoring

As the name suggests, technical monitoring is done by process administrators who can closely track processes, automations, and events to ensure they are executed successfully. From monitoring applications, embedded in the lobby, a process administrator can monitor execution of process steps, data flow, system integrations for external service calls, automation health, and events processing outcomes. The administrator or authorized person not only can track the processes but also take actions that provide resolutions to issues and ensure business continuity.

Let's consider an example in which there is a purchase requisition business process that has an approval flow, task automation to extract the requisition document, and a connection to an SAP S/4HANA system to post the approved purchase requisition data. Once the process is triggered, an administrator can see the execution flow of the process steps with its metadata, including the process data. If there are any errors, for example, while executing the automation or application programming interface (API) calls, then the process will be in an error state, and the administrator will be notified. Figure 11.2 shows an overview of the technical monitoring of the processes.

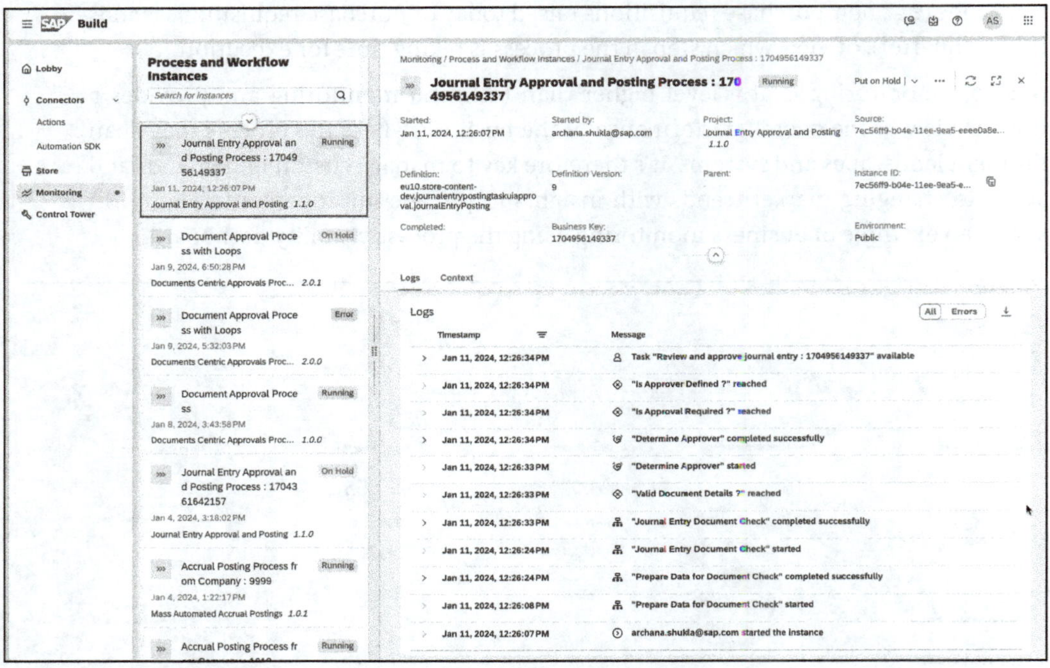

Figure 11.2 Technical Monitoring of Processes

11.3 Business Monitoring

In this section, we'll introduce the process visibility capability of SAP Build Process Automation and explain the following:

- How to create and configure visibility scenarios

- What options are available to add visibility into the end-to-end processes that spans across heterogenous systems (SAP or non-SAP)
- How to deploy and run the visibility scenario for real-time business monitoring

11.3.1 Overview of Business Monitoring in SAP Build Process Automation

The process visibility capability of SAP Build Process Automation provides an end-to-end, real-time dashboard for business users to monitor their business processes and to manage KPIs. Similar to a fitness watch that proactively monitors aspects of your health, process visibility monitors the health of your business processes, give insights into process performance, and informs the process owners of any deviations or anomalies.

Along with process monitoring, the process visibility dashboard can also be used to track milestones, KPIs, and various other service-level agreements (SLAs) that are critical to business. Process owners can see the complete overview of their processes and KPIs or can drill down to get deeper insights into process instances and data, which can be used eventually to define process improvement strategies. These strategies are key to a company's transformational journey during market evolutions and to increase customer satisfaction, thereby meeting market competition.

Each business project in SAP Build Process Automation can have its own visibility scenario, which can be created directly from the project **Overview** page, with few configurations and without the need for any coding. You can build an end-to-end dashboard for processes deployed in SAP Business Workflow, SAP Build Process Automation, SAP NetWeaver, and so on, as well as non-SAP systems and any custom processes specific to your organization.

Events play a critical role in process visibility. These events, together with the process context data, are pushed from processes across different systems (in a predefined format) to an events acquisition application. Process visibility then offers real-time monitoring by processing the event and the context data. For processes in SAP Build Process Automation, SAP Business Workflow, flexible workflows, SAP S/4HANA, and so on, this capability is available by default. This means you don't need to explicitly send events. Once the systems are configured per the guidance in the SAP Help Portal documentation, the events are automatically received and processed by the visibility scenario. For others, such as SAP Ariba, SAP SuccessFactors, or any third-party systems, the events must be sent using process visibility event APIs.

The general steps to enable process visibility for end-to-end processes are as follows:

1. For custom processes, implement logic to push events in a predefined format using public APIs.
2. Create and configure a visibility scenario.
3. Release and deploy the business project containing the visibility scenario.

4. Run the process in the system:
   - As the process is run, the events will be acquired by the visibility scenario in raw format.
   - These events will then be correlated and processed.
   - Information will be shown in the process visibility workspace or dashboard, thus enabling real-time monitoring for business users to track process execution and take corrective actions.

11.3.2 Creating and Configuring the Visibility Scenario

The visibility scenario is an artifact through which process owners can monitor end-to-end business processes across heterogenous landscapes. It's created inside the business project, and one or more of the processes inside that business project can be added to a visibility scenario to build the dashboard.

As stated earlier, the visibility scenario editor is a no-code tool, which doesn't require a coding specialist. With just a few configurations, you can easily create dynamic, real-time dashboards that can then be accessed by the business users or stakeholders to track daily business KPIs and process operations.

Before you start creating the scenario, here are a few guidelines:

- **Define the end-to-end process that needs to be monitored.**
  For effective process monitoring, it's important that you define the entire end-to-end process chain and identify which processes needs to be monitored and in which order, as well as the common condition between all the selected processes that will correlate them.

  For example, a purchase requisition process might span SAP Ariba, SAP S/4HANA, SAP Build Process Automation Process, and external non-SAP systems. In this case, the common correlation condition will be the purchase requisition ID, which will bind all the processes across different systems.
- **Align with the business users on the KPIs they want to track.**
  As business users own the processes and will track them in their daily routine, it's necessary to align with these users and get their requirements on business KPIs. These business KPIs should then be mapped with the available process data to ensure that stakeholder needs are fulfilled
- **Establish thresholds, cycle times, and alert mechanisms.**
  Clear objectives and performance levels are key to business monitoring. Defining process thresholds and providing permissible cycle time limits with actionable insights and mechanisms to alert the responsible users can derive the desired outcomes by taking early measures to avoid any unwanted delays in business operations.
- **Outline governance and authorizations.**
  Monitoring a business process requires the right roles and must be done only by

authorized users who are responsible for those processes. This may vary across organizations, regions, and business roles of an individual. For example, procurement processes and data in the Germany region can't be monitored by the procurement office of the India region. Governance plays a crucial role in ensuring data privacy and integrity. Defining the roles and responsibilities to provide secure access to the visibility dashboard is therefore important.

- **Maintain a continuous improvement strategy based on the data.**
  One of the main objectives of having such an analytical real-time dashboard is to continuously track the process performance and derive insights to improve the process. Any organization must have a robust strategy to regularly assess the process performance, implement optimizations, and adapt to changing market needs as the core goals for successful process monitoring.

Overview Page

Once you're ready with the visibility dashboard design, create the scenario by choosing **Create • Visibility Scenario** from the business project **Overview** page. Figure 11.3 shows how to create the visibility scenario in the business project.

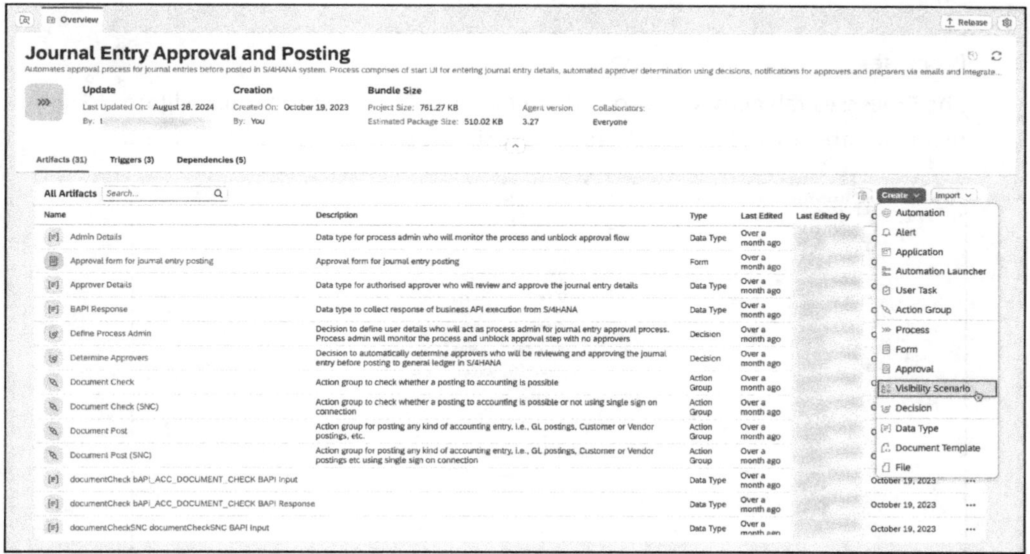

Figure 11.3 Creating a Visibility Scenario in a Business Project

This will create the scenario with the given name. Next you configure the scenario based on your dashboard requirements such as the processes to be monitored, the threshold, the performance KPIs that must be created, user notifications or alerts, and so on.

From the visibility scenario overview page, you can view the **Scenario Name** (or change it) and **Scenario ID**, which is a unique identification name for the scenario. This is automatically generated when you create the scenario and takes the scenario name without spaces. Once created, it can't be changed. Because the visibility scenario shows the process instance data in graphical forms, you can also give the business object name on which the process is based, such as purchase requisitions, sales orders, or journal entries. By default, the business object name will show **Instance/Instances**, but it's recommended to change this so that the relevant business entity on which this visibility scenario is based (e.g., sales order, purchase requisition, etc.) is shown to the business user while accessing the visibility dashboard (see Figure 11.4).

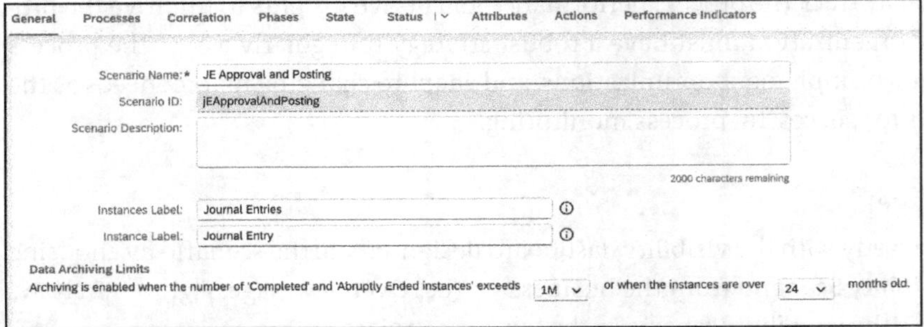

Figure 11.4 Configuring the Visibility Scenario

Processes

The **Processes** tab allows you to add the processes you want to track. Figure 11.5 shows processes area to add different process participants in the visibility scenario.

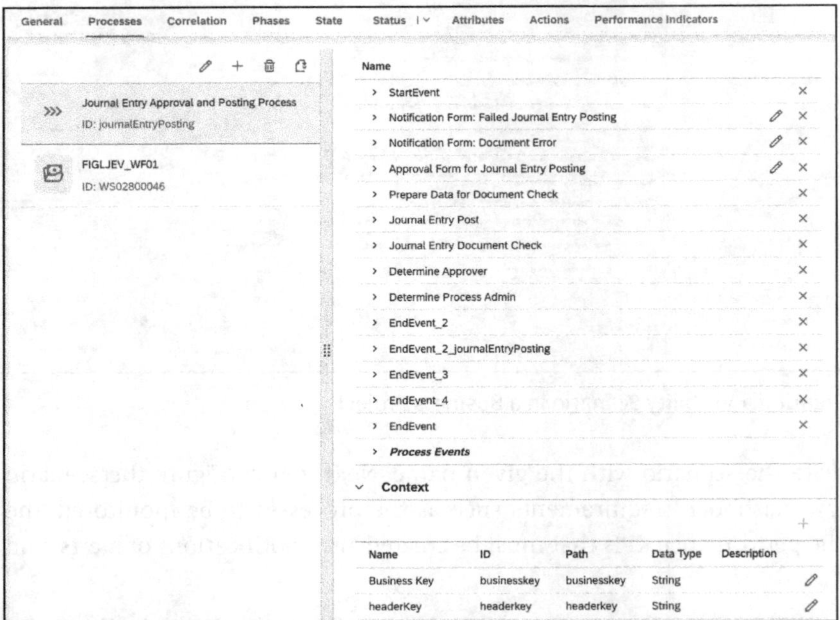

Figure 11.5 Overview of the Processes Tab

You can choose from the following processes:

- **Process in business project (where this visibility scenario is created) in SAP Build Process Automation**
  All you need to do is click on **Add Process**, select from the list of available processes, and import it to the visibility scenario for further configuration. As soon as the process is added, the respective events and context data are automatically imported. Keep the following points in mind:
  - For each activity in the process, there will be two events: **Created** and **Completed**, defining when this activity was started and completed during the runtime execution of the process.
  - The **Process Start** and **End** events define when the process was started and completed overall.
  - Apart from these, there are special events to capture other states of the process lifecycle such as **Suspended**, **Erroneous**, **Canceled**, and **Resumed**.

  Figure 11.6 shows options of processes from different systems such as SAP Business Technology Platform (SAP BTP), SAP S/4HANA, Qualtrics, SAP NetWeaver, and any third-party that can be added to the visibility scenario.

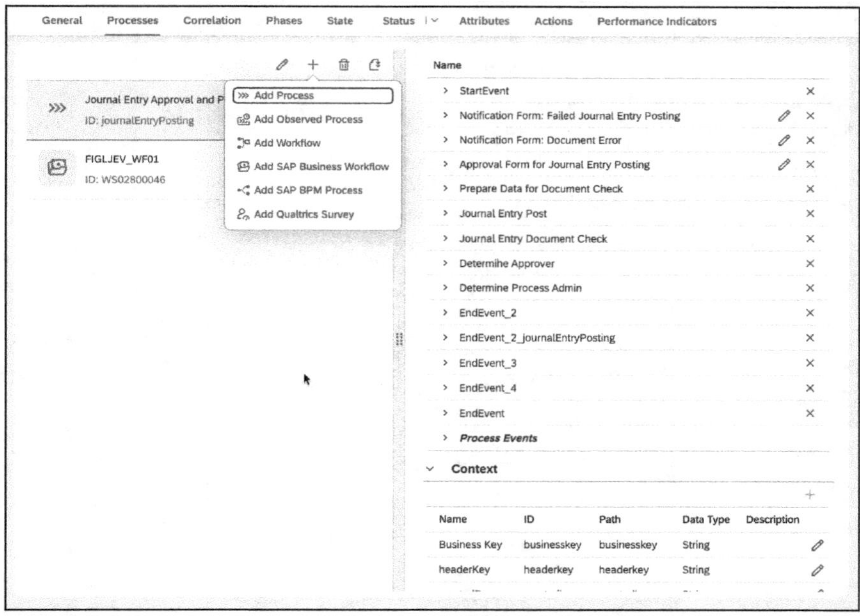

Figure 11.6 Adding Processes Available in the Business Project

Figure 11.7 shows the list of available processes in the same business project (where the process visibility scenario is created) after clicking the **Add Process** option.

You can then view the events added for different activities in the process, as shown in Figure 11.8.

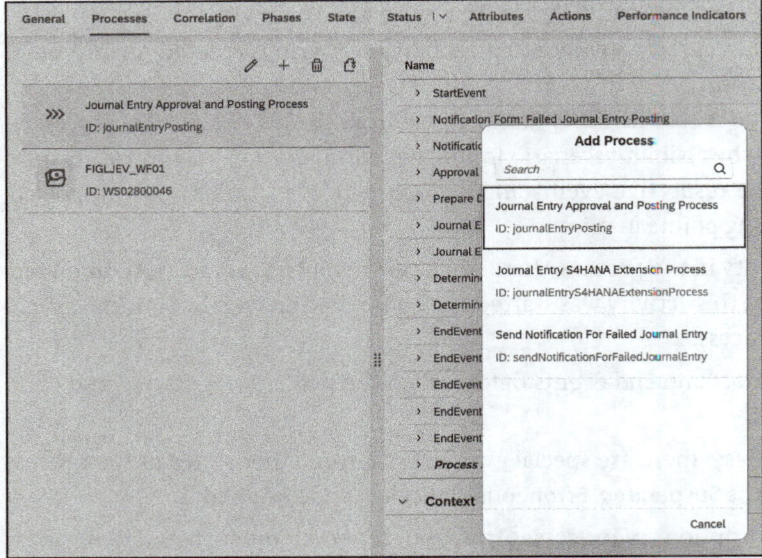

Figure 11.7 Selecting One of the Available Processes in the Same Business Project

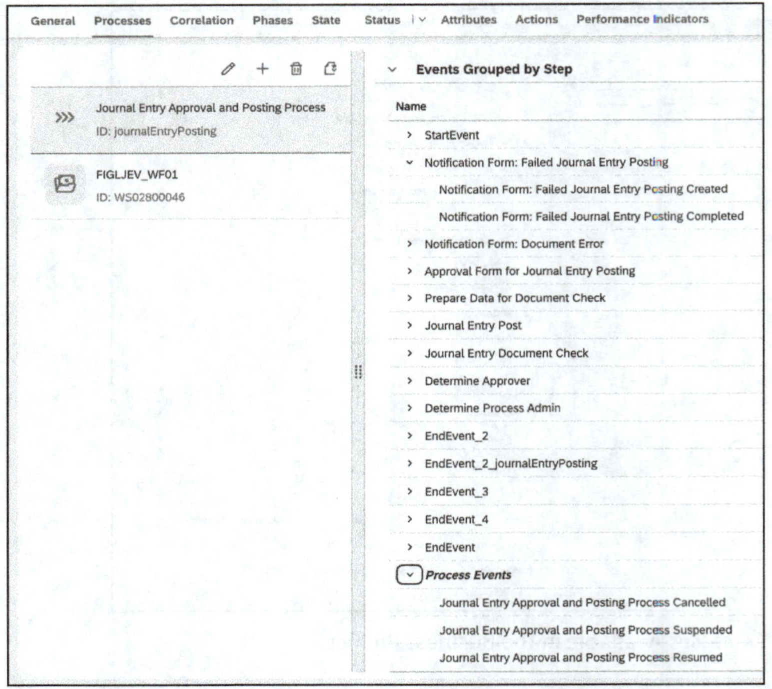

Figure 11.8 Viewing the Events Added for Different Activities in the Process

As an example, say there's a business partner approval process that has two approval flows, one decision to determine the approver and one action project to post the data in the backend system. While configuring the process, you've also selected the process context attributes that are exposed to the visibility scenario. When you add this process to the visibility scenario, there will be events and context data automatically imported. Table 11.1 provides a list of event names for different activities of the process as they are acquired by the visibility scenario.

Activity	Event Name	Event
Process Start	StartEvent	StartEvent Created
Approval Form	ApprovalForm1	ApprovalForm1 Created
		ApprovalForm1 Completed
Approval Form	ApprovalForm2	ApprovalForm2 Created
		ApprovalForm2 Completed
Decision	DetermineApprover	DetermineApprover Created
		DetermineApprover Completed
Action	BusinessPartnerPost	BusinessPartnerPost Created
		BusinessPartnerPost Completed
Process End	EndEvent_1	EndEvent_1 Completed
Process Events		BusinessPartnerApprovalProcess Cancelled
		BusinessPartnerApprovalProcess Suspended
		BusinessPartnerApprovalProcess Resumed

Table 11.1 List of Event Names for Respective Activities of the Process

- **Business workflows or flexible workflows available in the backend SAP S/4HANA system**
  Out-of-the-box visibility is available for business workflows and flexible workflows from the SAP S/4HANA system. To import a business workflow, both systems (SAP BTP and SAP S/4HANA) must be configured and connected. All the information is available in SAP Help Portal. Once configured, you'll be able to directly discover business workflows from an SAP S/4HANA system and add them to the process. Once the process is added, the events and context data will also be added into the visibility scenario. As shown in Figure 11.9, you can add a business workflow from SAP S/4HANA using the **Add SAP Business Workflow** option.

  In Figure 11.10, you can select the available business workflows in the backend SAP S/4HANA system, and in Figure 11.11, you can see the business workflows being added as process participants in the visibility scenario with events and context data that gets imported.

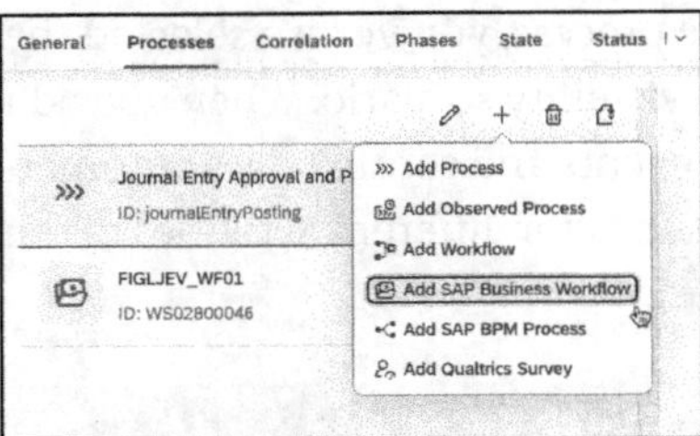

Figure 11.9 Adding a Business Workflow from SAP S/4HANA Using the Add SAP Business Workflow Option

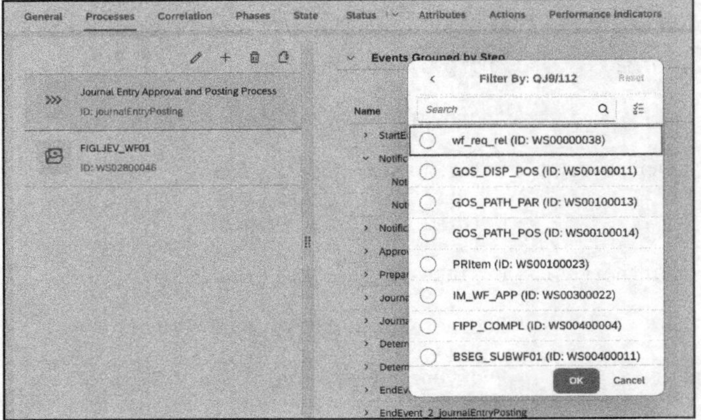

Figure 11.10 Selecting Available Business Workflows in the Backend SAP S/4HANA System

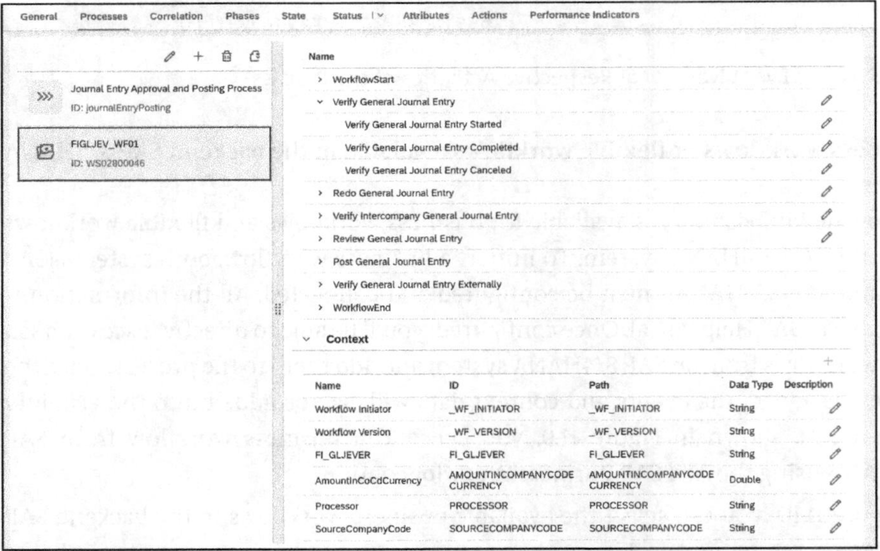

Figure 11.11 Business Workflows Being Added as Process Participants in a Visibility Scenario with Events and Context Data

- **Any third-party or line of business workflows that aren't available in the previous two options**
  This requires publishing relevant events from these systems and then creating and configuring the scenario. Using the **Observed Process** option, any custom processes can be created and configured. In such cases, you must manually add all the event and content data (as pushed from the backend system) to the observed process. Any mismatches in the event names may not be registered while processing the data.

  Figure 11.12 shows adding observed processes for custom workflows, and Figure 11.13 shows the events and context that are manually added in the observed process.

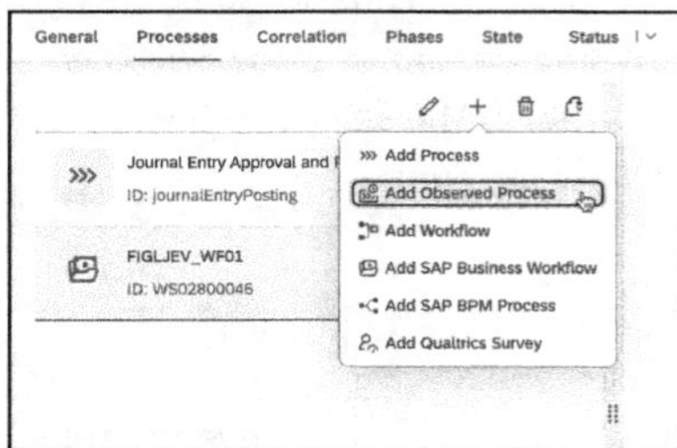

Figure 11.12 Adding Observed Processes for Custom Workflows

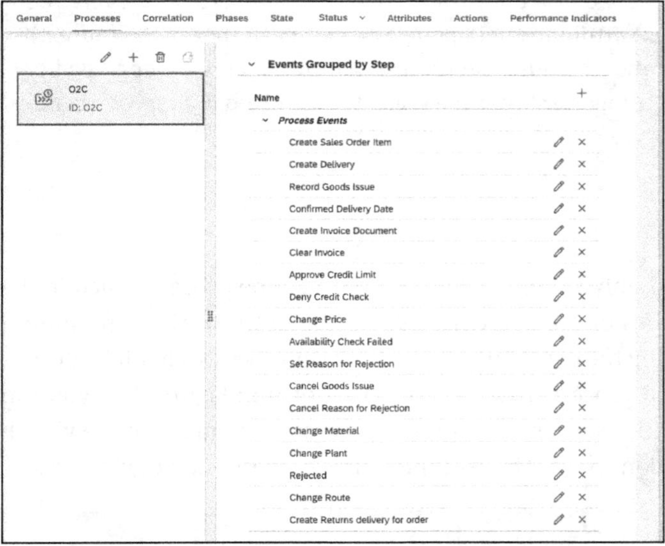

Figure 11.13 Manually Adding Events and Context in the Observed Process

For example, if you want to work with SAP SuccessFactors, then from the integration center of SAP SuccessFactors, you can push the event using visibility APIs on any operations in SAP SuccessFactors such as a new job application request, an update to a job application, and so on. These events will be collected in the visibility scenario and processed later on to show the analytical charts.

Note

If you have ongoing survey results that are needed by the business users to make informed decisions, then click the **Add Qualtrics Survey** option with which surveys can be embedded in the visibility dashboard.

Note

If you have process orchestration workflows from an SAP NetWeaver system, you can use the **Add SAP BPM Process** option to add them in the visibility scenario. To discover such workflows in the visibility scenario, configurations must be done in the SAP NetWeaver system, and a destination with cloud connector has to be set up in SAP BTP. You can get more details from SAP Help Portal (*https://help.sap.com*).

Note

From the **Processes** area, you can also delete and edit the process names, event data, and context information. If the process has changed, then you can reimport the process with two options: **Overwrite the names of events and context attributes** (meaning all events and context data of the entire process will be reimported) or **Overwrite the data type of the context attributes** (reimports only the data types of the context). This is a quite useful operation because the process will continue to be upgraded or modified with time, and it's important that the visibility dashboard is adapted accordingly.

Correlation

The **Correlation** area is used only when you have more than one process added in the visibility scenario. It provides options to define how different processes are interrelated. This helps in processing the data from across different processes in different systems to form an end-to-end process instance flow. As shown in Figure 11.14, you can configure correlation conditions when two or more processes are added in the visibility scenario to provide the context attribute of the source process and target process.

Note

During correlation at runtime, the content of the source process is compared with the target context. Sometimes, you'll notice that no instances are correlated, which may be because your source context doesn't have the data, or the context field that you're comparing is empty. It's recommended to keep the source context where you expect the data to be complete and available.

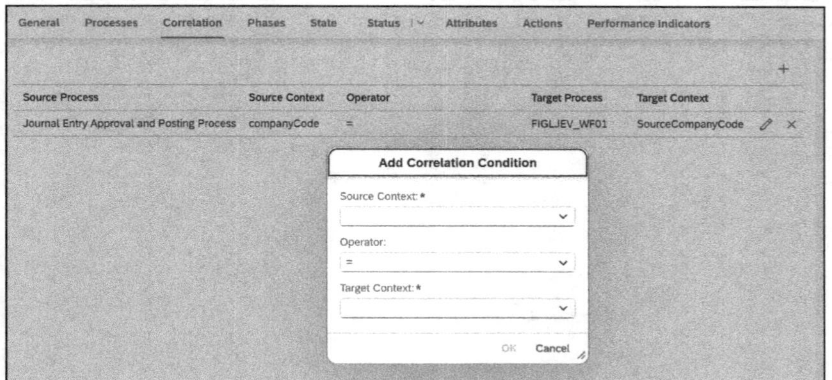

Figure 11.14 Configuring the Correlation Condition When Two or More Processes Are Added in a Visibility Scenario

Phases

The **Phases** area is one of the key configurations in the visibility scenario. As the name suggests, you combine multiple process steps into logical groups called phases. For example, a process might have 50 activities with scripts, actions, mail steps, and so on, and you want to present these activities in a simpler form that business users can understand and use to identify and resolve problems. Therefore, these activities can be grouped as Validating Journal Entry, Approving Journal Entry, and Posting Journal Entry, which are key phases in the process.

Each phase is created based on the events where you must configure from which event(s) this phase will start and end. If there are multiple events, then any one event that is reached first is considered for phase start and end. Besides this, you can also have a few nonmandatory properties such as defining navigational links to the phase, which can lead to direct access to the backend system or defining conditions to exclude showing phases in the dashboard (see Figure 11.15).

State

The **State** area is normally prefilled with relevant events from the process such as when a process instance is considered to be open or running, completed, or abruptly ended. Abruptly ended means that the process didn't follow the due course of completion by reaching the end event; instead, it was either canceled, suspended, or became erroneous due to an API call fail or backend system that couldn't be reached. It's not mandatory to change these configurations unless you have special events or requirements for state change.

State configurations are heavily used to lay out the visibility dashboard representations such as how many business partner processes are running, cycle time of completed business partner approvals in a month, or number of business partner approvals that were canceled across regions (see Figure 11.16).

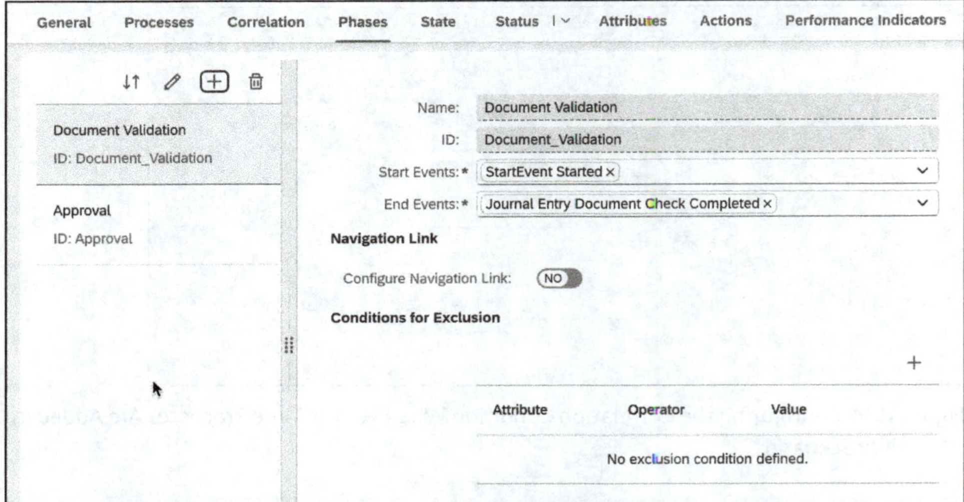

Figure 11.15 Phases Configuration Options to Add Start and End Events, Navigational Links, and Exclusion Conditions for the Phase

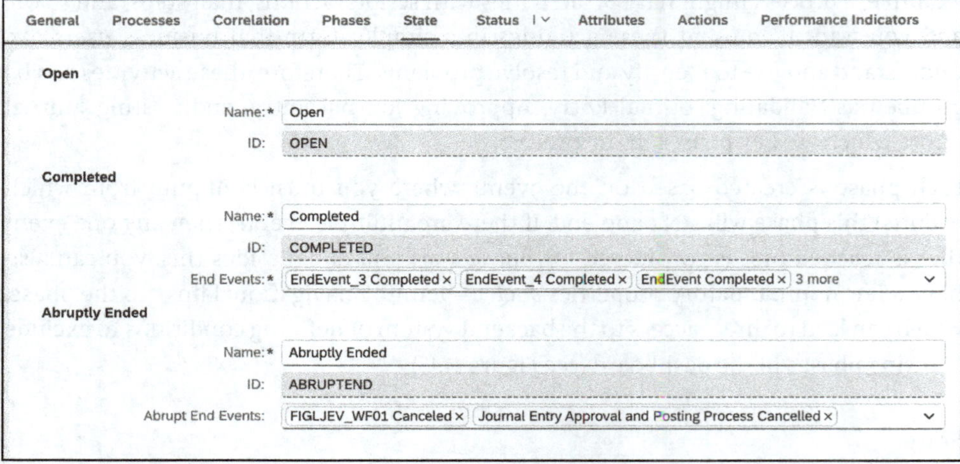

Figure 11.16 Configuration of State: All Needed Events Are Added by Default

Status

Status is another important configuration that determines the color codes in visibility dashboard charts. There are three statuses in the visibility scenario: **Critical**, **At Risk**, and **On Track**. **Critical** is represented in red, **At Risk** in yellow, and **On Track** in green in the visibility dashboard.

For example, if the target of the process is 4 days and the threshold is 80%, the process must be completed within 4 days:

- If the process completes or is still running *within* 4 days, then it's marked green for **On Track** status.
- If the process completes or is still running *after* 4 days, then it's marked red for **Critical** status.
- If the process completes or is still running *between* 80% of 4 days, which is 3.2 to 4 days, then it's marked yellow for **At Risk** status.

By default, the status conditions are predetermined based on the process state; for example, if the process fails or cancels, then it's regarded to be in critical state, but you can add new conditions based on the specific project requirements to define when a process instance can be considered as **Critical**, **At Risk**, or **On Track** (see Figure 11.17).

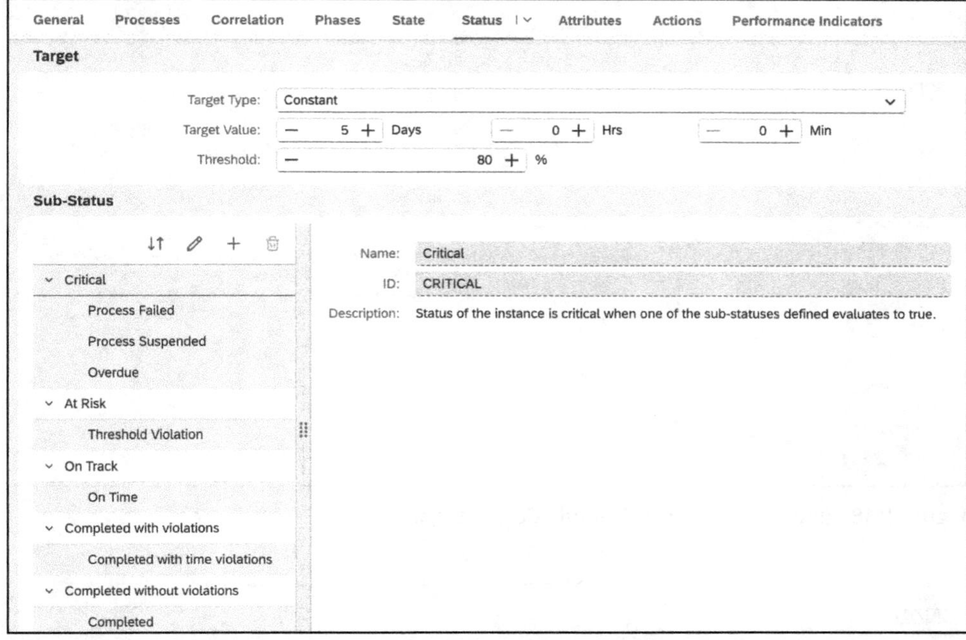

Figure 11.17 Defining the Target and Sub-Status of the Visibility Scenario

Attributes

Tracking KPIs and other important metrics in real time is a feature of proactive monitoring. These indicators can be created using **Attributes** and **Performance Indicators** areas, but we'll start with the **Attributes** section first. Attributes are the building blocks for KPIs. They are derived from various configuration properties and contexts of the visibility scenario. For ease of use, the visibility scenario comes with default attributes such as state, status, number of instances, completed steps, scenario instance ID, and so on.

Each attribute has certain properties that define them and how they can be shown in the visibility dashboard. The **Data Type** defines the outcomes of the attribute such as string, integer, date, Boolean, and so on. The attribute usage is defined by whether this attribute is **Dimension** or **Measure**. **Dimension** describes qualitative data such as the state of the process instance, start time of the process, completed steps in the process, and so on. **Measure** is the aggregation, or any formula applied on the data, such as average time taken to complete the phase (or phase cycle time), number of process instances running, duration between process start and end time, number of journal entry rejected by approvers, and so on (see Figure 11.18).

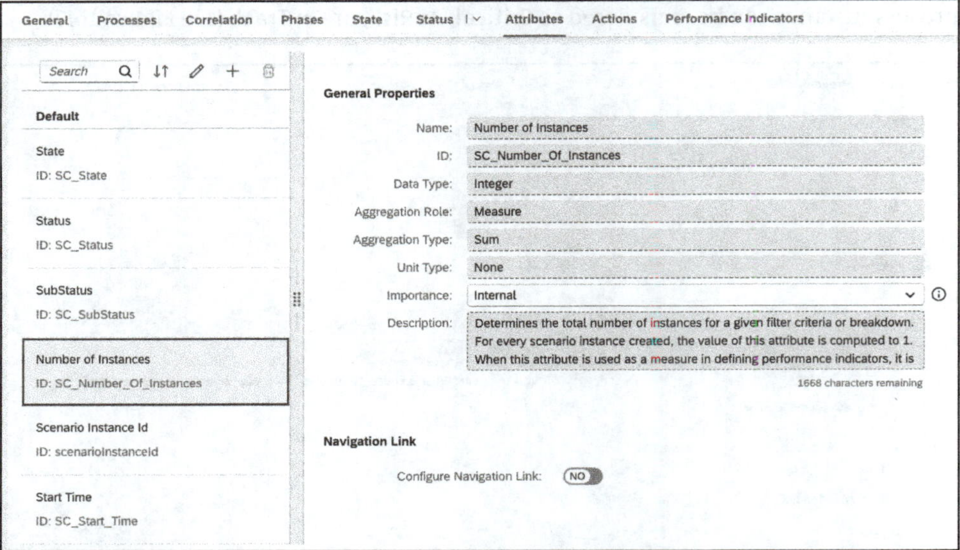

Figure 11.18 View and Configure Default, Context, and Calculated Attributes

Note

All context data is regarded as **Dimension**.

Based on the **Importance** property, the attribute is made visible in the dashboard (see Figure 11.19). Attributes are marked as the following:

- **Key**
  Shown as a column in the instances view.
- **Significant**
  Shown in the instance details view in the **Instance Information** area.
- **Internal**
  Used only during design time for KPI configurations. They aren't shown in the dashboard.

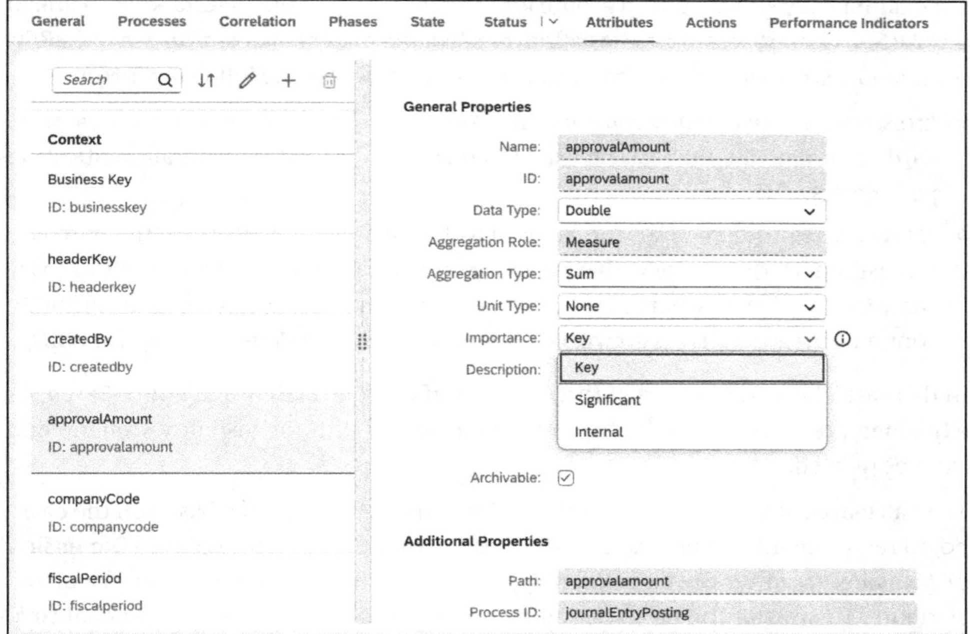

Figure 11.19 Configuring Importance Based on Which Attributes Are Viewed in the Visibility Dashboard

Note

When you click on the visibility dashboard charts, you are taken into the **Instances** view. It shows all the instances that contribute to that chart data. On clicking an individual instance in **Instance** view, it opens the **Instance Details** page, which shows complete information with the process flow of that instance.

Another interesting feature of **Attribute** configuration is **Navigational Link.** It gives you the provision to attach a URL to the attributes, which can be a deep linked URL to an SAP S/4HANA system, a URL to user data in SAP SuccessFactors, and so on. The connections to these systems happen via **Destination** in the SAP BTP cockpit.

To use this feature, first a destination must be created in the SAP BTP cockpit with additional property `bpmprocessvisibility.navigationLink = true`. Next, a destination environment variable must be defined from the project settings. This environment variable can then be chosen under the **Navigational Link** property, and a relative path to the actual URL must be defined.

For example, say you have an attribute for a purchase requisition ID and want to link to an SAP S/4HANA system with the URL *https://s4hana.ondemand.com/ui#PurchaseRequisition-displayAdvanced?PurchaseRequisition=12786*, where the following is true:

- *https://s4hana.ondemand.com/ui#PurchaseRequisition-displayAdvanced* is a standard common URL that can be used to create the destination with authentication parameters.
- *?PurchaseRequisition=12786* can be used as the relative path. Because the purchase requisition ID will change with each attribute, you can define it dynamically as `{purchaseRequisitionID}` where `purchaseRequisitionID` must be available as the visibility context. So, the relative path will be *?PurchaseRequisition={purchaseRequisitionID}*.

In this case, the attributes are of three types: **Default**, **Calculated**, and **Entity-Based**. As explained previously, default attributes are available with the visibility scenario and can't be deleted.

The **Calculated** attribute is the custom attribute that can be created based on the dashboard requirement and process design. These attributes can also be of type **Dimension** or **Measure** based on the **Expression Type** selected during creation of the calculated attribute. The rest of the property definitions such as **Importance**, **Navigational Link**, **Data Type**, and **Aggregation Role** are the same as in the default attributes.

Note

Expression types are set by predelivered formulas or functions to create the attributes that are then used to define the KPIs. They are a powerful means to define KPIs in a no-code approach without writing any code. A few examples of expression types are duration between events, which calculates the time taken between two events; number of times the event has occurred, which gives a numeric value of how many times the given event (e.g., business partner request rejected) occurred for all the process instances; month of the attribute, which can be used as a dimension to create KPIs (e.g., number of new business partner requests distributed over the months); and whether an event occurred to inform that an event has occurred and that some action based on this can be taken.

For example, as shown in Figure 11.20, an attribute called **Rejected Journal Entries** is created that calculates the number of journal entries rejected by the approvers using expression type **Event A occurred and not Event B** where **Event A** is **Approval Form for Journal Entry Posting Completed**, and **Event B** is **Journal Entry Post Created**. Based on the process design, the post event occurs only when the journal entry is approved or else a notification is sent to the requestor.

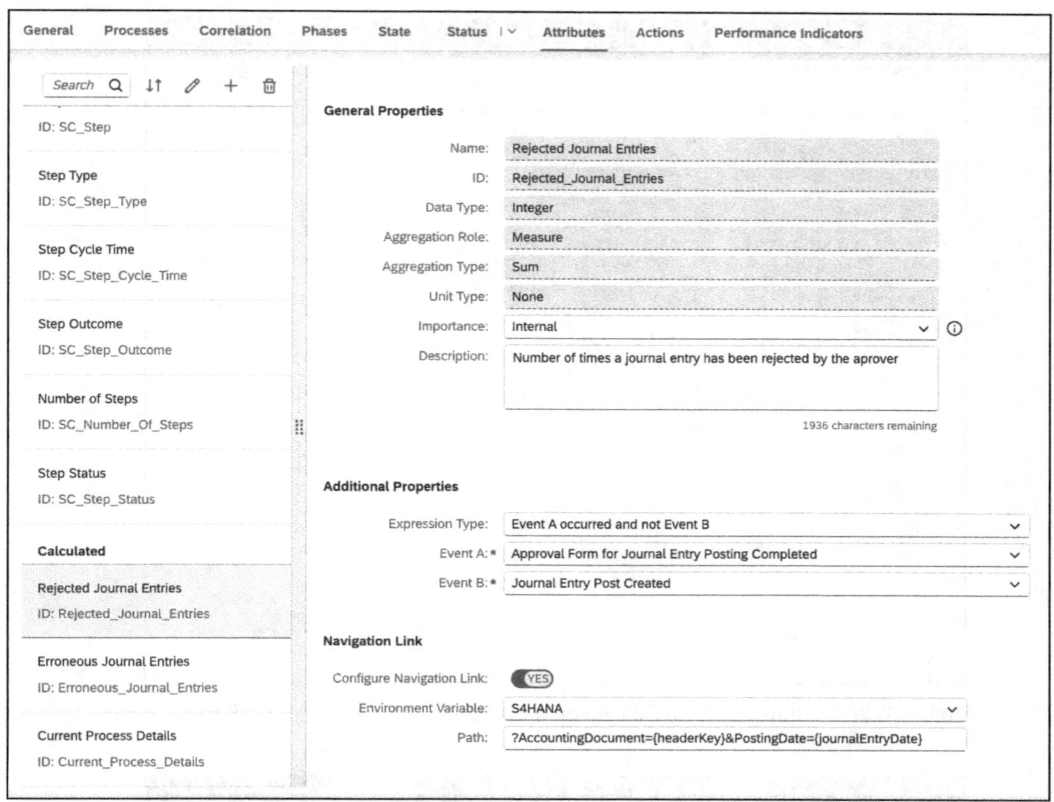

Figure 11.20 Calculated Attribute with Expressions and Navigational Link

The final attribute category is **Entity**. This category is used when there are multiple one-way or two-way associations in the process, such as a sales order is associated with multiple sales order items or a sales order is associated with an invoice where one invoice can have multiple sales order entries. Entities are good ways to define the link between one object and another in the end-to-end process, which means from a sales order header, you can navigate to order items, as shown in Figure 11.21. An entity can be regarded as a structure of calculated and context attributes of a visibility scenario. For associations where a business object is related with another business object, and you want to define the KPIs (Figure 11.22), then an entity can be used so that the KPI is viewed at an entity level. For example, if a purchase order is created with multiple line items, and you want to know which line items are associated with the purchase order, you can define an **Entity** called **PO** with a calculated and context attribute and then define a KPI based on this entity.

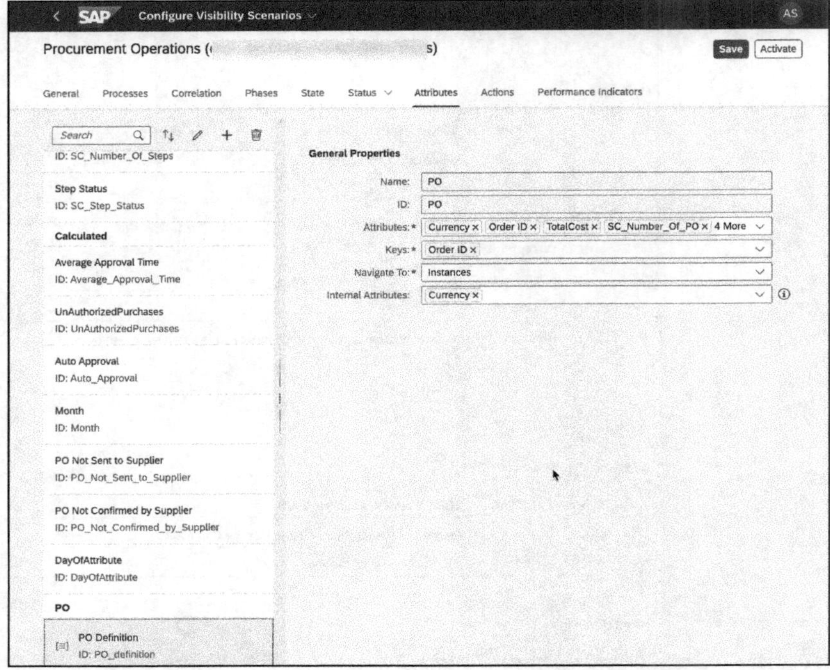

Figure 11.21 Defining an Entity Based on a Context and Calculated Attribute

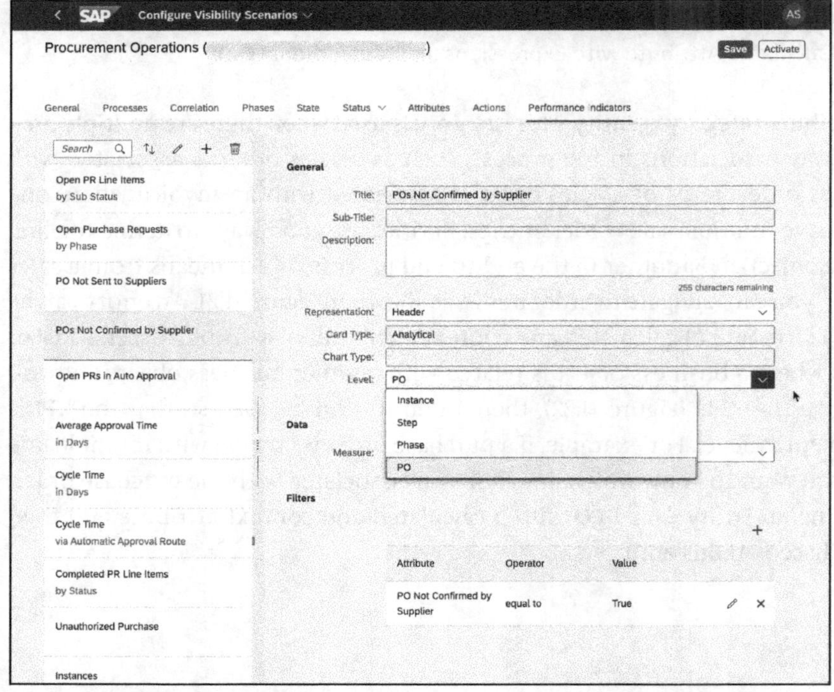

Figure 11.22 KPI Based on an Entity

When viewed in **Process Workspace**, such KPIs will provide navigation to the items or header based on the entity definition. This way, business owners can easily track both header-level and item-level process instance data from the same visibility dashboard. From the **PO Not Sent to Suppliers** tile shown in Figure 11.23, you'll see the detailed view of the purchase orders not yet sent to the suppliers, as shown in Figure 11.24. In Figure 11.25, you'll see the line-item details of the purchase order when you click a particular order. This shows all the line items associated with that purchase order.

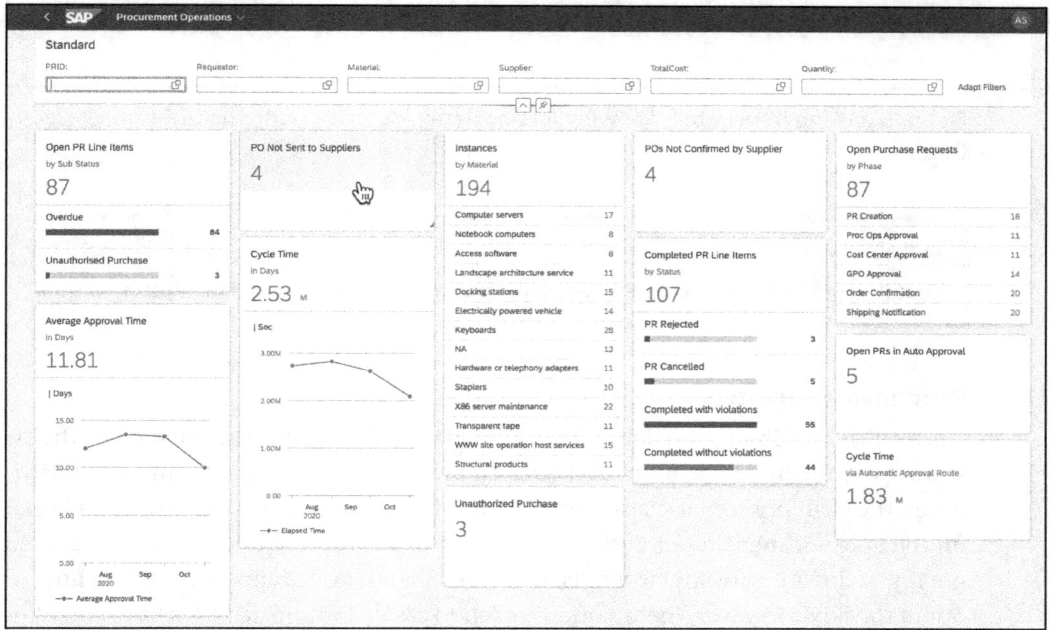

Figure 11.23 Process Visibility Dashboard with an Entity-Based KPI

Procurement Operations

PO

Supplier: | Order ID: | TotalCost: | PO Not Sent to Supplier: =1 | Adapt Filters (1)

Supplier	Order ID	TotalCost		PO Not Sent to Supplier	PO Not Confirmed by Supplier	Auto Approval
MuseComm Pvt Ltd	2300578437	3,000.0	EUR	1	0	0
Autohaus Krauth GmbH	2300578438	28,118.08	EUR	1	0	0
Xerox GmbH	2300578440	2,000.0	EUR	1	0	0
ARP GmbH	2300578439	1,200.0	EUR	1	0	0

Figure 11.24 Detail View of Purchase Orders Purchase Orders Not Sent to Suppliers

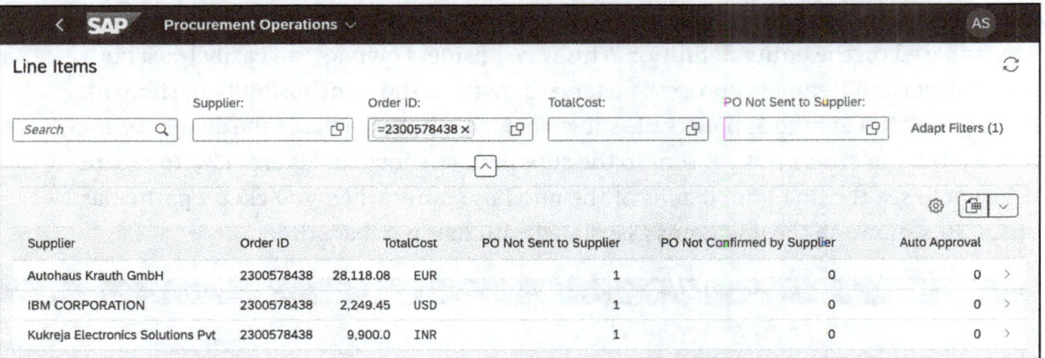

Figure 11.25 Line-Item Details Showing All Line Items Associated with the Given Purchase Order

> **Note**
>
> Details on how to access the **Process Workspace** are provided in later chapters.

Performance Indicators

Once all the attributes are defined, you can define the **Performance Indicator** that is shown as a tile in the visibility dashboard represented in the chart form. As for **Attributes**, the visibility scenario also comes with default KPIs such as **Open Instances** based on the **Status** dimension or **Cycle Time** of the process represented as a header and so on. If you don't create any new KPIs, then these default ones appear in the dashboard out of the box. However, there's often a need to create customized KPIs based on individual business needs, which you can do with the visibility scenario (see Figure 11.26).

For example, say you need to track a KPI for rejected journal entries because any rejection requires rework and that directly affects the overall cycle time of the process and may affect customer satisfaction or incur greater cost to the company. In Figure 11.26, you can see a KPI to capture rejected journal entries, which will be represented as a donut chart so that the total number of rejected journal entries is shown across different company codes for an organization. This kind of insight can help business process owners identify the region with problems and take the necessary measures to improve the situation. You can further refine the data by using different filter options to show more accurate and reliable data to the process owners.

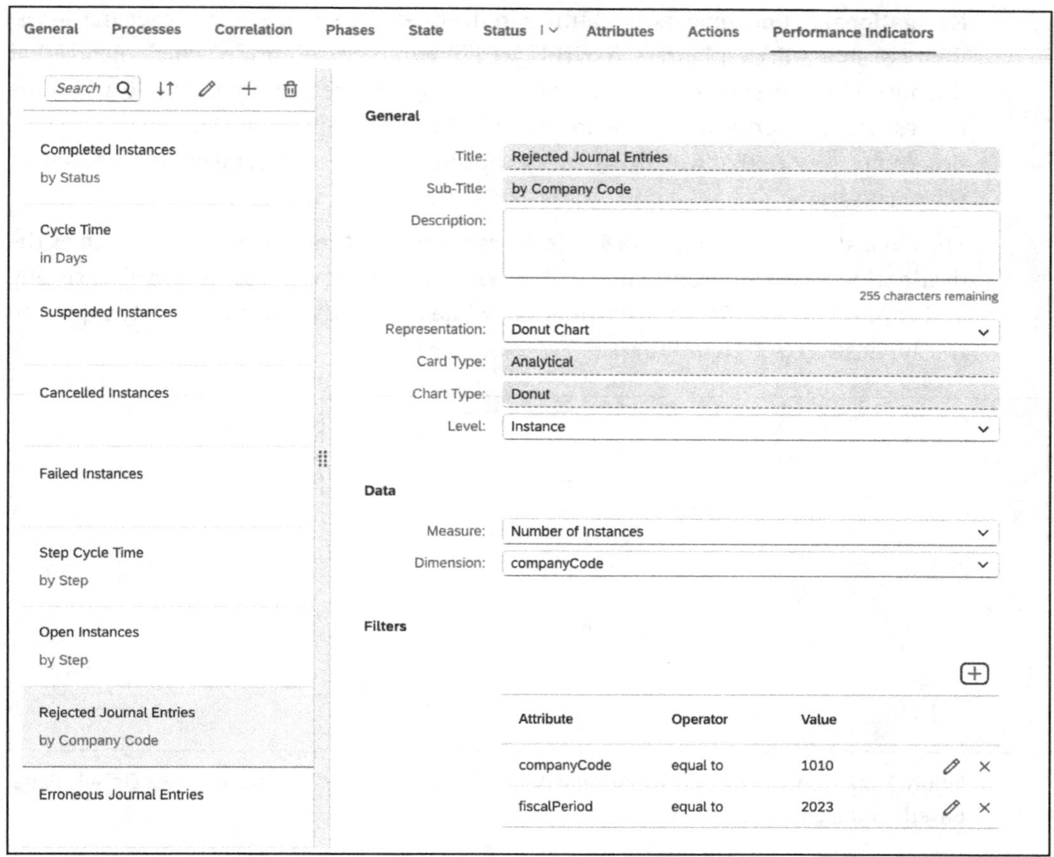

Figure 11.26 Configuring KPIs That Will Be Tracked by Business Users in the Visibility Dashboard

Actions

Identifying the process problems and proactively resolving them before they can disrupt the business continuity is one of the primary benefits of business monitoring. Visibility scenarios can be configured to alert business users of such inconsistencies and delays in process execution. The **Actions** section gives two options: **Navigational Action** and **Trigger Process**. These options allow business users to respond quickly to the process problem by (1) navigating to the backend system or application via URL to look deeper into the cause of the issue or (2) starting another process, which, for example, can notify the business user or administrator of the issue over email, automatically recover from the failure, or handle failures based on the process steps.

Navigational Action appears as a button in the process instance details page in the visibility dashboard. Configuration of this action requires a condition, which means that the button will be shown only when all the conditions are met; for example, when the process status is **Critical** and the **Company Code** is **1720**, then call a SAP Cloud Application Model service that will broadcast the notifications to all the stakeholders based on the process data.

The navigational link configuration is the same as for other link properties such as the destination-based environment variable, which is created as a business project setting and connects via an SAP BTP destination, while the relative path is the actual path to the application (see Figure 11.27).

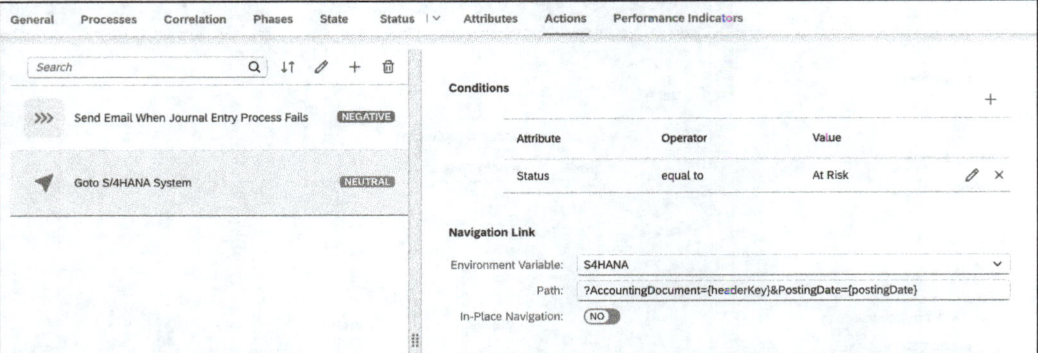

Figure 11.27 Configuring a Navigational Action That Will Appear in the Instance Details Page Based on the Conditional Criteria

The **Trigger Process** action has two options: user-based and system-based. For user-based actions, a **User** button appears in the process instance details page of the visibility dashboard and needs to be clicked explicitly, whereas system-based actions are triggered automatically, as soon the conditions are met. For both options, you must also select from the available processes in the same business project that will be triggered (see Figure 11.28).

Note

Only the processes that have API trigger as the start activity can be used for the trigger process action.

At runtime when this action is triggered, it will internally call start process APIs and will pick all the needed authentication details from the SAP Build Process Automation instance created under the **Service and Instances** area in the SAP BTP cockpit. As you already know, the start process API needs an input payload, which varies from one process to another. The same applies while configuring trigger process actions where you must enter the input payload. In the input payload, a dynamic value can be provided using curly braces ({}) where the actual values will be picked from the visibility scenario context data at runtime.

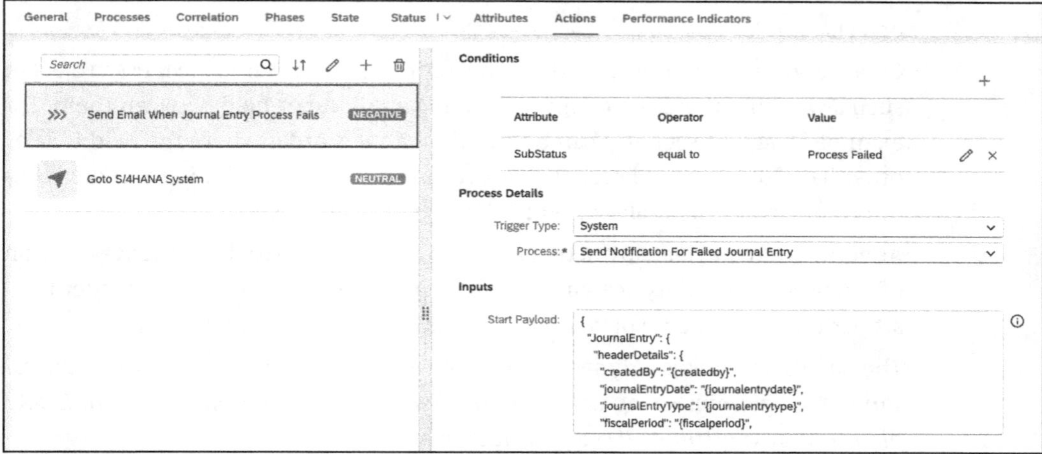

Figure 11.28 Adding a Trigger Process Action That Will Appear as a Button or Trigger Automatically Based on the Trigger Type

> **Hint**
>
> You can get the sample start payload from **Control Tower • Environments • Public or Private Environment • Triggers**. Select and click the process trigger to view the trigger details.

Advanced Features

With this, all the basic features of creating and configuring the visibility scenarios are covered. However, there are a few advanced features that are worth understanding. In this section, we'll learn about the following features:

- **Correlation properties for 1:N or N:1 relationships**
  The cardinality of a business object relationship with its entities is *1:N*; for example, one sales object is associated with multiple sales order items, and one invoice is associated with multiple purchase orders. This cardinality is very common in establishing the relationship between different business objects and provides how their respective process instances can be accessed. In the process visibility dashboard where a business process owner wants to navigate from one entity to another, there is a need to show the dashboard at both the header and item level with a navigation link from the header to item level information. In Figure 11.29, you'll see details of the header level **Request ID** and can navigate to its items via **View Approvals** navigation links. Figure 11.30 shows the visibility scenario of the item level for 1:N relationships.

To achieve this kind of navigation, there is a set of simple configurations:

- From the business project **Overview** page, create a process for the visibility scenario. This scenario will be based on the header level information of the business object.
- Create another visibility scenario at the item level in the same business project in such a way that there is a common binding attribute or field between them. For example, the first scenario can be based on a sales order where the field is sales order ID, and the second scenario can be based on a sales order item where it has information of which sales order ID that item belongs to.
- As you want to navigate from the header to item level, go to the **Attributes** section of the process visibility scenario, create a calculated attribute in the header level scenario with the name of your choice, and configure a navigation link.
- The navigation link is a dynamic URL such as *<Site URL>/site#pv-showInstances?sap-app-origin-hint=&sap-ushell-navmode=inplace&scenarioId=<scenarioID>&/filter=<context attribute ID> eq '<Value>'* where
  - *scenario ID* is the ID of the item level visibility scenario.
  - *context attribute ID* is the ID of the common binding context attribute, and *Value* can be dynamically picked such as *filter=SalesOrderID eq {GivenSalesOrderID}*.

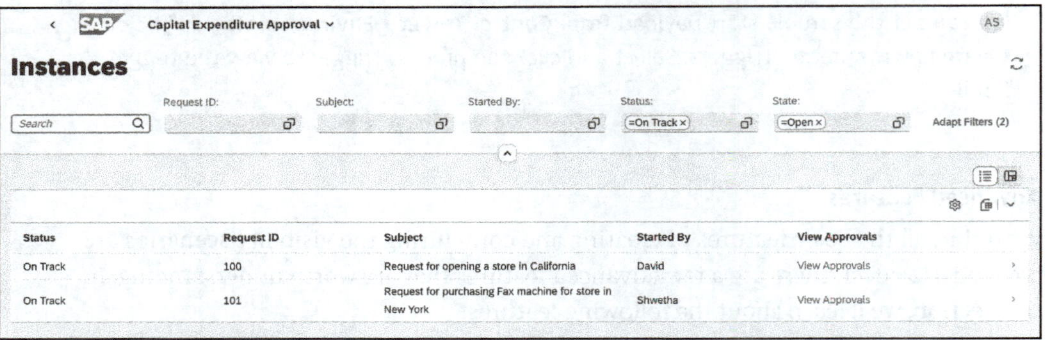

Figure 11.29 Visibility Scenario at the Header Level for 1:N Relationship with Navigational Link for Item Level Entity

The *N:1* cardinality is like 1:N but the entity relationship is reversed, which means multiple line items or business objects correspond to one entity, for example, multiple purchase order items can be linked to one invoice, or multiple billing documents are associated with a single invoice. In such cases, the correlation condition can be defined between these different process instances with a cardinality of N:1 or 1:1, and so on.

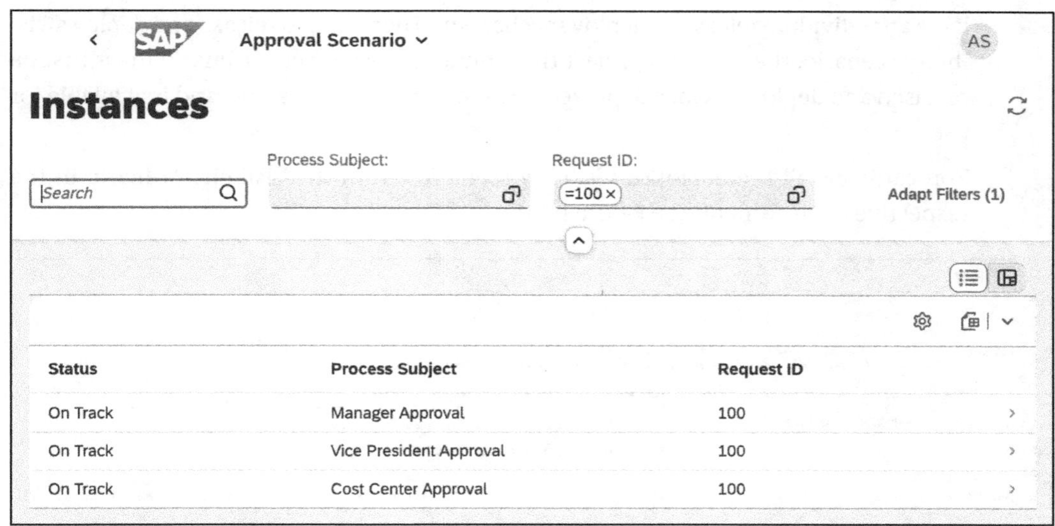

Figure 11.30 Visibility Scenario for the Item Level for 1:N Relationship

- **New expression type for attributes with step processing**
  In process visibility, each activity of the process is called a step. For example, if the sales order approval process has three activities, such as decision for approver determination, an approval form, and a post action, then each of these are regarded as steps in process visibility. Now let's assume that you want to view who is the current processor of the approval form, or if there are multiple levels of approvals, then you want to see which step the approval is on and who approves the form. In such cases, where you want to show such step-related information at the process visibility instance page or details page, then you can add a calculated attribute with the step-related expression type.
- **New target for step processing**
  Setting up the target in process visibility is an important configuration, based on which an early warning is given to the process owners for any inconsistencies or delays. There was already a target configuration at the process level, and customers requested a target at the step level as well so that business owners can track the progress of the individual activities.

11.3.3 Lifecycle of the Visibility Scenario

Visibility scenarios inherit the lifecycle of the business project and other artifacts inside the business project. Like other artifacts, the visibility scenario once developed and saved must be released and deployed before it can be consumed by the end user. In SAP Build Process Automation, all the artifacts are contained inside a project and don't have an individual release or deploy mechanism. Therefore, to release and deploy a visibility scenario, the business project that contains the visibility scenario must first be released and deployed. Once deployed, the scenario becomes active and is available for use.

You can view all the deployed visibility scenarios from the **Visibility Scenario** in the respective environment (see Figure 11.31).

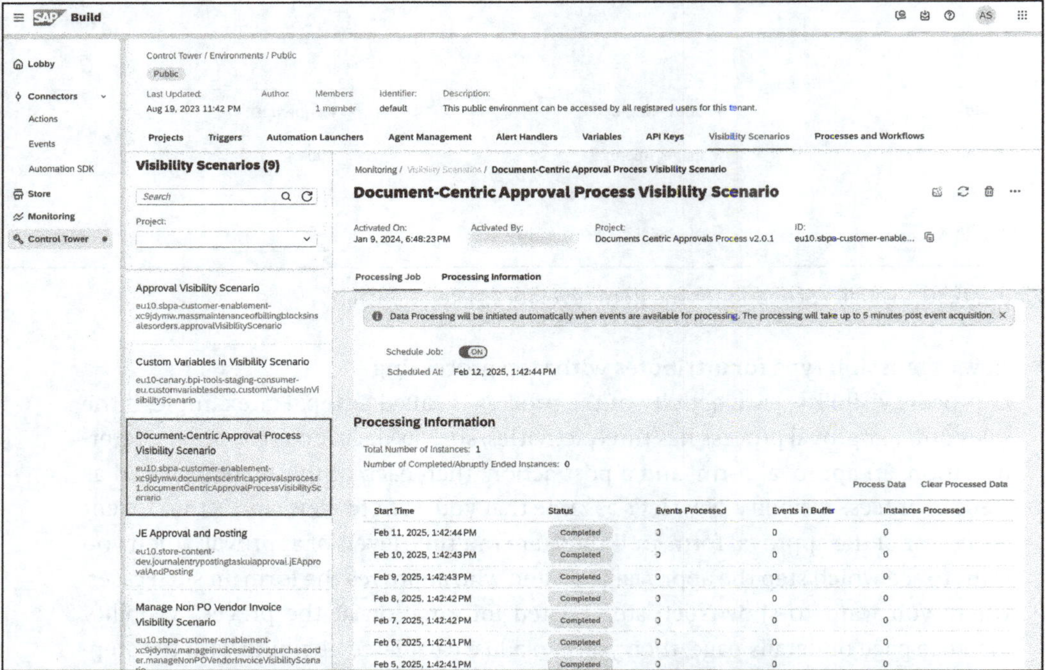

Figure 11.31 View of Manage Visibility Scenarios Showing All the Deployed Scenarios

Editing and deleting visibility scenarios can be done directly from inside the business project. Once edited, the scenario must be released and deployed for the changes to reflect in the dashboards such as new KPIs, updates to existing KPIs with new dimensions, or new columns in the instance view.

Note

Deleting the scenario from a business project is only a design-time change until you deploy the business project again. Once the scenario is deleted, it can't be accessed anymore from the Monitoring application or from end-user process workspaces. Additionally, all the events and processes data will be immediately deleted, with no options for recovery. With this in mind, be cautious when deleting scenarios.

Any organization has a landscape architecture for development, quality, and production. There are guidelines and best practices for continuous integration and continuous delivery (CI/CD) to streamline the development lifecycle. One regular practice in CI/CD is to transport content from the development landscape to quality and finally to production. SAP Build Process Automation is integrated with the SAP Cloud Transport Management service in SAP BTP. As explained earlier, the visibility scenario can't be transported individually as it's bundled in the business project. Using the SAP Cloud Transport Management service, a business project can be transported from one system to another. The transport is based on versioned content, which means you can select a released/versioned business project and choose to transport to another SAP BTP account such as transferring content from a development tenant to a quality tenant.

Note

Whenever any process that is associated with the visibility scenario is changed, it's mandatory to update the scenario by reimporting the respective process. If an SAP Build Process Automation-based process (in the same business project) is changed, the visibility scenario immediately starts showing a design-time error and forces you to update the scenario before you release or deploy.

Apart from this, other operations, such as export, import, publish to library, manage members, and so on, are derived from the business project lifecycle. These operations are applicable to all the artifacts, including the visibility scenario, that are bundled in that business project.

11.3.4 Tracking and Optimizing Processes

The primary goal of business monitoring is to track the processes in real time, detect problems in the process execution, removing bottlenecks early enough in the process flow, and maintain the threshold and SLAs to meet customer demands. All of these proactive measures help companies stay ahead in the competitive market by reducing the overall operational costs, creating faster and more efficient way to solve the problem, maintaining high-quality company standards. meeting business KPIs with greater customer satisfaction, and eventually improving business outcomes.

Accessing the Visibility Scenario Dashboard

Visibility scenarios can be accessed by the business user or process owners from the **Process Workspace** application, which is integrated with the SAP Build Work Zone service, as shown in Figure 11.32.

Each information tile in the **Process Workspace** can be clicked to get more details on what process instances are aggregated to show that information. For example, clicking **11 Critical** tasks in the **Open Invoices** tile leads to another instances application that shows the 11 process instances that are in a critical state.

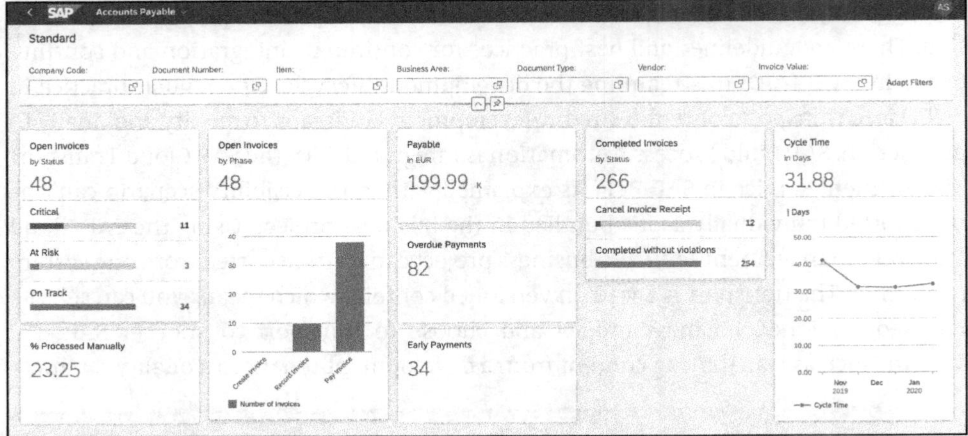

Figure 11.32 Process Visibility Dashboard

By using the **Export To Spreadsheet** icon (below **Adapt Filters**), the entire data can be exported in a Microsoft Excel sheet, which can then be used for audit purposes, generating custom reports that can then be circulated internally or used for other analytical purposes (see Figure 11.33).

Accounts Payable Dashboard

Invoices

Company Code: | Document Number: | Item: | Business Area: | Document Type: | Vendor: | Invoice Value:

Status: =Critical | State: =Open | Adapt Filters (2)

Status	Company Code	Document Number	Item	Business Area	Document Type	Vendor	Invoice Value	
Critical	2000	5705	2	9900	KR	Infix Corporation	4,363.180	EUR
Critical	2000	23925	2	9900	KR	Phunix GmbH	20.900	EUR
Critical	3000	24871	2	9900	RE	American Business Services	5,428.940	EUR
Critical	3000	190571	2	7000	KR	Phunix GmbH	20.900	EUR
Critical	1000	118023	2	8889	KR	C.E.B. BARCELONA	5,883.420	EUR
Critical	1000	297687	2	9900	KR	Phunix GmbH	4,155.300	EUR
Critical	3000	117051	2	1	KR	Tiefland Glass AG	4,394.910	EUR
Critical	1000	298916	1	9900	KR	American Business Services	358.240	EUR
Critical	2000	244085	2	9900	KR	Jotachi Deutschland AG	20.900	EUR
Critical	2000	118627	2	8889	KR	American Business Services	4,473.980	EUR
Critical	3000	116644	2	3000	KR	Phunix GmbH	8,338.550	EUR

Figure 11.33 Process Instances in a Critical State

There is one more level of information that can help business process owners get more insights, such as how many steps have been executed, the duration of completion, why a process instance is in critical state, which process step is blocked, and how long the delay will be. Figure 11.34 shows the details of a process instance execution, and Figure 11.35 shows the path of the process execution with steps that are blocked.

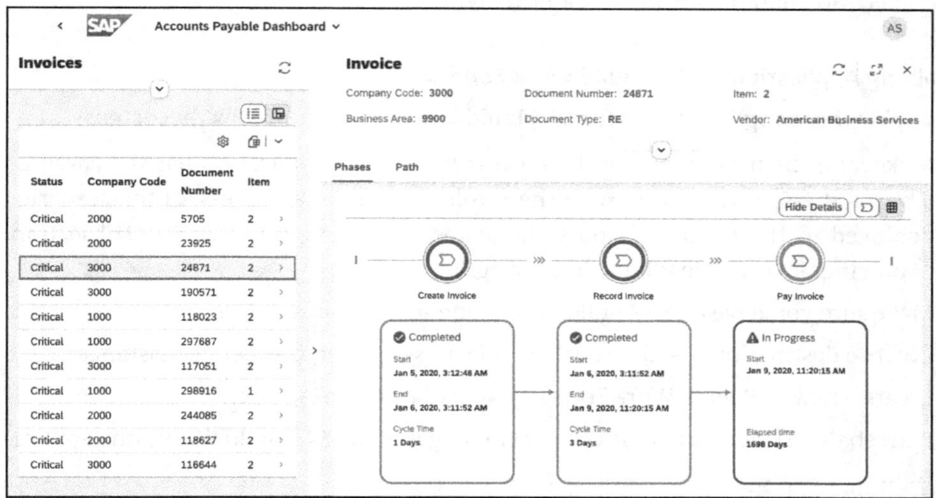

Figure 11.34 Details on a Process Instance Execution

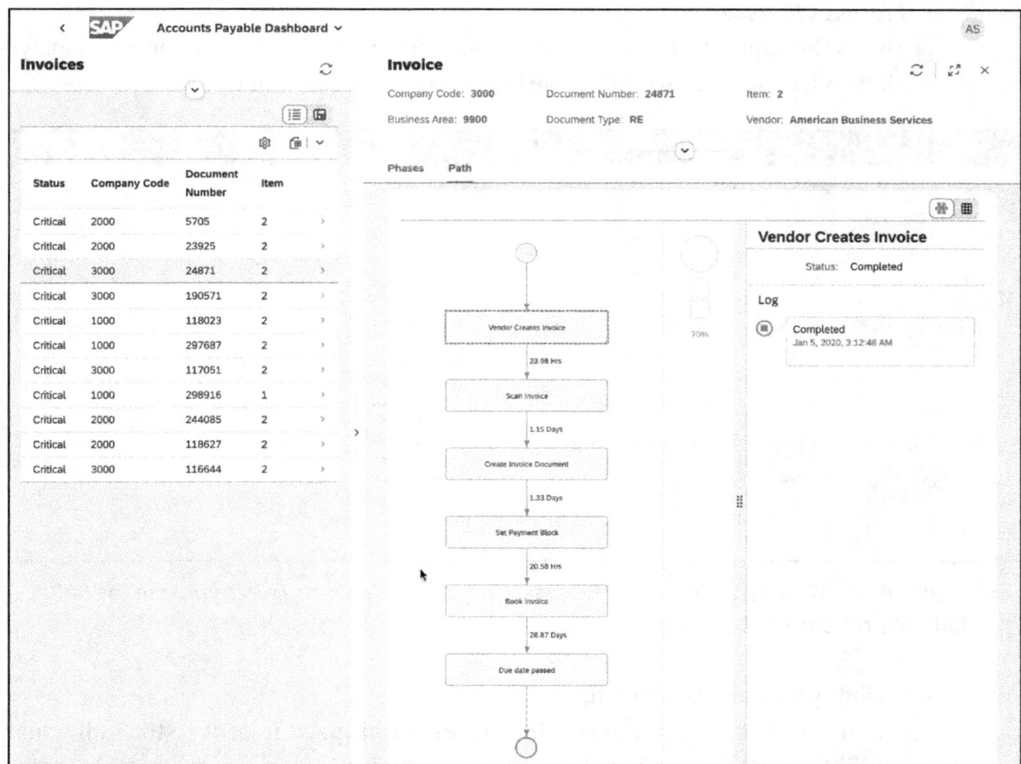

Figure 11.35 Path of the Process Execution Showing Which Step Is Blocked

With such a comprehensive dashboard having three levels of drilldown options, process visibility provides effective real-time business monitoring with consolidated process execution data, business KPIs, and analytics.

Enabling Applications in SAP Build Work Zone

To enable these applications in the SAP Build Work Zone site, follow these steps:

1. Make sure the user accessing the **Process Workspace** application has the `AdvancedUser` and `PVOperator` roles. With these roles, the user can access all the scenarios deployed in the tenant. If you want to restrict access, see the next subsection, "Applying Governance While Monitoring."
2. Make sure you have an SAP Build Work Zone service subscription.
3. Create a destination based on SAP Build Process Automation service instance.
4. Create a new SAP Build Work Zone site, or use any existing one.
5. Refresh the **HTML Apps** in the **Channel Manager** of the SAP Build Work Zone application.
6. Once you refresh, there will be three applications related to the visibility scenario in the **Content Manager • Content Explorer • HTML Apps** (see Figure 11.36):
   - **Process Workspace**
     This is the application that will appear as a tile in SAP Build Work Zone. It displays all the visibility scenarios that the logged-in user is entitled to see.

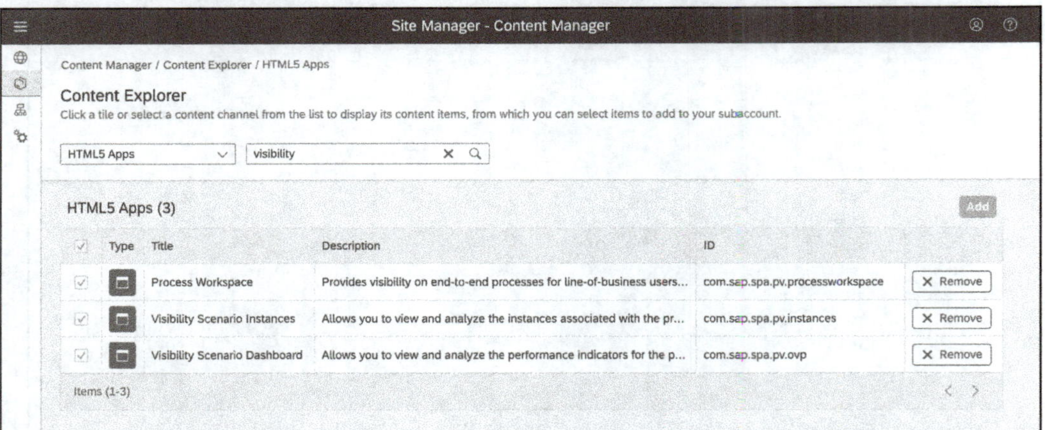

Figure 11.36 HTML Applications for Process Visibility Scenarios as They Appear in the SAP Build Work Zone Content Explorer

   - **Visibility Scenario Dashboard**
     When you click on a scenario in the **Process Workspace**, it leads to the individual visibility scenario dashboard, which is used by business users to track processes and KPIs in real time.

- **Visibility Scenario Instances**
  This application gives you a complete view of all process instances that contribute to the aggregated data shown in the dashboard. It opens when you click on any chart in the dashboard to drill down to next level of information.

7. From **Content Explorer**, select the three applications, and add them to **Content Manager**.
8. In **Content Manager**, click on the **Everyone** role, and add the **Process Workspace**, **Visibility Scenario Dashboard**, and **Visibility Scenario Instances** applications (see Figure 11.37).

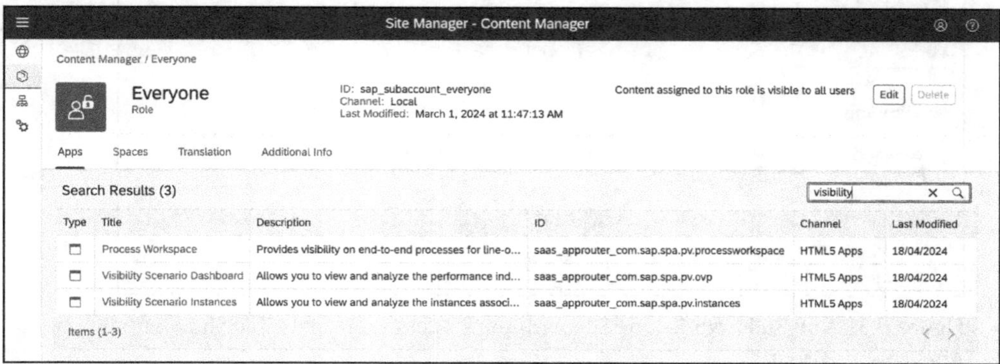

Figure 11.37 Visibility Application Added to the SAP Build Work Zone Role

9. To add the visibility dashboard as a tile, open the group where you want to add it. Click on the group of your choice, and add only the **Process Workspace** application to the group, as shown in Figure 11.38. The other two applications aren't meant to be embedded as a tile as they need additional information to be loaded. This information comes when you select the respective scenario or process instance.

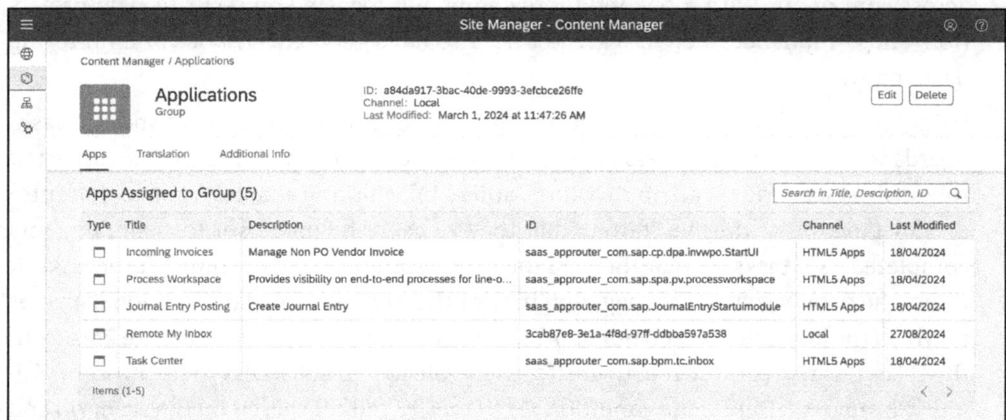

Figure 11.38 Process Workspace Added to the SAP Build Work Zone Application Group to Be Shown in the Site

10. Once you save and open the site, you'll see the **Process Workspace** tile (see Figure 11.39). If there are any issues while opening the application, relevant error message(s) will be shown. If you understand the error message, you can fix it yourself or contact your IT administrator.

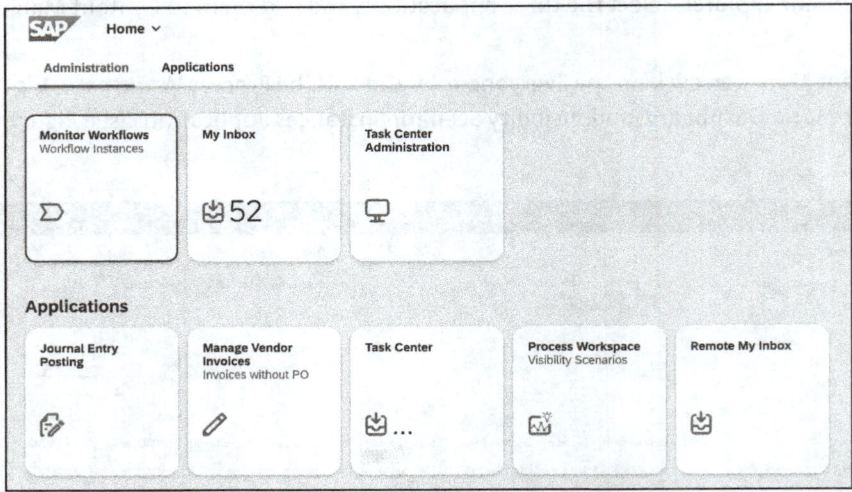

Figure 11.39 SAP Build Work Zone Site with the Process Workspace Tile Shown under the Applications Group

Applying Governance While Monitoring

The visibility dashboard shows sensitive data that can be misused if not governed appropriately. Establishing the right access with the right roles based on the user responsibilities is critical in business monitoring. For example, the general ledger process data must only be accessed by the finance team of that region and organization, and purchasing data must not be shared across different company subsidiaries or even across continents. With more and more companies focusing on security, data protection, and compliance, there is a greater need to have good governance in monitoring dashboards.

Using the `PVRestrictedOperator` role, the business user is given access to specific dashboards with limited data access per the conditions defined while creating the restricted role such as scenario ID, attribute values, and so on. This role enables users to track process instances that they've started, that they've been the processor for regrading any completed user tasks, or that they are the processor for of any upcoming user tasks. It filters the data based on the logged-in user or the organizational or functional groups to which the logged-on user belongs. As shown in Figure 11.40, click on **Create Role** from the `PVRestrictedOperator` role to create a new role to control access to the process visibility dashboard. Define the condition for the restricted role based on the scenario and the attribute values (Figure 11.41). Once the role is created, it can then be added to any role collection and assigned to the user, as shown in Figure 11.42.

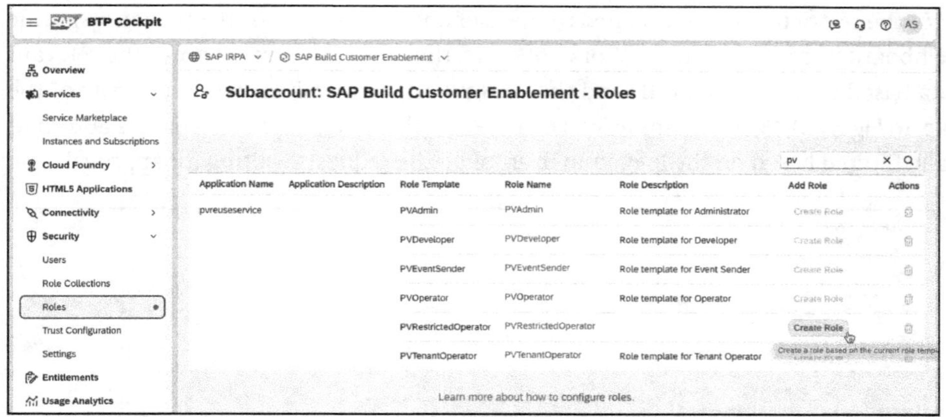

Figure 11.40 Creating a New Role to Control Access to the Process Visibility Dashboard

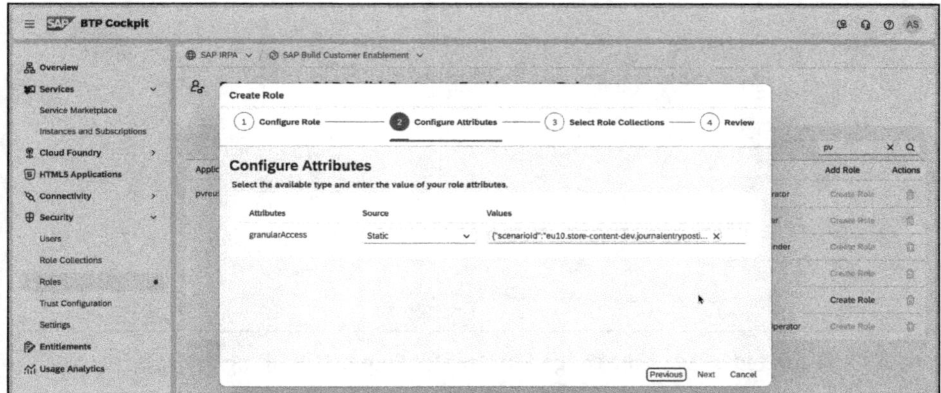

Figure 11.41 Defining Conditions for the Restricted Role Based on the Scenario and Attribute Values

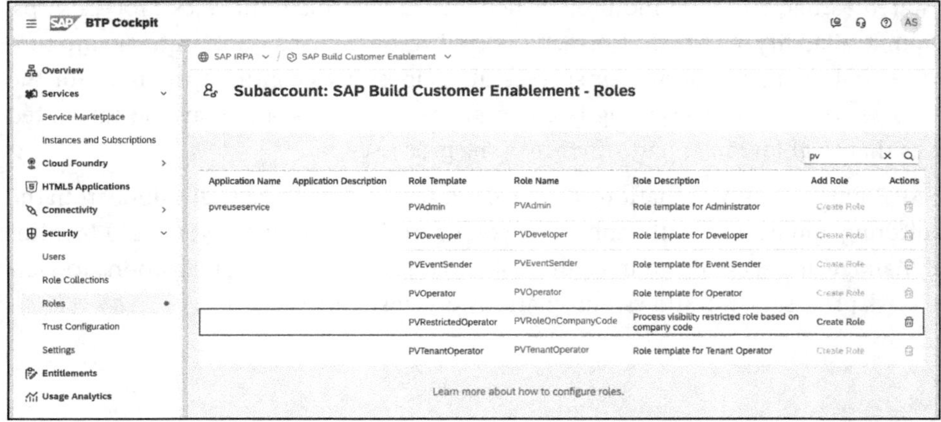

Figure 11.42 Adding a Role to Any Role Collection and Assigning the Role to the User

Once the restricted role is assigned to the user and the user opens the scenario-specific dashboard from the SAP Build Work Zone site, then the user is only shown the relevant data based on the scenario ID and the context attribute values defined as part of the role. In Figure 11.43, for example, you can see that the data in the workspace is automatically filtered based on the logged-in user for all the requests waiting for approval.

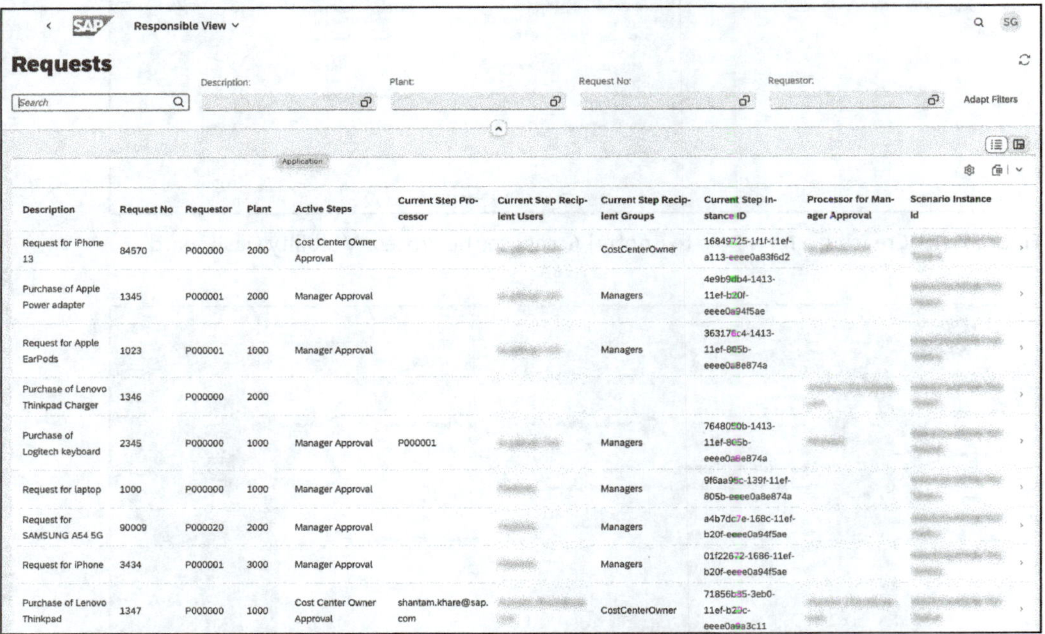

Figure 11.43 Restricted View of the Logged-In User for Requests Waiting for Approval

11.4 Technical Monitoring

As stated earlier, technical monitoring helps an IT administrator track process steps, system availability, execution correctness, and resource consumption to ensure process reliability. It's important for successful businesses to continuously monitor the processes to check if everything is going per plan and that there are no unwanted issues that might hamper performance of the processes.

In SAP Build Process Automation, technical monitoring can be centrally done from the **Monitoring** application in the lobby. As you open the application, you'll see **Monitor** and **Manage** areas, as shown in Figure 11.44, that provide further applications to view and track processes, workflows, automations, and visibility scenarios.

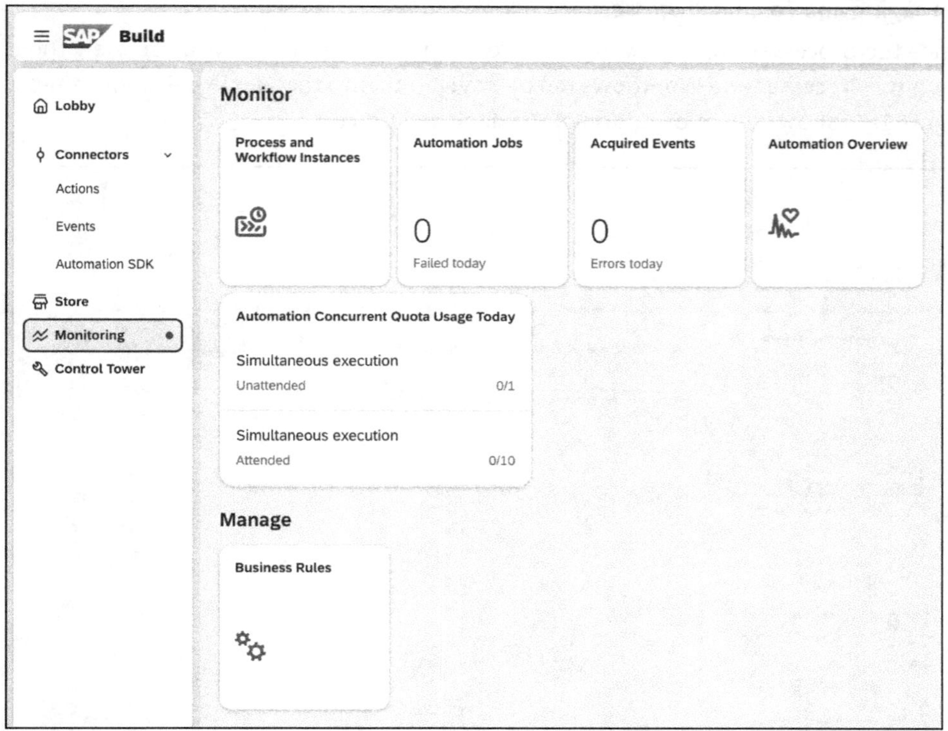

Figure 11.44 Technical Monitoring View

> **Note**
>
> Users need to have the `ProcessAutomationAdministrator` role collection to access the Monitoring application. This role collection has all the roles required to monitor processes, automations, events, and visibility scenarios. You may also create a new role collection and add specific administrator roles based on the project artifacts.

In the following sections, you'll be introduced to different technical monitoring applications of SAP Build Process Automation with which an administrator can track the health of processes, automations, and visibility scenarios.

11.4.1 Processes and Workflows

Organizations need a robust process monitoring capability to ensure business stability by taking rapid corrective measures in case of errors. In this section, you'll learn about managing and monitoring processes and workflows. There will be detailed descriptions of the respective applications and how an administrator can use these applications to track processes, handle errors, and take appropriate actions to maintain process efficiency.

Managing and Testing Processes

Once the processes are deployed in the respective environment, you can view them from the **Processes and Workflows** tab by navigating into that deployed environment. Figure 11.45 shows the information of the deployed process.

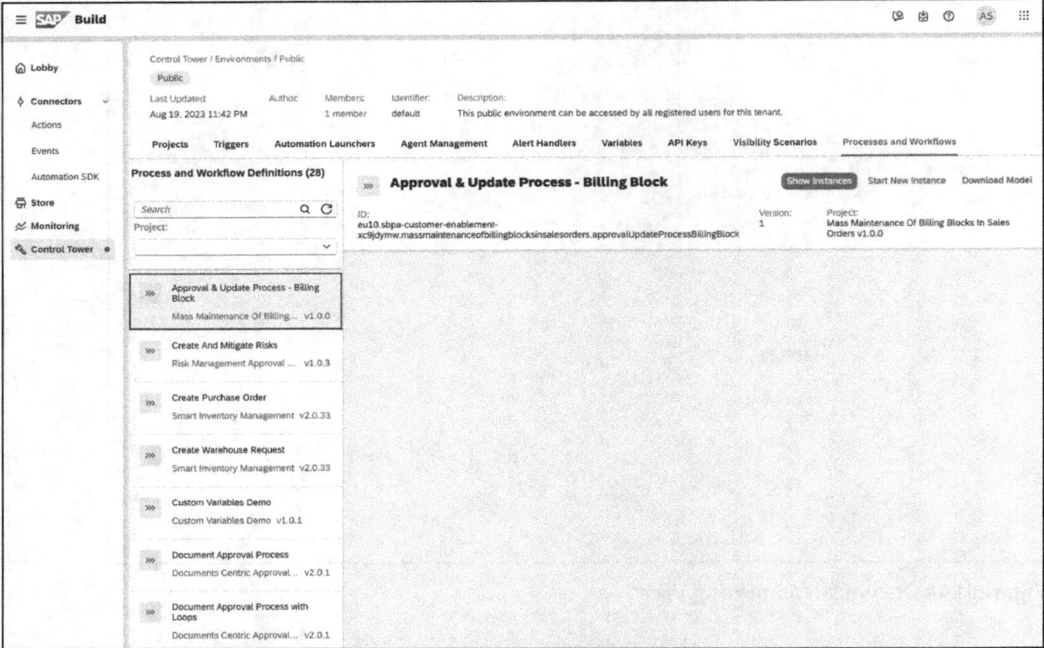

Figure 11.45 Metadata of a Deployed Process

For each deployed process, you can see the following:

- **ID**
  The unique process definition ID for the process. This will be used while working with SAP Build Process Automation APIs.
- **Version**
  The version of the process that is deployed.
- **Project**
  The business project name that contains this process.

You can not only view the process metadata but also test the process using the **Start New Instance** option. You can also download the Business Process Model and Notation (BPMN) model for the process using **Download Model** or directly navigate to all the process instances via **Show Instances** (see Figure 11.45). These options help an IT administrator manage the processes, and they help developers analyze and fix process issues.

> **Note**
>
> You might wonder why process and workflow terminology is used in SAP Build. Because SAP Build Process Automation combines capabilities from previous process management and automation services, this new terminology is used to differentiate between the old and new deployments where *process* represents the business process deployed from the latest SAP Build Process Automation, and *workflow* represents processes deployed from the old service.

Monitoring Processes

SAP Build Process Automation offers a centralized tool to technically monitor all processes and workflows deployed in the same SAP BTP account. It offers many benefits such as a unified and centralized application, real-time process execution logs, and detailed insight into process data and metadata.

Navigate to **Monitor • Processes and Workflows Instances** to get a complete list of processes based on the status filters. You can adapt the view using filter options such as **Project, Definition, Environments,** and so on (see Figure 11.46) to get a detailed view of a particular process instance.

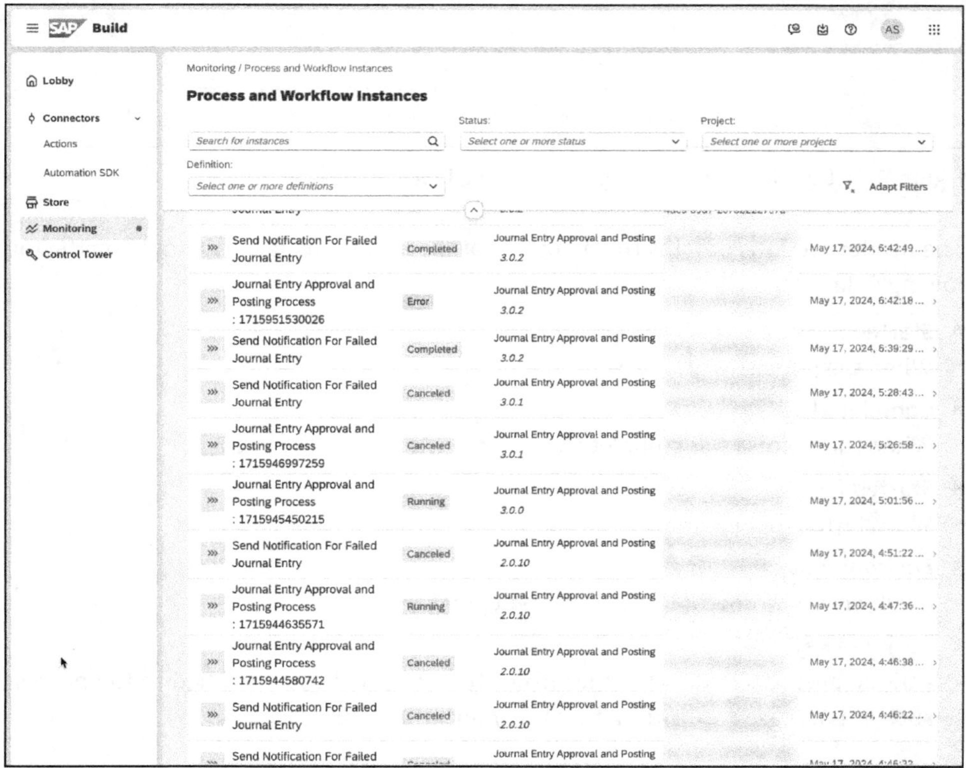

Figure 11.46 Overview of Process Instances

For example, let's say you want to see all failed or erroneous process instances for the **Journal Entry Approval** process that are deployed in a public environment. When you select a process instance, it takes you to the details page from which an IT administrator can access all the information for a process instance (see Figure 11.47).

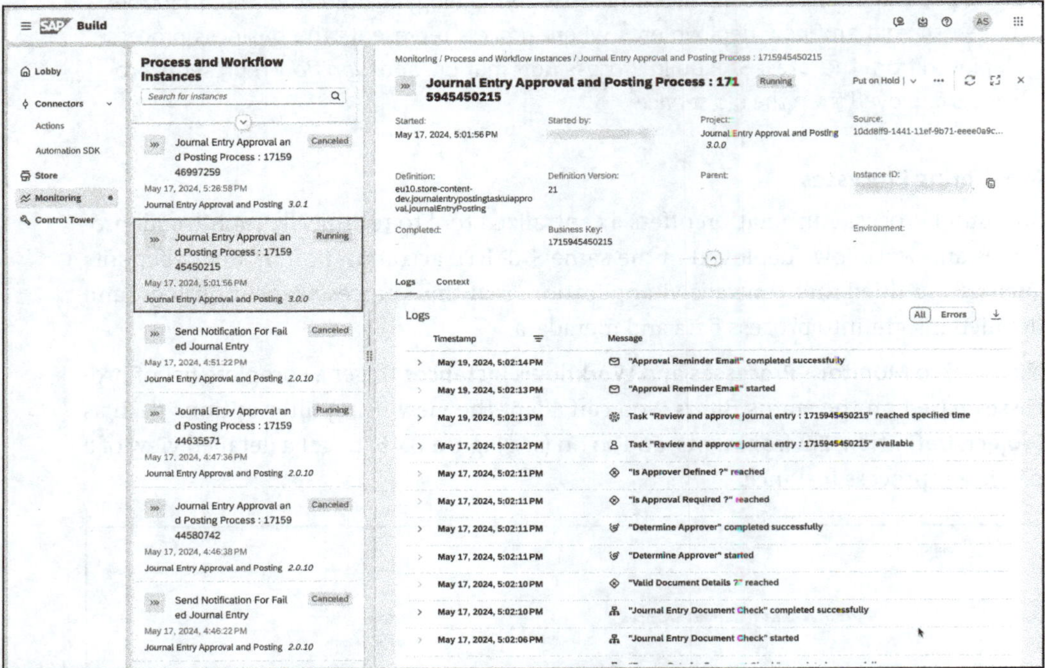

Figure 11.47 Detailed Execution Log of a Running Process Instance

The header section contains the metadata of the process instance denoted by the following fields:

- **Started**
  When the process was started.
- **Completed**
  When the process was completed.
- **Started By**
  Who started the process.
- **Environment**
  In which environment this process is deployed.
- **Business Key**
  Unique business key that comes from the functional data attached to the process such as purchase requisition ID, sales order, and so on.

- **Definition**
  Definition ID of the process that belongs to this process instance.
- **Parent**
  Parent definition ID of the main process in the monitored process instance of a subprocess.
- **Instance ID**
  Unique instance ID used while accessing the process automation ID such as to fetch the status of the process, to get a complete trace of execution logs, and so on.

Administrators have the authority to put the process instance on hold or cancel the process as the business demands. They have full access to the execution logs, which gives the entire trace of every step of the process execution with respective metadata based on the step type. For example, a user task will contain who initiated the process, who the recipients of the task are, and a unique task instance ID. On the other hand, a decision will contain a rules service ID, rule service version, and so on.

The **Context** section in the process **Monitoring** page (see Figure 11.48) shows the entire process data as the process is executed from one activity to another.

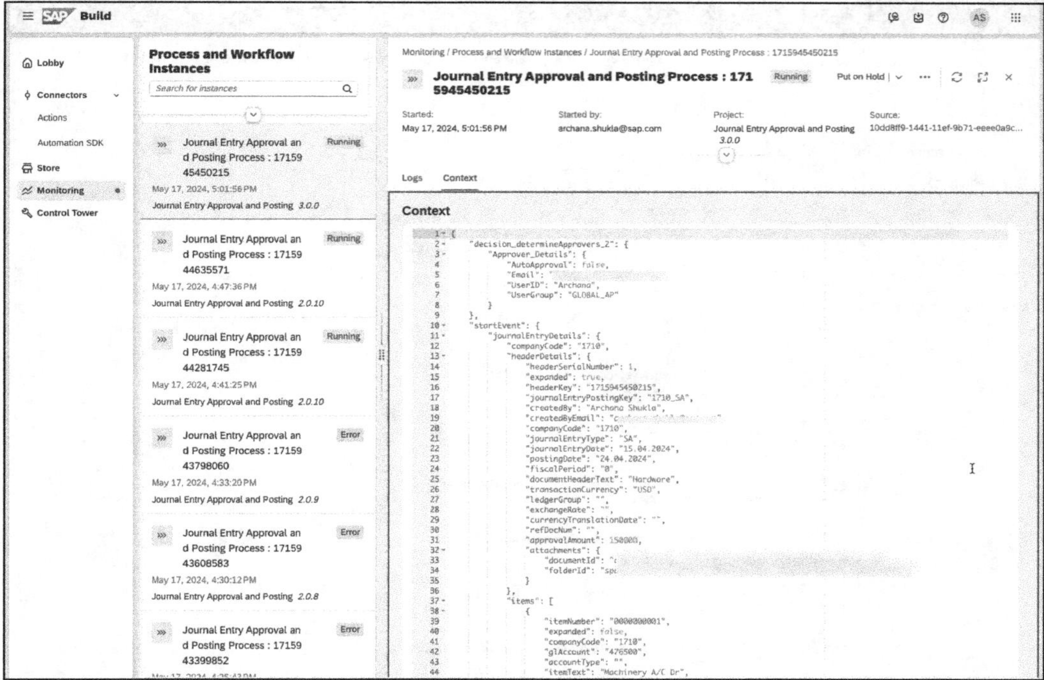

Figure 11.48 Process Context View of a Process Instance

This data contains all the information in a structured manner, including users, roles, business objects, API call outcomes, events, and so on. Looking at the context, an administrator can validate the data and detect problems to maintain data integrity and compliance. Due to the sensitive nature of information and adhering to data protection policies, only authorized users who have defined roles such as `WorkflowContextAdmin` can view the context. This provides security by protecting the data against leaks and unauthorized access.

If you have the requirement to create your own custom monitoring application with added governance and security, SAP Build Process Automation provides a library of RESTful APIs for workflow instances that can be used to access the data (with the right authentication and authorization). These and other related APIs are available in **SAP Business Accelerator Hub • SAP Build Process Automation • Workflow** (see Figure 11.49):

- **/workflow-instances/{workflowinstanceId}/execution-logs**
- **/workflow-instances/{workflowinstanceId}/context**
- **/workflow-instances/{workflowinstanceId}/error-messages**

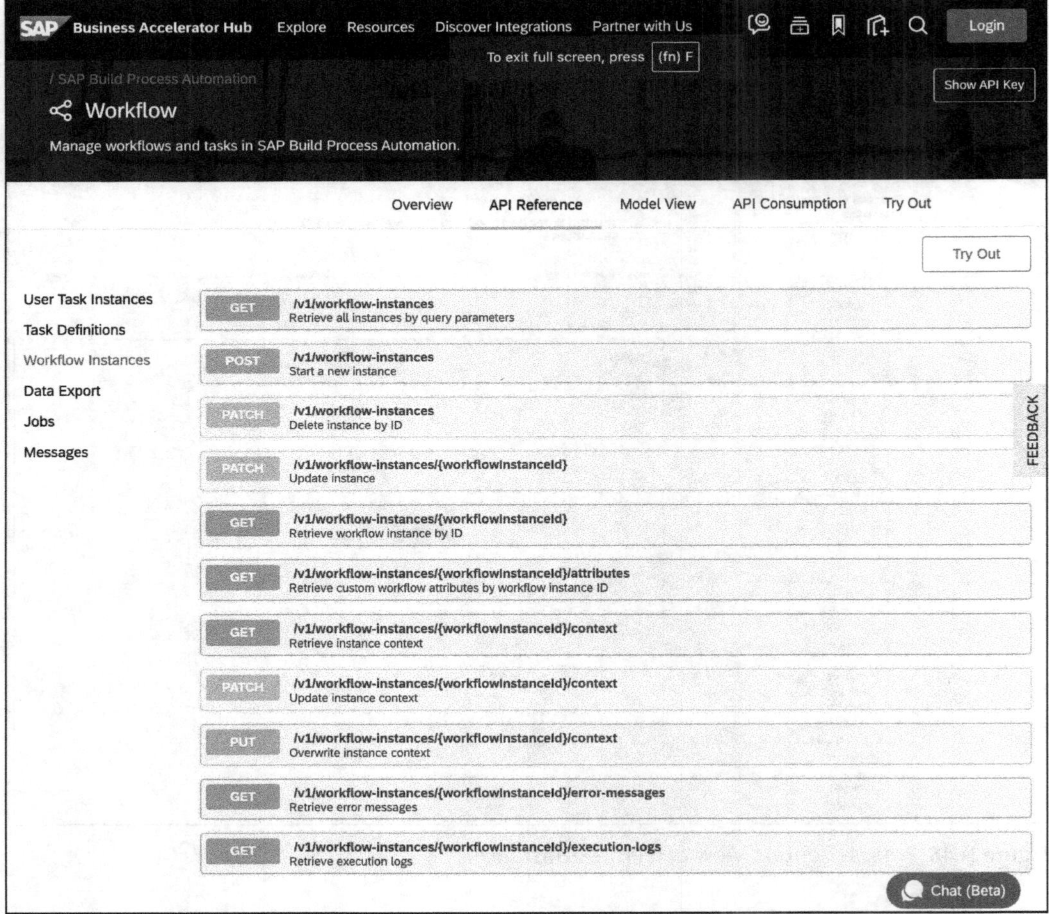

Figure 11.49 Workflow APIs to Get Process Context, Execution Logs, and Error Messages

11.4.2 Automation

The automation capability, also known as a robotic process automation (RPA) bot, is provided in SAP Build Process Automation. Like processes, automations also consist of multiple activities that are executed in attended or unattended mode by a robot or bot via a desktop agent. Monitoring automation execution is important to provide transparency and improve the ability to identify defects.

In SAP Build Process Automation, automations are run as scheduled jobs triggered by the system or attended jobs triggered by the user or via API. These jobs run on a desktop machine called a desktop agent. You can go to **Monitoring • Automation Jobs** to view all the jobs running or completed via all the agents configured in that tenant. Filter options help you search for relevant automations across different business projects, statuses, environments, desktop agents, and so on, as shown in Figure 11.50.

11

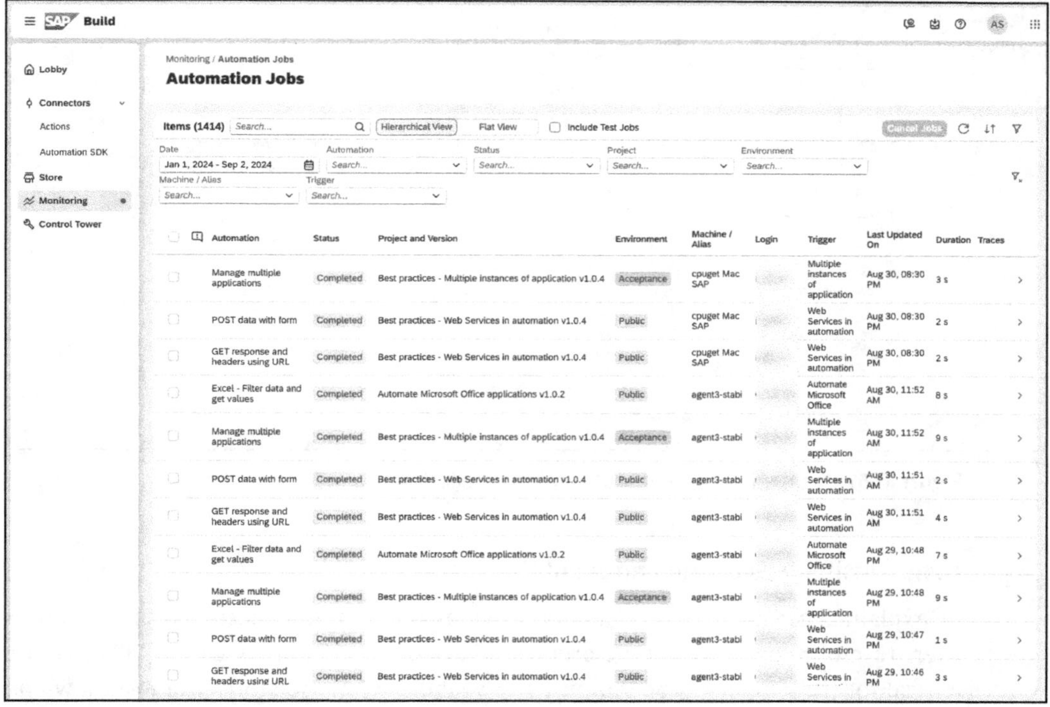

Figure 11.50 View of Automation Logs for a Specific Period

> **Note**
>
> Like the process Monitoring application, automation Monitoring application access is controlled by roles. Users must have the `ProcessAutomationAdministrator` role collection or individual automation admin-specific roles such as `IRPAOfficer` or `IRPAPersonalDataAccess`, to monitor automations.

From the monitoring view, administrators get the complete overview of the bots that are running, completed, or failed across public and private environments. They can measure the performance of the bot to keep track of execution time as one delay may lead to delays in subsequent executions, resulting in a cluster of problems that affect the business (see Figure 11.51). To avoid delays, administrators have the authority to cancel jobs or investigate detailed traces and execution history to identify the operational or functional issues that may have caused the bot to fail or get stuck.

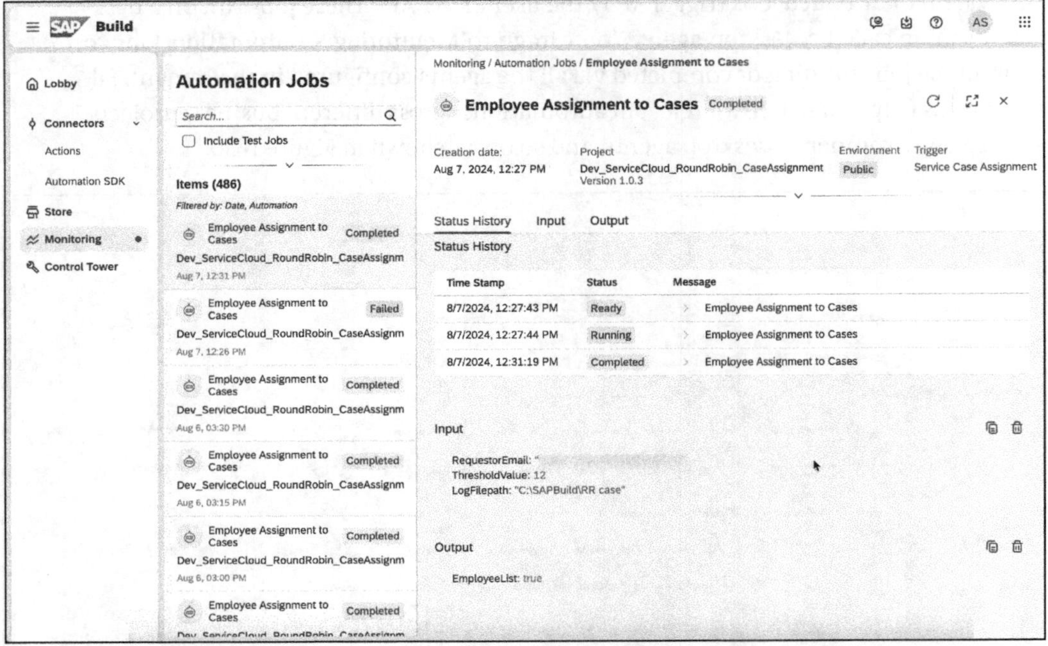

Figure 11.51 Detailed View of an Automation Execution Log

11.4.3 Visibility Scenario and Events

As explained earlier, process visibility provides an end-to-end dashboard for business users to monitor their business processes in real time and manage the KPIs that are critical for their business operations. The visibility scenario stores the workflow and event information, processes the information, and presents it in an intuitive graphical dashboard.

Administrators can view all the deployed visibility scenarios from the **Visibility Scenario** tab of the respective environment by navigating to **Control Tower • Environments**. On selecting a visibility scenario, admins get an overview of scenario metadata such as when it was activated or deployed, who deployed it, which business project it belongs to, and the unique scenario ID.

The critical activity is to manage the event and data processing information. Using this information, an administrator gets a complete list of event processing information indicating if the visibility dashboard is correctly receiving the data or not. Information such as how many events got processed, how many events are waiting to be processed, and how many workflow instances are processed are all directly related to each other.

For example, one workflow can have four steps. Each step will have two events (one for started and one for completed). Let's say there are four workflows that are started, and each one is at a different step. Events will be based on where each workflow is. If the first workflow is in the first step, then it's one event; if the second workflow is in the third step, then it's five events; and so on. Likewise, you can get information for all ongoing event processing.

You'll also notice the **Process Data** and **Clear Processed Data** options in Figure 11.52. As and when the process is executed, the events and process data are collected in the database by the visibility service. By default, all the data is processed every five minutes. All processed data is stored in another database. Let's imagine that due to some reason, such as system downtime, network issues, or faulty executions, the events weren't received on time or were received incorrectly. IT administrators can then clear the processing data once the issues are fixed and manually process the data using the **Process Data** option, bringing the dashboard back to the correct state.

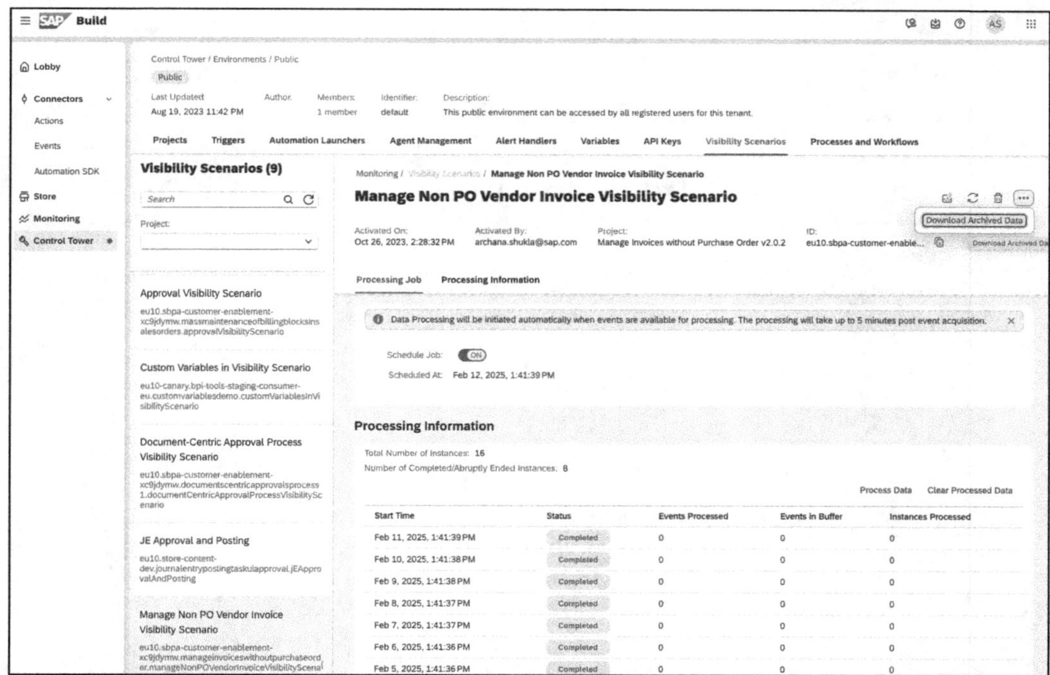

Figure 11.52 Processing Information of a Visibility Scenario

Two other noteworthy functions for administrators are as follows:

- **Download Archived Data**
  A huge amount of process and event data is stored every day in the database based on the rate of workflow execution. Customers are charged for the storage. It's therefore advised to archive the data from time to time to free the resources and save money. All such archived data can be downloaded for audit and analysis purpose using this option.
- **Navigate to Dashboard**
  This option provides the direct access to the visibility dashboard for administrators to ensure data correctness, so they can control any deviations or misinformation from predefined parameters. With this option, administrators are saved from creating SAP Build Work Zone sites and configurations to access the visibility scenario dashboards.

To manage events data, process visibility has an event-driven approach where events are acquired every time an activity is started and completed in the process. The event is therefore the key to data viewed in the dashboard, as shown in Figure 11.53. A wrong event or missed event can jeopardize all the business KPIs, leading to incorrect analysis by the business user who depends on this data to identify process bottlenecks.

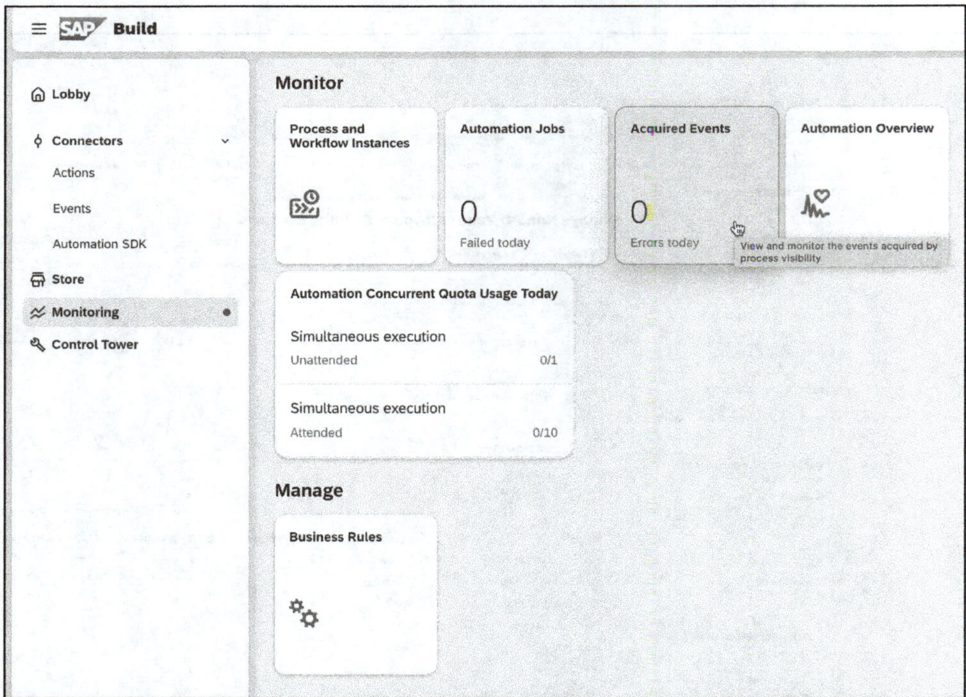

Figure 11.53 Application to View Events Acquired from External Systems for Visibility Scenario

All the visibility and business events that are pushed by processes or workflows can be studied from **Monitoring • Event Acquisition**. Due to the high volume of event data, you must explicitly filter based on **Process Definition ID**, **Process Instance ID**, **Timestamp**, and so on to see the events data, as shown in Figure 11.54. Notice the event types grouped by process instance ID and arranged by timestamp.

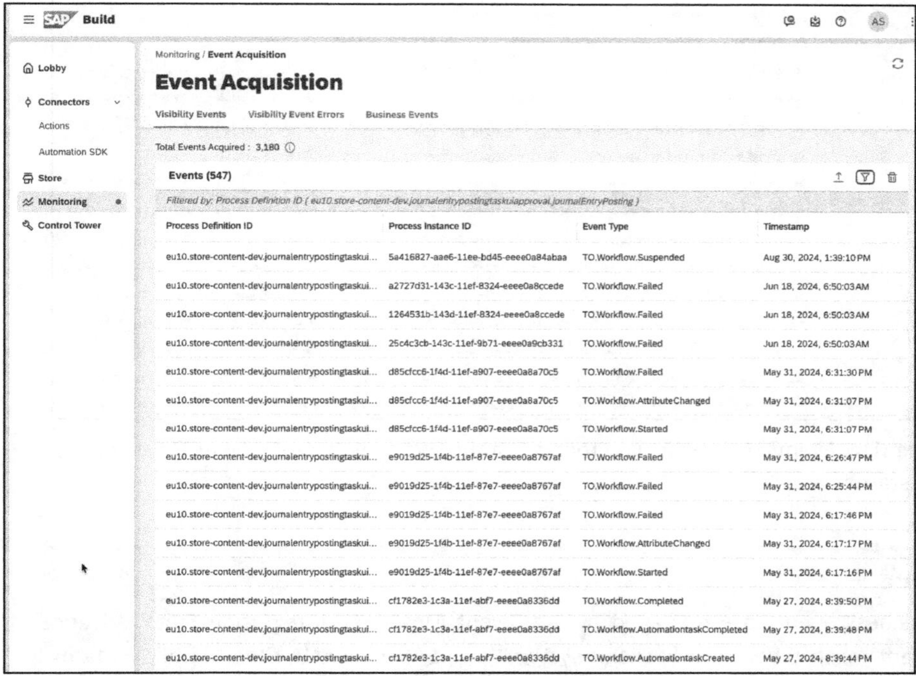

Figure 11.54 View of All the Events Acquired for a Process

For example, let's study the events data for process **JournalEntryApproval** with instance ID **cf1782e3-1c3a-11ef-abf7-eeee0a8336dd**, as shown in Figure 11.55.

When the workflow is started, it throws event **TO.Workflow.Started** at 8:37 pm. Then, at 8:37:18 pm, immediately after the workflow starts, an automation is triggered, and event **TO.Workflow.AutomationtaskCreated** in thrown by the workflow, which is completed within a couple of seconds at 8:37:24 pm. Then, after one more automation execution, a decision is called at 8:37:28 pm, and the process throws event **TO.Workflow.RuletaskCreated**.

An administrator can read the entire event acquisition flow with the timestamp and can find any event anomalies that may damage the business user experience with malformed data. Along with the visibility-based events from the same application, any business events published by external systems such as SAP S/4HANA that are used to trigger the process can also be monitored.

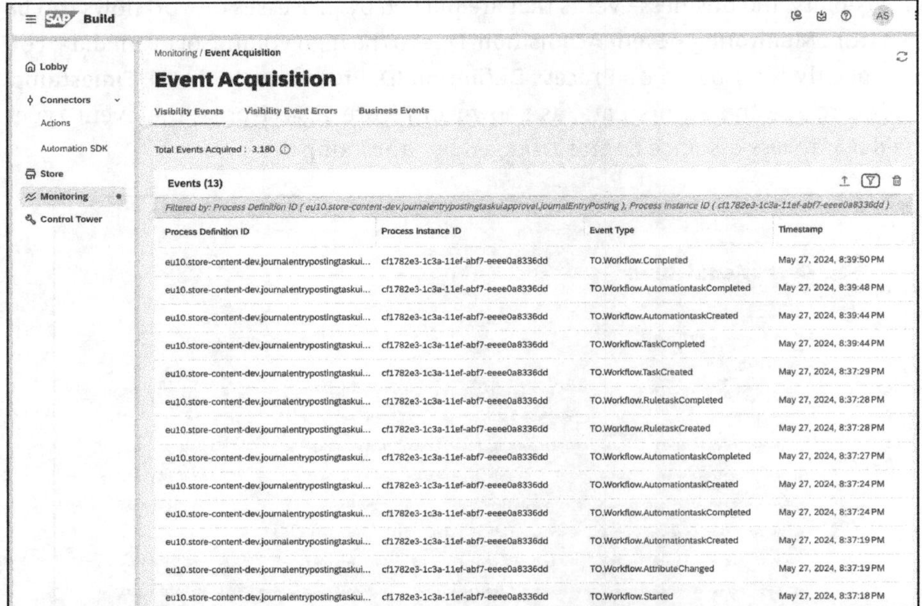

Process Definition ID	Process Instance ID	Event Type	Timestamp
eu10.store-content-dev.journalentrypostingtaskui...	cf1782e3-1c3a-11ef-abf7-eeee0a8336dd	TO.Workflow.Completed	May 27, 2024, 8:39:50 PM
eu10.store-content-dev.journalentrypostingtaskui...	cf1782e3-1c3a-11ef-abf7-eeee0a8336dd	TO.Workflow.AutomationtaskCompleted	May 27, 2024, 8:39:48 PM
eu10.store-content-dev.journalentrypostingtaskui...	cf1782e3-1c3a-11ef-abf7-eeee0a8336dd	TO.Workflow.AutomationtaskCreated	May 27, 2024, 8:39:44 PM
eu10.store-content-dev.journalentrypostingtaskui...	cf1782e3-1c3a-11ef-abf7-eeee0a8336dd	TO.Workflow.TaskCompleted	May 27, 2024, 8:39:44 PM
eu10.store-content-dev.journalentrypostingtaskui...	cf1782e3-1c3a-11ef-abf7-eeee0a8336dd	TO.Workflow.TaskCreated	May 27, 2024, 8:37:29 PM
eu10.store-content-dev.journalentrypostingtaskui...	cf1782e3-1c3a-11ef-abf7-eeee0a8336dd	TO.Workflow.RuletaskCompleted	May 27, 2024, 8:37:28 PM
eu10.store-content-dev.journalentrypostingtaskui...	cf1782e3-1c3a-11ef-abf7-eeee0a8336dd	TO.Workflow.RuletaskCreated	May 27, 2024, 8:37:28 PM
eu10.store-content-dev.journalentrypostingtaskui...	cf1782e3-1c3a-11ef-abf7-eeee0a8336dd	TO.Workflow.AutomationtaskCompleted	May 27, 2024, 8:37:27 PM
eu10.store-content-dev.journalentrypostingtaskui...	cf1782e3-1c3a-11ef-abf7-eeee0a8336dd	TO.Workflow.AutomationtaskCreated	May 27, 2024, 8:37:24 PM
eu10.store-content-dev.journalentrypostingtaskui...	cf1782e3-1c3a-11ef-abf7-eeee0a8336dd	TO.Workflow.AutomationtaskCompleted	May 27, 2024, 8:37:24 PM
eu10.store-content-dev.journalentrypostingtaskui...	cf1782e3-1c3a-11ef-abf7-eeee0a8336dd	TO.Workflow.AutomationtaskCreated	May 27, 2024, 8:37:19 PM
eu10.store-content-dev.journalentrypostingtaskui...	cf1782e3-1c3a-11ef-abf7-eeee0a8336dd	TO.Workflow.AttributeChanged	May 27, 2024, 8:37:19 PM
eu10.store-content-dev.journalentrypostingtaskui...	cf1782e3-1c3a-11ef-abf7-eeee0a8336dd	TO.Workflow.Started	May 27, 2024, 8:37:18 PM

Figure 11.55 Events of a Particular Process Instance with Instance ID, Event Type, and Timestamp

Note

Business events are managed by SAP Event Mesh/SAP Integration Suite, advanced event mesh where the backend system (e.g., SAP S/4HANA Cloud Public Edition) is extended for event-based scenarios with SAP BTP. SAP Event Mesh is configured to receive events from backend systems and provides a subscription for these events to SAP BTP. On receiving an event, the respective tasks are performed as configured in SAP Event Mesh, such as starting the process when a sales order is created in the SAP S/4HANA system.

11.5 Summary

Effective monitoring is key to business success, and it's important for any organization to diligently observe and report any potential bottlenecks in the process execution that might affect business outcomes. By identifying inconsistencies and fixing them, you'll not only gain smoother process execution but also greater customer satisfaction and improved KPIs. In this chapter, you learned how technical monitoring and business monitoring can be enabled while automating processes with the no-code monitoring applications of SAP Build Process Automation.

The benefits of the **Monitoring** area introduced you to the importance of monitoring such as centralized access, enabling governance, and real-time dashboards, which help business users effectively track processes in their daily work.

When discussing business monitoring, you learned how process visibility works and how to create and configure a visibility scenario using a no-code editor. Configuring the visibility scenario involves various areas such as (1) **General**, where you define how the scenario name and business instance name will appear in the dashboard; (2) **Processes** to select the available processes across different systems, which can be SAP BTP, SAP S/4HANA, SAP NetWeaver, or any third-party; (3) **Correlation**, where a mapping is defined between processes to establish an end-to-end relationship between their process instances; (4) **Phases**, which are created for logical grouping of the entire process and are relevant for business users; (5) **State**, which defines the events that contributes to open, completed, and abruptly ended process states; (6) **Status**, which can be used to configure the process thresholds and substatuses for critical, at-risk, on-track, and completed process instances; (7) **Attributes and Performance Indicators**, which are used to create and configure the actual business KPIs that will be tracked by the business users in real time; and (8) **Actions**, such as navigation and trigger workflows that can be added in the process instances to help business users mitigate bottlenecks by performing appropriate activities.

After creating the visibility scenario from the lobby, you learned how business users get access to **Process Workspace** for tracking and optimizing processes. This section provided a detailed view of different tiles and applications in the process visibility dashboard such as the **Open Instances** tile, **Specific KPI** tile, **Instances View** tile, **Individual Instance View Details** with phases and path diagram. It also helped to understand how to use the SAP Build Work Zone service to embed the **Process Workspace** tile in the portal.

Moving from business user monitoring, we discussed technical monitoring and how an administrator or a technical process owner can monitor all the different kinds of artifacts such as processes, workflows, automations, events, and visibility scenarios from a centralized application. In process monitoring, you learned about (1) admin operations such as put on hold, suspend, resume, and so on, which can be performed by administrators on process instances; (b) logs, which gives a detailed trace of how the process has progressed with runtime metadata of the individual step; and (c) the **Context** area, which has process context details after each activity is completed. In automation monitoring, you learned how to track automation jobs and get details on the automation performance. For visibility scenarios and events, you learned to view the visibility scenario as it runs in real time with running jobs and processing information. You also learned how to archive the processed data of the visibility scenario and how an administrator can directly navigate to the visibility dashboard without the need for SAP Build Work Zone configuration. Besides visibility, there were also details on how to manage events data that is captured by the visibility scenario and process events.

Using the monitoring applications and the process visibility dashboard, both IT administrators and business users can work toward detecting problems earlier in the process execution, getting insights into the issues, and deriving plans to optimize the process for better on-time delivery, resource utilization, overall operational cost, and gaining better opportunities for increased business outcomes.

In the next chapter, you'll learn about governance in process automation and how different governance techniques and rules can be applied to process automation artifacts to align with a company's business policies.

Chapter 12
Governance

Governance in process management and automation establishes a set of rules that defines the boundaries and checks needed while performing various operations. This chapter will introduce you to different governance techniques in SAP Build Process Automation and how they can be applied while working with process automation artifacts.

Every organization has some form of governance in place. It's important that companies ensure their processes are aligned with the company's business strategies and policies, as well as that they are abiding with external regulations. The outcome of good governance is efficient processes, proactive management, continuous compliance with business policies, effective decision-making, and greater accountability. All of this yields better customer satisfaction and business financial results.

Governance involves varied consistent activities at different stages of the business object lifecycle, which includes people with different authorities, roles, and responsibilities. The role and responsibilities play an important part in timely completion of these activities within the constraints of the compliance. Implementing governance rules, defined by executives and policy makers, mainly lies with managers or process owners who must educate their employees on governance processes and policies, auditing operations to find governance lapses, reporting on compliance, and providing corrective measures on violations.

Violation of governance rules discovered during audits or otherwise may incur heavy fines and thus exhaust valuable resources of any organization, which may result in loss of credibility, brand value, and revenues. Effective governance is therefore the vital requirement for any small, medium, or large organizations to sustain in the market and handle competition.

In this chapter, we'll introduce various governance frameworks in SAP Build Process Automation and how you can use them for access management, collaborative working, segregating deployments based on organizational structure, managing transports, and controlling project lifecycle operations.

12.1 Establishing Governance in Business Processes

In today's world, just about every company has automated its manual processes whether in finance, supply chain, procurement, sales, or HR management areas. These automated processes run on their own with minimal human intervention. Organizations have realized the need of strong governance in such automated processes to handle internal and external risks that might create problems at later stages.

Governance in business processes requires clearly defined process steps and roles that can access those steps, rules, and policies on how a process and its steps can be performed, clear responsibilities of the personas handling the process steps, and process lifecycle. The governance rules are applied at many levels of process lifecycle in SAP Build Process Automation such as while accessing business projects from the lobby, deploying projects to different environments, working in collaboration with other teams or members, and transporting projects from one cloud landscape to another.

Business solutions with process automation and governance included have several benefits:

- Only authorized users can access the projects and the steps across departments; for example, developers for finance projects can only access respective projects and can't see projects from the procurement department, and specific members of the finance department can only view or edit the specific projects.
- Business rules are used to automate policies and compliance rules such that it's automatically applied during different stages of process execution, and there is no need for human interaction to review manually. This brings the agility in rolling out the policies and changing them when needed.
- Efficient enforcement of compliance and mitigating the internal and external risks are needed because humans aren't required to review and follow the governance guidance manually.
- Continuous tracking via embedded monitoring and governance tools helps to assess the violations early in the process and gives sufficient time to managers for taking appropriate actions.
- Audits become smoother and less costly because auditors get all the information such as who made the changes, when the changes were made, how the content is securely transported, what governance policies are implemented, and so on.
- A centralized governance model with different governance techniques is applied to SAP Build projects and their artifacts to maintain a unified approach and consistency while implementing governance.

As stated earlier, a governance framework is the foundation to manage and maintain process efficiency, support transparent decision-making, and reduce audit risks with enhanced compliance. SAP Build Process Automation provides governance at different stages of the business project lifecycle. Here are the governance approaches available with process automation:

- **Access management**
  Defines user roles and responsibilities to access viewing and editing of projects.
- **Collaboration and sharing**
  Provision to add team members and share the project with developer, administrator, or viewer privileges to work collaboratively while implementing the process.
- **Environments**
  Isolating business project deployments with environments that are created to provide separations based on departments, line of business (LoB) operations, industries, regions, continuous improvement landscapes, and so on.
- **Transport management**
  Embedded use of the SAP Transport Management service to transfer business projects across landscapes, without the need to export and import.
- **Approval flows while creating projects**
  In many organizations, any new process or change in critical business functions such as finance or life science must go through an approval flow where the authorized persona reviews the changes based on corporate policies.

These techniques can be achieved via different options in SAP Build Process Automation that aims to help in defining and delivering effective governance such that all the processes are correctly executed and aligned with the organizational strategy. In subsequent sections, we'll look at these different governance techniques in detail.

12.2 Business Process Access Management

With increased digitalization and use of cloud technologies and applications, robust access management has become a necessity for many organizations. Access management is one of the basic and first steps toward governance. It relies on the mechanism of managing and controlling user access to the business project for viewing, development, and administrative activities based on departmental hierarchies. By implementing access governance, companies aim to ensure that the right individuals have the right access to designated resources while preventing unauthorized access and maintaining compliance.

In SAP Build Process Automation, access management has several components for effective user access. This includes *access control* to restrict access to business projects based on roles and responsibilities, *role-based access* to assign access to users based on their roles in the organization (e.g., developer, administrator, etc.), and *managing audit* for regularly monitoring the audit log service to ensure that all activities are being performed within the regulatory guidelines.

Access control is explained in detail in upcoming chapters where you're introduced to various roles and role collections that can be assigned to users based on the access requirement for different process automation artifacts. These granular authorizations elevate the security measures and industry regulations, empowering organizations to efficiently enforce governance.

Role-based access is also used to restrict users to view or edit applications or data to ensure that employees or business users can only access the information they are authorized to work on and is needed to perform their jobs. It's not about assigning individual roles to the users but on providing access based on organizational roles and responsibilities. In many ways, role-based access provides better security because it's implemented at the company level and saves time from tedious individual role assignments.

Implementing role-based access requires (1) understanding the business needs to ensure that only the users of specific job functions are permitted; (2) establishing policies, guidelines, and industry standards to develop a robust governance framework; (3) defining clear roles and responsibilities to authorize people for a specific chain of activities; (4) closely monitoring lapses in adherence to governance rules and rectifying them; (5) promoting enterprise-level objectives on governance; and (6) educating the teams on the activities that people involve in process management have to follow.

A variety of roles—owner, editor, deployer, and viewer—are defined in SAP Build Process Automation that grant view, share, edit, publish, develop, deploy, and delete privileges to users. They are assigned based on user responsibility in the organization; for example, the administrator and the owner/business project creator are assigned all the roles, the developer is assigned the editor and deployer roles, and so on.

To enable role-based access in SAP Build Process Automation, select the **Manage Members** option of an individual project, as shown in Figure 12.1, and then add individual members or user groups to give them developer, viewer, or administrator privileges. These privileges are applied to the entire project, including all of its artifacts such as processes, forms, decisions, automations, and so on.

Here's the list of privileges available for a business project that can be given to individual users or groups of users to control access to the business project:

- **Viewer**
  With this privilege, the user can only view the projects and deploy them. No editing is allowed.
- **Administrator**
  With this privilege, the user can perform admin tasks such as sharing the project, deleting the project, and editing the project. The administrator has full access to perform all lifecycle operations.
- **Developer**
  With this privilege, the user can create, edit, deploy, and manage project artifacts

and their dependencies. This is mainly for IT developers who create, deploy, test, and manage projects.

- **Everyone with access/Only project members**
  These general privileges are assigned at the project level so that all the members of the team/user group that have access to the lobby or are added to the individual project get the respective access.

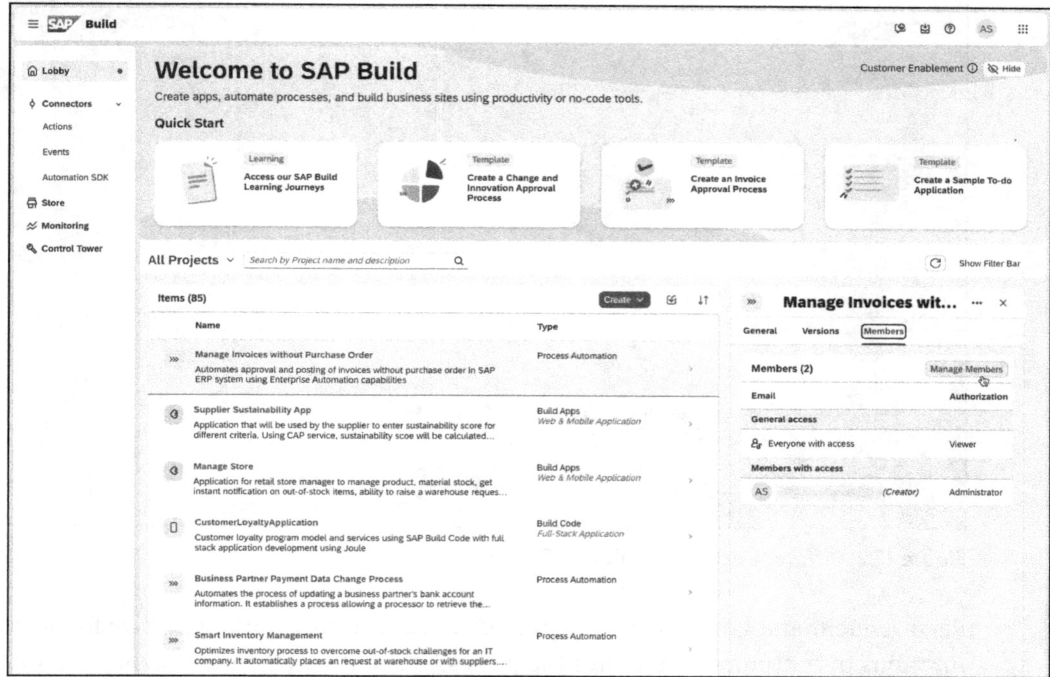

Figure 12.1 Managing Project Access for Users

Figure 12.2 shows these different privileges and how are they granted to different users.

As the privileges are assigned, the user access is restricted to the business projects in the lobby. For example, when a user named Lorin logs in to the lobby application, she'll see only the projects that are available to the finance department as she only has view permission for finance projects. Similarly, a user James who belongs to the HR department can both view and edit the projects related to HR because he has the developer permission for HR projects.

> **Note**
>
> The project creator is by default the administrator of the project and has the rights to define role-based access. The access is specifically needed when multiple users of a team work on the same project during the implementation phase or when restricted users are allowed to modify the project in production.

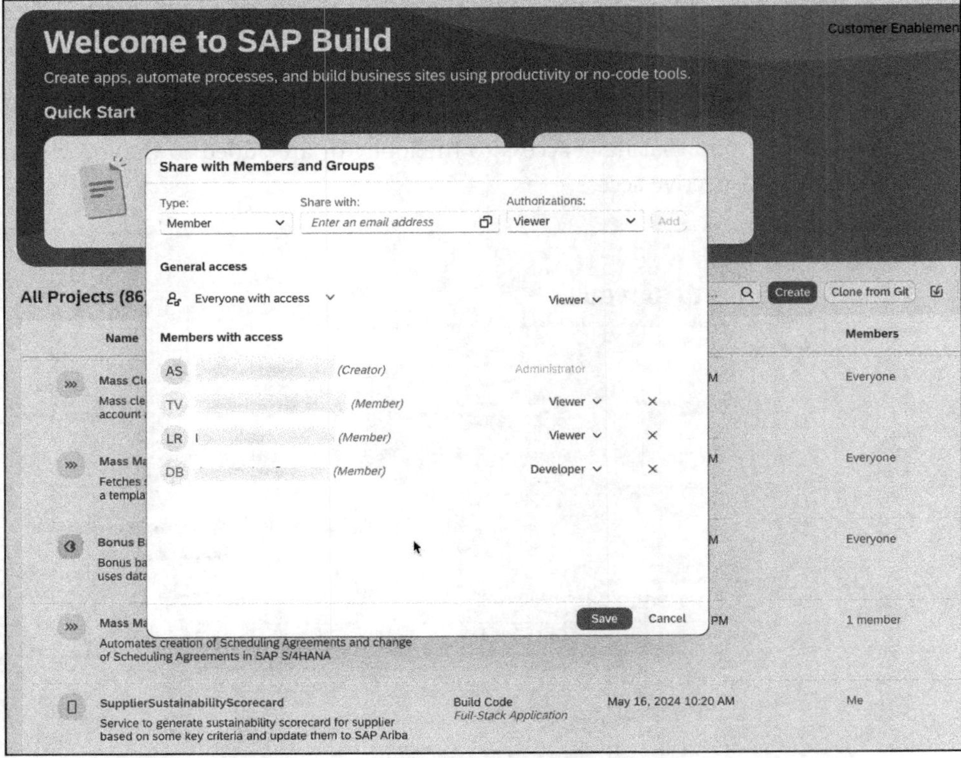

Figure 12.2 Different Access Granted to Different Users

Regular audits and assessment are key for effective governance. It's important for organizations to continuously monitor the processes and discover inconsistencies in following the governance policies. Such proactive approaches help to optimize the processes, maintain external or internal regulatory guidelines, and enhance productivity. In Chapter 13, we'll introduce you to the audit logging capability that can be used to study security events.

12.3 Environments

An environment is a space in which a business project is deployed and run. It contains information on runtime configurations such as variables, triggers, API keys, alerts, and so on, as well as a list of physical resources that are needed for automation execution such as agents.

Environments were introduced to provide the following benefits:

- An isolated container for deployments and execution of projects across different LoBs

- Reuse of resources of the environment by projects deployed in that environment rather than creating duplicate variables, agents, and so on
- Controlled access for viewing logs, deploying, monitoring, and executing business projects at runtime with sensitive information and data

Implementing process governance is difficult, and it's impossible if the design time and runtime tools don't enforce governance rules at different process lifecycle stages. SAP Build Process Automation took the opportunity to provide a holistic approach to manage governance via Environments to establish secure settings of running the processes in an isolated area. Figure 12.3 gives a glimpse into the environment application where you can see all the different shared environments that are created together with the assigned members who created the environments and last updated information.

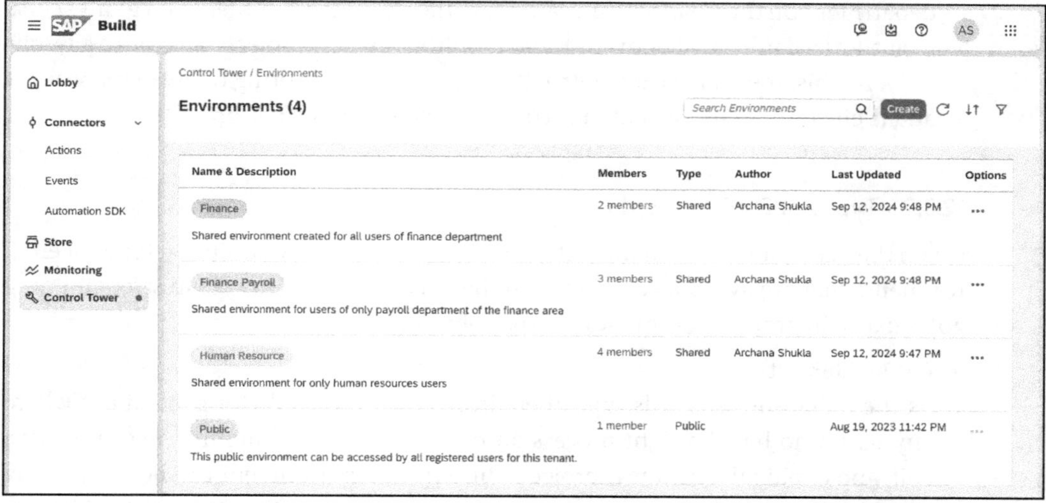

Figure 12.3 Overview of Environments

In this section, you'll learn about environments in a bit more detail to understand why they are needed, what different types of environments are available, what environments contain, and how to work with environments in SAP Build Process Automation tools.

12.3.1 Why Do Organizations Need Environments?

There are often questions on why environments are needed when organizations have development-quality production setups to segregate project lifecycles at different stages of project development. Here are some of the factors that explain why environments are needed:

- Enabling lightweight containers to deploy, test, monitor, and access control compared to creating different SAP Business Technology Platform (SAP BTP) accounts that incur extra costs for execution. The recommendation for continuous improvement and continuous development (CI/CD) is still to have three separate accounts for development, quality, and production. Using an environment helps specifically when you want to share business projects and resources within the same organization such as in the finance area having payroll, planning, treasury, audits, and operations departments and not across LoBs (e.g., agents maintained by the finance area shouldn't be used by procurement).

- Deploying project versions in different containers with different configurations such as deploying the project in one environment with a different destination configuration than in another. IT admin can create production environments for developers to test the implementation before pushing to production.
- Segregating deploying and monitoring across LoB projects. For example, a payroll department on the finance team can create their own environment and grant access to authorized users to deploy, test, and monitor business projects specific to payrolls only. By this, the teams can control the access and monitoring of business projects and their configurations that may contain exclusive information.

12.3.2 Types of Environments

Given the need for environments, let's now understand the two different types of environments that are available and how can they play a significant role in strengthening governance in process automation–based solutions:

- **Public (default)**
  As the name suggests, this type of environment is available for general use where any user who has the right process automation roles can monitor, execute, view, edit, and deploy the business project. This type of environment can be accessed by all users of an SAP BTP account, irrespective of their organizational role and hierarchy. This environment exists by default and needn't be explicitly created. For example, when you deploy a project, you'll automatically see the public environment even if you haven't created any shared or private environment.
- **Shared**
  This environment is created to restrict the access of the users for project lifecycle operations such as deployments, editing, monitoring, and so on. Only process admins (users who have the `ProcessAutomationAdmin` role) can create such environments and give access to users/developers who have the `ProcessAutomationDeveloper` role. By introducing shared environments in a corporate landscape, you can do the following:
  – Deploy the same project version in different environments.
  – Deploy different versions of the same project in different environments.

- Give access to different users in different environments.
- Temporarily create environments for a specific time frame for testing or handling any bug fixes, organizational changes, and so on.
- Create different environments for different departments of the same organization or for different LoBs.

Let's understand various use cases with examples, as illustrated in Figure 12.4. Consider that there are three projects, Project A, B, and C. Project A is deployed in the public environment, which means it can be accessed by all the users who have the required permission. Project B is deployed in a shared finance environment where users who have the permissions for that environment will be able to access it. Project C is deployed in another shared environment that is created for the HR department employees. With different shared environments, there is a restricted access to the business projects that establishes the required accountability and control to the respective LoB teams.

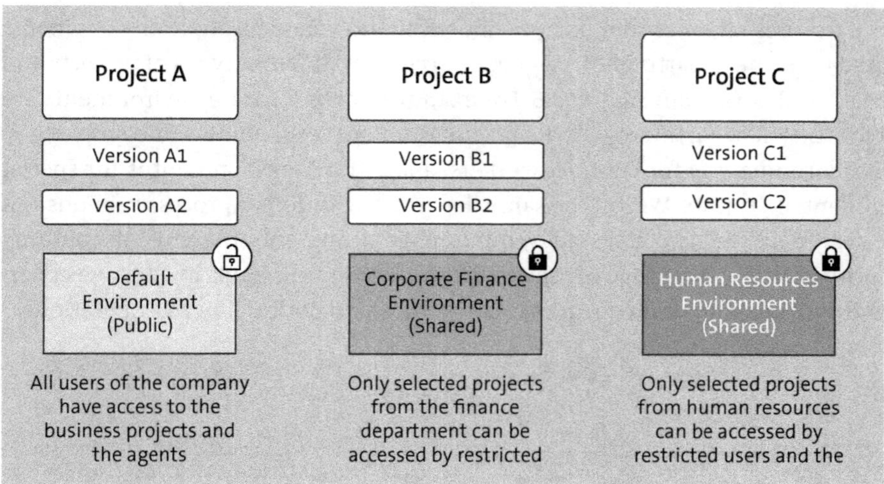

Figure 12.4 Pattern Showing How Public and Shared Environments Are Accessed across Different Projects

Another pattern is where different versions of the same project are deployed in different environments. In this scenario, shown in Figure 12.5, two shared environments are created for the HR department where version 1 is deployed in one environment, and version 2 is deployed in another environment. There can be multiple requirements where such a pattern can be helpful such as when a special project has to run for a short duration, or there is a change expected in the department for which the main project has to be updated, and both the old and new projects must run in parallel.

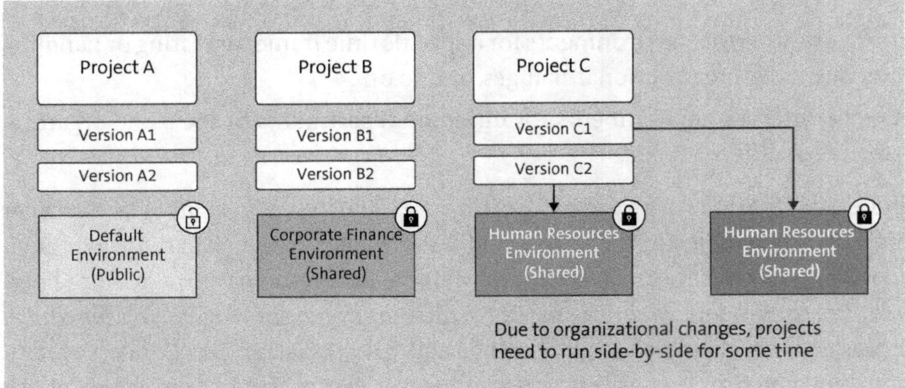

Figure 12.5 Pattern Showing How a Shared Environment Can Be Used for a Project with Different Versions

Another pattern is an extension to a previous scenario where multiple environments are created for same department to segregate access for different types of projects and employees, as illustrated in Figure 12.6. For example, three separate environments are created for the HR department where one environment is for all full-time employees, another environment is for contract workers, and a third environment is for special and confidential projects. Within the same department, different process versions can be selectively deployed and accessed by different environments, thereby safeguarding any security breach and leakage of sensitive information. This helps in effective governance and control of process execution, monitoring, and critical lifecycle operations.

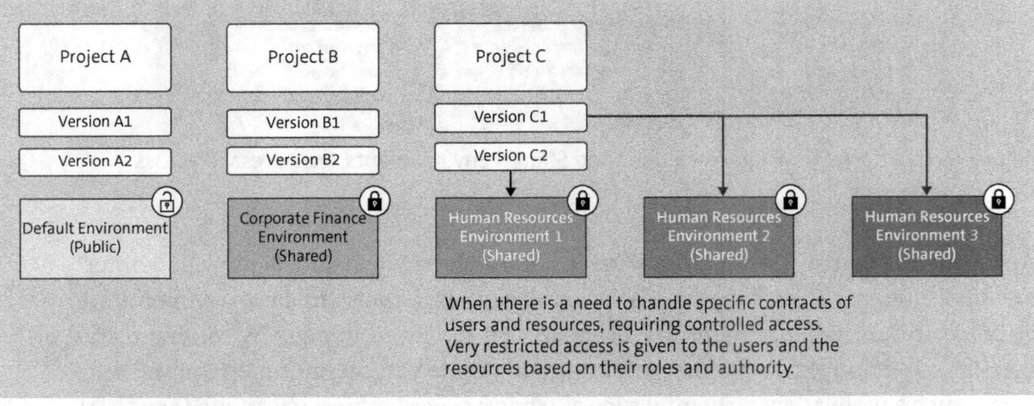

Figure 12.6 Pattern Showcasing Another Variant of Shared Environment Usage for Different Versions of the Same Project

12.3.3 What Do Environments Contain?

Governance requires that a process and its artifacts are created, tested, managed, and monitored within the defined industry rules by authorized users. There can be many variations on how the changes can be made to the artifacts, and it's important to carefully monitor these changes for imposing compliance. With environments, organizations can easily and effectively create boundaries within which the users can operate by defining a specific set of configurations to run the processes and automations, such as the following:

- **Projects**
  Set of business projects that can be accessed from an environment.
- **Automation Launchers**
  To start the automation manually.
- **Variables**
  To reuse automation variables across projects.
- **Agent Management**
  To allocate agent machines for running automation.
- **Alert Handlers**
  To configure email alerts in automation steps.
- **Variables**
  List of variables created in the project deployed in the environment.
- **API Keys**
  Unique access code that can be used to execute public API endpoints for security reasons where the service instance details can't be exposed.
- **Triggers**
  List of API triggers and scheduled triggers that are deployed in the environment.

Figure 12.7 shows a view of a public environment and its entities where you see the **Name & Description**, **Version**, **Author**, and **Last Updates** information of the deployed business project in public environment. By default, only the **Active** version of the project is shown, but you can toggle the **Show Inactive Version** button to also see all active and inactive versions of the project in the public environment. There are other entity sections such as **Triggers**, **Automaton Launchers**, **Variables**, and so on, which can be clicked to view the respective artifacts that are available within the environment during runtime execution of the project.

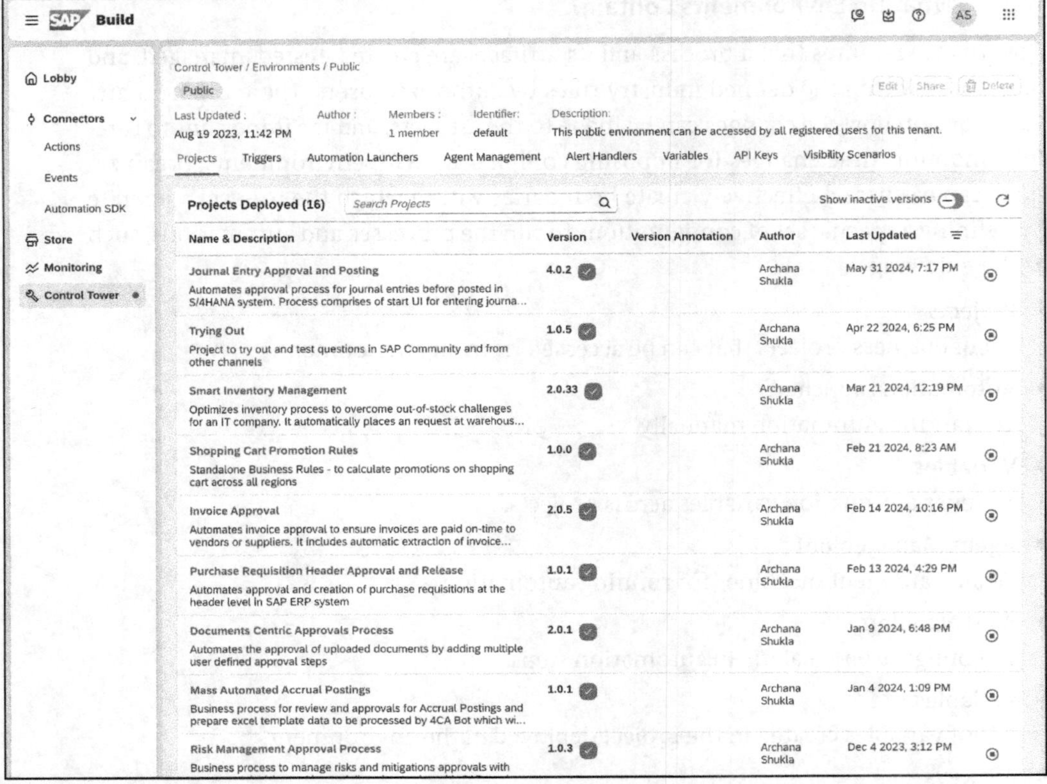

Figure 12.7 View of a Public Environment

12.3.4 Creating and Using Environments

An environment can be created and managed via the **Environments** tile in **Control Tower** under the **Tenant Configuration** section, as shown in Figure 12.8.

The **Public** type of environment exists by default, which can be used if no other environment is required. The **Shared** type of environment can be created, edited, deleted, and shared from the Environments application. As explained in the previous sections, multiple shared environments can be created and managed by the IT admin based on the governance policies of the organization or a particular department. Figure 12.9 shows the environment application with a list of all environments that are created in this tenant.

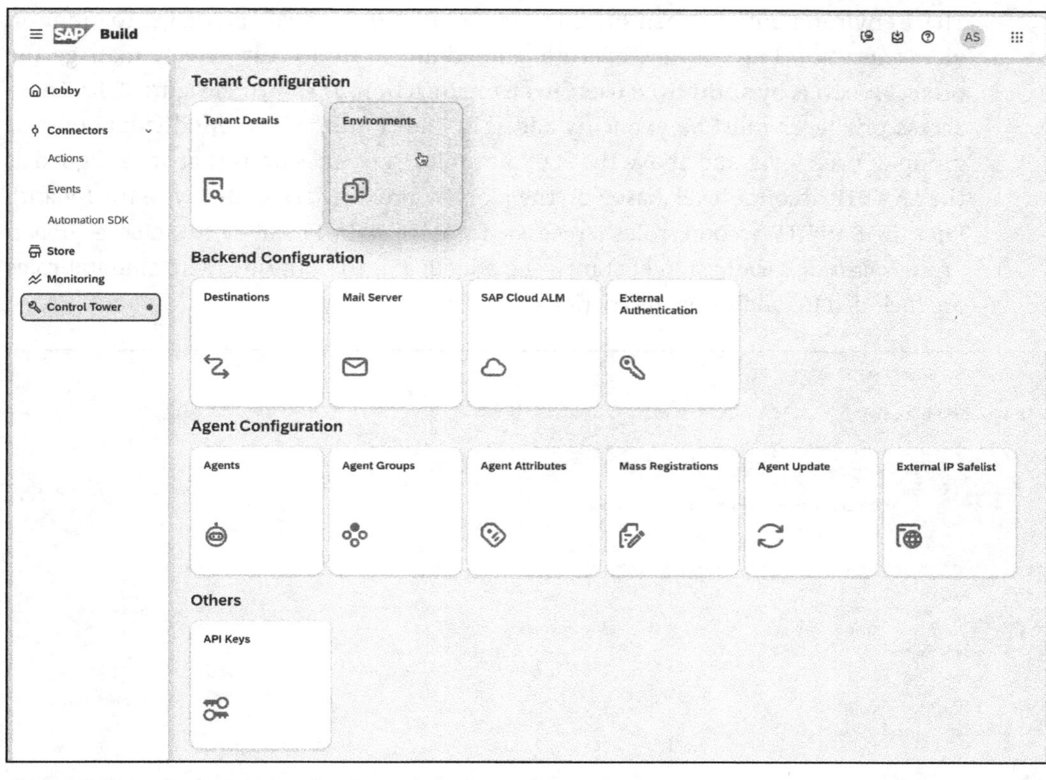

Figure 12.8 Environment Application Tile in Control Tower

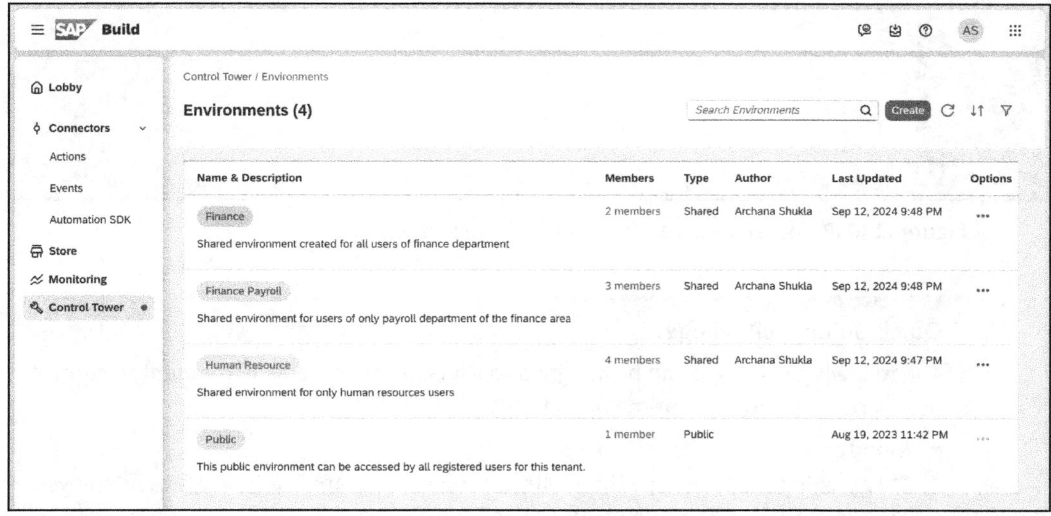

Figure 12.9 View of All Environments Created for This Tenant

These environments are then shared by the users with adequate privileges to monitor, view execution logs, deploy, execute, and administrate to restrictively manage the project lifecycle by authorized users in an organization, as shown in Figure 12.10. These access privileges must be explicitly added by the IT admin to the individual user or group of users over and above the standard roles/role collections that are assigned at the SAP BTP account level (based on the identity provider associated with the tenant). First the SAP BTP account roles (`ProcessAutomationAdmin`, `ProcessAutomationDeveloper`, `ProcessAutomationDelegate`, etc.) must be added, and then these environmental roles are included for additional restrictions.

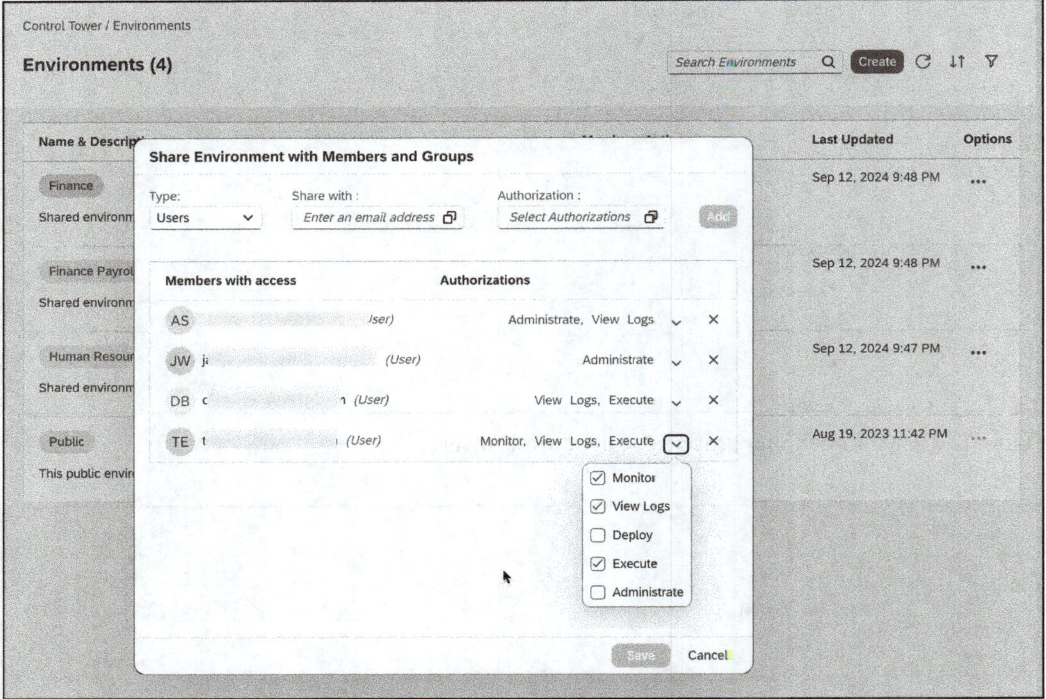

Figure 12.10 Permissions Given to Users for Environment Access

Quick Guide on Privileges

The following privileges can be assigned to users in the shared environment to control user access for various runtime operations:

- **Monitor**
  This privilege is to access the technical monitoring application to view deployed projects and inspect deployment issues. This doesn't have view execution logs permission.
- **View Logs**
  This privilege is to access technical monitoring applications to access detailed exe-

cution logs, which may contain sensitive organization or user information with view-only permission.

- **Deploy**
  This privilege is to manage deployment and undeployment of projects from the given environment.
- **Execute**
  This privilege is an end-user permission that allows the user to execute the process and access the forms or visibility dashboard at runtime. This doesn't include monitor, view logs, and deploy permissions.
- **Administrate**
  This privilege is an admin permission given to users who are authorized to add agents and other resources to the environment, as well as create and manage environments. It inherits permissions from the **Monitor** and **Deploy** privileges but excludes the **Execute** and **View Logs** privileges.

Once the environment is created and user/user groups are assigned, it becomes available during deployment of the projects from the lobby. All the projects that are deployed in the respective environment can be seen from the **Environments** page in **Control Tower**. Besides this, you can also see triggers, automation launchers, agents, and so on that are available for a particular environment. Figure 12.11 shows an overview of the projects deployed in a shared environment created for an HR department, and Figure 12.12 shows an overview of a shared environment created for a finance department.

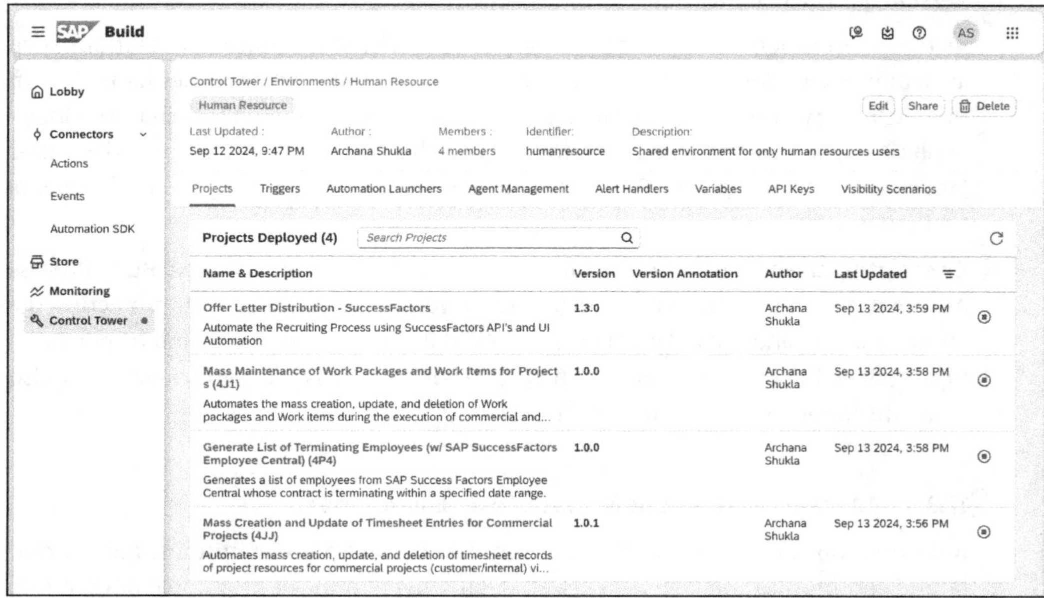

Figure 12.11 Overview of the Projects Deployed in an HR Environment

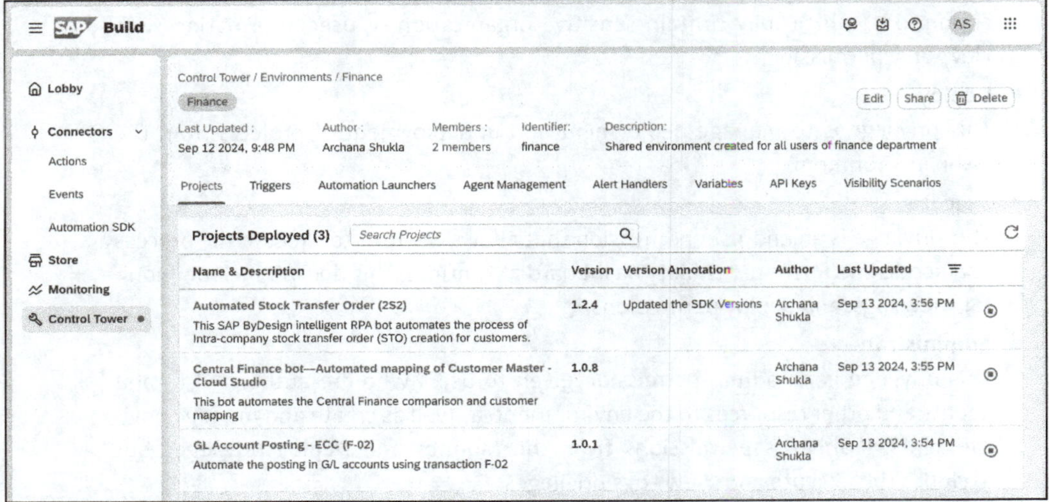

Figure 12.12 Overview of the Projects Deployed in a Finance Environment

From the **Projects** section, you can **undeploy** a project using ⦿. The undeployment operation will undeploy all the versions (active and inactive) of the project. In a shared environment, only the latest deployed version of the business project will be available and active for use, but in public environments, all the versions are available for access. There will always be one active version of the project, and the rest of the versions will be inactive. By default, only the active version is shown and is available for deployment in the public **Environment** page. If you want to see all versions, toggle the **Show Inactive Version** button. Refer back to Figure 12.7, which shows the toggle button.

As the environments are set, the user access is restricted for runtime operations such as deploy, execution, monitoring, and so on. For example, when a user named Lorin (having deploy permission to Finance and Payroll environments) logs in to the lobby application, she will be able to deploy only in those environments. Similarly, a user James who belongs to the HR department, can deploy projects only to environments that are related to HR.

By elevating the governance with an introduction of environments, SAP Build Process Automation aims to provide a holistic framework to organizations in protecting the strategic assets and related changes by profiled users with clear roles and responsibilities. Such stable practices not only help in having a consistent performance but also generate opportunities for improvements.

12.3.5 Allowed Destinations Mapped to an Environment

A destination is meant to connect to other cloud or on-premise systems for any outbound communication. The connection details given in the destination configuration can be technical user (via client credential, SAML bearer protocols, client certificates, etc.) or named user (via user token, password, etc.). For security reasons, the destination that uses the technical user is vulnerable and can be misused to extract sensitive information if it gets unauthorized access. Managing unauthorized access to a destination can therefore be seen as the subdivision of access management to ensure that the communications to internal or external systems are secure and done by the right team within the right procedural guidelines.

To secure destination usage, destinations can be mapped to a given environment. For example, destination that connects to finance system can be restricted to be used while deploying finance projects and to be available only to employees of the finance department. To achieve this, navigate to the **Destination** tile from **Control Tower**, and then click on the **Add** button to add the destination from the list of available destinations in the tenant, as shown in Figure 12.13.

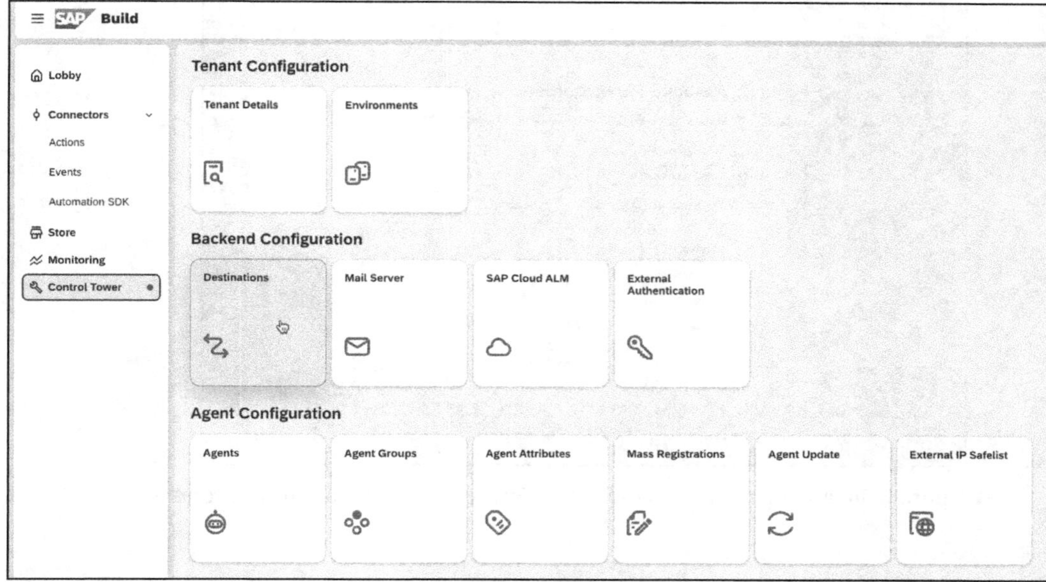

Figure 12.13 Opening the Destination Tile from the Control Tower

All destinations available in the SAP BTP cockpit are listed (which aren't yet added to any environment), and you can then select any destinations and associate them to all environment/public usage or to a list of specific environments by selecting **All Environments** or **Specific Environments**, respectively, as shown in Figure 12.14.

Maintaining destinations in the Environments application is quite easy. Here are some related tips:

- Destinations that aren't used in the projects can be removed by the IT admin.
- Destinations that are removed or renamed from the SAP BTP cockpit are shown as errors and can't be used while deploying the project.

- By default, all destinations are mapped to the public environment unless explicitly added to any specific environment by the IT admin.
- IT admins should be able to change the environment to public or to any other specific environment based on the change in the organizational hierarchies.
- Destinations available in the SAP BTP cockpit can be managed directly by clicking the **Open in BTP Cockpit** button from the **Destinations** page in **Control Tower**. IT admin must have the `IRPAOfficer` role to navigate to the SAP BTP cockpit to handle whitelisting destinations.

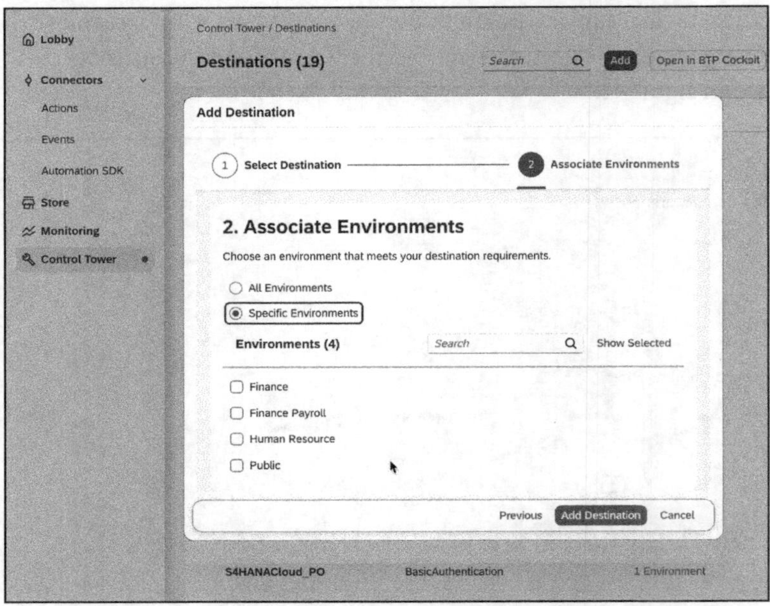

Figure 12.14 Adding the Destination to All Environments or Specific Environments

12.4 Transport Management

Continuous integration and continuous delivery , popularly known as CI/CD as mentioned earlier, is an established practice used in project development for reliable deliveries. In this practice, implementations are done by the developers in one landscape, which is associated with a central repository. Business projects are developed and tested before being deployed to the production landscape. The project team can do frequent changes and keep merging them to the production landscape. With CI/CD, organizations can build, test, and deploy software projects more often, with high-quality security. They can release project versions in small iterations, give early access to customers for feedback that eventually helps in early detection of bugs, and improve overall software productivity. However, CI/CD pipelines have potential risks such as unauthorized access, malicious attacks during transport, and poor credentials or weak passwords that can lead to leakage of sensitive data and insufficient control of core project lifecycle operations.

Governance is key to control such important operations in the CI/CD pipeline that are vulnerable to security risks. The goal is to ensure that operations such as creating, updating, or transporting the project are monitored and approved by authorized users only. Governance provides a set of guidelines, procedures, and policies for the process development and quality delivery that must be consistently followed by developers, IT admins, and business users. With this, organizations can create a reliable CI/CD pipeline with lower risks of exposing vulnerabilities and timely meeting of delivery timelines to address the market demands at a faster pace.

Transport management is a critical step in project delivery where the project developed or updated in one landscape, say a development system, is delivered to another landscape for testing or productive usage, say a quality or production system, respectively. While transferring the content from one system to another, there is a high risk of security breaches that may result in leaking customer data, causing companies huge losses in terms of money and brand trust. SAP Build Process Automation integrates with the SAP Transport Management service to provide a secure tunnel for transporting content from one SAP BTP account to another. This way, developers can confidently and consistently deliver projects from development to quality to production landscapes with reduced risks.

The IT administrator has full control of managing content transports from one system to another. The admin owns the lifecycle of the transport operation and decides how the systems are arranged in the CI/CD pipeline, which users should have what access, which backends are to be connected, and how the content is transported across systems. This streamlines the transport process and safeguards the overall release management procedure from any unwarranted security threats.

In this section, you'll learn how to transport business projects from one tenant to another in the different lifecycle stages of the transport.

12.4.1 Transporting Business Projects

Before transporting the content, there are some preparations to be done by the IT admin:

- Create separate SAP BTP accounts for development, testing, and production.
- Classify the account as development, quality, or production from the **Tenant Configuration** page in **Control Tower** of SAP Build Process Automation.
- Create a common account with the SAP Transport Management service enabled, and then create a service instance of this service, and assign the right roles to the user who will transport the project (check SAP Help Portal documentation to learn more about which roles or role collections to use: *https://help.sap.com*).

Figure 12.15 shows a destination that is created to connect to a production system from the SAP Transport Management service.

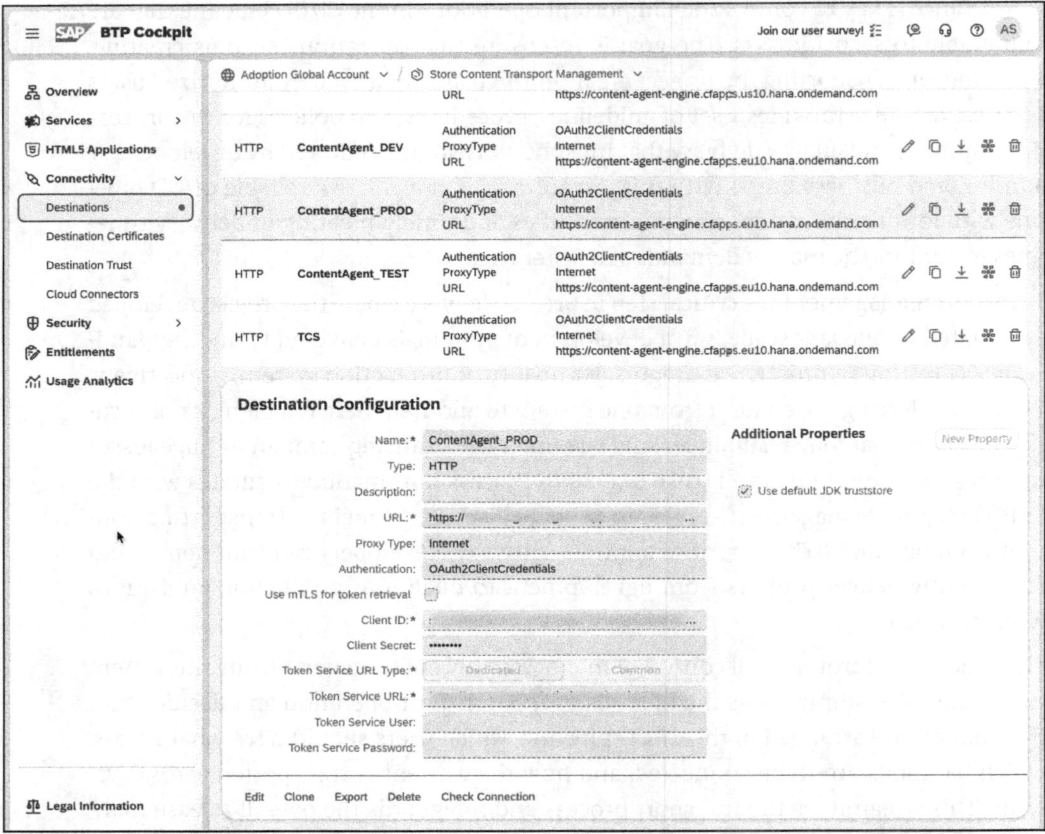

Figure 12.15 Destination to Connect to a Production System from SAP Transport Management Service

Allowed Operations in a Classified Tenant

A tenant that is classified as **PRODUCTION** or **TEST** *can't* perform the following operations: create new project, add project from store, create project from template in store, import editable version of the project, save as new project, rename project, release project, and publish project.

A tenant that is classified as **PRODUCTION** or **TEST** *can* perform these operations: import a versioned project, deploy or undeploy the project, delete the project, and manage members. Additionally, the **TEST** tenant can also use the transport option for transporting a business project from quality to production.

The **DEVELOPMENT** tenant has no restrictions, and all operations are allowed.

After the initial setup and configurations are done, the IT admin then does the following in the development, quality, and production accounts (these systems will later act as source and target systems during content transportation):

1. Creates a service instance and service key of the **Content Agent Service** from SAP Support Portal in development, quality, and production accounts based on the landscape setup at the respective organization.
2. Creates destinations in respective development, quality, and production accounts based on **OAuth2ClientCredentials** authentication such as `dev_contentagent`, `prod_contentagent`, and so on. The authentication details such as client ID, client secret, token service URL, and so on can be fetched from the service keys of the **Content Agent Service**.
3. Creates similar destinations in accounts where the SAP Transport Management service is installed for each of the development, quality, and production accounts such as `tms_contentagent_dev`, `tms_contentagent_prod`, and so on. These destinations are also **OAuth2ClientCredentials** based on where the credential details can be picked from the **Content Agent** service instance (created for each of the development, quality, and production accounts in the preceding step).
4. Opens the Transport Management application (from the account where the SAP Transport Management service is setup), and creates nodes, one each for development, quality, and production systems, from the landscape visualization menu.

   Figure 12.16 shows an overview of the Transport Management application with configuration options for transport nodes and routes.

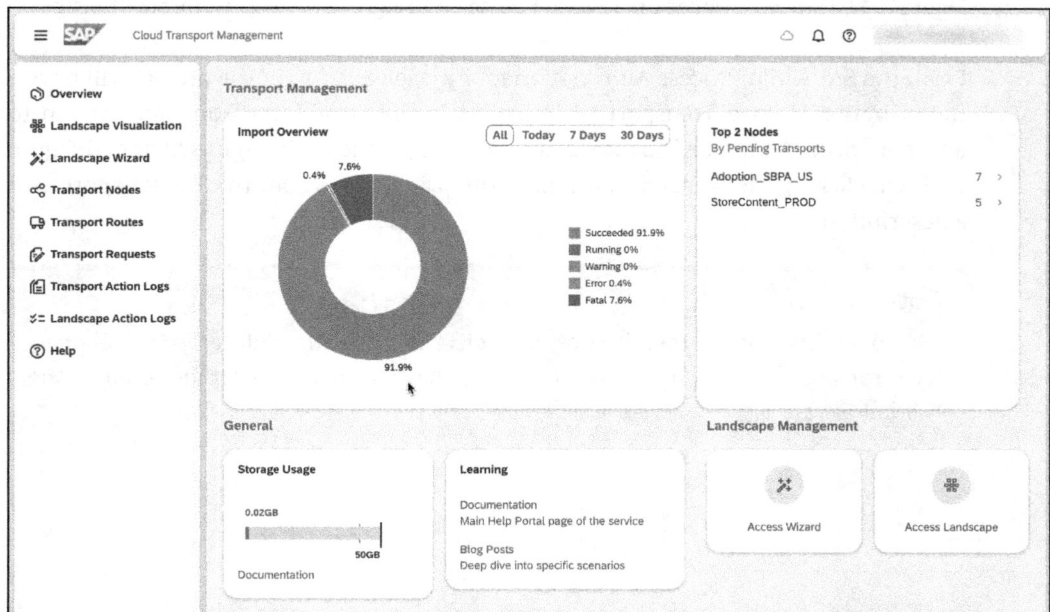

Figure 12.16 Transport Management Application View

5. Configures the node with the respective destination created in the Transport Management application account, such as **tms_contentagent_dev** for the development account and **tms_contentagent_prod** for the production account. Once the nodes are created, a link must be established between the nodes based on the transport strategy of the organization such as development to quality, quality to production, or directly from development to production.

   You can get all the latest information on SAP Transport Management from the SAP Help Portal guide at *http://s-prs.co/v592803*. Figure 12.17 shows transport nodes created between source and target tenants for content transfer.

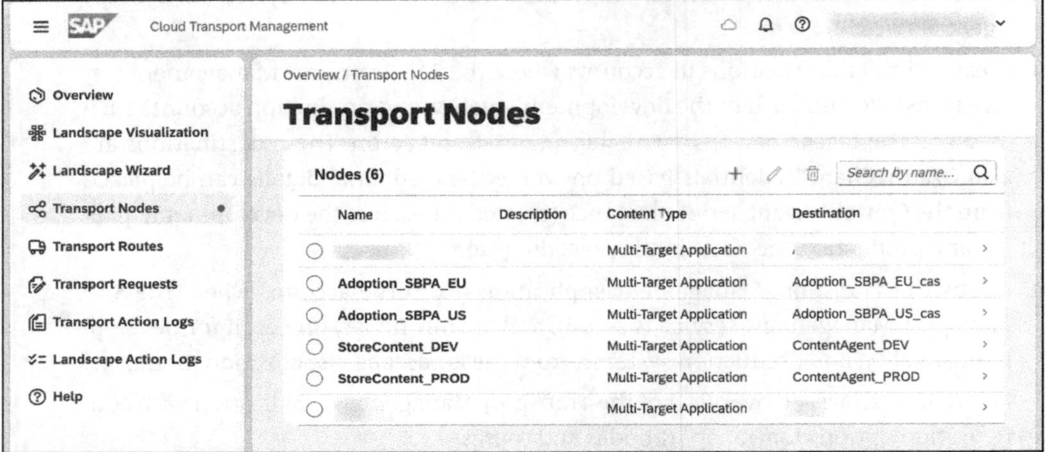

Figure 12.17 Transport Nodes Created for Source and Target Tenants for Content Transfer

From the SAP Build Process Automation lobby, each version of the project will have a menu option named **Transport** to transport business projects from one system to another without the need to export and import, as shown in Figure 12.18 and Figure 12.19. On clicking the **Transport** option, a popup appears to confirm the transport with a description.

Note

Only the users who are members of the project and have the right project level privileges can use the **Transport** option. If you don't have the permission, then contact your IT administrator.

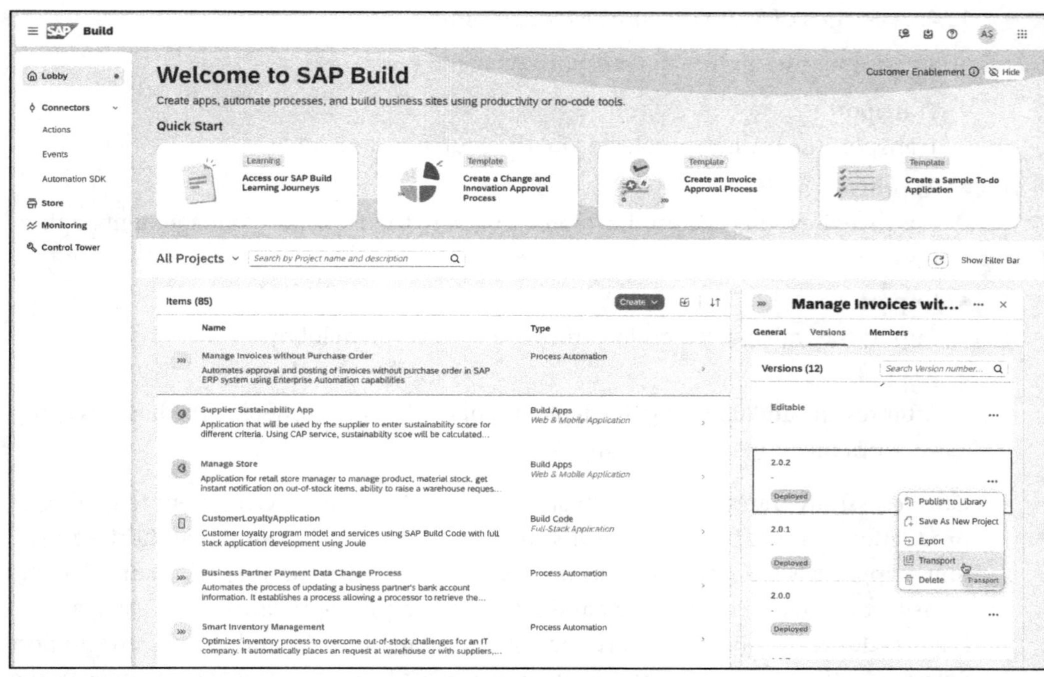

Figure 12.18 Transport Option for a Business Project from the Lobby

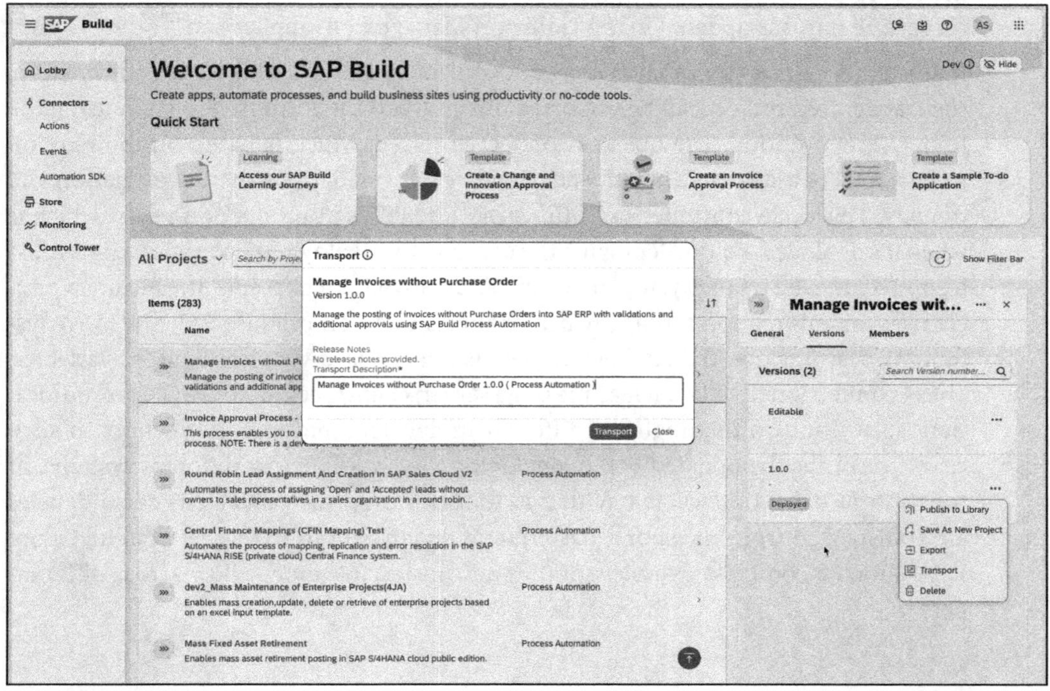

Figure 12.19 Confirming the Transport with a Description

12.4.2 Stages of SAP Transport Management

There are four phases in SAP Transport Management:

- **Transport**
  Transport the business project from the source system lobby.
- **Approval**
  Approve the request from the **Import Queue** in the Transport Management application.
- **Import**
  Import the transported project from the target system lobby.
- **Deploy**
  After testing and updating the deployment configuration, deploy the business project from the target system lobby.

On successfully transporting the business project from the development to quality or production system, the project waits for manual approval from the authorized user. These approvals are introduced as part of the governance process to ensure that the transport procedure is compliant and secure. To do the approval, the user navigates to **Import Queue** of the Transport Management application and checks the **Import Selected** option, which moves the business project from the transported queue of the development system to the import queue of the quality or production system. When the import is successfully completed in the target system (quality or production), the respective status is updated in the Transport Management application.

In the target system (i.e., quality or production) lobby of SAP Build Process Automation, the transported project can be imported directly into the respective system using the Import menu option. The business project is tested, which is based on the requirements. If there are errors, then the developer is informed for further investigation and remedy. This cycle continues until the project qualifies as production-ready, which is when it's deployed. Before deploying the content, the deployment configurations must be updated based on the system in which it's deployed. For example, the business project uses an automation artifact, which connects to an on-premise SAP S/4HANA system. As best practice, the connection details are exposed via environment variables. These connection details, such as message server, client, Secure Network Communications (SNC), and so on, might be different in development and production. In such cases, they must be updated before deploying the project across different systems, or the runtime execution will fail. With this, the business project and all its dependencies are transported from the source system and deployed into the target system. Figure 12.20 shows a complete overview of the transport management flow across different personas.

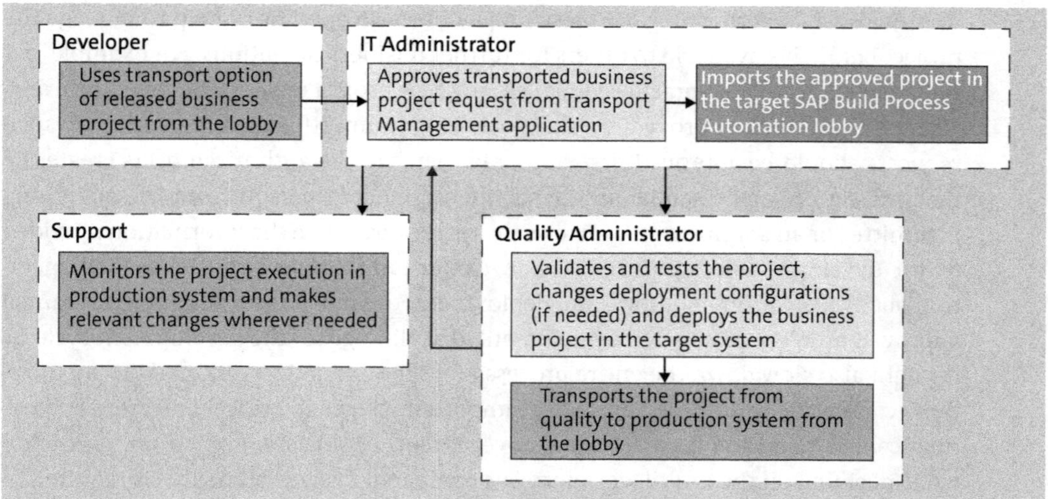

Figure 12.20 Complete Overview of the Transport Management Flow across Different Personas

12

One of the challenges in governance during CI/CD implementation is controlling the user access to the right systems leveraging the right tool. SAP Build Process Automation provides a unified workbench and tools to enable a holistic governance framework (e.g., environments, Transport Management application, granular permissions, etc.) that sprawls across different artifacts and systems, thus ensuring consistent and robust implementation of governance policies across all business processes for selected operations. These centralized applications provide various reliable access management capabilities to control the lifecycle of the business process across organizations of a company, which also includes granular level permissions and privileges. Not only this, but an IT admin can also monitor the business processes to identify any unauthorized access, control the user access for different operations and systems (or environments), and carefully inspect the production system to handle any security incidents.

12.5 Approval Flow for Controlling the Project Lifecycle

Governance aims to help organizations function within regulatory boundaries defined by the organization or external bodies such as government authorities or other regional governing departments. Some of the key components of good governance are (1) clear roles and responsibilities, (2) well-defined policies, (3) effective monitoring, and (4) a continuous improvement strategy. When these components are implemented during process development stages and delivery, they encourage the sense of responsibilities and inculcate good governance in the teams, resulting in higher efficiency and customer satisfaction.

Approvals while performing business project lifecycle operations help to manage the project before it's available to others for further business operations. For example, per the organizational best practice guidance, any new project creation in the finance area isn't allowed until approved by the governing committee. This means that such requests should be approved first, and only then can the authorized users create the new project. In such manual approvals, the designated person from the governing committee or an administrator reviews the request based on the information provided by the developer. Once approved, the developer can then perform the needed operation such as creation, modification, deployment, or transport of the project. Manual approvals aren't necessary to be implemented at all stages of the development cycle, as it might also slow down the entire process and speed of delivery to market. Organizations can decide to impose approval at important stages in critical LoB areas such as finance, life science, manufacturing, government, healthcare, and so on. Flexibility while deciding on the multilevel approvals compared to strict standards is required to ensure good delivery speed and innovations.

SAP Build Process Automation enables an approval flow for **Create** and **Transport** operations for a business project. An IT administrator enables or activates the approval flow from the **Governance Rule** page in **Control Tower**, as shown in Figure 12.21 and Figure 12.22, where governance rules are enabled for **Create Project**. Once activated, a default process template is enabled and imported in the lobby for create and transport operations. The lifecycle operation is paused until the governance rule is executed and someone manually approves it.

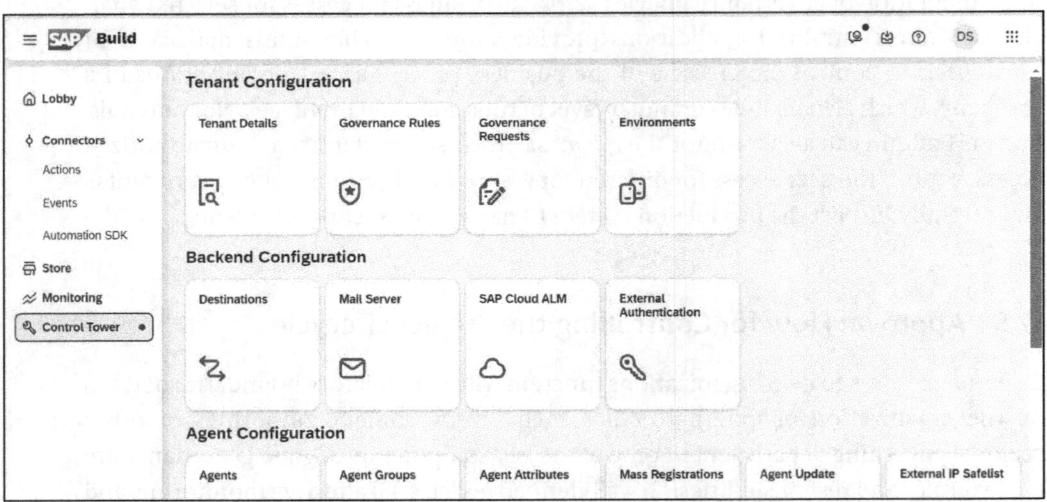

Figure 12.21 Governance Rules Options in Control Tower to Enable Approvals for Create and Transport Option

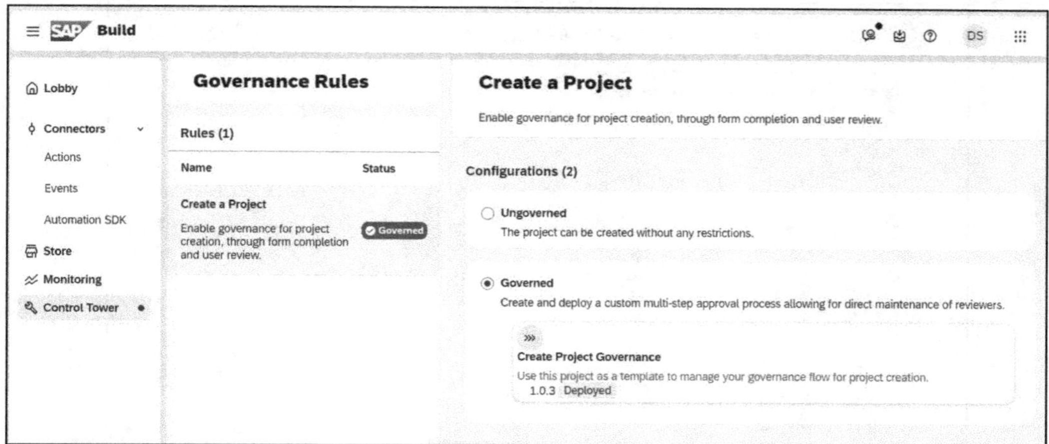

Figure 12.22 Governance Rules Application from Where the Create Project Option Is Enabled for Governance

Note

Governance rule is activated at the tenant level, which means that once the governance is enabled, let's say for the create operation, then all the business projects created from the lobby will have to follow the governance process. Exceptions can be defined, however, in the process model created for governance. For example: if the approval isn't needed for certain departments, then a conditional block can be added in the process to skip the approval.

Figure 12.23 and Figure 12.24 give an overview of the governance rule process with approval flow for creating a project. Members of the governance committee will update the template to add steps needed for governance, such as the requestor fills the questionnaire via the form activity in the process, and this form is then reviewed and approved/rejected by the predetermined approvers. Once modifications are done, the process must be deployed and published to the library. This process act as a governance rule that must be fulfilled by the developers who are creating or promoting the business project from the lobby. It means that when the governance rule is enabled, and a developer tries to create a business project, it will start the governance process, and the create operation will be paused. Once the request is approved, the project will be created and shown in the lobby like any other project.

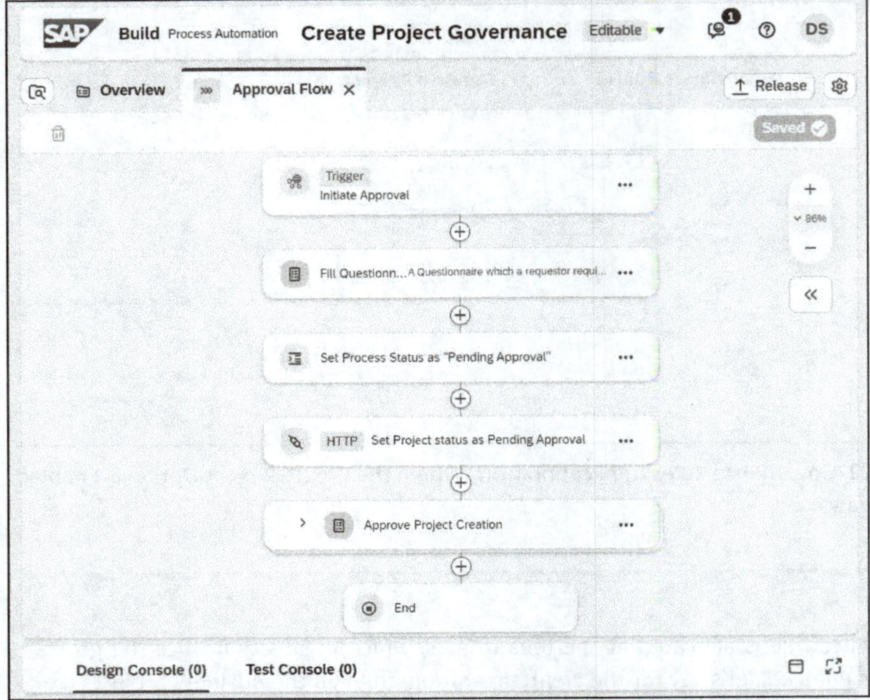

Figure 12.23 Overview of the Governance Rule Process for Creating a Project

In the governance process, the first step is a form with a questionnaire. The questionnaire is decided by the governance committee based on whether it's designed for a create or transport flow and organizational policies and procedures that are unique for each company. The questions can be similar to these for the create operation:

- What is the name and description of the project?
- What is the value proposition?
- Which functional area of a given LoB does the project cater to?
- Which backend systems are used?
- How many business users are expected to use this project?

For the transport operation, the questions can be framed like a detailed test report, as follows:

- Are you using accessing personal data?
- Are you following the product standards guidelines?
- Will you show the report for product standard approvals?
- Is this deployment to a production or quality landscape?

This questionnaire will be available in the developer's inbox and must be filled by the developer.

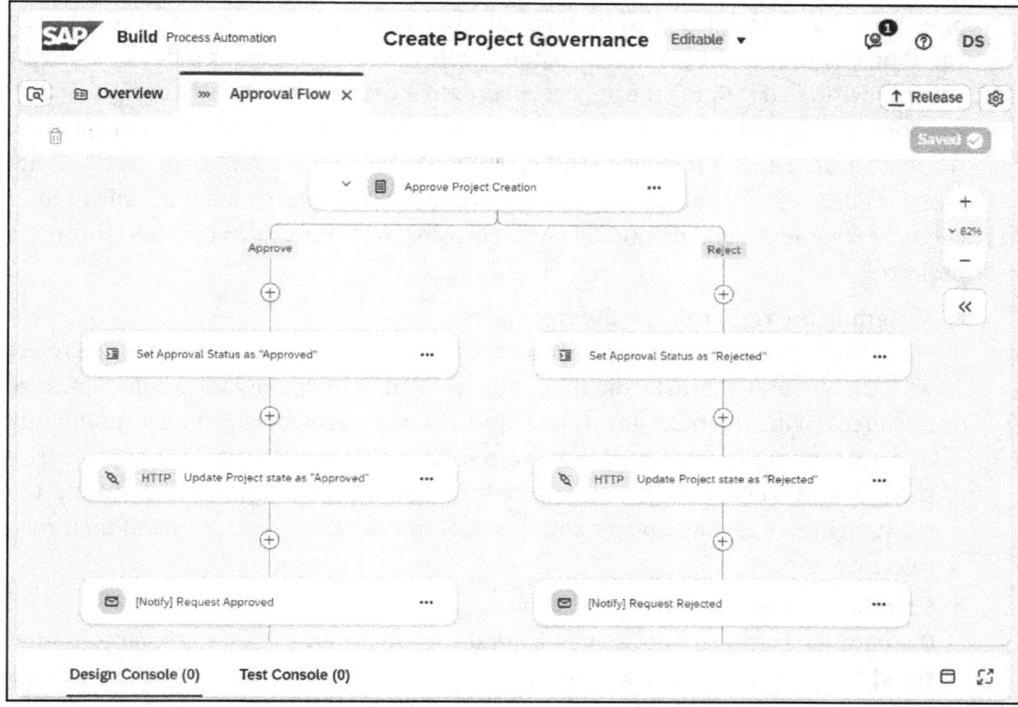

Figure 12.24 View of the Approval Flow Showing How the Process Will Execute Once the Project Creation Request Is Approved or Rejected

Once submitted, the filled form reaches the approver's inbox. Approvals are done by the IT administrator who navigates to the **Governance Requests** page from **Control Tower**, reviews the request form submitted by the developer, and approves or rejects based on the governance rules. The governance process has the flexibility to create the approval strategy based on the governance procedure of the organization. These approvals can be single-step or multistep approval, parallel approval, or sequential approval, and they can be regularly reviewed and changed with changing governance policies. This makes the entire business project lifecycle more efficient and secure.

An administrator can check the request logs from the **Governance Requests** page. It lists all the projects that are under review, approved, or rejected. In addition, it also shows other details such as the request type (create or transport), request creation date, who created the request, status of the request (**Pending**, **Approved**, **Rejected**), date when the status was updated, and so on.

12.6 Maintaining Business Projects

Effective project maintenance is an important activity in the project lifecycle. Once the project is created and deployed in the productive system, it often needs to be updated based on business user or end user requirements and changing market demands. Right governance in maintaining the project ensures that the right users with the right authorizations are making informed decisions while modifying the project, managing the scope, and prioritizing the changes. The approach to define a robust project maintenance guideline is critical for project success and overall improvement in time to market. Here is some governance guidance that can help in effectively maintaining business projects:

- **Assigning the right roles to the right people**
  This ensure that the most appropriate people are assigned to modifying the project who understand the user requirements, as well as helps organizations optimize resources. With the right governance process, the developers can also confidently update, test, and deploy business projects, avoiding any disruptions and security risks. Such access management procedures inculcates accountability among the team members as they understand the importance of these changes and their relative outcomes.
- **Stakeholder engagement**
  Changes in the project are driven by end-user requirements. It's critical to understand these requirements and devise a plan to update the project; otherwise, the entire project can be derailed, causing risks of regression. The governance framework ensures that there are proper people who are deployed to understand stakeholder requirements and to plan and implement the changes. The deployment of these project changes can be further controlled with an approval process, especially for critical LoBs. This means that any change must be first approved by the governance committee, and then it can be deployed to production. Such a centralized and unified governance framework strengthens the quality of the delivery and ensures effective project maintenance.
- **Monitoring and regular reviews of the control process**
  The same governance rules don't always apply. The rules will change with changing market demands, company strategies, introduction of new technologies, and updates in key deliverables and key performance indicators (KPIs). Monitoring the governance process is important for detection of early delays in the project maintenance and cost overflow. These regular reviews help in timely updates of the governance procedures, which ensures effective resource allocation, on-time delivery of the changes to the market, and satisfied stakeholders.

Throughout the business project lifecycle, SAP Build Process Automation has several governance features, such as access management, environments, transport management, and approval flow, which provide a holistic governance framework for effectively creating and maintaining business projects. Access management defines permissions such as Viewer, Developer, and Administrator, which can be assigned to users based on their responsibilities while managing the project. For example, the Developer role is given to employees who will implement the changes, the viewer role is given to business users who can view the existing process and define the plan to update the project based on stakeholder requirements, and so on. Environments define the boundaries for users to deploy, monitor, and execute process runtime artifacts. For example, a developer who is modifying the business project for the finance department must have the Deploy permission to the environment related to the finance department. The developer might also need permission to monitor and view logs to see the test results after process execution to validate the changes from the Monitoring application. Before transporting the changes to the production tenant, an approval flow can be enabled for the transport operation so that the changes are again reviewed and approved for the business users and domain experts.

Embracing governance during project maintenance leads to improved speed and quality of the updates, better resource management, and enhanced security. This eventually helps to improve time to market for the updates leading to better outcomes and happy stakeholders. SAP Build Process Automation offers centralized, no-code tools to enforce governance while managing and maintaining business projects.

12.7 Summary

Governance is demanded by all organizations across all LoBs. Small companies informally introduce governance in their processes, while large companies have strict governance policies, and all business processes are enforced to be built and executed only within these boundaries. Among them, many companies also deal with external sources. Such companies must apply governance rules and procedures even more broadly to ensure secure transactions and regulatory compliances. Apart from this, there are often audits where reports are needed to prove to internal and external authorities that the company is adhering to the governance policies. These reports help to establish trust and credibility with customers and stakeholders. With this in mind, you can understand the importance and need for governance in the industry.

This chapter started with an introduction about the various benefits of governance through which (1) only authorized users can access the project, (2) rules are applied to automate policies, (3) there is continuous monitoring to enforce compliance, and (4) smoother audits occur. The governance framework of SAP Build Process Automation provides access management, project sharing, environments for segregating deployment of projects, transport management, and approval flows while performing project lifecycle activities. Following are various governance techniques that you learned about in this chapter:

- With role-based access during business project access management, privileges such as Viewer, Administrator, and Developer can be given to individual users or groups of users to control unwanted access while creating and managing processes.
- With environments, an isolated container is provided for deployments where access can be further controlled by permissions such as View Logs, Deploy, Monitor, and Execute. Environments are public (default) or shared. By default, the project gets deployed in the public environment. You can choose to create shared environments based on the organization or subdivisions within the organization. Shared environment is a powerful concept where different versions of the same project can be deployed to different shared environments, and different users can get access to different environments to perform runtime activities. In environment applications from Control Tower, users can create new shared environments and assign privileges to monitor, deploy, execute, and administrate business projects. From each environment application, you can view deployed projects, automation launchers, agents, variables, and triggers.
- Destination configuration in a specific environment from **Control Tower** enables secure connectivity to external systems for each organization and ensures that the system data isn't accessed across departments. All destinations available in the tenant are listed while enabling the whitelisting of one or more destinations for a particular environment. You can add/remove the destination from shared environments or make it available for all environments.
- With transport management using the SAP Transport Management service, there is a provision to move the business project from development to quality to production tenants without any manual download or upload. This helps developers confidently deliver projects and establishes a strong CI/CD pipeline. You learned how to (1) set up the SAP BTP cockpit to enable the SAP Transport Management service and configure the destination, (2) configure transport nodes and routes in the Transport Management application to define source and target systems, and (3) transport a version of the business project from the SAP Build Process Automation lobby using the Transport option. You also learned about the four stages of SAP Transport Management, which are transport, approval, import, and deploy. Each business project goes through all the stages as the project moves from one tenant to another.
- With the introduction of an approval flow during the project lifecycle operations of create and transport, these operations are paused until the governance process flow is executed and approved. From **Control Tower**, the governance rules can be enabled for create and transport operations. Once enabled, a default process template is created in the lobby, which can then be modified by the members of the governance committee based on the organization's guidance. The governance process flow is finally published to the library and applied every time a project is created or transported.

Applying right roles to right people, understanding stakeholder requirements correctly, and regularly monitoring the governance process are key to effective business process maintenance. Throughout the entire business project lifecycle, there are numerous options to enforce governance during process management and maintenance, which can be efficiently applied to strengthen governance while creating, monitoring, and executing the business processes. These functionalities embedded in no-code tools with enhanced user experience provide easy configuration options to include governance in various business project lifecycle operations. Process governance guides the developers, administrators, and business users of the procedures regarding creating, deploying, and accessing the business project. It also ensures security in accessing data, assigning roles, defining runtime environments, and monitoring the process execution.

In the next chapter, you'll learn in detail about various authorization options, authentication techniques, and audit logs information supported by SAP Build Process Automation.

Chapter 13

Authentication, Authorization, and Audit Logging

This chapter discusses authorizations for SAP Build Process Automation, including what roles to use and when to use them. You'll also learn how audit logging in SAP Build Process Automation ensures that all critical actions within the platform are recorded for security, compliance, and troubleshooting.

Authentication, authorization, and auditing provide security for a distributed cloud environment by allowing any user with the proper credentials to connect securely to protected application servers This allows organizations to track who did what in their systems, which is essential for investigating and preventing security incidents.

In this chapter, you'll learn about authentication and authorization mechanisms in SAP Build Process Automation, secured ways to connect to cloud and on-premise environments, and how audit logs are captured in the platform.

13.1 Approach to Authentication and Authorization

Authentication refers to verifying the identity of users or systems before allowing access to the SAP Build Process Automation platform. SAP Build Process Automation doesn't have an implicit user management and authentication mechanism. It leverages the user management and authentication mechanism of SAP Business Technology Platform (SAP BTP). Authentication in SAP Build Process Automation is ensured via Single Sign-On (SSO; the default identity provider), which integrates with Identity Authentication. SAP Build Process Automation uses role-based access control to manage permissions.

Authorizations are done by role collections that are maintained in SAP BTP. Users are assigned specific role collections that grant them permission such as developing, deploying, or monitoring the process. To assign roles, SAP Build Process Automation relies on the standard functionality of SAP BTP.

Let's understand the role collections for SAP Build Process Automation:

- `ProcessAutomationAdmin`
  This role has the highest level of access in SAP Build Process Automation and provides

comprehensive administrative privileges within the platform. This role is assigned to IT administrators and is crucial for maintaining the stability, security, and proper functioning of SAP Build Process Automation across an organization.

The user with this role has implicit manage privileges on all objects (projects, packages, environments, agents) and handles key tasks such as these:

  - Manage process configuration, permissions, and authorizations.
  - Monitor, troubleshoot, and manage automation processes.
  - Access and manage audit logs for compliance and security.
  - Schedule triggers for the business process.
  - Access the **Event Acquisition** and **Monitor Visibility** scenario tiles.
  - Access the Document Information Extraction interface for annotating templates.
  - Create and act upon environments depending on privileges.

- `ProcessAutomationDeveloper`

  This role is typically assigned to users who are responsible for creating and configuring automated workflows, integrating external systems, and maintaining the logic behind business processes.

  Key responsibilities include the following:

  - Create, edit, deploy, and publish the processes and automations.
  - Access the Document Information Extraction interface for annotating templates.
  - Create all the artifacts, such as forms, decisions, and automations, and configure visibility scenarios in the process.
  - Administrate the content that has been granted in admin mode.
  - Monitor jobs based on environment permission.

- `ProcessAutomationParticipant`

  This role is typically assigned for business users who need to interact with automated processes but don't require design, development, or administrative access. This role participates in the business process and performs the following tasks:

  - View tasks in My Inbox.
  - Use forms to complete their tasks by either submitting or approving them.
  - Perform task operations such as claim and release.
  - Register and run agents.
  - Start workflow instances.

- `ProcessAutomationDelegate`

  This role is assigned to a developer who doesn't have rights to create or import a project. However, this role allows a user to perform operations on projects where the corresponding privileges have been defined by another developer or administrator. Key responsibilities include the following:

- View or edit existing projects, packages, environments, agent groups, and registered agents.
- Monitor the execution for a set of agents or environments.
- Administrate the content that has been granted in admin mode.
- Monitor jobs based on environment permission.

The SAP Build Process Automation roles can be viewed in SAP BTP under **Security • Role Collections**, as shown in Figure 13.1.

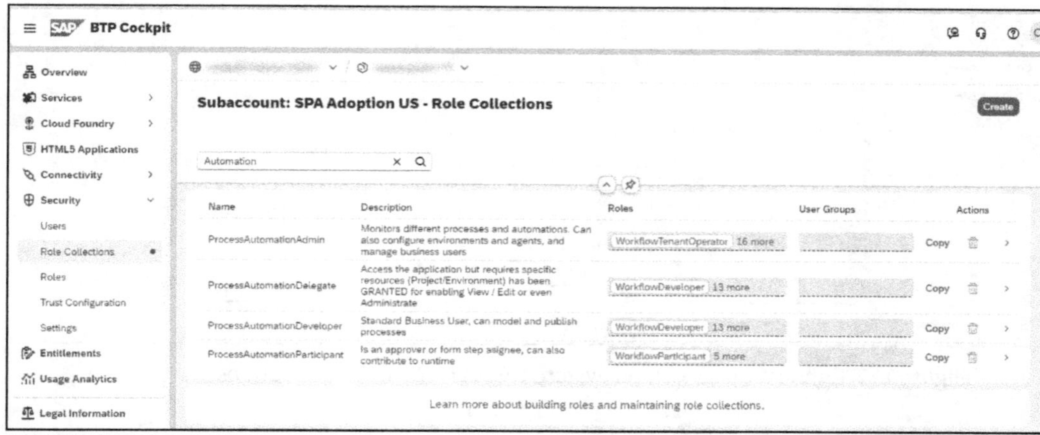

Figure 13.1 Role Collections for SAP Build Process Automation in SAP BTP

13.2 Securely Connect to Cloud and On-Premise Systems

SAP Build Process Automation connects to cloud and on-premise systems securely using **Destinations** in SAP BTP. You can use a destination to connect your application to the internet (via HTTP, remote function call [RFC], or WebSocket RFC) or to an on-premise system (via HTTP or RFC). As shown in Figure 13.2, you can create a destination to connect to an SAP S/4HANA Cloud system by choosing subsection of **Connectivity • Destination** in SAP BTP. Make sure to add the destination parameters **sap.application-development.actions.enabled** as **true** and **sap.processautomation.enabled** as **true**, which allows you to access the destinations in SAP Build Process Automation.

You can also send and retrieve emails using the mail artifact in SAP Build Process Automation by configuring a mail destination via **Connectivity • Destination** in SAP BTP, as shown in Figure 13.3.

If you're connecting to an on-premise system, you must use cloud connector, which serves as a link between cloud applications and on-premise systems. In cloud connector, the subaccount of the ABAP system is connected so that cloud connector can be used within a communication system or destination in the subaccount.

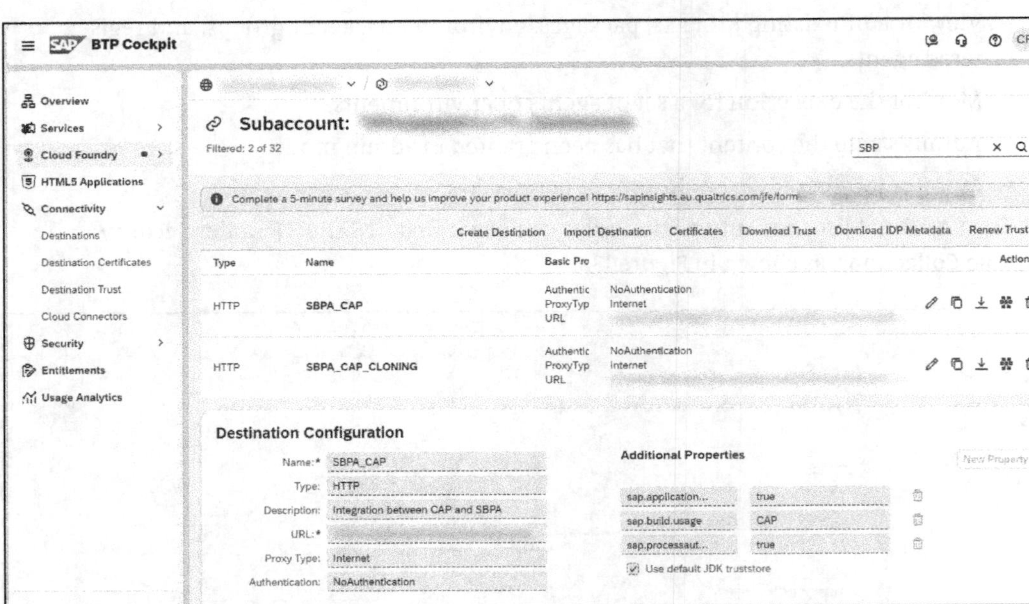

Figure 13.2 Connection to a Cloud System through SAP BTP Destinations

Destination Configuration

Name:*	sap_process_automation_mail
Type:	MAIL
Description:	Mail Server Configuration for SAP Process Automation
Proxy Type:	Internet
Authentication:	BasicAuthentication
User:	processbuilder-mails
Password:	••••••••

Figure 13.3 Connection to the Mail Server through SAP BTP Destinations

13.3 Audit Logs

An *audit log* is a record of events and changes. SAP software creates logs based on events. Audit logs are records of these event logs that provide documentary evidence for an event or activity. SAP Build Process Automation leverages the SAP Audit Log service in SAP BTP. After successful subscription to the SAP Audit Log service in the same subaccount of SAP Build Process Automation, the subscriptions can be seen under **Services • Instances and Subscription** in SAP BTP, as shown in Figure 13.4.

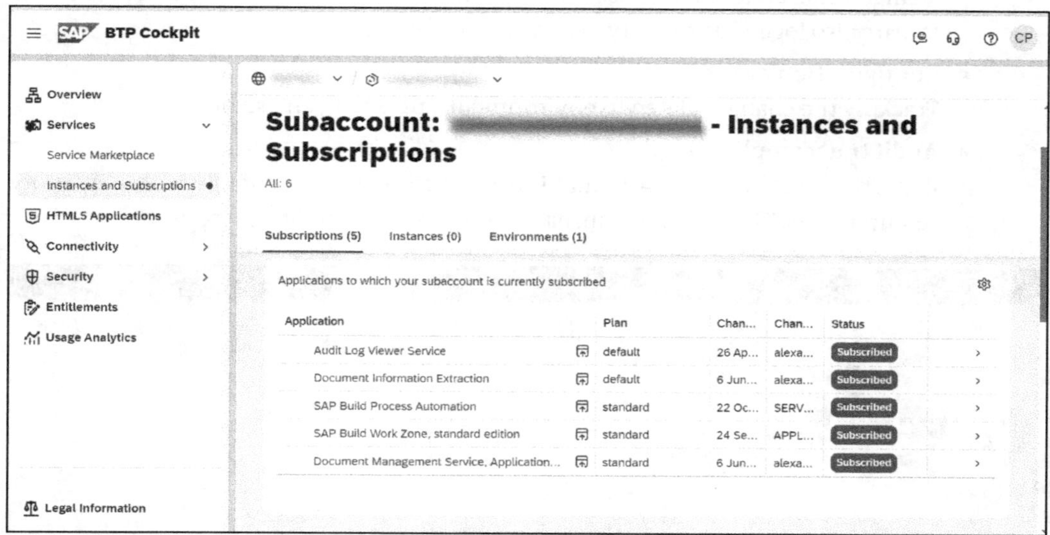

Figure 13.4 Audit Log Viewer Service in SAP BTP

The SAP Audit Log service stores audit logs representing different actions taken over your account and/or data. Predefined audit categories represent these actions:

- **data-access**
  Read-access logging records for access to sensitive personal data.
- **data-modification**
  Data modification logging records for sensitive personal data.
- **security-events**
  Logging of general security events such as login, logout, and others.
- **configuration**
  Logging of critical security configuration changes.

The **Auditlog Viewer** service can be accessed by clicking [icon] under the **Instances and Subscriptions** section in SAP BTP. You can access the different audit categories in the **Auditlog Viewer** application, as shown in Figure 13.5.

Audit logging in SAP Build Process Automation ensures that all critical actions within the platform are recorded for security, compliance, and troubleshooting purposes, including the following:

- **User activities**
  Events related to account data changes, which are logins, logouts, and password changes.

- **Security events**
  Security-related events, such as login attempts (successful or failed), password changes, role assignments, user attempts to access a service they're not authorized for, are also logged as security events for the subaccount.
- **Configuration events**
  Tracking the change related to environments, packages, and so on.
- **Audit trail compliance**
  Exporting audit logs in SAP Build Process Automation to comply with regulatory requirements. This ensures transparency and accountability in the system.

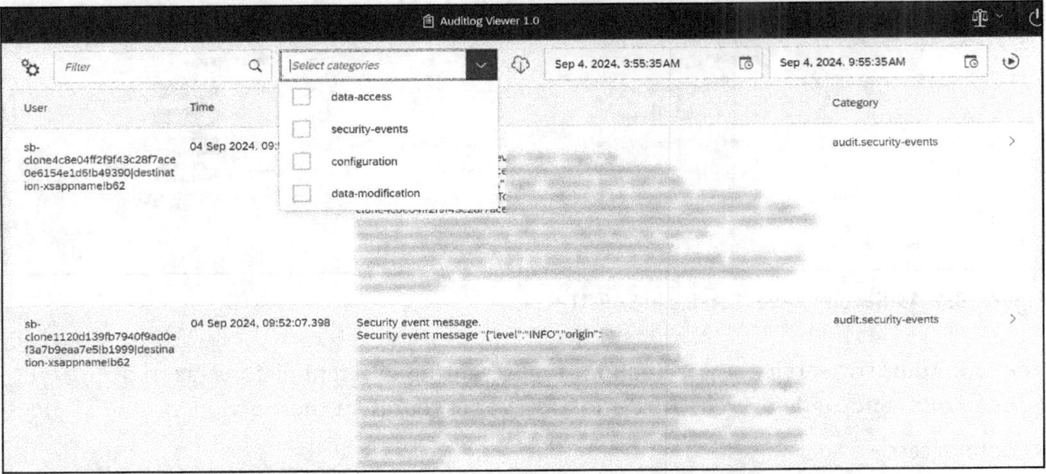

Figure 13.5 Logs in the Auditlog Viewer Service

The audit log message captures the whole context of the reported event. The message itself depends on the event generated. For more information about the message detail, you need to read the message and cross-verify the information with the corresponding SAP Help Portal information for the SAP Audit Log service and the events within the corresponding application provider. As shown in Figure 13.6, the first message in the **Auditlog Viewer** application explains that the user has modified the content of a process artifact.

Apart from the SAP Audit Log service, if the business process project has a process visibility artifact, you can build dashboards to monitor processes in and across SAP and third-party applications. You can view the logs of the process by selecting the **Visibility Scenarios** tile in the **Manage** area under **Monitoring** in SAP Build, as shown in Figure 13.7.

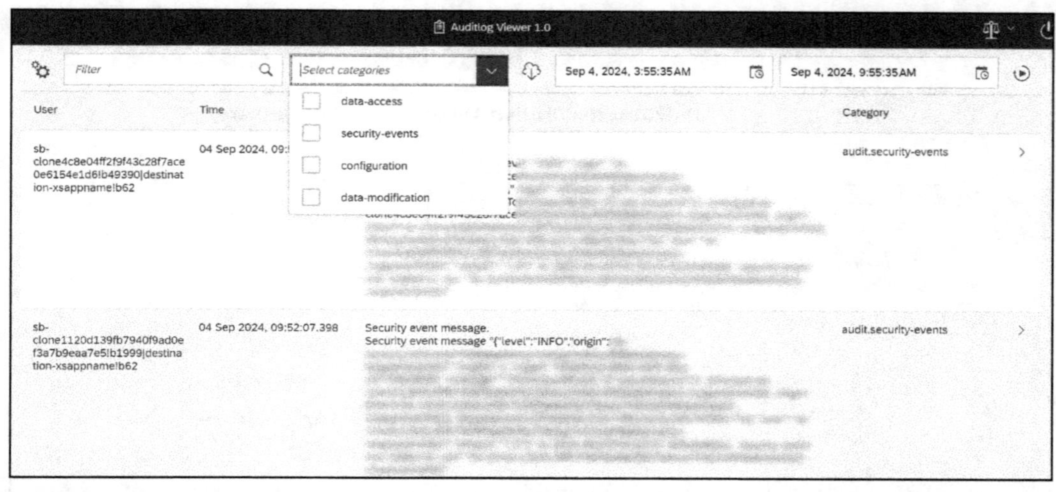

Figure 13.6 SAP Build Process Automation configuration logs in Auditlog Viewer

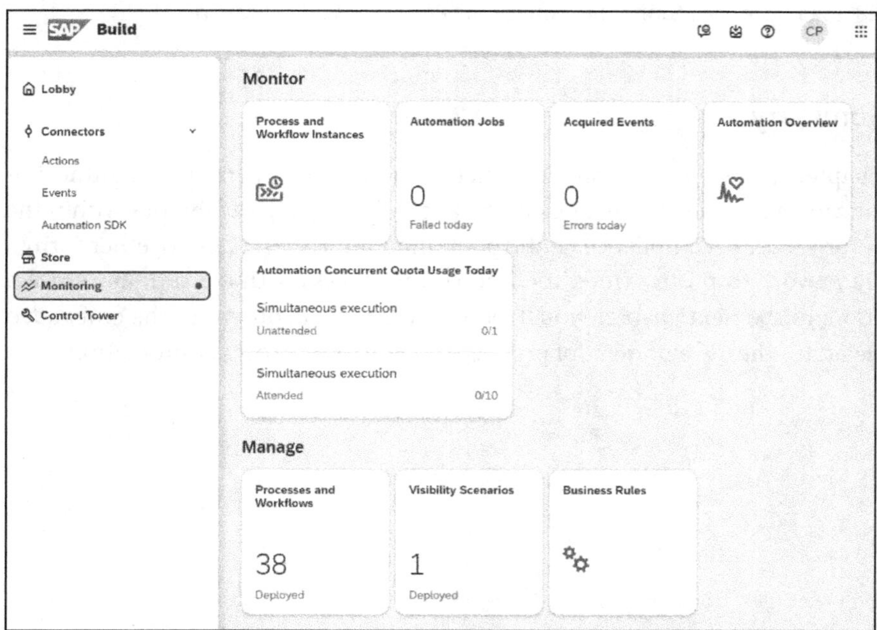

Figure 13.7 Accessing Visibility Scenarios under Monitoring in SAP Build

As shown in Figure 13.8, you can monitor processes as well. Refer to Chapter 11, Section 11.3, which explains in detail about designing and configuring visibility scenarios.

Figure 13.8 Logs from the Monitor Section for the Process Visibility Scenario

13.4 Summary

In this chapter, you've learned about authentication mechanisms, rights authorizations that can be assigned to business users, and audit logging capabilities within the platform for security, compliance, and troubleshooting issues. They together form a robust framework to protect, monitor, and control access within SAP Build Process Automation. In the next chapter, you'll learn how prebuilt content in the embedded store accelerates the development of processes in SAP Build Process Automation.

Chapter 14
Working with Store Content

This chapter introduces the SAP Build content store, which is an application to discover and deploy business projects from a plethora of enterprise-ready content packages that are built on common industry use cases. We'll discuss prebuilt content, various types of content, and how you can access and deploy them.

With low-code/no-code tools, SAP helps organizations accelerate their developer efficiency, thereby addressing the challenges faced by scarcity in skilled developers and empowering citizen developers or non-IT developers to contribute to application and process development. To further accelerate the solution development, SAP Build has published a variety of prebuilt content packages across different industries and lines of business (LoB) on common identified whitespaces in the SAP Build content store. These content packages can be easily imported, configured, and deployed in customer-managed SAP Business Technology Platform (SAP BTP) accounts, without the need of a professional developer.

In this chapter, we'll introduce the SAP Build content store where you'll learn about different kinds of stores and how to discover prebuilt content from these stores. We'll also explore how to publish and commercialize the content.

14.1 What Is a Store?

Content stores are applications that allow users to access enterprise-ready, industry-based prebuilt content. It's like a store with ready-made business projects that just need a little setup and configuration to be running in a respective SAP BTP environment. The content packages in the store are designed to address common business problems across different industries and LoBs, with the idea of reducing the development time and bypassing long development cycles in building the project from scratch.

There are three types of content stores—SAP Business Accelerator Hub, the embedded store, and SAP Store—as described in below sections:

14.1.1 SAP Business Accelerator Hub

SAP Business Accelerator Hub is a public website (*https://hub.sap.com*) and central hub for all application programming interfaces (APIs) and business content published by SAP. Users can explore and test APIs such as OData, REST, Simple Object Access Protocol (SOAP), remote function call (RFC), and so on provisioned by SAP products and LoBs such as SAP S/4HANA, SAP Ariba, SAP Concur, SAP SuccessFactors, SAP Customer Experience, SAP Fieldglass, and others. Along with APIs, you can also discover and study events, graphs, adapters, core data services (CDS) views, and business content.

Navigating to the **Domains** section in SAP Business Accelerator Hub takes you to Business Content for Integrations and SAP Build, as shown in Figure 14.1. The SAP Build content store is a one-stop shop for discovering and exploring all SAP Build business content packages to fit your automation and extension needs. Content can be further filtered based on different categories such as **Industries**, **Products**, **Lines of Business**, and so on. Figure 14.2 shows a detailed page of an SAP Build content package in SAP Business Accelerator Hub. For more information on the content package, refer to Section 14.3.1.

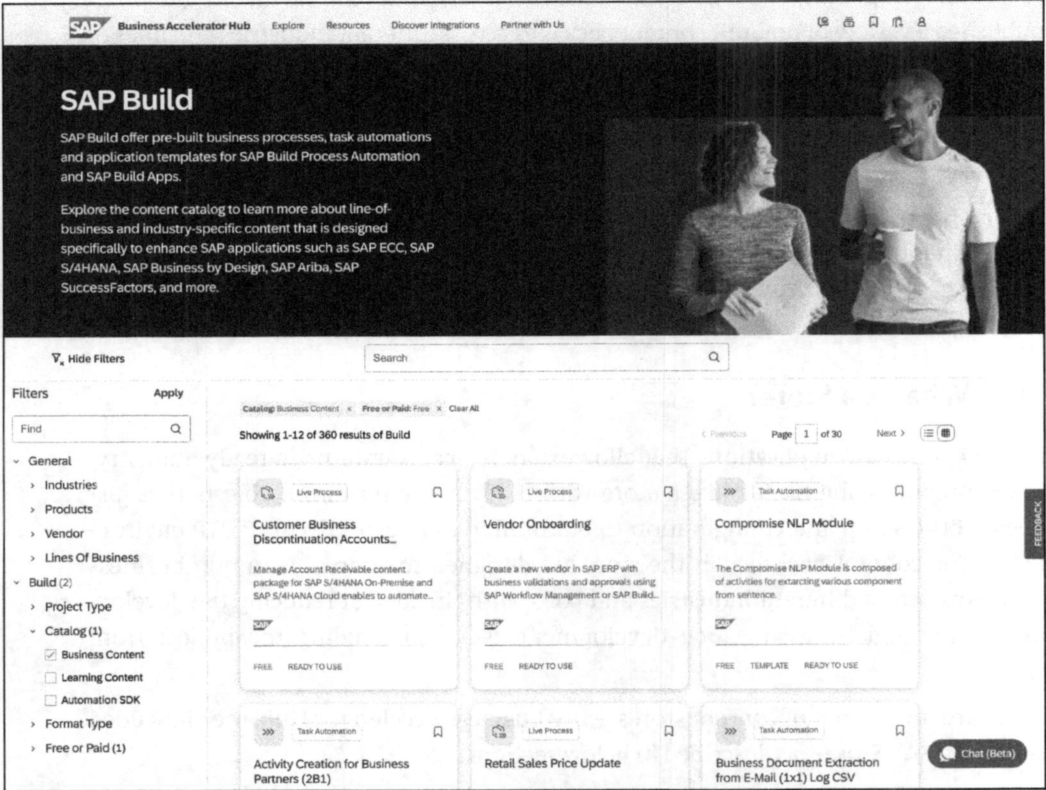

Figure 14.1 Overview of the SAP Build Page in SAP Business Accelerator Hub

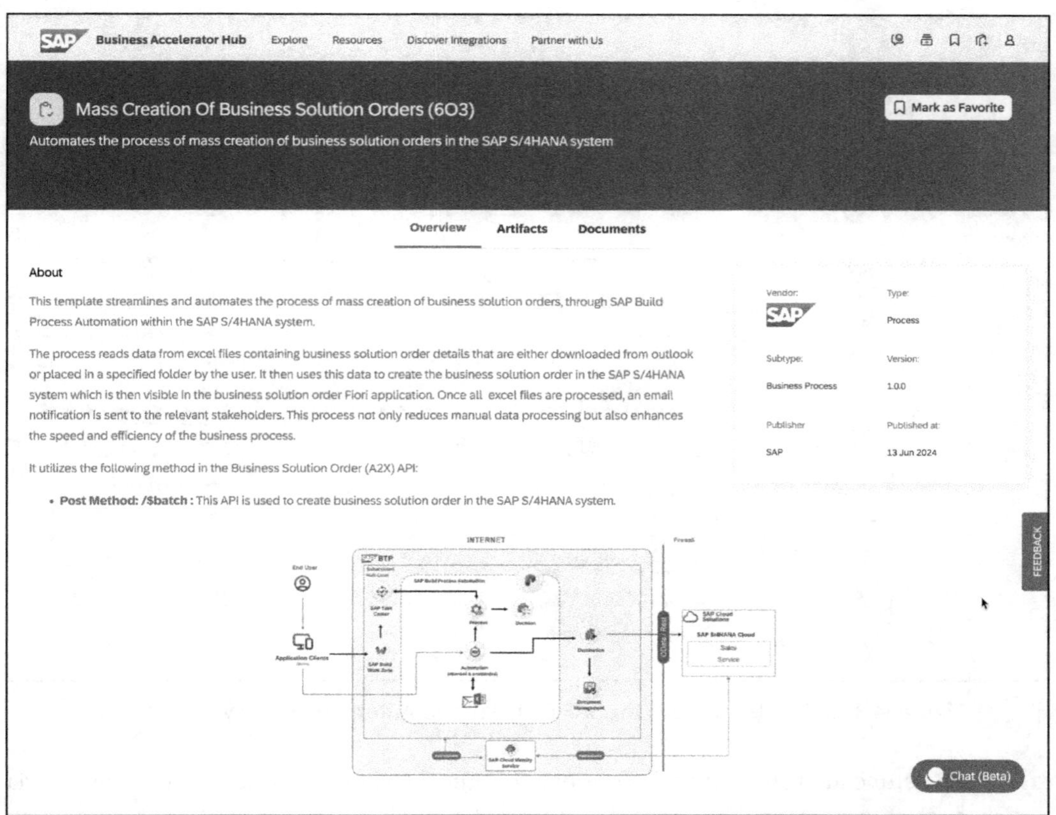

Figure 14.2 Content Package Detailed Page in SAP Business Accelerator Hub

14.1.2 Embedded Store

The *embedded store* is a cloud-based application that is available with the SAP Build Process Automation service. This embedded application is the most significant feature of the service, which provides a range of content packages curated for different needs, embedded in the design studio. While SAP Business Accelerator Hub is only for discovery of content, the embedded store gives a unified-integrated experience to discover and import the content directly into the lobby. The other difference is that you can explore the SAP Build content packages in SAP Business Accelerator Hub without the need for any service license, but you do need an SAP Build Process Automation license agreement to explore and import the content from the embedded store. The idea is to let nonlicensed customers and stakeholders also explore and experience the content packages.

The embedded store can be accessed by clicking the **Store** option from the **Lobby**, as shown in Figure 14.3.

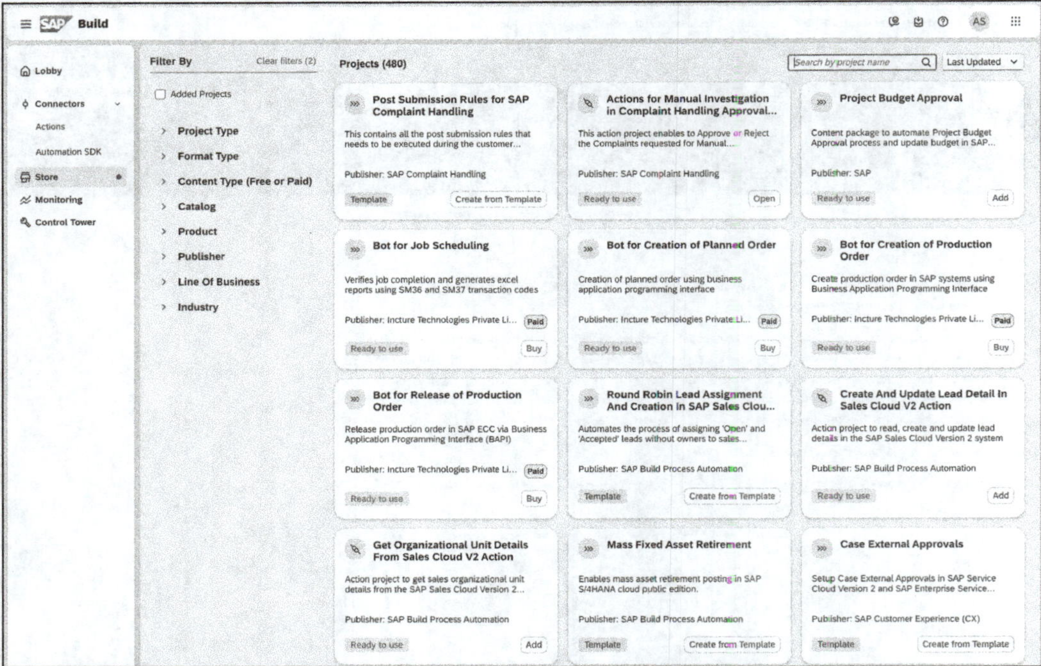

Figure 14.3 Embedded Store in the SAP Build Process Automation Lobby

The embedded store provides options to filter the content based on multiple criteria such as project type, free or paid content, LoBs, and industries as in SAP Business Accelerator Hub. Among hundreds of prebuilt content items, these filter criteria allow users to explore the right content based on their requirements and understand the benefits of the content that provides all the needed artifacts in one package.

Each content package is specially crafted to offer a suite of capabilities that work seamlessly together to realize the specific use case. For example, let's consider a use case involving the mass maintenance of billing blocks in sales orders. To build this use case, we need capabilities such as automation to fetch and update sales orders in Microsoft Excel sheets using an SAP S/4HANA API, process for approval, decision to automatically determine approvers, monitoring dashboard, and so on. The content package for this use case contains automation, data types, processes, forms, triggers, decisions, and environment variables.

Apart from discovering and viewing the content, the embedded store provides options to directly import the content in the lobby. The free content, which means the content that doesn't have any commercials attached, can be directly added in the lobby. There is also paid content published by SAP Partners or other stakeholders, which have to be first purchased from the vendor and then downloaded, as instructed by the respective vendor. Such content packages have a **Buy** option that takes you to SAP Store, which has the information regarding the content and details of the purchase.

14.1.3 SAP Store

SAP Store (*www.sap.com/store.html*) is a marketplace for a variety of solutions and applications from SAP and its partners. Customers across the world can browse the solutions based on different industries and categories. These content items can be tried for free or need to be purchased before you can use them. SAP Build–based content packages or commercialized solutions published by SAP or its trusted partners are available in SAP Store. These packages can also be explored from SAP Business Accelerator Hub and the embedded store, as shown in Figure 14.4, but they must be purchased before using them.

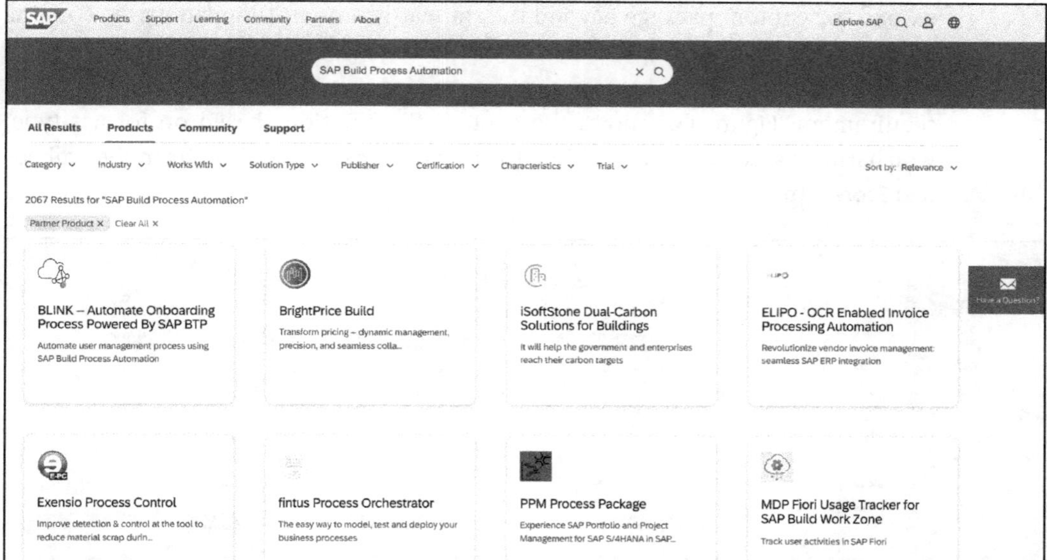

Figure 14.4 View of Content Package and Solutions of SAP Build Process Automation in SAP Store

14.2 Types of Content

While exploring any of the content stores, you'll notice different types of content categories. These categories classify the content into groups and help in searching the right content. Let's understand what these content categories and their relevance in the store.

14.2.1 Format Type

Following are different types of content format that are available in the embedded store in SAP Build:

- **Ready-to-use**
  Prebuilt content is a collection of various artifacts to provide packaged solutions for different use cases. Ready-to-use content refers to complete, out-of-box packages that are meant to be used as is with little or no configuration. You can import the content from the store, configure it based on the instructions attached to the content package, and use it immediately. Such content can't be extended or modified and is upgrade safe.
- **Template**
  Content packages are evolving and gaining market share as they significantly accelerate developer productivity and save resources. There's often a need to change or extend the content package beyond the standard content. Template-based content packages cater to these requirements. Such packages are added as new business project in the lobby and can be changed or extended completely based on the project requirement. Figure 14.5 shows the content filter options based on format type, which can be accessed by expanding the **Format Type** filter option from the embedded **Store** application.

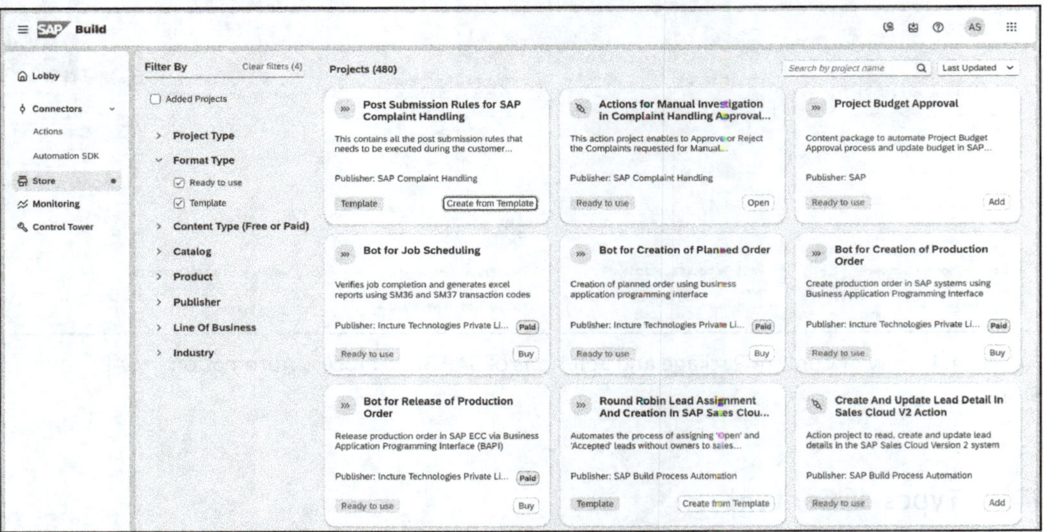

Figure 14.5 Content Filter Options Based on Format Type

> **Note**
>
> Some content packages are both template and ready-to-use. Users are given options to import them as-is or create a new project. Based on the requirements, you can import, configure, and deploy a ready-to-use version or create a new project from the template to modify it.

14.2.2 Project Type

SAP Build comprises the following solutions: SAP Build Apps, SAP Build Process Automation, SAP Build Work Zone, and SAP Build Code. The *project type* represents the type of business project based on these services. SAP Build Process Automation project types are categorized as follows:

- **Business Process**
  These template content packages are created using the SAP Build Process Automation toolkit. This type of content package is a complete set of processes, decisions, automations, data types, visibility scenarios, forms, and other artifacts that are required for the use case, along with SAP Integration Suite. They can be directly added into the lobby and configured using no-code editors for respective capabilities.
- **Task Automation**
  These are ready-to-use or template automation content packages created from the old SAP Intelligent Robotic Process Automation (SAP Intelligent RPA) service and later migrated to SAP Build Process Automation. As the name suggests, theses content packages contain only automation and supporting artifacts such as data types, Microsoft Excel template files, and so on.

Note

You may notice another type called **Live Process**. These are the ready-to-use content packages that were built using the legacy workflow management service. These content packages are also updated so they can be imported to the SAP Build Process Automation lobby, but they can be configured or extended using the old editors such as SAP Business Application Studio (for workflow changes), the manage rules editor (for business rules), and so on.

- **Actions**
  These are standard connectors to enable external connectivity to SAP and non-SAP systems or applications. For example, if you want to use actions created based on workflow APIs in your SAP Build Apps project, then you can use the **Process Actions** package. Figure 14.6 shows the content filter options based on the project type, which can be accessed via the **Project Type** filter option on the **Store** application.

Note

Besides **Process** (from SAP Build Process Automation) and **Action** project types, you'll also find many other project types based on the respective services such as **Build Code**, **Build App**, and so on.

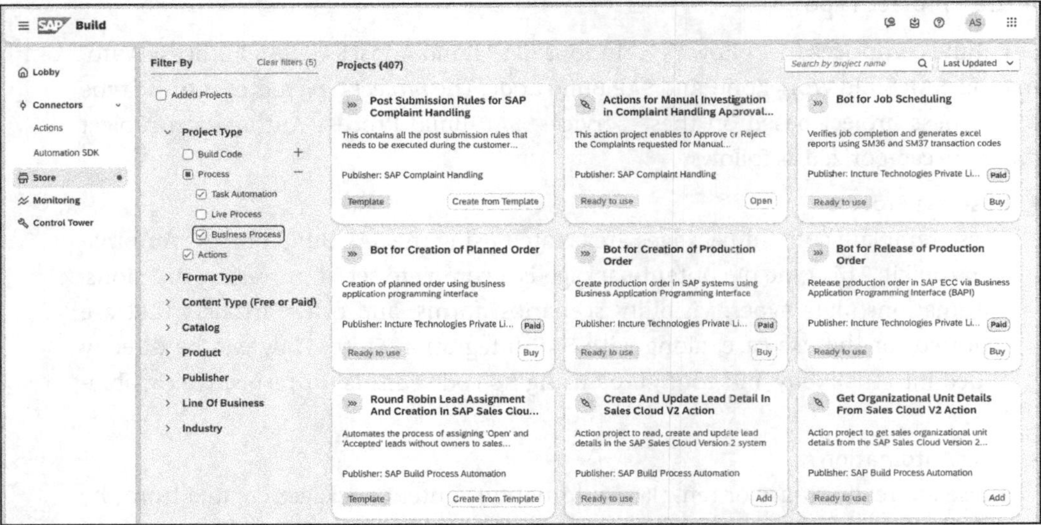

Figure 14.6 Content Filter Options Based on Project Type

14.2.3 Catalog

The **Catalog** is another categorization of content packages where the content is presented based on the usage, which is a great option for users who want to scroll through learning or productive content:

- **Business Content**
  These are enterprise-ready content packages that can be installed and used with few configurations. All the scenarios are valid identified whitespaces across SAP LoBs catering to a variety of industries. Based on the type of the business content (ready-to-use or template), they can be easily deployed with simple configurations or extended based on additional requirements.

  For example, **Journal Entry Posting Approval**, which is a template content package, can be set up and configured using the attached documentation in a few simple steps. Then, you can choose to deploy it as is or extend it with more artifacts based on your use case requirements.

  Another example is the **Automated Purchase Order Upload** ready-to-use content package, which means that you can't extend this content package. It must be configured and used as instructed in the documentation.

- **Learning Content**
  There is a whole list of learning content packages that can be used for references, learning, and demo purposes. These accelerators help train users on how to use SAP Build Process Automation features and capabilities based on common business cases and extension scenarios. Examples include how to use SAP Document

Management Service (DMS); how to use SAPUI5 applications in your processes; and best practices when using Business Application Programming Interface (BAPI), Microsoft PowerPoint, Microsoft Excel. You can select from a range of learning use cases and expand the learnings in building real use cases for your organization.

Figure 14.7 shows the content filter options based on catalog types, which can be accessed from the **Catalog** filter option in the **Store** application.

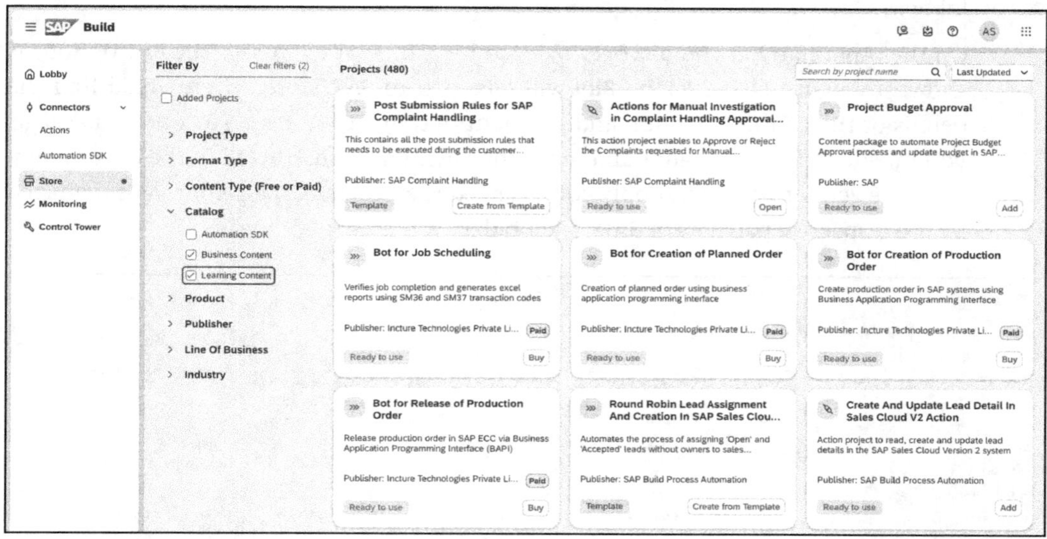

Figure 14.7 Content Filter Options Based on Catalog Types

Note

Automation software development kits (ADKs) are also shown in the **Catalog** type. This is a list of functions delivered by SAP to be used in the automation capability of SAP Build Process Automation. They facilitate the developer to create automations and work with different frameworks such as Microsoft Excel with the Excel SDK, send emails via Outlook using the Outlook SDK, use Google Documents for storing information using the Google Document AI SDK, and so on. These SDKs aren't meant to be imported. To use an SDK, you have to add it as a dependency in your project, and then consume the respective functions in the automation.

14.2.4 Free or Paid

SAP Build provides both free and paid content. Content published by SAP is mostly free but requires an SAP Build Process Automation service subscription to use them. Paid content is created by SAP and its trusted partners, which are available in SAP Store and SAP Business Accelerator Hub.

When using free content, you just go to the embedded store and add the content into the lobby. For instance, you want to use the **Mass Fixed Asset Retirement** content package. Navigate to the embedded store application, select the **Content Type** filter option, and then select **Free**. Search for "Mass Fixed Asset Retirement" content package, select the content package tile to get an overview of the content package, read the documentation to ensure that you've done all the necessary setup, and then click on **Create from Template option on the overview page**. This will import the content directly into the lobby.

On the other hand, if you discover paid, content then you'll notice the **Buy** option, as shown in Figure 14.8. Clicking **Buy** will take you to SAP Store, where you'll find the details of the content, vendor, and payment options. To use such a content package, you must contact the vendor and make the payment as instructed in the documentation. The vendor will then supply you the needed artifacts. All paid content is managed by the respective partners or SAP stakeholders.

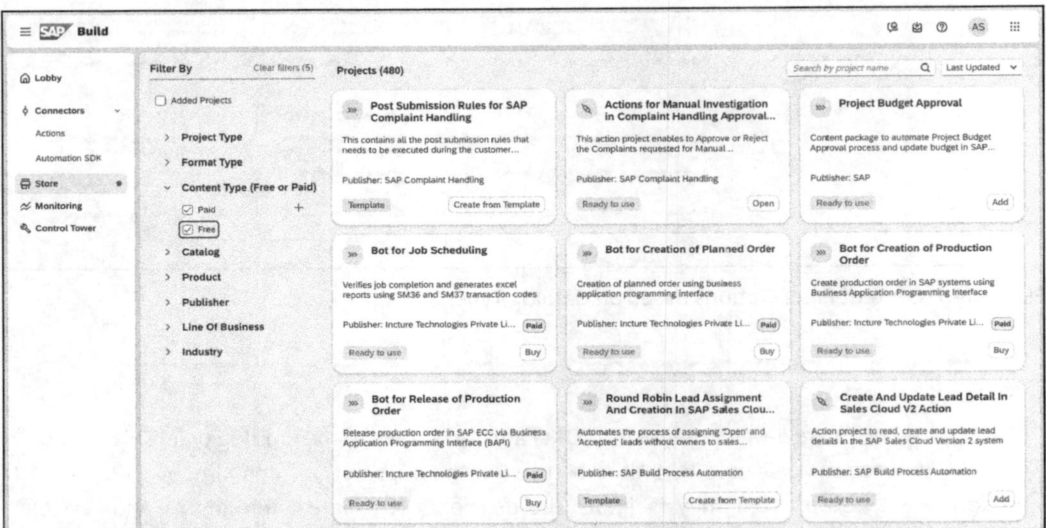

Figure 14.8 Content Filter Options Based on Commercialization

14.3 Discovering and Working with Content

In this section, you'll learn how to discover and explore the content from the stores, how to filter and search to find the right content, how to import the prebuilt content in the lobby, and how to configure, extend, and consume the content for your requirements.

14.3.1 How to Find the Right Content

Prebuilt content helps organizations accelerate their development by saving resources, saving time to market with a modularized approach to the problem, and reducing waste with less coding and testing efforts. Often customers aren't able to find the right content and leverage the benefits of these enterprise-ready content packages that align with SAP best practices and give a full scope of customization and extension.

The first step toward finding the right content is knowing the need or problem that you're trying to solve. Then, you can think strategically about what type of content will help you achieve those needs while also resonating with your organization.

Next, start discovering the content via the SAP Build embedded store or the SAP Business Accelerator Hub store. The experience in both stores is same.

> **Note**
>
> The SAP Business Accelerator Hub store is free to explore, which means you can navigate through the hub without the need for an SAP Build or SAP Build Process Automation service license. For the embedded store option, however, you'll need an SAP Build or SAP Build Process Automation service subscription.

Let's see how to discover the content in SAP Business Accelerator Hub; the same steps will apply for the embedded store as well.

Open the SAP Build content domain (*https://hub.sap.com/build*) in SAP Business Accelerator Hub. In the left panel, you'll see search and filter options. Using these options, you can filter the content or directly enter the search term to find the content you're looking for, as shown in Figure 14.9.

For example, let's say you want to search for procurement content, which is related to purchase orders. To do so, expand the **Lines Of Business** item on the left, and select the **Sourcing and Procurement** checkbox. This will show you the list of related content. This is a type of search where you know exactly what your requirement is.

Another example is when you have a rough idea and want to explore all the content related. In that case, you can directly enter your search terms in the search box (e.g., "fixed asset" or "purchase orders"), and you'll be presented with all the content that matches those terms. To refine the search, keep adding more filter criteria or search terms, as shown in Figure 14.10.

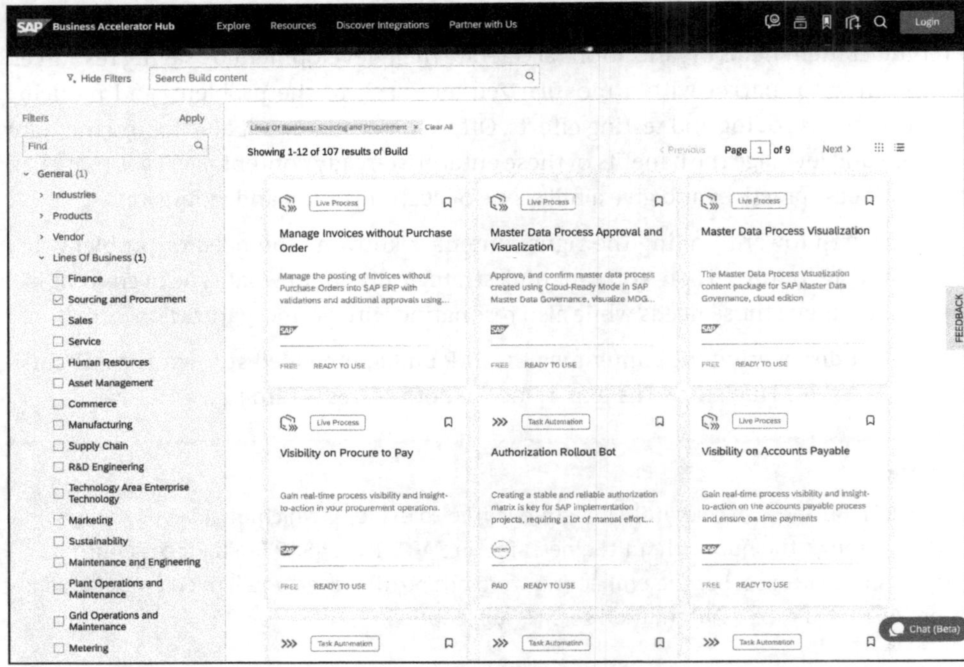

Figure 14.9 Filtering the Content Based on a Specific Line of Business

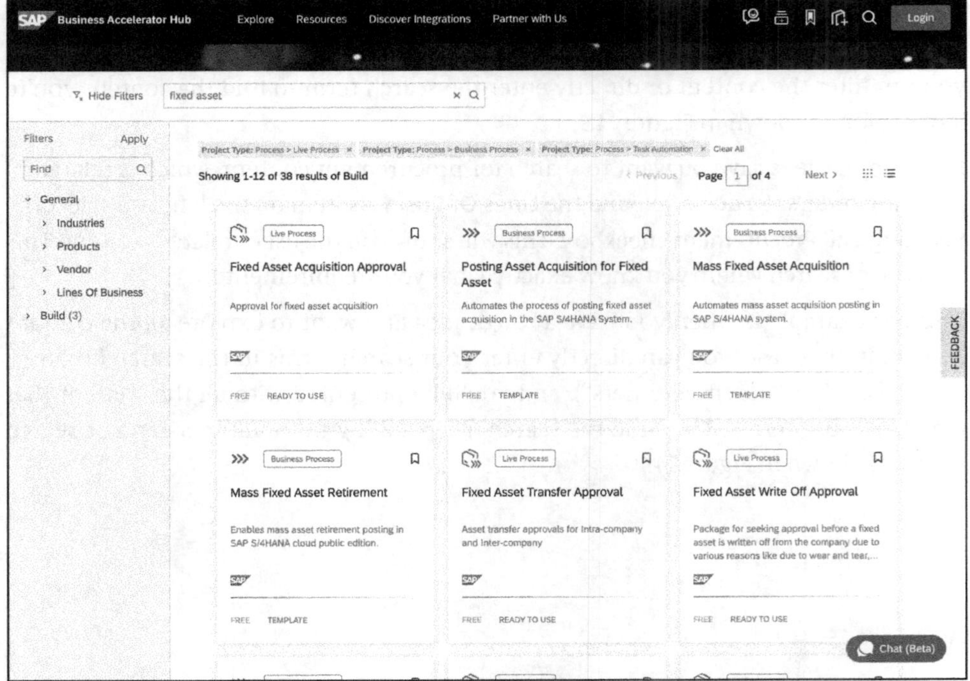

Figure 14.10 Filtering the Content Based on Search Terms

Each content tile in the search or filter results has basic properties such as the content name, short description, content type, free or paid content, and content publisher. This gives you a quick view of the content. To explore the content in detail, click on the tile, and you'll be led to complete information about the content package, as follows:

- **Overview**
  In this area, as shown in Figure 14.11, you'll find the long description of the content, which explains what this content does, its value proposition, and how it extends the standard LoB solution. A solution diagram is also included that explains the interactions of different components and services used in building the content.

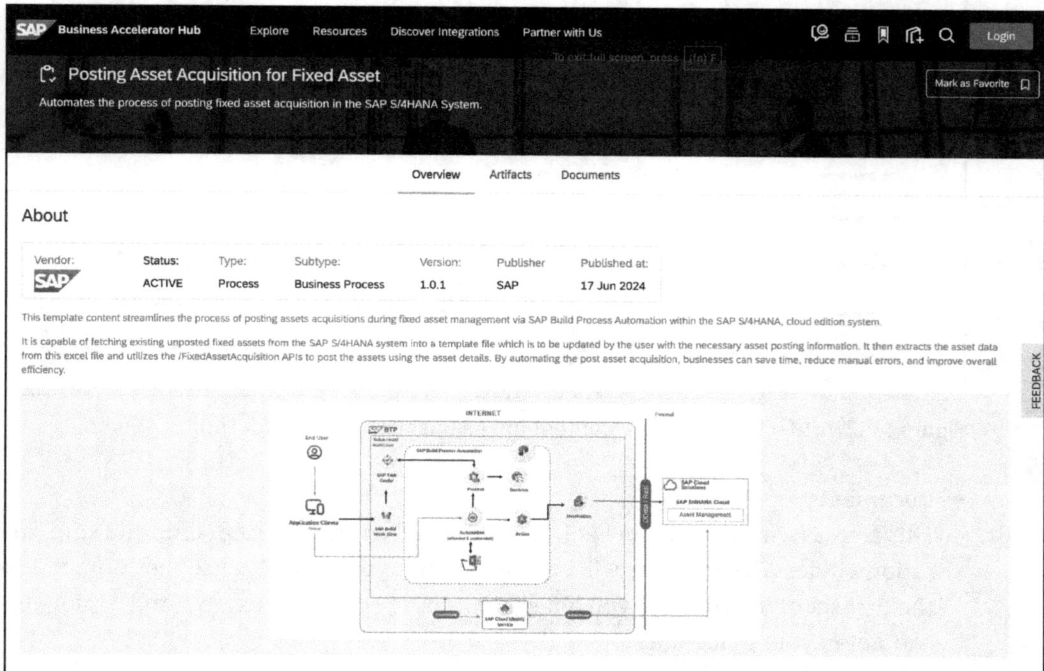

Figure 14.11 Overview Page of the Content in SAP Business Accelerator Hub

- **Artifacts**
  This area, as shown in Figure 14.12, takes you to a complete list of prebuilt artifacts (with short descriptions) that are bundled with the content package. A content package contains a range of artifacts that are required to build the use case, such as processes, decisions, automations, data types, forms, a visibility dashboard, SAPUI5 applications, iFlows, other files, and so on. These are the artifacts that will be configured or extended to consume the content.

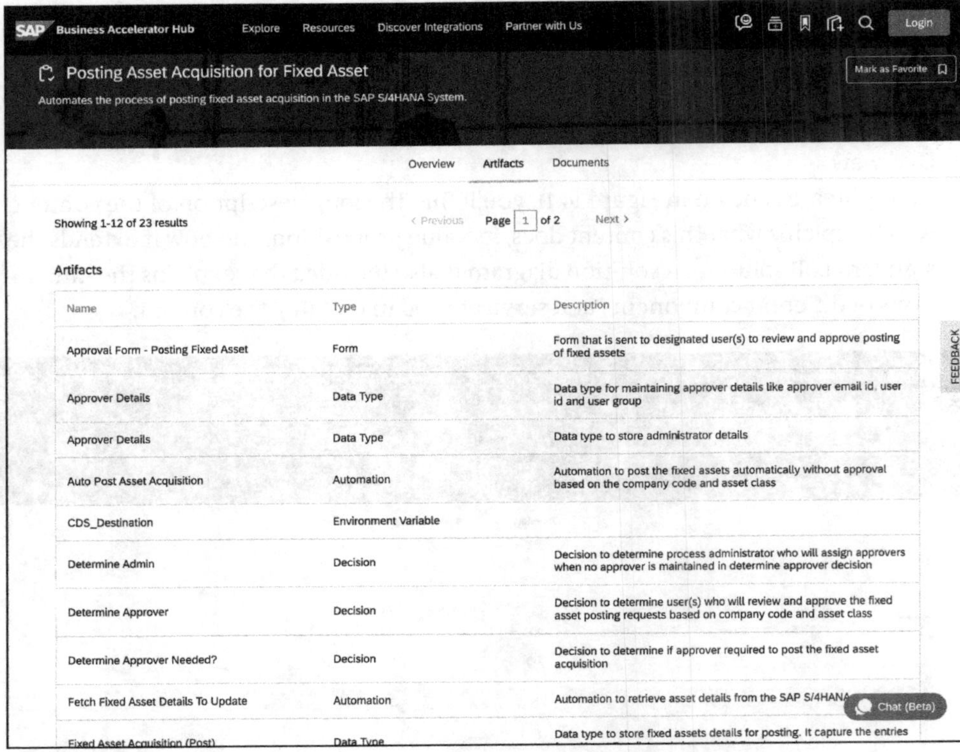

Figure 14.12 Artifacts View of the Content in SAP Business Accelerator Hub

- **Documents**
  This area, as shown in Figure 14.13, provides detailed step-by-step setup and configuration guides. These guides will be needed once you decide to use the content. It's the first set of instructions you'll need to import, configure, extend, and deploy the content in your landscape.

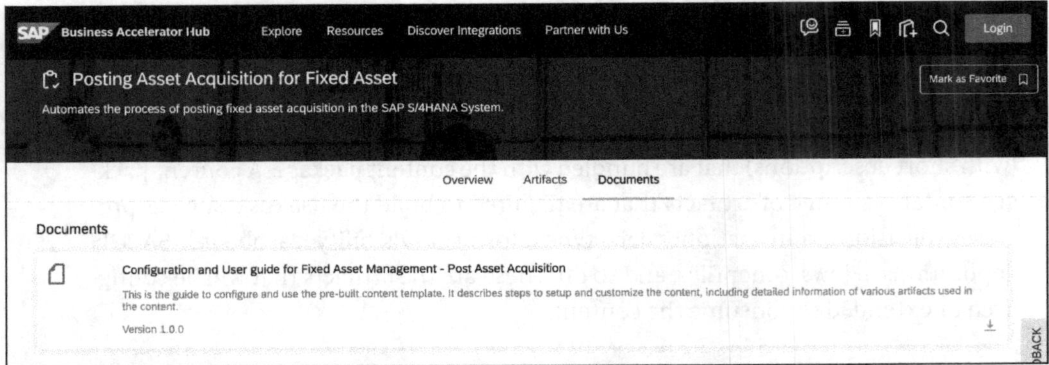

Figure 14.13 Documents Area of the Content Package in SAP Business Accelerator Hub

14.3.2 How to Import and Use Content

Once you've identified the right content and decided which prebuilt content package you want to use, then you prepare for setting up the content. Prebuilt content items are very easy to set up. The first step is to open the setup guide and follow the prerequisites to ensure that you have the right platform and service installations needed for the content

After you've installed the SAP Build Process Automation service, follow these steps:

1. Open the SAP Build Process Automation application, which will take you to the **Lobby**.
2. From the **Lobby**, click the **Store** application from the left panel.
3. In the embedded store that opens on the right, search for the content you want to import using various filters such as **Project Type**, **Catalog**, **Content Type**, and so on.
4. From the filtered or searched content package tile, click the content tile you want to explore.

You'll notice button options such as **Create from Template** and **Add** (sometimes, both options appear). Choose from these options based on how you want to use the content:

- **Create from Template**
  This button is shown for content packages that are meant to be extended, as shown in Figure 14.14. A new business project will be created from the template content, and you get the full scope for any customization or extension.

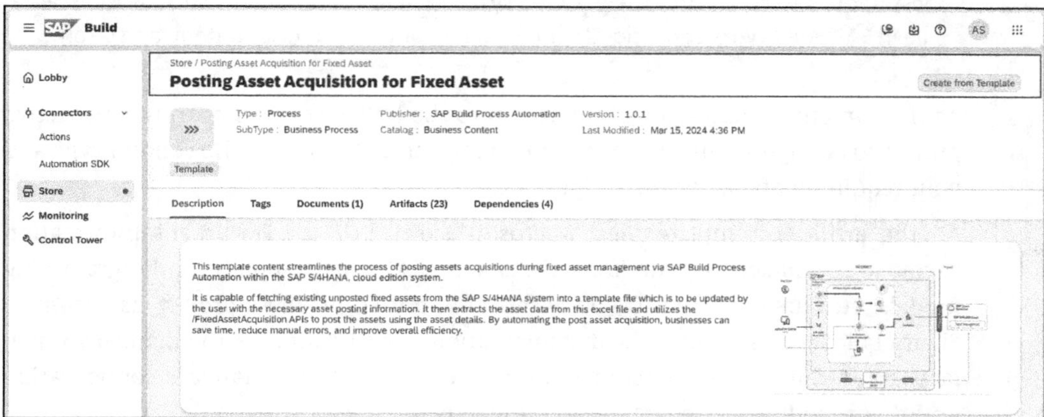

Figure 14.14 Importing the Content in the Lobby via the Create from Template Option

- **Add**
  This button is shown for ready-to-use content, which means these content packages are imported as is with only minimum configurations allowed (see Figure 14.15).

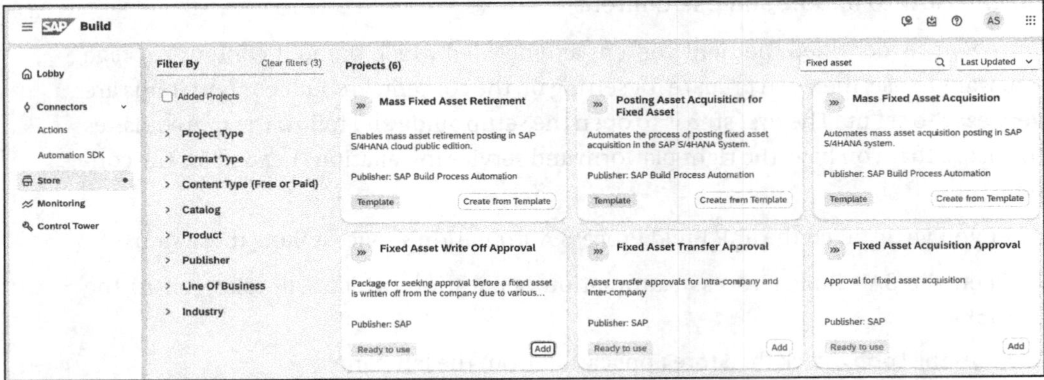

Figure 14.15 Ready-to-Use Content with the Add Option to Import the Content into the Lobby

> **Note**
>
> A few content items will have both options, which means that they can used as ready-to-use content or template content based on your requirements.

> **Note**
>
> The setup and configuration guide (attached with each content item) must be followed to ensure that the content is setup with all the needed configurations such as destinations to connect to the backend system, SAPUI5 applications such as the **Start** tile in SAP Build Work Zone, cloud connectors for any on-premise content, and so on.

As the content is successfully imported into the lobby, you can open the respective project to configure, customize, or extend the content based on the content type and your requirements, as shown in Figure 14.16.

All the artifacts, template files (Microsoft Excel, PDF, or Document Information Extraction), and source code (for SAPUI5 apps, SAP Cloud Application Programming models, etc.) are provided to you with the imported business project, as shown in Figure 14.17. You can follow the documentation to first configure the mandatory artifacts, as explained in the documentation, and use it with the standard configuration options or modify it based on the additional requirements.

Prebuilt content helps you develop the end-to-end solutions with ease. It's available in various categories and can be configured and extended without knowing the complex IT technologies through the no-code tools. Special care has been taken to ensure that the content packages can be installed directly from the embedded workbench giving a unified experience in various phases of solution development. It's always recommended to find appropriate content to kick-start the development rather than starting from scratch, which may lead to the need for more skilled developers and higher costs.

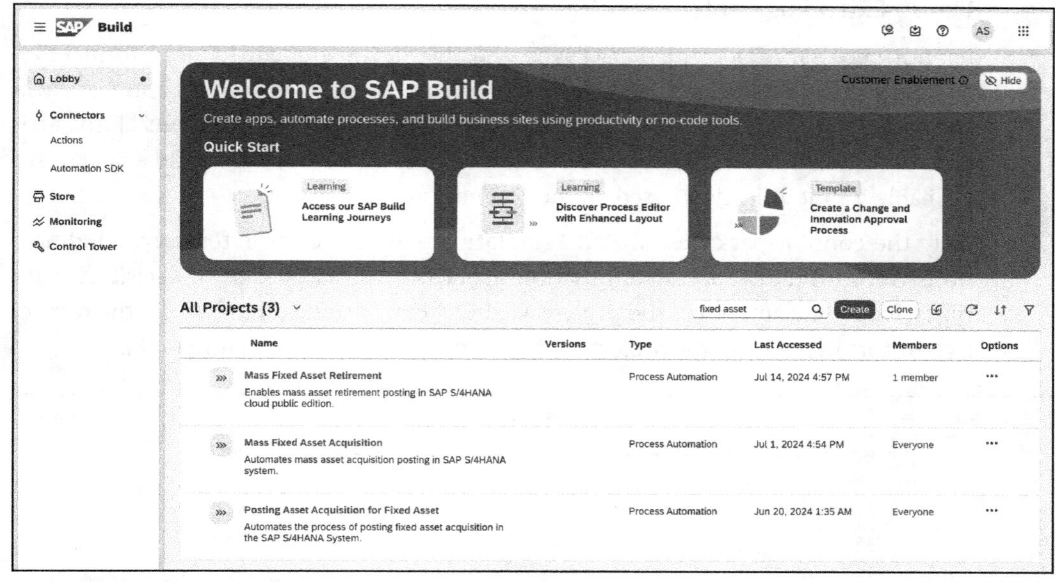

Figure 14.16 Content Successfully Imported into the Lobby

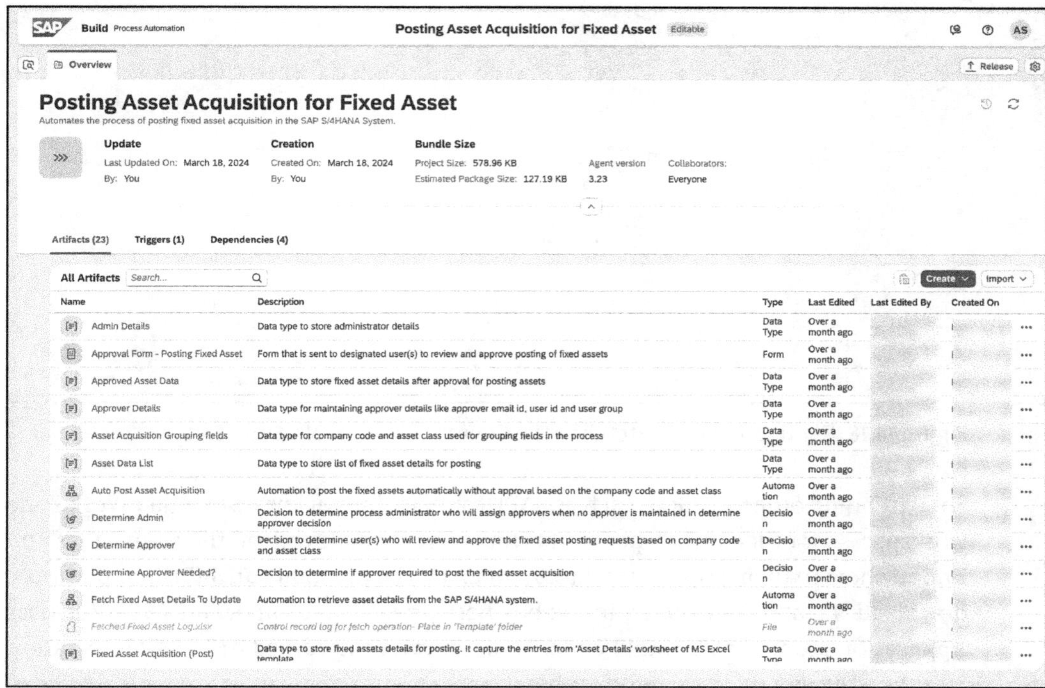

Figure 14.17 Overview of the Business Project Created after Importing the Content

14.3.3 Extending Content

It's not necessary for content packages available in the embedded store to meet all business needs. In such scenarios, you can extend the existing prebuilt content package based on the additional requirements. Each content package contains all the artifacts and their source files. As you add the content in the lobby, all the artifacts are available for modification based on the content category.

Only the content packages labeled **Template** are available for extensions. Modifications to SAP Build Process Automation artifacts such as processes, decisions, data types, forms, automations, and process visibility can be done directly from the respective editors in the business project of the imported content (see Figure 14.18).

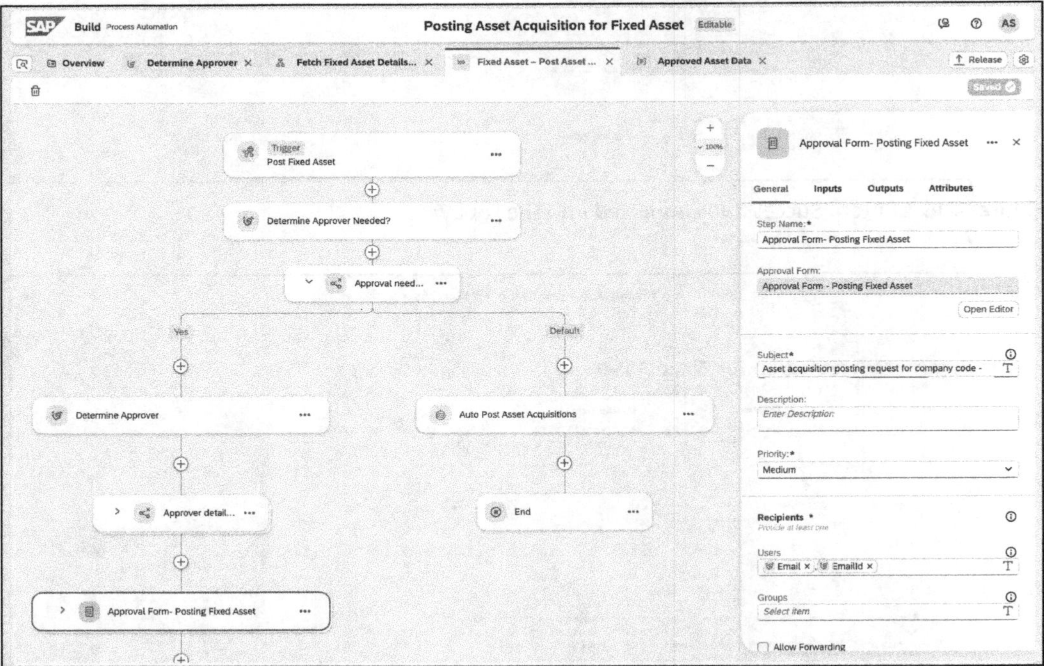

Figure 14.18 Configuring and Extending the Content from the No-Code Editor

For extending other artifacts, such as SAPUI applications, iFlows, SAP Build Work Zone workspaces, or other such dependent artifacts, you can download the source files and import them in their respective editors. For example: to modify an SAPUI5 application, download the attached zip file in the business project, and import it in SAP Business Application Studio or SAP Build Code. An iFlow can be modified from SAP Integration Suite and the iFlow editor, among others.

You can not only modify existing artifacts but also add new artifacts and dependencies that are needed to solve the business case. Once the content is extended or modified, you can release and deploy the project and consume it without having to create it from scratch. With this flexibility, developers can now deliver end-to-end solutions through prebuilt out-of-the-box content by adapting them with custom implementations based on the unique requirements of their projects.

14.4 Creating and Publishing Content to the Store

Until now, we focused on consumers who want to explore and use the content packages. Now let's talk about the journey of SAP Partners and how they can build content packages and publish them in the SAP Store, together with SAP Business Accelerator Hub and the embedded store. The SAP Partner ecosystem contributions to SAP Build content store for a wide range of industry use cases is important to expand the opportunities and scale the content adoption.

SAP Build powered by SAP BTP provides low-code/no-code and secure platform that allows users to create content using their specialized expertise and experience with greater market reach to meet their client demands.

If you're wondering why you should consider publishing SAP Build Process Automation prebuilt content, here are some of the reasons:

- **Simplified development experience**
  Partners can build accelerator content at reduced cost with the Test, Demo, and Development (TDD) license available for SAP Build Process Automation. The simplified developer experience with a drag-and-drop approach gives the needed flexibility and agility to create template content.
- **Co-create business value with best practices**
  Collaborating and co-designing the solutions for common business problem based on SAP's best practice guidelines and having a deeper insight into customer demands can generate immense value for customers. Easier consumption and configuration models, options to bundle other SAP BTP services (e.g., SAPUI5, SAP Cloud Application Programming Model, SAP Business AI, etc.), and extension options can derive innovative solutions.
- **Commercialize solutions**
  Monetization is one of the most attractive strategies to create and sell content packages on the SAP platform. Partners can create tailor-made content packages to cater to their customer's business problems. They can then choose to sell the content by publishing it in SAP Store so that many other customers and stakeholders benefit from it. Build once and sell many is the key benefit of commercialization and can drive new revenue and opportunities.
- **Scale demand generation**
  When you host a content package in SAP Store, you can greatly increase your reach. It gives you access to attract thousands of SAP customers and leverage the digital

SAP marketplace to scale demand generation. Additionally you can also get the content certified by SAP to generate more traffic to your content. There is no better direct way to reach out to the market to establish your brand and start monetizing.

As such, content created using the SAP Build Process Automation tools can earn you benefits when published in SAP Store. Publishing the content in SAP Store is a simple process, as shown in Figure 14.19, where the partner first needs to have a valid SAP PartnerEdge agreement, SAP Store participation agreement, and TDD license. Once the prerequisites are met and your content is created, you can follow the guidance on SAP Solution Hub to publish the content in the store. When you publish the packaged content from SAP Solution Hub, it's made available in all the stores: SAP Store, SAP Business Accelerator Hub, and the embedded store in SAP Build After the review of your content by the store moderators, the content is approved and published in the store. This may take a couple of days depending on the content quality, but you'll be notified with a direct link to the store once the content is successfully available in SAP Store.

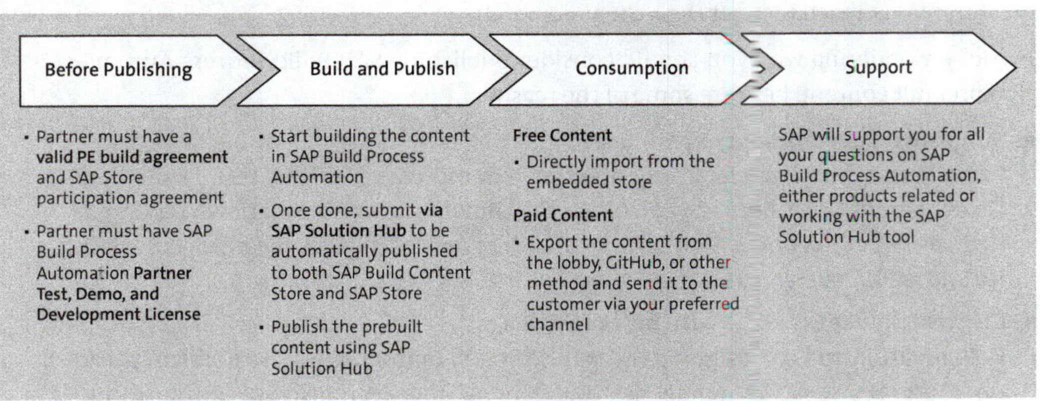

Figure 14.19 Partner Publishing Journey Overview

Partners can publish both reference and paid content, as shown in Figure 14.20. Usage of the content is decided based on the content type. For free or reference content, it can be directly download from the store and used by the consumers, but for paid content, the consumers approach the partner to make payment and clarify any queries, and then the partner provides them the content package to be downloaded via their preferred channel. For example, a partner can upload the content in any document repository and share it with the customer provide a setup service as part of the content contract, give access to their onboarding portal with documentation, and so on.

This is how SAP Build Process Automation content packages are created, published, and monetized by SAP Partners. We've touched on the basics of content monetization, and how it can benefit you. For more detailed instructions and discussions, you may reach out to your partner manager from SAP.

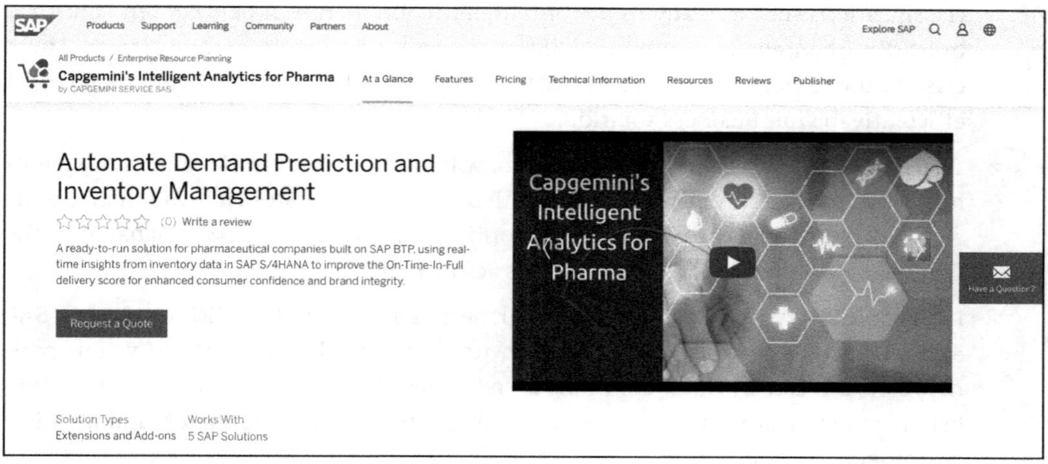

Figure 14.20 Partner Content in SAP Store with Monetization

If you're just getting started, you can visit SAP Store and explore various free and paid content. Gather inspiration, start your journey with either type of content, and create value for your customers and the ecosystem.

If you're planning to start implementation on any new processes or want to upgrade existing processes in finance, procurement, human resources, sales, asset management, manufacturing, and so on, then SAP recommends exploring various free and paid industry-based prebuilt content from SAP Business Accelerator Hub and SAP Store, respectively. Download the content from the SAP Build Process Automation lobby, and then configure, extend, and integrate the content based on the use case. This way, you save time and resources, gain the flexibility to adapt changes, get built-in SAP best practice models, and have a no-code user experience as well.

14.5 Summary

This chapter started with an introduction to the industry-based prebuilt content of SAP Build Process Automation, which is used to accelerate solution development by technical or nontechnical users of IT departments. SAP Build Process Automation has published multiple content items across different industries and LoBs. The chapter then progressed to introduce SAP Business Accelerator Hub, the embedded store, and SAP Store as means to discover, import, and publish content packages, for both SAP stakeholders and partners. You learned how to explore the content in SAP Business Accelerator Hub using the **Overview**, **Artifacts**, and **Documents** areas; how to import the content packages in SAP Build Process Automation lobby using the embedded store; and how to commercialize the content via SAP Store. Each content package has various classifications such as ready-to-use or template content, free or paid content, business or learning content, and content based on processes, automations, and actions. These classifications were explained in detail to help you discover the right content and use it efficiently in your business solution.

There are often questions on where and how to find the content. Section 14.3 explains how to search and filter the content in SAP Business Accelerator Hub and then use the embedded store in SAP Build to import, configure, and extend the content based on the unique requirements from organizations across regions.

In Section 14.4, you learned how SAP Partners can build and publish content in SAP Store. This content can be free of cost or with commercials. With this ecosystem, partners can co-create business solutions based on SAP Build Process Automation to publish them in SAP Store, leverage commercial benefits, and expand their brand with SAP customers.

The next chapter will take you on a journey of extending standard business processes of SAP. It will help you understand various standard processes across SAP's LoB solutions and how they can be extended using SAP Build Process Automation.

Chapter 15
Extending Standard Processes

In this chapter, we'll dive into the world of process extensions. You'll see why there is sometimes a need to extend standard processes running in your SAP applications, get an overview of different SAP process extension possibilities, learn what process variants are, and explore some example extension scenarios.

As mentioned in Chapter 3, business processes are the lifeline of every organization. All the activities and tasks in everyone's business day belongs to a process. Well, at least the majority of them do. Often, we don't even realize it during the actual doing, but others do. The process owners know, for example, because they need to ensure that at the end of this predefined set and flow of tasks, the desired result will be achieved, also known as the business outcome.

Going back in history, when the first attempts took place to support these tasks (aka business processes) with software, it was quite specific and very much tailored to the single use case and situation at the organization. There wasn't a kind of standard software available that organizations could buy, customize to their needs, and—voilà—have the process up and running. Instead, it was an individual program, specific implementation, and each time it was done from scratch again. As you can imagine, this is quite cumbersome, a lot of effort, invites failure, and is costly. Couldn't it be done better?

Five IBM employees also had this question in the 1970s and came up with an answer. You might have heard of them: Dietmar Hopp, Hasso Plattner, Claus Wellenreuther, Klaus Tschira, and Hans-Werner Hector. They are now known as the founders of SAP. On April 1, 1972, they founded the System Analyse Programmentwicklung (System Analysis Program Development) company with this revolutionary idea in mind: create standard enterprise software that integrates all business processes and enables data processing in real time, that is, a *standard* for all *business processes* in *real time*.

Starting with R/1, the first financial accounting system, evolving to R/2 in 1981, followed by R/3 in 1992 and with the innovation of the in-memory database SAP HANA in 2011, it was just a matter of time to release SAP S/4HANA in 2015 and start the journey with SAP customers into the cloud. Standardized software could enable every enterprise and organization to become intelligent, networked, and sustainable while bringing together the needed solutions, the latest technology, and best practices from diverse industries and lines of business (LoBs) to run any business process, integrated and digitalized in the cloud.

Although this could be a great and perfect ending of a book or a Disney movie, in reality, there still might be some challenges. Probably not everything is covered via the standard business applications, supporting the standard business processes. The following sections acknowledge this fact and discuss the need and possibilities to do business process extensions. We'll end the chapter with two techniques to take care of process extensions, with process variants, for example, or how to realize extensibility in SAP S/4HANA.

15.1 Standard SAP Processes and the Need for Process Extensions

The answer the SAP founders gave to the market was and still is revolutionary: a solution with standard functionality to support business processes in any industry or LoB while including best practices of all the decades of learnings with SAP customers and also providing the flexibility to either configure these processes and systems to meet business requirements or even extend that standard functionality in a smooth way. In short, that's the secret of the success of the SAP business applications. But where will you find these standard processes, and why are extensions needed? We'll explore this in the following sections.

15.1.1 Standard Business Processes

Standard business processes refer to a set of preconfigured workflows and functionalities, translated into transactions in a standard business application, such as an enterprise resource planning (ERP) system. They cover the aspects of recruit to retire, source to pay, design to operate, or lead to cash. The main goal is to ensure the smooth execution of these transactions, supporting the most common business operations. For a better understanding of this, we've listed a few examples of the standard business processes supported by SAP S/4HANA Cloud (this isn't a complete list, as that would deserve a book on its own):

- **Finance**
  Accounting and financial close, accounts payable, accounts receivable, asset accounting, contract accounting, customer payments, financial planning and analysis, general ledger allocation cycle
- **Asset management**
  Maintenance management, resource scheduling
- **Human resources**
  Time recording, employee integration

- **Manufacturing**
  Emissions management, planning to fulfillment, just-in-time supply to customers, make-to-order production, make-to-stock production, material replenishment, quality management
- **R&D/engineering**
  Project control
- **Sales**
  Complaint handling, sales order creation, customer returns, external billing, free-of-charge delivery, invoice correction, sales quotation
- **Service**
  In-house repair, service contract management, service quotation, warranty claim management
- **Sourcing and procurement**
  Central purchasing, central requisitioning, invoice processing, procurement of services, returns from store, supplier classification
- **Supply chain**
  Core inventory management, external transportation planning, retail logistics, return to supplier, transportation execution, warehouse physical inventory

Additionally, SAP delivers specific solutions for LoBs, such as human capital management (SAP SuccessFactors), spend management and business network, supply chain management, and customer relationship management (see Figure 15.1).

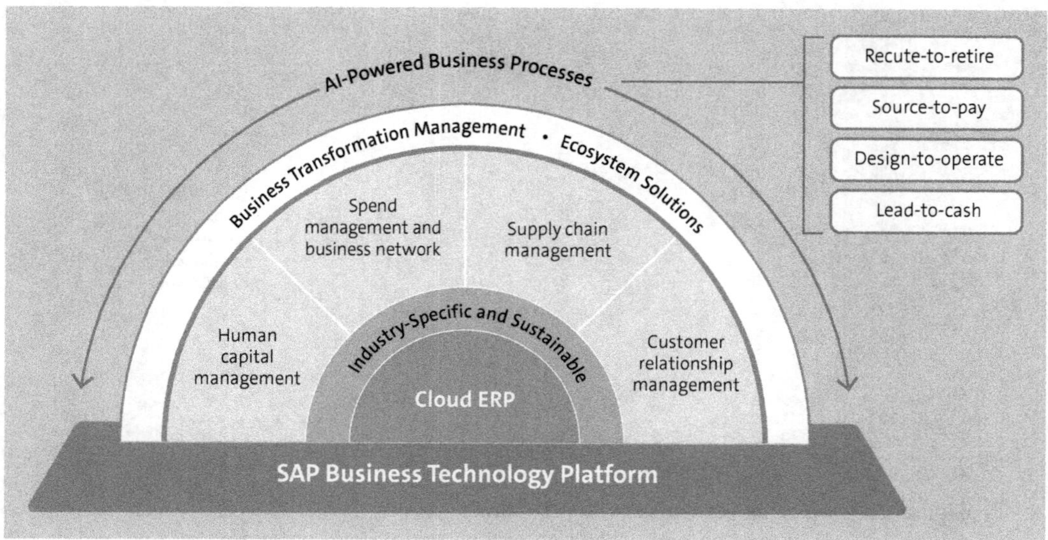

Figure 15.1 Cloud ERP Surrounded by Industry-Specific and LoB Solutions

SAP also provides industry-specific solutions, for example, automotive, banking, chemicals, government, insurance, mining, retail, and so on. SAP supports standard business processes that can be customized.

The covered business processes can be configured along best practices, based on the tremendous business knowledge SAP has gained through the past decades. You'll find the most efficient, robust, consistent, and compliant (industry or LoB) way to get support for the business processes in your organization.

15.1.2 The Need for Process Extensions

Although so many business processes, industries, and LoBs are covered, sometimes it's not enough. Additional flexibility is required also due to some limitations that might be revealed while using standard business processes. In rare cases, there could be very specific requirements in an industry that aren't covered yet or evolving business models that aren't reflected. Sometimes, unexpected and rather sudden new regulations and modifications in laws, particularly in finance, data privacy, or trade regulations, call for changes.

To bring in a more positive note, there are also conditions that call for extensions of the standard (we touched on them slightly in Chapter 2). For example, you may want to run an innovative enterprise and differentiate yourself from your competitors, as shown in Figure 15.2. Having the flexibility and agility to do so is key and must be considered when it comes to the execution of business processes.

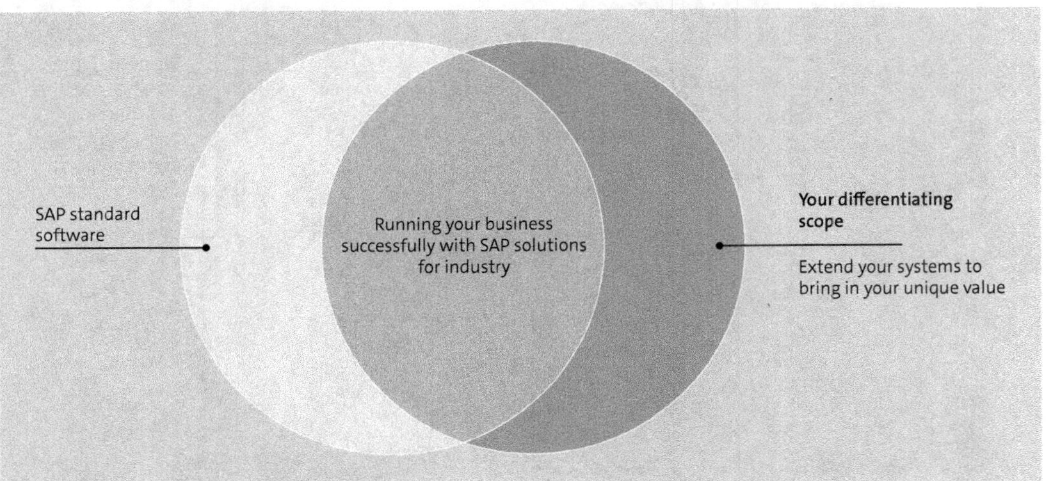

Figure 15.2 Balancing Innovation with Standard

Of course, you don't want to reinvent the wheel every day, and sometimes, you'll just be happy to improve the performance of your invoice approval process, for example, to get the invoices clearly and directly assigned to the ordered goods or services and have them paid in time, to benefit from any discounts, or just to avoid costly dunning procedures. This is not only costly monetarily but also can cost your reputation. Remember Chapter 1, where we discussed the topic of enterprise automation, using SAP Signavio solutions to figure out where processes need to be improved and using SAP Build to implement these improvements. This could either be done by creating extensions of certain business processes or by automating dedicated tasks within the standard process itself.

You might also want to streamline the user experience (UX) or include specific corporate design guidelines across all user interfaces (UIs). Although the business applications provide capabilities to adapt them, the usage of an external UI development environment might bring the desired result. Again, SAP Build, as part of SAP Business Technology Platform (SAP BTP), is useful here. In Chapter 1, called SAP Build *the* extension solution.

Another argument for extensions is to keep the core of your business applications clean. In fact, it should also be the goal of every extension project to keep the core clean, meaning untouched or unchanged. In other words, this means keeping the core of the standard business application as close to the standard as possible, while extending the functional scope based on the business requirements. With "clean core," you'll follow a concept and strategy to extend cloud systems, particularly SAP S/4HANA Cloud systems in a safe way, especially to ensure that any upgrade can be done smoothly. Therefore, you can make use of two extensibility options: on-stack extensibility or side-by-side extensibility. Side-by-side extensibility is done with SAP BTP. We'll explore the differences later in Section 15.2.

Let's revisit another part of enterprise automation, which is also a push to extend standard business processes: the integration with third-party systems or other SAP systems. Imagine a business process spanning different LoBs or even different organizations or enterprises. Of course, this can't be handled completely in the standard business process, as it's normally also not in their scope. Regarding scope, all software is delivered with a certain (functional) scope. If you want to go beyond this scope, you're doing an extension. Integration in this case could also mean you want to orchestrate the activities covered within the business processes. Or, you want to include other stakeholders in the process execution who are probably not even part of your organization, don't have access to the standard business application, and should never have access. Rather, you want to include them for specific tasks, notifications, approvals, and so on. This also leads to an extension of the standard process, as you wouldn't share the login credentials to your business-critical ERP system with every customer or supplier, right?

15.1.3 Risks of Extensions

Extensions are making a lot of sense and also provide several benefits, but there are also some risks, which we don't want to neglect. More extensions lead to more complexity that you need to handle by using the right business technology platform that is best suited for extensions in SAP environments. If you're unsure, please read Chapter 2 again.

Extensions also include additional costs for the development, testing, and maintenance that you need to be aware of. On the other hand, the needed extensions will also provide concrete benefits for your business. Just think about the example of the properly managed invoice approval process mentioned before, with using discounts and having happy suppliers. Make sure you've chosen the extension platform that helps you reduce this total cost of ownership and provide smart ways to accelerate the development of process extensions. Again, refer to Chapter 2.

Taking care of security aspects, user management, testing, governance, lifecycle management, and seamless updates of your core business applications also needs to be considered. But you already know which chapter to revisit for this (Chapter 2!).

15.1.4 Benefits of Extensions

Process extensions help improve your operational efficiency because you can digitalize and or even automate business processes end to end. No manual steps occur in between, so you can handle every connection from one to the other system elegantly. Having processes that are really tailored to business needs will allow faster decision-making and more streamlined operations.

Fully digitalized, improved, and automated business processes, meaning the standard and their extensions, will lead to an improved UX. With the right toolset, it will even be seamless and delightful for your business user. Get them an intuitive experience and only include them wherever it's necessary; let a bot do the more tedious work.

A stable core and a rich set of standard business processes are advantages on their own. But we've learned now that process extensions are required here and there. And at the end of the day, they provide the required scalability and flexibility to grow your business, adapt faster on changing business models, and to not only react to changing business requirements but to spearhead them.

Transfer your innovative ideas as fast as possible into your business processes. Don't let the standard hinder you from becoming the intelligent enterprise you want to be. If you do your process extensions right, you'll achieve a faster time to market. You can use the process extensions not only as an option to enhance the standard but to turn it into a competitive advantage to outperform.

Now that you have a better understanding of what standard business processes are and why there's a need for process extensions, it's time to explore the different process extension possibilities. We'll also explain on-stack and side-by-side extensibility.

15.2 Different SAP Process Extension Possibilities

Let's now focus on how to ensure the unique requirements are fulfilled that can't be met by a standard business process. We need an extension. In general, you have two options if you want to build an extension for SAP S/4HANA:

- On-stack extension
- Side-by-side extension

We'll discuss these two options, their differences, how they relate to your business process, and which approach you should choose in the following subsections, followed by a real-life example.

15.2.1 On-Stack Extensions

On-stack extensions are tightly coupled to data, transactions, and applications. You just leverage the tooling integrated in the business application to do the extension, while the extension is technically running on the same stack. Examples are to adapt a standard UI, add a custom field to a standard UI, or adapt a standard business process, for example, with one additional approval step. Here, this is *key user extensibility*, which requires an understanding of the business without the need for too many technical details. We've mentioned this before as low-code or no-code development. On the other side, you would create a custom, full-stack application, which is called *developer extensibility* and pro-code development.

But let's focus on the process part, which also falls under the category of key user extensibility. In the on-premise world, you would have used SAP Business Workflow, as part of SAP ERP, for this or leveraged the flexible workflow extension via the Manage Workflows app, which is also available in the cloud. Here you make use of the workflow tooling embedded in the business application. You can also build on preconfigured workflow scenarios and make smaller changes to them or use them as is.

Remember the scope of software. The embedded workflow tooling's only scope is workflows directly coupled to this application because it's a local process extension. The tooling doesn't consider processes spanning different systems, outreach to users who don't have access to this application, usage of other technologies (e.g., bot automation, business rules, AI), or central maintenance of all process extensions. If you're wondering why we need SAP Build Process Automation when there are also workflow tools in certain business applications, this is it: they have another, more narrowed scope and singular purpose that won't change.

Tip

If you're more interested in the workflow technology as part of SAP S/4HANA, there are also books available. There are also migration recommendations regarding what to do with custom workflows in SAP ERP and whether to move to SAP Build Process Automation. For example, check out *Workflow for SAP S/4HANA* (SAP PRESS, 2024).

15.2.2 Side-by-Side Extensions

Side-by-side extensions, in contrast, are loosely coupled to the standard business application. The extension could even be a standalone application, which is developed with a broad spectrum of diverse development preferences involving different skillsets and programming languages (e.g., ABAP, Java, or JavaScript). You can also choose a low-code/no-code approach for this. You might also want to include external stakeholders and users who won't have access to your core business application. Data changes are independent from the data in SAP standard tables or fields, which means no transaction consistency is necessary. To make this more concrete, let's revisit the improved invoice approval process. We won't change the data of the invoice itself in the business system but instead accelerate the processing. Typically, side-by-side extensions run on an independent infrastructure, which isn't directly coupled to the business applications so it has a different lifecycle and update cycles, and it fosters scalability and openness.

As you're familiar with the capabilities of SAP BTP, as outlined in Chapter 2, especially SAP Build, so if you now compare the characteristics described for side-by-side extension, there's clearly only one possible conclusion: SAP Build is *the* extension solution. With SAP Build, you can create full-stack applications as extensions, create software as a service (SaaS) applications, automate processes and workflows, go for a low-code approach, and design efficient digital workspaces, as shown in Figure 15.3.

SAP Build gives you the flexibility to choose between various development models to support fusion development, provides templates and prebuilt content for faster time to value, supports the clean core approach, and is enterprise-ready. The latest technologies are included, for example, generative AI capabilities and constant updates in an independent update cycle.

Whatever extension you want to create, SAP Build provides the needed toolset to extend with digital workspaces, pro-code or low-code apps, and, of course, automated processes (Figure 15.4).

There are also scenarios where it's not an either-or question, so you would use a combination of on-stack extensibility and side-by-side extensibility. Both approaches complement each other. You might also need to extend an on-premise solution with SAP

BTP, which is supported. SAP Build ensures the evolving business requirements are met and that you can drive your digital innovation agenda with SAP BTP as your innovation layer next to the stable and clean core.

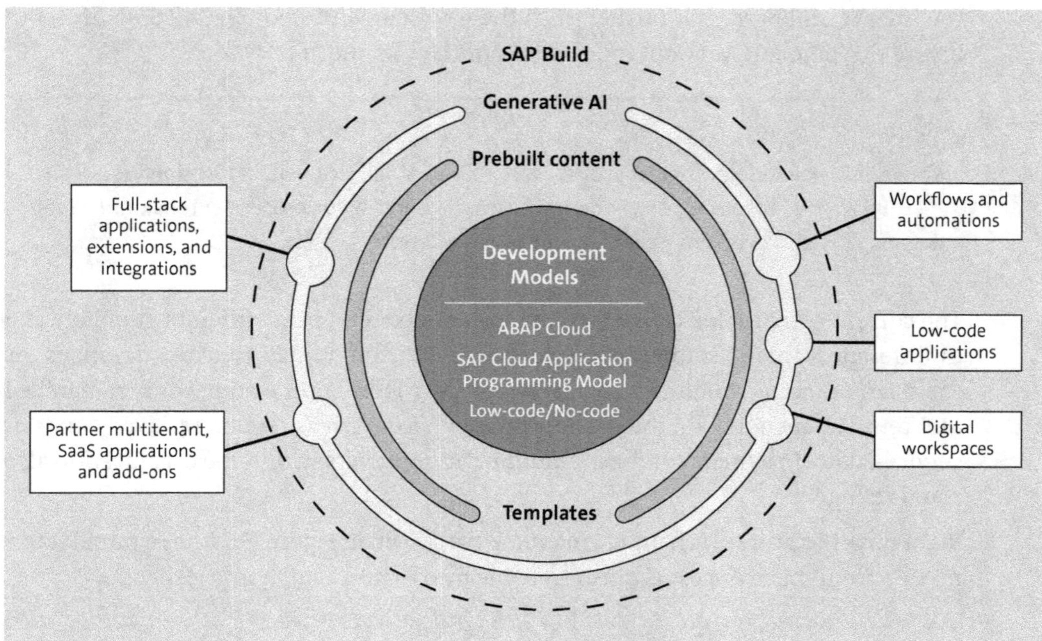

Figure 15.3 SAP Build to Extend Cloud ERP and Other Cloud Business Solutions or On-Premise Systems

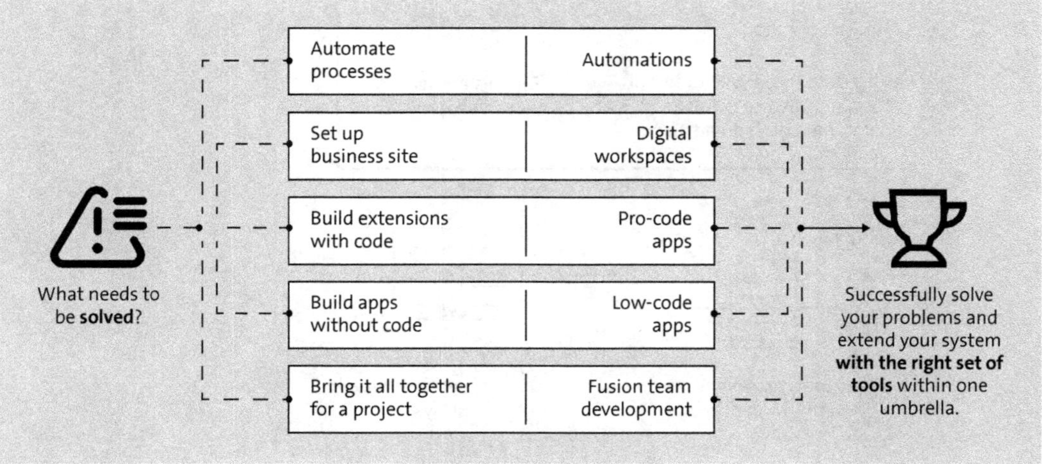

Figure 15.4 Extension Possibilities with SAP Build

15.2.3 Invoice Approval Process Example

Let's now walk through a concrete business process extension example: the invoice approval process we've mentioned before. The standard business process is to create the invoice and process it further, start the payment, and so on in SAP S/4HANA. But before this happens, we want to include some checks and approvals.

> **Note**
>
> If we only needed another approval step, the on-stack extensibility option would suffice. But in this case, we need some more intelligence in our process extension. The following scenario is common across many organizations.

The current scenario looks like this: Invoices are coming from different suppliers and in different formats: sometimes paper-based and sometimes attached to emails. As there are different conditions for the suppliers, the invoices also need to be handled differently. Depending on the details of the invoice, such as sender, received goods or services, day of payment, or gross amount, the approvers within the organization also could change.

Right now, the process is handled manually, as shown in Figure 15.5. The required information from the invoices is transferred one by one into a huge spreadsheet.

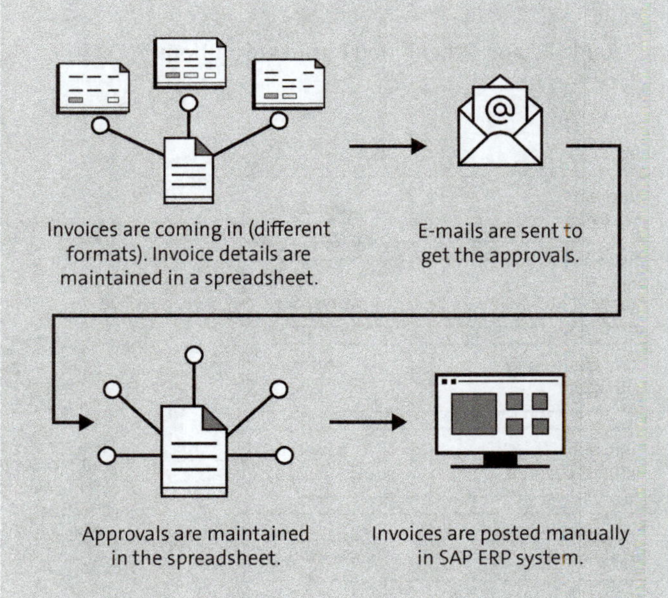

Figure 15.5 Invoice Approval Process

The spreadsheet will be sent around via email to get the needed approvals. Once the approval is given, the invoice will be posted into the ERP system. Needless to say, this approach is time-consuming, cumbersome, costly, and prone to failure and wrong postings. We want to change this with the help of SAP Build.

The invoices are scanned with an application developed with SAP Build Apps and uploaded into the portal, which is defined with SAP Build Work Zone. This triggers a process (extension) in SAP Build Process Automation. The information from the invoice is automatically extracted using AI capabilities and the Document Information Extraction feature and then stored in the process context. If an auto-approval isn't possible, for example, due to the amount of the invoice, and a dedicated approval is necessary, the selection will start. With decisions, the applicable approver will be selected based on certain criteria. The approver gets all the required information in an approval form in the My Inbox application. Once approved, the invoice gets automatically posted into SAP S/4HANA, and the standard business process for payment and so on will start, as shown in Figure 15.6.

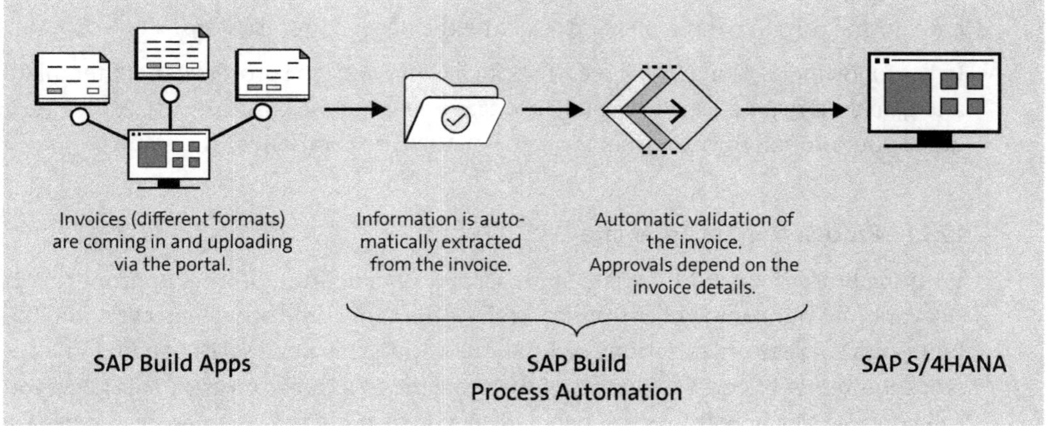

Figure 15.6 Invoice Approval Process with SAP Build

We could now even expand this process and leverage SAP Integration Suite to store the invoices on a shared file drive for audit purposes or to send notifications in Microsoft Teams to inform the finance department of the upcoming payments.

With this process extension, we've automated big portions of the whole activities, streamlined the process, increased the throughput times, ensured everything is based on the same and accurate data, and reduced the potential of errors dramatically.

> **Tip**
>
> This invoice approval process example is also available as a mission in SAP Discovery Center (*https://discovery-center.cloud.sap/*), listed as **Process and approve your invoices with SAP Build Process Automation.** Feel free to try it on your own.

15

This has been one possibility for creating business process extensions. Let's visit some other options by using process variants in SAP Build Process Automation, have the starting point in SAP S/4HANA, and involve the extensibility wizard.

15.3 Using Process Variants

SAP Build supports the notion of fusion development; that is, pro-code developers as well as citizen developers, business process experts, or key users being able to collaboratively work on extensions. Within SAP Build Process Automation, there is another capability available that enables pro developers to create a process variant template, and then a key user or business process expert can define a new process variant based on this preconfigured process variant template. The pro developer has full flexibility and all the known capabilities available to create a process variant template. On the other hand, the creation of the process variant is rather limited and only focused on the configuring certain parameters and selecting the needed process steps.

In the following sections, you'll see what can be done with the process variant template and gain a more practical view with a use case you might be familiar with. In fact, we'll guide you through the necessary steps to have a process variant as the result.

15.3.1 Process Variant Template

With the process variant template approach, you can define a process up front that is relevant for your complete enterprise while at the same time provide pockets of flexibility for different organizations or locations within this process. Depending on specific requirements, certain variants of the same process can be created. This gives you full control over how things are handled, as you also define which process steps are available in a process variant template itself, but acknowledges some varieties. These variants might be necessary due to different conditions in different countries, different handling or complexity of a process based on certain parameters, or because some subsidiaries must handle things their own way, but still according to the global guidelines and priorities.

The process variant template, which is available for the business process expert or key user to adapt, is also constrained. Only what has been "allowed" by the creator of the template can be changed, for example, the number of similar process steps where you only allow one step to start the process, the minimum number of approval steps, and so on. These guardrails might also be helpful for the ones creating the process variant, as they don't need to deep dive into all the possibilities provided by SAP Build Process Automation but are limited and safeguarded to only touch the artifacts and activities relevant for their business case. Because it's also reducing the number of needed technical details, they make changes and variants quite easy without greater knowledge needed regarding the capabilities of the tool, the process editor, and everything else you already know from the chapters in this book.

Using process variant templates also puts you in the position to follow a "process factory" approach. As in a factory, you have some raw materials, put them together in a smart way, and let them leave the assembly line. But in your case, the raw materials are the artifacts of a business process, which will be assembled by a developer, as a process variant template, and leave the assembly line to be configured by the business process expert or key user to reflect their business needs and support their daily tasks in a process variant.

15.3.2 Capital Expenditure Approval Process Example

Let's make this more tangible with a concrete example along a capital expenditure (CAPEX) process. Within this process, you aim for approvals for everything that leads to capital expenditures. Whenever funds are needed to buy, upgrade, or maintain physical assets, they are recorded in the balance sheet of the enterprise. Typically, everything invested to grow or at least maintain the current business is included, for example, buying machines, constructing buildings, investing in long-lasting equipment (useful life more than one year), or purchasing software licenses. This is not to be confused with operating expenses, which are short term, part of the income statement, and used for day-to-day operations.

For this CAPEX process, there is a defined set of process steps needed to fulfill the business requirements. But, depending on the nature of the funding request and the type of investment (project, buildings, software licenses), different levels of approval are necessary. In this case, the possible process steps are assembled in a process variant template and can then be configured within a process variant in SAP Build Process Automation. How does this look in the product? We'll explore this in the following sections.

Business Process Project and Some Prework

First, you create a business process project, as you've seen in Chapter 4. Let's name it "Capital Expenditure". Now, you need to do some conceptual work and segmentize the overall process in consumable units of independent processes, which you want to provide as process steps within the variant for the business process expert or key user, comparable to the concept of subprocesses (refer to Chapter 5, Section 5.3.3). So, you create a set of independent processes, and these will be the process steps in the variant template later, for example, like this:

1. **Initialization**
   To start the process variant, include a trigger and probably a form to notify the requestor about the start of the process.

2. **Approval**
   This is where the music plays in your variant template, where the different checks are included, who should approve the request, the concrete approval forms, notifications, and so on (see Figure 15.7).
3. **Finalization**
   To end the process variant, include a notification for the requestor.

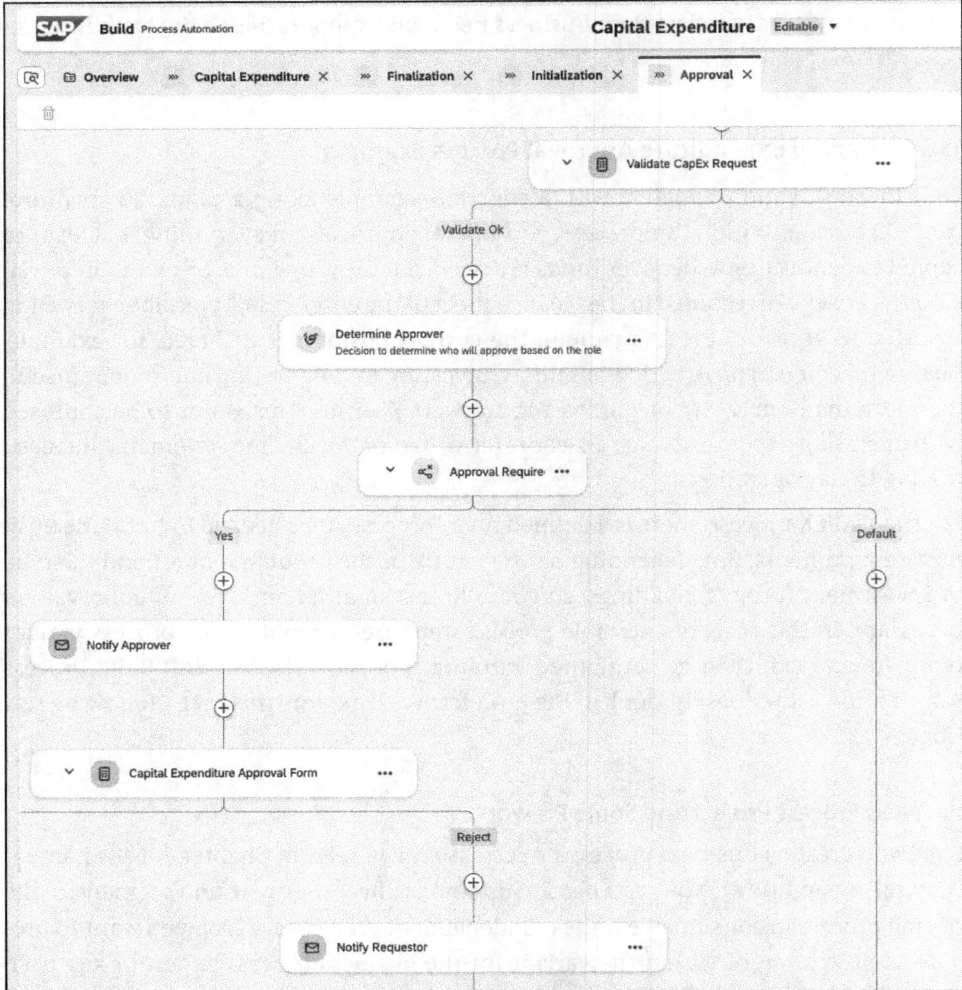

Figure 15.7 Approval Process (Step) of the Capital Expenditure Variant Template

Once this prework is done, you're ready to create the actual process variant template. No surprises here—you just follow the procedures you've already seen and learned in the previous chapters.

Add and Define a Variant Template

After opening the business process editor of the Capital Expenditure process, you might want to define a form to start the process (here called **Capital Expenditure Start Form**), as described earlier in Chapter 5. Afterward, click the **+** icon to add another artifact or step, here the **Variant Template**, as shown in Figure 15.8.

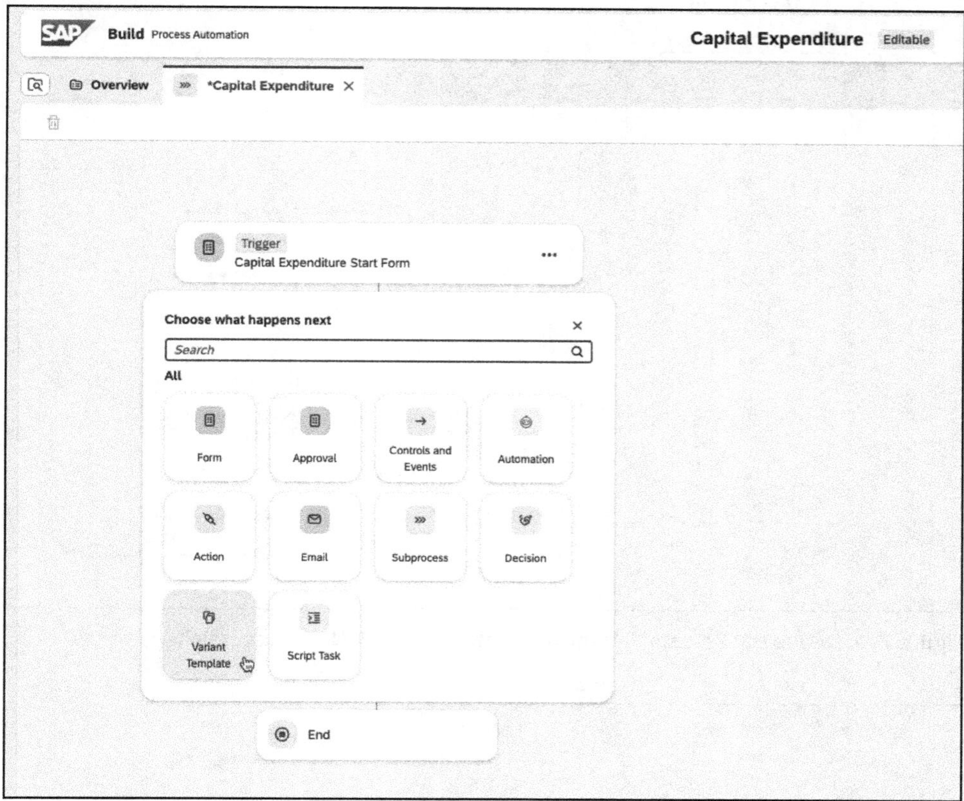

Figure 15.8 Adding a Variant Template to a Process

The variant template is empty and requires adding some process steps, as shown in Figure 15.9. These process steps are the processes created before, such as initialization, approval, and finalization.

Once the variant template is opened, again you'll see at this stage an empty space (see Figure 15.10). This is great because you now have the freedom and flexibility to define and add which steps (processes) should be part of this variant template and which pockets of flexibility you want to give to the business process expert or key user later.

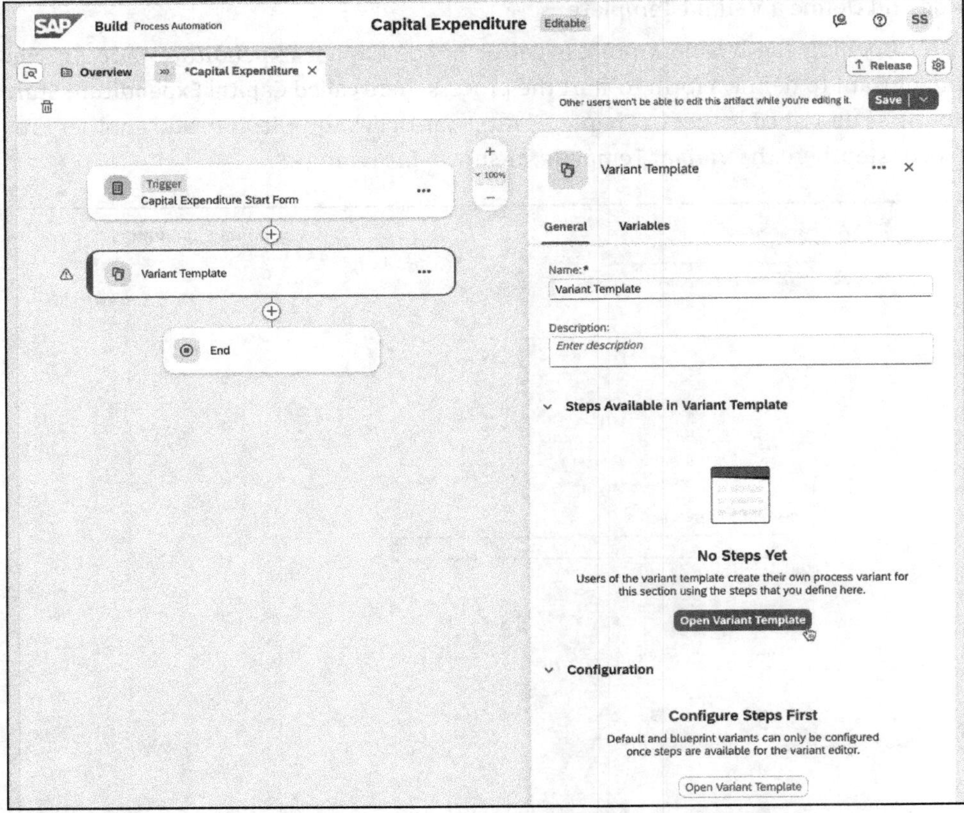

Figure 15.9 Define the Process Variant Template by Adding New (Process) Steps

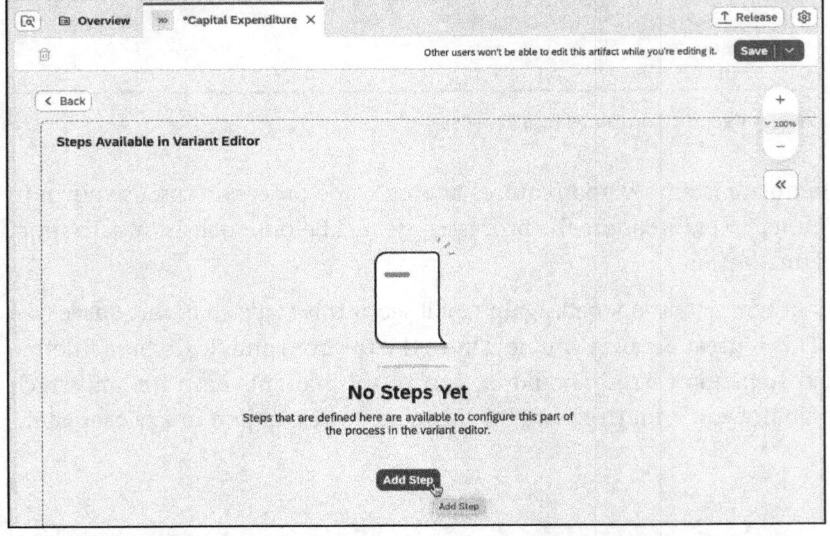

Figure 15.10 Adding Steps (Processes) to the Process Variant Template

Now you also realize why it has been important to do the conceptual prework at the start. After clicking **Add Step**, you can only select existing business processes that are part of the same project and add them as steps, as shown in Figure 15.11.

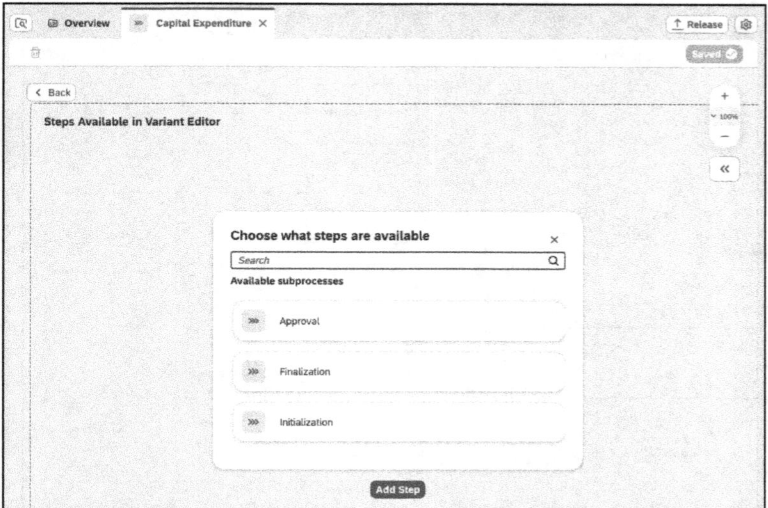

Figure 15.11 Selecting Available Processes to Act as Steps in the Process Variant Template

Don't be concerned regarding the order in which you add these processes as steps. Right now, you only define which of them should be part of the variant template.

Once you've added the steps required in the variant, you move on with the fine-tuning of these steps. As you know from other artifacts, you define the inputs and outputs, which you can choose from the process context. But there is one big difference here: you configure the behavior of the process step in this variant.

As shown in Figure 15.12, you configure the **Approval** step next. First, define whether it's a required step and if so, define the minimum number of uses. Because it's a **Required Step**, at least **1** is the right entry here. When needed, you can also define the maximum number of uses. Normally, you only have one initialization step. The same is true for the finalization step in our example, as it also just appears once. We also want to make this more compelling and easier for consumption by our target audience. Remember, we're creating this variant template for the business process experts and key users, so let's select recognizable icons and a nice color for the **Step Appearance**. We've discussed reducing the needed technical details in our process variant template, which is what we do in the **Step Inputs**. For example, define whether it should be possible to assign a role as an input to this process step.

Once everything is configured for this set of process steps, it's time to go back to define a default variant under **Configuration** in Figure 15.13. This variant will always be used if none of the other possible variants created by the business process expert or key user later aren't chosen, based on the given conditions.

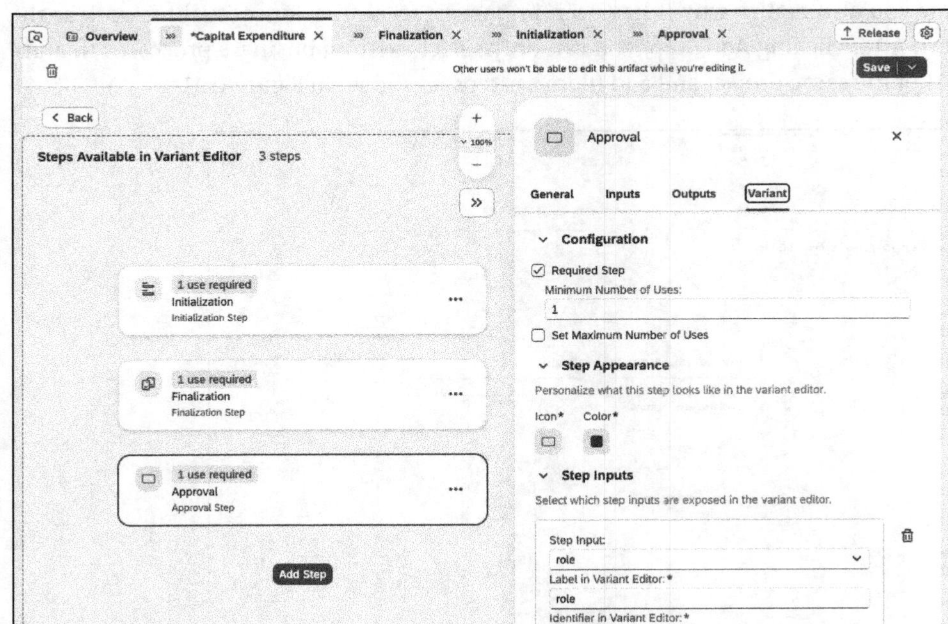

Figure 15.12 Configuring the Process Steps and Their Behavior in the Variant Template

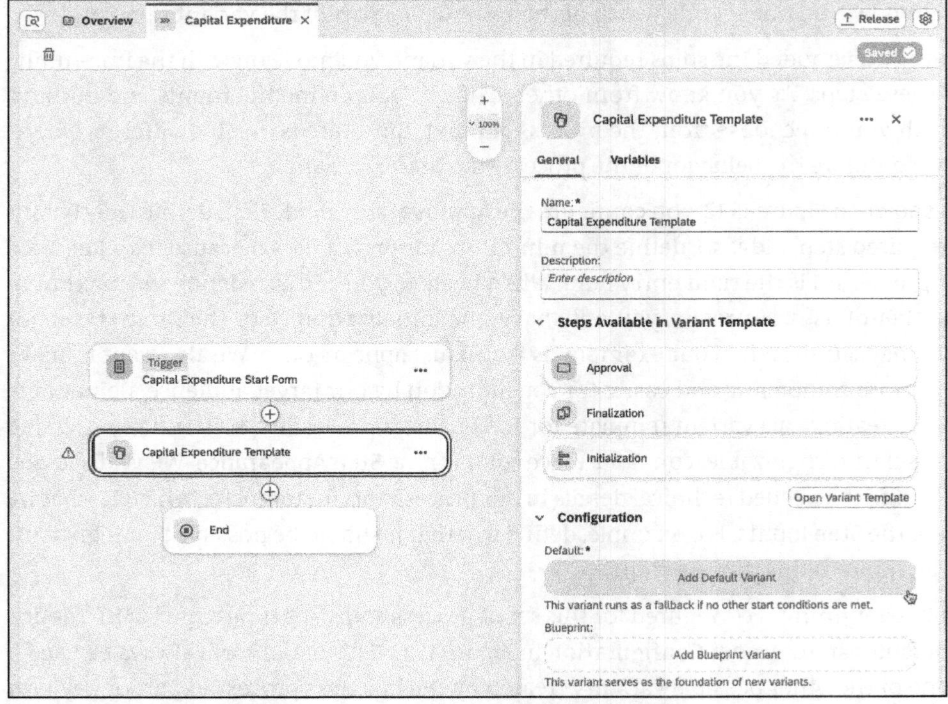

Figure 15.13 Adding a Default Variant to the Process Variant Template

Before, in the variant editor, the order of the steps wasn't important because we only defined a kind of container of possible process steps. Now, it's different. Here, you give a concrete order of process steps to be executed in the default case, as shown in Figure 15.14. Additionally, you can assign conditions for this step or provide some parameters for **Step Configuration**. Here, we see again the **role**, defined as a step input before. Let's assign **Local Manager** with a concrete role, but again, only for the default variant—our backup so to speak.

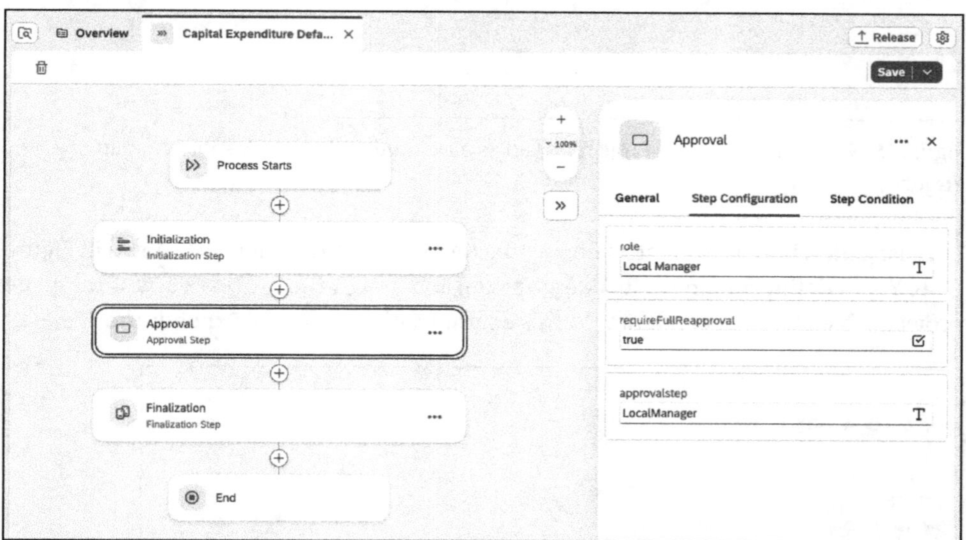

Figure 15.14 The Default Variant with a Flow of Steps and Further Configurations

Besides the default variant, you might also want to add a **Blueprint** variant. All the options that apply to the default variant are also available for this one.

To finalize this whole procedure, you need to release and publish this project. But we're not completely done yet on this side. Our goal was to provide this process variant template, containing the default template and blueprint template, as a very easy-to-use environment for the business process experts or key users. For this, two things need to happen, as we'll discuss in the next sections.

Create a Variant Management Project and Add Dependencies

First, create a new business process project. In this new project, you add a dependency to the variant template project created before. This is the binding between these two projects that allows you to use the variant template here. There's nothing spectacular to do. In the newly created business project called *Capital Expenditure Variant Management*, for example, open the **Project Properties**, and underneath **Dependencies**, just add the business project containing your variant template. In our example, this has been named **Capital Expenditure**, as shown in Figure 15.15.

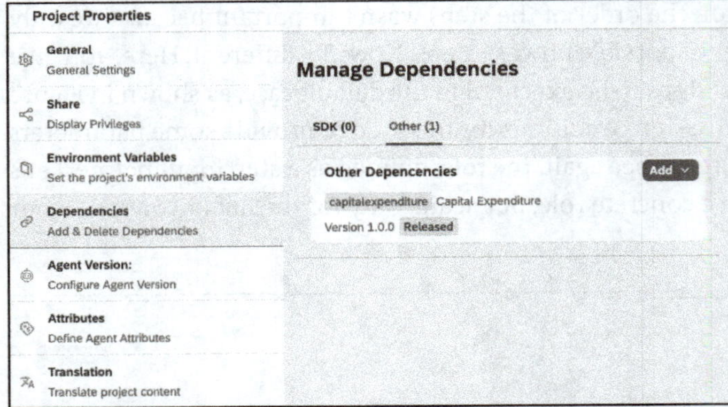

Figure 15.15 Variant Template Project Added as a Dependency to the Variant Management Project

The last action here is to create a new artifact, **Variant Management**, as shown in Figure 15.16. You'll define a name, add a description, and assign the process variant template project, which has a dependency. In this example, it's the **Capital Expenditure** project.

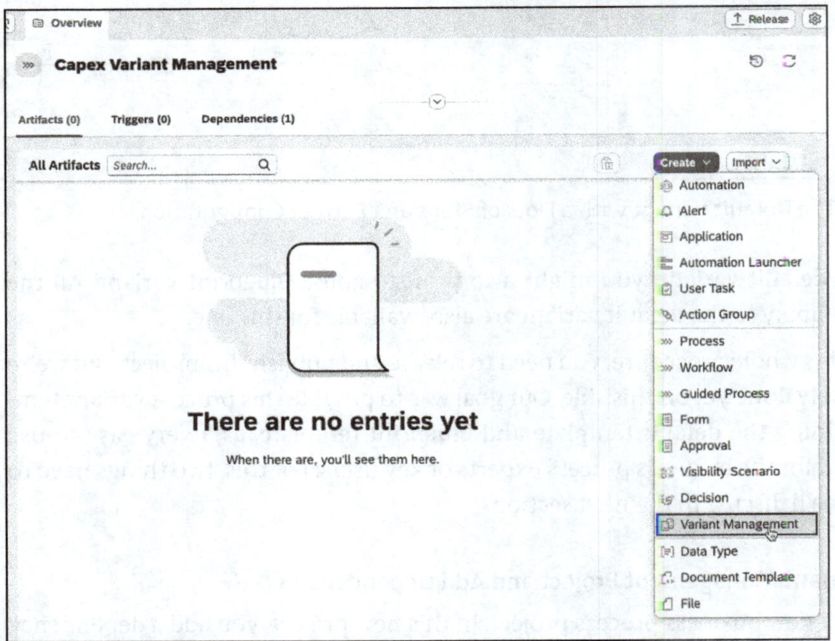

Figure 15.16 Creating a Variant Management Artifact

Note

You can only add one variant management artifact in a project.

The stage is ready to be entered by the variant user, referred to as the business process expert or key user before. The one who wants to create a new variant of the process, where we've created the process variant template before is the variant user. We're switching roles and are ready to define the process variant.

Create a Process Variant

Once you've opened the created variant management artifact, you can create process variants based on the template, as shown in Figure 15.17. Within the constraints but also with the given flexibility, as a business process expert or key user you can reflect the business needs in your very own process variant in SAP Build Process Automation. In our example, there are already some variants of the capital expenditure process created. You'll find a variant for infrastructure that includes two approval steps, and a variant for investments in machinery and plants, which has a three-step approval process.

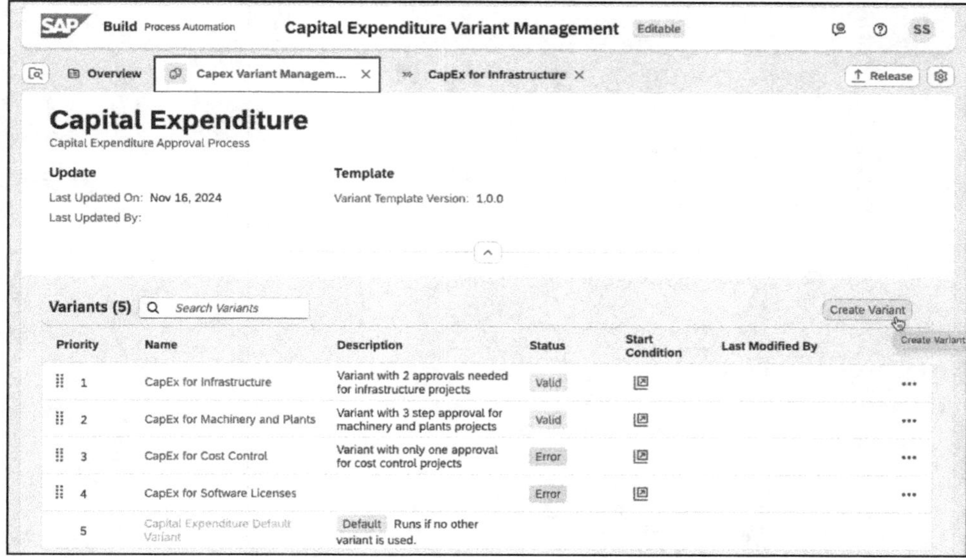

Figure 15.17 List of All Created Process Variants and Starting Point to Create a New Variant

There are two process variants listed with the **Status** as **Error**. No need to panic here. There are just some configurations not finalized yet. Every process variant can include dedicated start conditions. Finally, you'll find the default variant at the end of the list.

The order of this list is also important but can also be adapted. Once this project is released and deployed, it will start from the top to the bottom. And, if none of the defined conditions are met to execute any of the created process variants, there's only one option left: to run the default variant. This variant is also different from the others because you can't change it here. That's why we've also named it as our backup option. This variant will be executed when all the others aren't applicable.

If we take a closer look at one of these process variants in Figure 15.18, we find two types of steps only once: the initialization step and the finalization step. This is because, as we said in the process variant template, each of these step types should only appear once in a process. On the other hand, we can add several approval steps. In this variant, we have three different approval steps or approval layers from a business point of view. After the initialization, the cost center needs to approve, followed by the purchasing department, and finally the sustainability head has a say.

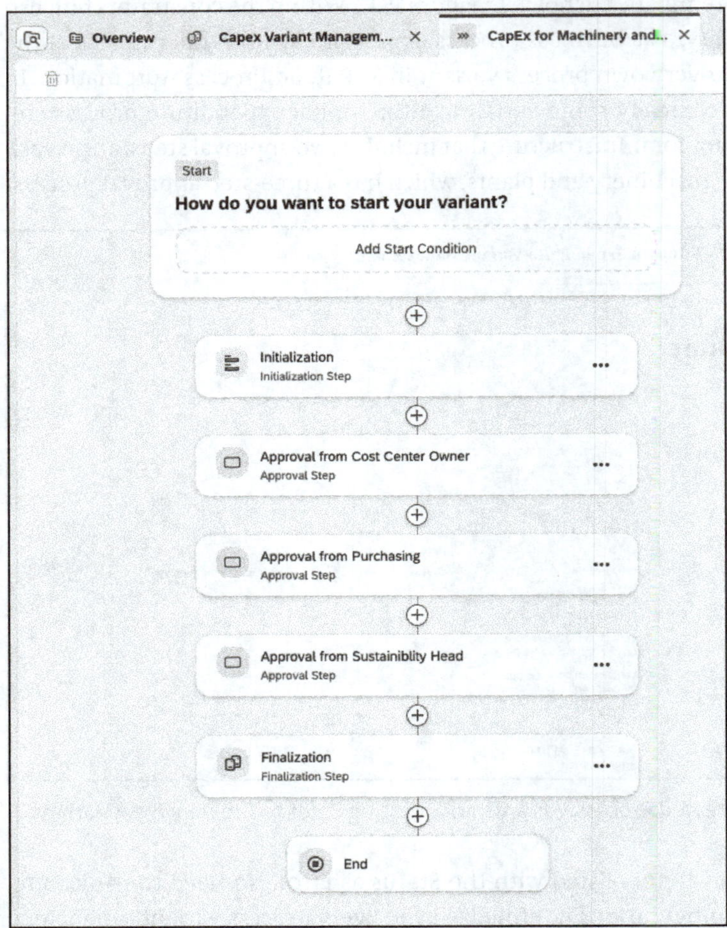

Figure 15.18 Variant of the Capital Expenditure Process for Machinery and Plants

With the concept of process variants, you have a very flexible option available to provide templates to the business process experts or key users in your organization. Though their overall flexibility in creating different shapes of a process is constrained, there are sometimes good reasons to do so.

Now that we're finished with this option to create process extensions, let's visit another one, which is very close to SAP S/4HANA in the next section. So far, we were focused on driving the process extensions with SAP Build Process Automation as our starting point. We'll change that focus next.

15.4 SAP S/4HANA Extensibility and Creating a Process Extension for SAP S/4HANA Cloud

To accommodate varying business needs and preferences while adhering to a clean core strategy, SAP S/4HANA Cloud customers use SAP BTP to extend the core applications, as you know. To reiterate, *clean core* is a concept and strategy to extend SAP S/4HANA Cloud systems in an upgrade-safe way.

Using SAP Build for extending the core capabilities of SAP S/4HANA Cloud is clean-core compliant. Figure 15.19 shows the different options for extending SAP S/4HANA Cloud with SAP Build. SAP Build supports key user extensibility with low-code applications such as SAP Build Apps, SAP Build Process Automation, and SAP Build Work Zone. SAP Build supports developer extensibility with pro-code applications SAP Build Code and SAP BTP, ABAP environment.

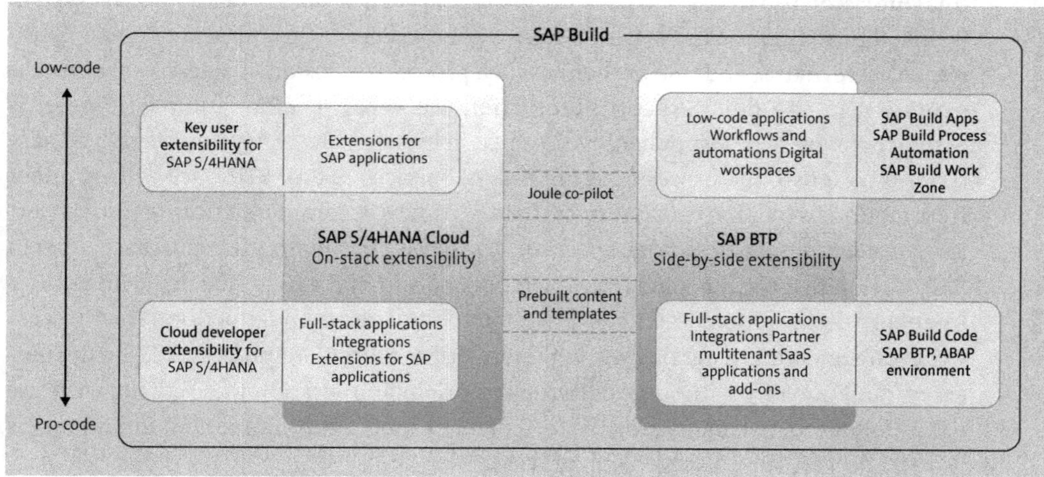

Figure 15.19 SAP S/4HANA Cloud Extensibility with SAP Build

The set of available applications to extend SAP S/4HANA Cloud may be a bit overwhelming at first sight as you wonder how to find the right application for your use case and how to find the right API or extension point for the application you want to extend. You can refer to various documents from SAP that provide guidance, but searching and finding all that information is tedious. In a cloud environment where there are constant innovations and recommendations, even experts find it difficult to identify the correct tool. This is where the new SAP Build extensibility wizard or assistant comes in.

The SAP Build extensibility wizard, which is available in SAP S/4HANA Cloud Public Edition, aims at simplifying the creation of custom extensions. It's designed to streamline on-stack extensions in SAP S/4HANA Cloud as well as side-by-side extensions in SAP Build by providing step-by-step help to the end user. The wizard can be launched from any application in SAP S/4HANA that supports key-user extensibility such as Manage Sales Order, New Supplier Invoice, Manage Journal Entries, and so on. It also filters the extension points (APIs and events) based on the application it was launched from, thereby helping users easily create the extension they want. The wizard can be used to create side-by-side process extensions using SAP Build Process Automation.

Prerequisite

The SAP Build extensibility wizard is available with any of the SAP Build services, including SAP Build Process Automation. By subscribing to any SAP Build services, the wizard can be launched in SAP S/4HANA Cloud, and extensions can be created. The wizard understands the SAP Build service entitlements and roles to show only those extensions that are possible.

Imagine a scenario where you need to ensure that each invoice generated in the system is linked to a purchase order with minimal user intervention. There are several methods to achieve this, and one of them is via a process extension. A side-by-side process extension in SAP Build Process Automation can verify whether a purchase order is assigned for every new supplier invoice created in SAP S/4HANA system. If no purchase order is assigned, the process can use SAP AI Services in SAP BTP, such as Document Information Extraction (DOX) service, Data Attribute Recommendation, Invoice Object Recommendation, SAP AI Core, and so on, to identify the appropriate purchase order in SAP S/4HANA. If the machine learning model fails to find a right match, it can notify a member of the procurement department for further manual action. This process extension operates in parallel with the standard procurement process, adding the necessary functionality without modifying the original process. In this section, we'll use the SAP Build extensibility wizard to create a process extension in SAP Build Process Automation for such an extension process.

The SAP Build extensibility wizard can be launched from any application that allows key-user extensibility. Because we want to create an extension process for invoice processing, let's try to open an app to create an invoice by clicking the **Create Supplier Invoice** tile in the SAP S/4HANA launchpad, as shown in Figure 15.20.

This opens the **New Supplier Invoice** page shown in Figure 15.21. Click the user icon on the top-right corner of the screen, which opens a *user menu*. Choose the **Create Extension** option to launch the SAP Build extensibility wizard.

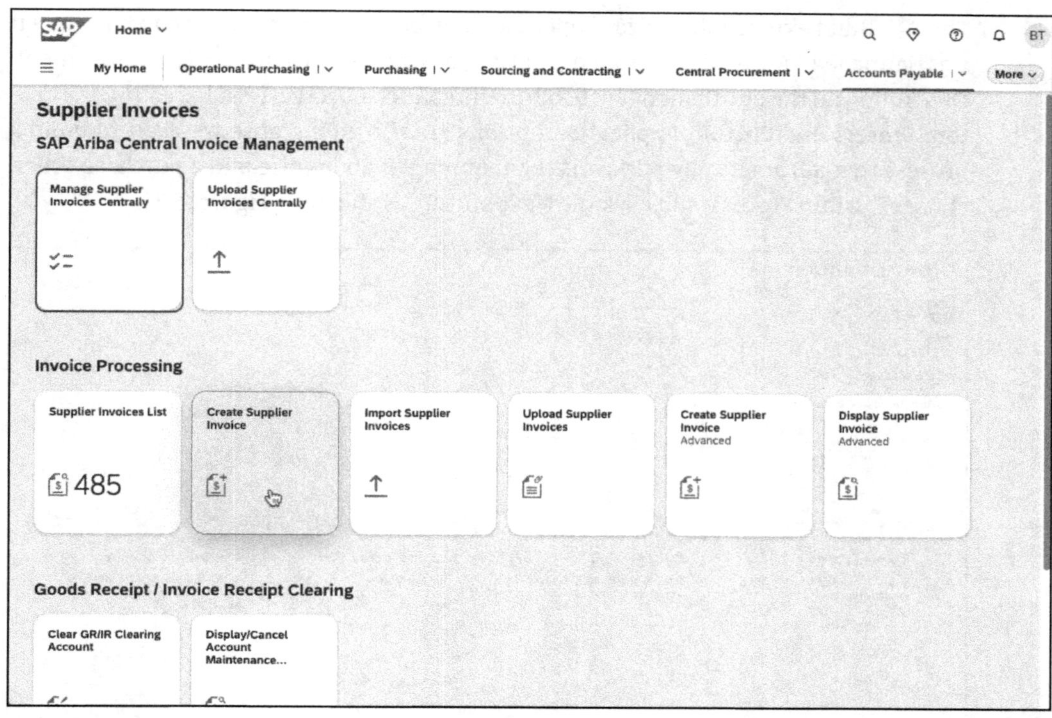

Figure 15.20 SAP S/4HANA Launchpad for Accounts Payable

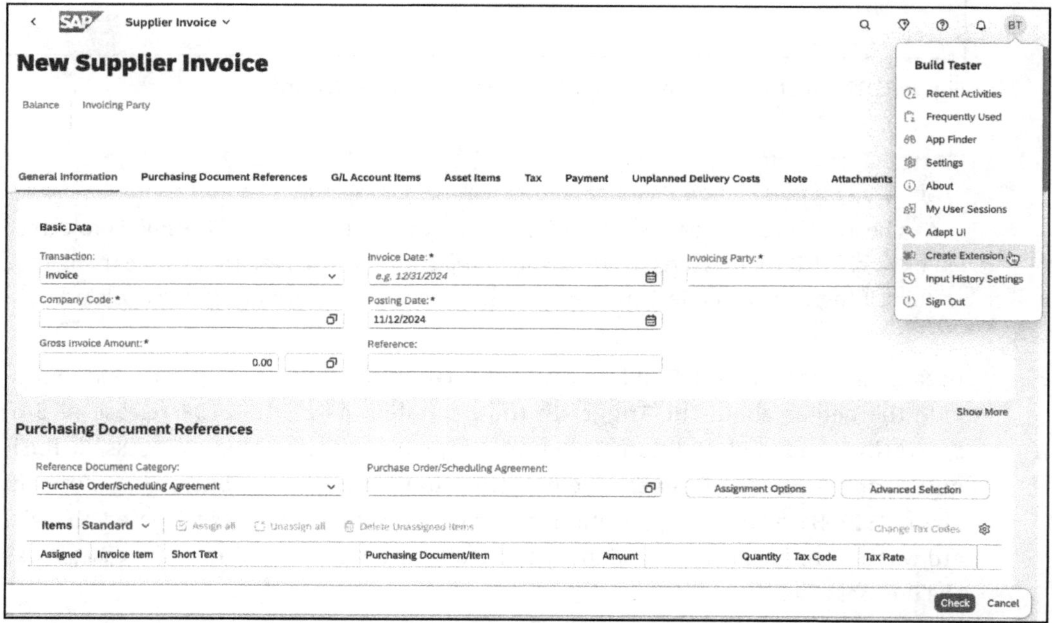

Figure 15.21 Launching the SAP Build Extensibility Wizard from the User Menu

The SAP Build extensibility wizard opens as an overlay of the application from where it was launched. All available extensibility types are shown in a tile arrangement. Depending on the entitlement available in the connected SAP BTP subaccount, the **Custom Process** and **Custom Application** options are shown. Because we want to build a process that automatically performs the mapping of an invoice and a purchase order, choose **Custom Process**, and click the **Next** button, as shown in Figure 15.22.

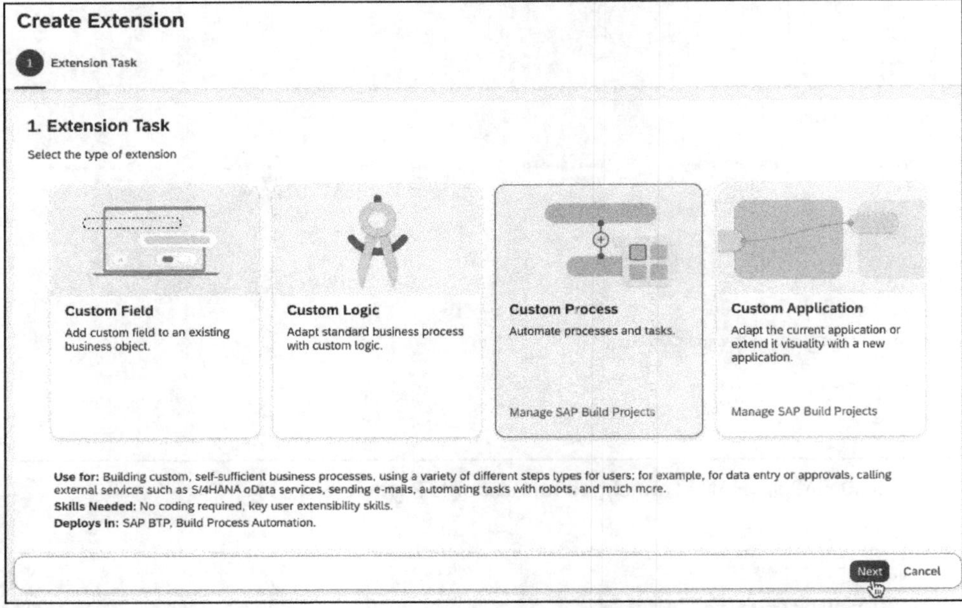

Figure 15.22 Creating a Custom Extension Process from the Wizard

> **TIP**
>
> As a best practice, navigate to the SAP Build lobby using the **Manage SAP Build Projects** link to check if there is already an existing project for the use case that you're trying to build. If there are no projects that help with your use case, continue with the wizard.

Because a process in SAP Build Process Automation requires a trigger, we need to provide the details about the **Trigger** in the next step. As of the 2408 release of SAP S/4HANA Cloud Public Edition, the wizard only supports triggering a process via **Business Event**. Select the **Business Event** option, and click the **Next** button, as shown in Figure 15.23. Based on the application from where the extension was triggered, the wizard was automatically able to identify the available extension point. In the case, it was the **Business Event**.

The wizard also recognizes the **Event Object** that is relevant for the **New Supplier Invoice** application and automatically applies a filter on the **Business Context**. This automatic filtering helps the user effortlessly find the right event object instead of searching SAP Business Accelerator Hub or other documentation. If you had launched the wizard from a different application, then the default filter could be different than for **New Supplier Invoice**. Of course, you can override these filters and search any event that is needed for your use case.

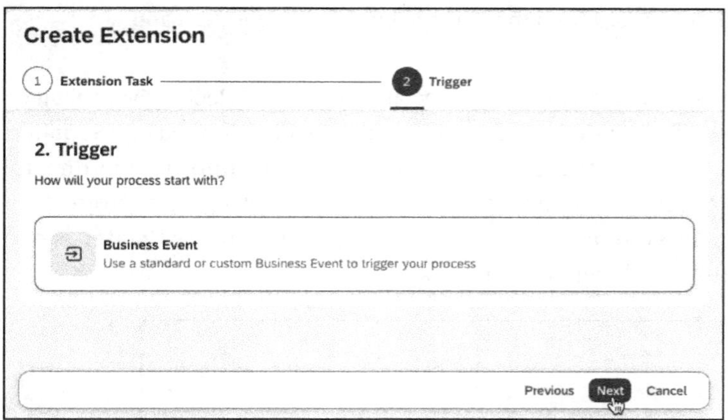

Figure 15.23 Choosing a Trigger for the Custom Process Extension

For the use case of automating the assignment of a purchase order during invoice processing, choose **Supplier Invoice Events**, and click the **Next** button, as shown in Figure 15.24.

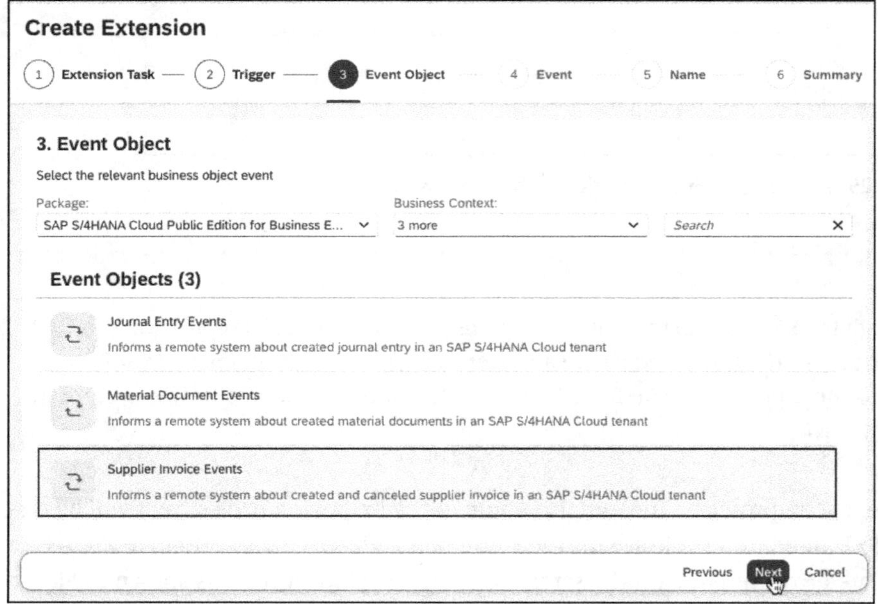

Figure 15.24 Selecting Event Objects in the Wizard

Tip

The wizard helps to find the right extension point based on the application from where it was launched. To make use of this capability, always launch the wizard from an application that is semantically closer to the extension process that you want to build. For example, if you want to build an extension process for purchase requisition approval, then launch the wizard from the Manage Purchase Requisition app.

In the next step, the wizard lists the **Events** associated with the selected **Event Object**. The **Supplier Invoice Events** contains two events: **SupplierInvoice Created** and **SupplierInvoice Canceled**. The wizard gets the event objects and event information directly from the SAP S/4HANA system. For our use case, we want to start a process from every new invoice created in the system. Therefore, select the **SupplierInvoice Created** event, and click the **Next** button, as shown in Figure 15.25.

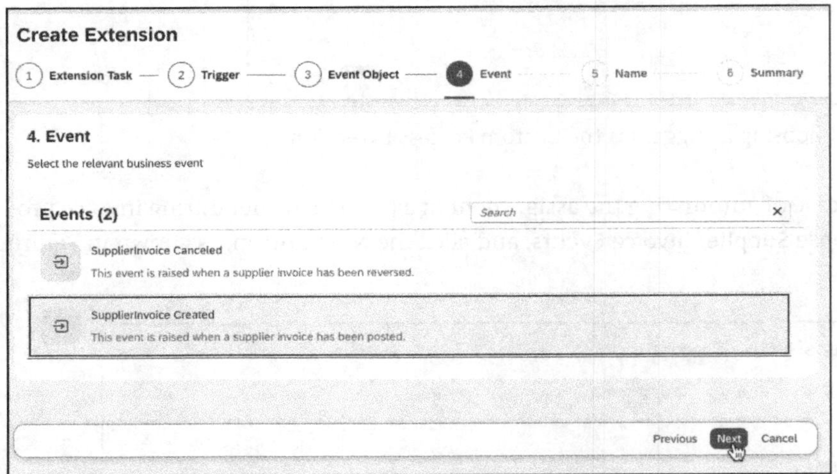

Figure 15.25 Selecting an Event to Trigger the Process Extension

Note

Although the SAP Build Extensibility Wizard requires any SAP Build service, such as SAP Build Process Automation, to create an extension, you need entitlement to SAP Event Mesh to run a process in SAP Build Process Automation based on a business event in SAP S/4HANA.

In the next step, provide a meaningful name to the extension project that will be created in SAP Build. If a project with the same name already exists, then you'll get an appropriate error message on the screen. Click the **Review** button, as shown in Figure 15.26.

In the **Summary** screen, you're shown a summary of all the selections made in the previous steps. Click the **Create** button to complete the project creation from the wizard.

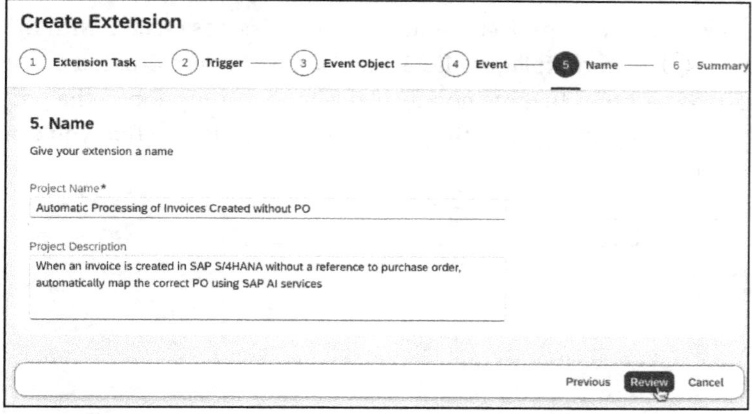

Figure 15.26 Naming the Extension Project

> **Note**
>
> Every time you complete the wizard, a new project will be created in SAP Build. You can't use an existing project and add additional artifacts to it.

When the project creation is successful, the wizard shows a success message with an option to open the created extension process in SAP Build Process Automation, as shown in Figure 15.27. You can either use the **Manage SAP Build Projects** link or the **Open Extension in SAP Build** button to navigate to the lobby for further development of the process extension.

Figure 15.27 Success Message after Project Creation

When the process extension is opened in the SAP Build lobby, you can see that a process with event trigger is available. Based on the selection made in the wizard, an event project was created and published to the library. A process automation project was created, and a dependency to the event project was added. A process was created with the business event as trigger, as shown in Figure 15.28. The **SupplierInvoice Created** event has a few invoice-related attributes that are now available for consumption in the process. The wizard does the heavy lifting in setting up the overall project so that you can start modeling the process.

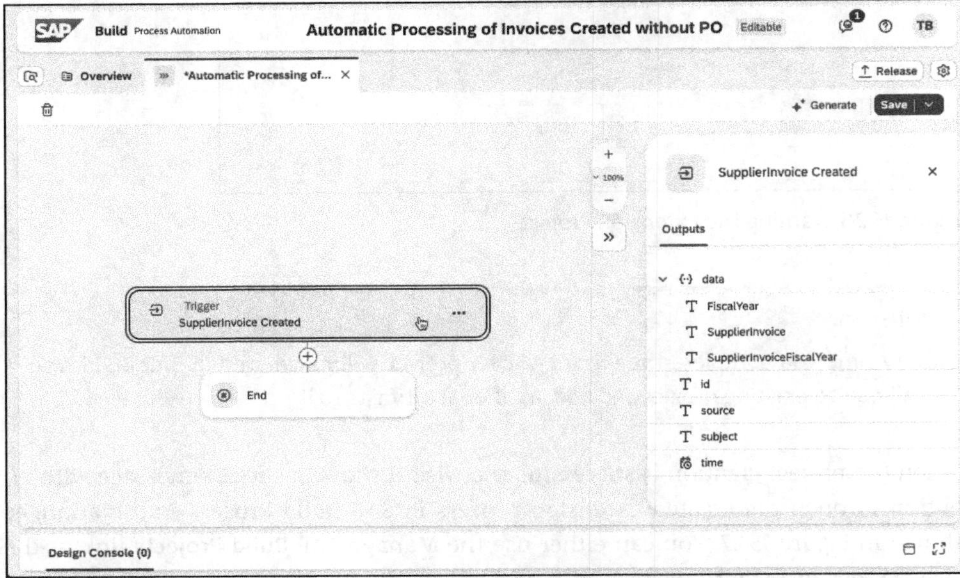

Figure 15.28 A Process Created with a Business Event Trigger

You can continue to model the process, as shown in Figure 15.29; release a version; and deploy it in SAP Build Process Automation. When an invoice is created in SAP S/4HANA, the **SupplierInvoice Created** event is observed in SAP Event Mesh, which then starts the process in SAP Build Process Automation.

Now you're familiar with both ways—either starting the extension activities with SAP Build as your entry or using the extensibility wizard particularly for extension in SAP S/4HANA Public Cloud Edition.

Before we close the chapter, we want to inspire you with some example scenarios around doing extensions with SAP Build Process Automation under different circumstances.

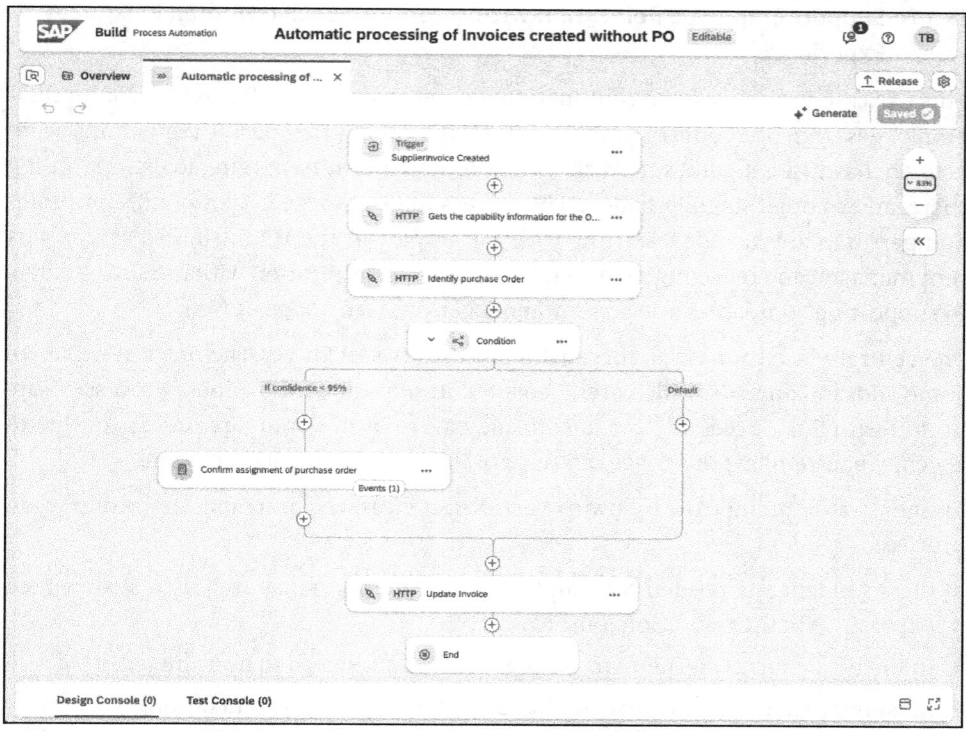

Figure 15.29 Process That Automatically Assigns the Purchase Order to an Invoice

15.5 Example Extension Scenarios

At this point, you're most likely already thinking about possible extension scenarios in your organization, including what kind of standard business processes you have, in which business solutions they are running (e.g., SAP S/4HANA, SAP SuccessFactors, SAP Ariba, and non-SAP applications), and where extensions would make sense and become a game-changer.

To stimulate this thought process, we'll discuss a few examples in the following sections where process extensions made the difference. These are in addition to the ones we've described, discussed, or shown in SAP Build Process Automation in the chapters so far.

These example extension scenarios are from different LoBs, based on real customer use cases, supported by several products of SAP Build, and created by fusion development teams. We'll see an employee relocation example, a use case related to bid management, and a sustainability scenario.

15.5.1 Support Employee Relocation with a Comprehensive Human Resources Experience

This use case is often seen at multinational enterprises. They offer relocation to their employees from one country to another. Just think about the specific regulations every country has when it comes to work, taxation, work permits, moving logistics, settling there, and so on. It's a fully-fledged HR process to be managed. A lot of different roles and personas are included, starting with the employee, the HR business partner, current manager, and receiving manager. Then, the communication with the stakeholders in supporting companies and government agencies is included as well.

The central place to manage these HR processes is SAP SuccessFactors. But there are some additional needs, which aren't covered in the standard business processes supported with SAP SuccessFactors, due to the nature of this complex process or due to specific requirements or service offerings of this multinational enterprise.

The goal was to provide the following benefits, services, and outcomes in the relocation process:

- Clear guidance is needed for employees and managers, as well as a streamlined experience in the relocation journey.
- In the host country, settlement support and logistics need to be arranged.
- Most important is to ensure compliance with regulations and taxation.
- The partner network should be used to involve them for relocation support functions.

By extending SAP SuccessFactors for this scenario, the solution included the following products:

- **SAP Build Process Automation**
  Executing the unified relocation process with task assignments, and schedule to bring together employees, HR business partner, and the receiving manager.
- **SAP Build Apps, SAP Build Code, and ABAP environments**
  Developing customized apps for employees to complete assignments related to the relocation and apps for the HR business partner to maintain data and provide requirements.
- **SAP Build Work Zone**
  Hosting a relocation workspace supporting a guided experience for employee information and a separate HR workspace for the HR business partner.

As you can see, the standard activities remain untouched in the SAP SuccessFactors system but have been surrounded by SAP Build, adding specific functions to realize a comprehensive HR experience for this relocation process.

15.5.2 Build Specialized Workspaces and Processes for Bid Management

Professional services companies are receiving a large numbers of requests for proposals (RFPs) from their customers every day. These requests need to be handled in a timely manner and, most of the time, afford cross-team collaboration and a straightforward evaluation process. The standard solution covered with SAP S/4HANA already handles a big portion of these efforts to manage the RFPs, but there is a desire to automate portions of this process and come to a more comprehensive solution via extensions.

In this use case, the company was aiming to fulfill these requirements:

- Automate the response creation for the incoming RFPs.
- Ensure the cross-team collaboration will be as smooth as possible—in the best case, seamless.
- Make use of standard business processes and use SAP S/4HANA Professional Services solutions to manage the responses.
- Strive for an optimized response management underpinned with a specialized workspace.

The extension of the standard business process that is hosted in SAP S/4HANA Professional Services has been realized with these developments:

- **SAP Build Process Automation**
  Covers and orchestrates the end-to-end bid process, including automation starting from an uploaded RFP document until the approval of a bid submission.
- **SAP Build Apps, SAP Build Code, and ABAP environments**
  These apps support the different activities in the process, starting from the submitted RFP to the review of the RFP requirements. They provide recommendations for solutions, teams, and consultants, which results in a high-level plan that needs to be approved.
- **SAP Build Work Zone**
  Home of the workspace for the bid management team, hosting the apps, and linking to knowledge articles and forums where specific bids can be discussed.

The extension in this case provided process automations, different apps during the process, and a central place for collaboration.

15.5.3 Create Dashboards for Company and Idea Management

Sustainability is one of the hot and important topics for almost every company these days. The efforts in this area are increasing and so is the demand to get transparency in the related activities. Are they really paying out, triggering the needed changes, and contributing to the sustainability goals? A lot of data is needed and must be maintained for this activity. As you can imagine, asking the departments to report their monthly utilities usage and expenses in a big, complex spreadsheet is likely a cumbersome and error-prone activity. Yes, we often come back to spreadsheets and how to use them more effectively or even avoid them. As for every process, this also holds true for processes related to sustainability: make the usage as simple and delightful as possible for the end user, and it's more likely to be used to drive the business outcome.

In addition, the standard procedures are covered in SAP S/4HANA. Most of the required data is already stored there, so let's use this data wisely and increase the transparency. Therefore, the following outcomes have been on the project list:

- Improve the efficiency and accuracy via a streamlined data maintenance process.
- Simplify collection of the monthly utilities' usage and expense data.
- Get rid of errors or redundancies in these data collection procedures.
- Improve data quality for more accurate analyses.
- Push the transparency to another level.

We see again a combination of various SAP Build solutions to ensure this list becomes reality:

- **SAP Build Process Automation**
  Before the data is maintained in SAP S/4HANA, it needs to be approved. The related process extension is triggered by a sustainability app and sends notifications to the involved managers.
- **SAP Build Apps, SAP Build Code, or ABAP environments**
  The utilities data is maintained in a mobile and desktop app before being submitting for approval. The data can be entered manually or, more elegantly, by scanning the utilities bills.
- **SAP Build Work Zone**
  This is the central place with a sustainability dashboard, including even gamification elements for cross-group competitions. In the dashboard, you're connected to aggregated data in the SAP Analytics Cloud solution. Furthermore, here you launch the app and access the approval process for incoming data submissions.

As you can see, these extensions could also live and work on their own and already provide more value and benefits. Although the combination leads to a comprehensive solution, it doesn't mean you need to have this solution already in mind when you start your own extension scenarios. Often customers start small, for example, by automating the process. Then, they evolve the data entry with an app extension and bring everything together for a smooth UX with a consistent look and feel and central entry points. You can also start with an app for a common data entry activity and have process automation as the next milestone. Another example is to first bring all data together in one central workspace and then have app or process extensions as the next step.

Don't feel overwhelmed when it comes to extensions and the questions of where to start and how. You don't need to boil the ocean at the very start. An incremental approach will most often be more beneficial, showing faster results and creating appetites for more. Collaboration, good knowledge of the process and the required improvements, and how to fascinate the people behind these processes are the keys to success. Always have these people in mind—they work the process, live the process, and influence the success of any process extension approach. But we're sure, you now have many ideas of where to start extending your standard business process and creating successful use cases like the ones we've introduced in this book.

15.6 Summary

In this chapter, we've started with a small journey through the history of SAP and learned how revolutionary it was to provide software solving daily business problems in a standardized way in a standard business application while also providing the flexibility to customize these standard business applications or even extend them. You know now what standard business processes are but also that there are cases where possible customizations aren't enough to fulfill diverse business requirements. The need for process extensions is clear, and you're now aware of the potential risks, how to handle them, and which benefits are in store for you and your organization.

When someone mentions on-stack or side-by-side extensions, you can explain the differences and know the combination of both approaches will work. In addition, the tool of choice for side-by-side extensions is no question anymore: it's SAP Build, as part of SAP BTP. We've made the topic of process extensions more concrete with the example of the invoice approval process, as an extension of SAP S/4HANA.

Fostering the idea of fusion development, we've introduced the concept and creation of process variant templates in SAP Build Process Automation. Though the process variant template is rather constrained for the business process expert or key user, it provides the right depth of flexibility to configure variants of business processes or process extensions in a very straightforward way.

We've also changed the starting point for a process extension approach. So far, the focus has been on how to create processes, especially process extensions, starting in SAP Build Process Automation and connecting to the underlying standard business solutions as you go. Now, with the extensibility wizard in SAP S/4HANA, you could also use this as point of departure for your process extension journey.

The process extension scenarios showed that there is huge potential and no limiting factor for any industry or LoB regarding process extensions. You could also start small and enlarge depending on the level of maturity or comprehensiveness you want to achieve.

In the next chapter, we'll focus on how to include different stakeholders more elegantly into SAP Build Process Automation, for example, the end user via SAP Task Center. You'll also learn more about fusion development. So far, we've only mentioned it here and there, but there's more to it.

Chapter 16
Scaling Process Automation

To scale process automation, we'll focus on two different personas. On one hand, we'll discuss how to involve the end user with SAP Task Center. On the other hand, we'll see how fusion development, including developers with different skill levels, is supported and which benefits you'll gain here.

Having reached this chapter, you've already gained some deep expert knowledge around SAP Build Process Automation and seen how this product complements the others by its nature. In Chapter 1, we talked about enterprise automation, combining SAP Build Process Automation with other solutions. There we mentioned SAP Signavio solutions to get insights into the business process and even run deeper analysis with process mining capabilities. Another aspect has been to leverage SAP Integration Suite to connect to any SAP or non-SAP applications, whether they are running in the cloud or on premise. Chapter 2 explored SAP Business Technology Platform (SAP BTP) and the several services offered on this platform to help your business run better and be able to extend any standard business application, particularly by using SAP Build. The whole process extension topic has been vibrant in the book in almost every chapter and detailed previously in Chapter 15.

In other words, there are several ways to scale SAP Build Process Automation and to involve other solutions, products, services, or capabilities, which complement each other to solve and support a specific business need. Now we want to focus on how to broaden this scale, mainly in two opposite directions—but we won't stretch it too much.

First, we want to get a better understanding of how to reach the end user, that is, the business users who are dealing every day with the business processes in standard business applications and also the processes or extensions created with SAP Build Process Automation. Determining how to involve this target group is also essential for the success of any business process so they can do everything possible and they are comfortable with it.

Second, we look at the other side of involved personas and people dealing with SAP Build Process Automation, that is, the developer. More precisely, we'll look at the different types of developers, as we'll talk about fusion development, including professional and citizen developers, key users, and business process experts.

16.1 SAP Task Center

Human tasks are core to business process management use cases. A human task needs human intervention to take required actions to complete the process step. Human task actions include approving or rejecting the purchase requisition, reviewing the invoices, or forwarding the sales order to different departments for verification. These human tasks are sent to the task inbox where authorized users can view the tasks assigned to them and take appropriate actions to complete the tasks. These tasks can be created from various SAP or third-party business process systems and solutions such as SAP SuccessFactors, SAP S/4HANA, SAP Ariba, SAP Concur, and so on.

Consider an order-to-cash scenario where a purchase requisition is created and approved in an SAP Ariba system, followed by purchase order generation in an SAP S/4HANA system, and the delivery process handled in third-party systems. In a typical setup, there will be reviews and approvals needed at all the stages of the business scenario across different systems. There can be situations where a business user from the procurement department must approve the purchase requisition/purchase order request and review the delivery note sent by the supplier. In such cases, the business user must navigate to different inboxes of different systems to do the approvals, which can be cumbersome, especially when performed in daily business.

A unified inbox is therefore needed where all the tasks from different systems can be viewed and operated from one user interface (UI) with a unified experience. SAP Task Center is the service in SAP BTP that provides an interactive application to manage tasks from different systems in a single inbox. SAP Task Center federates tasks from different systems such as SAP Ariba, SAP Concur, SAP Fieldglass, SAP Build Process Automation, SAP S/4HANA, and so on and shows them in a single inbox so that business users can view them and work on the tasks from one place, eliminating the need to work with multiple inboxes, as shown in Figure 16.1.

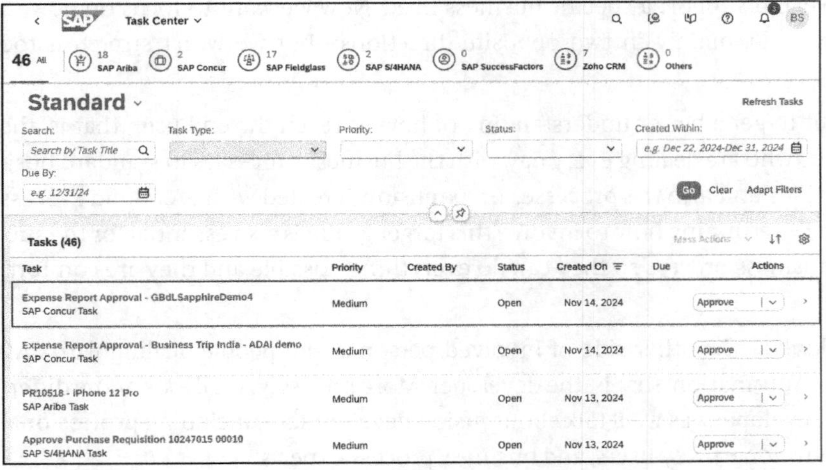

Figure 16.1 View of the SAP Task Center Application

In this section, you'll learn how SAP Task Center is used with SAP Build Process Automation where the human tasks can be viewed and operated by business users. It will cover only the features of SAP Task Center that are relevant to SAP Build Process Automation.

16.1.1 Benefits of Using SAP Task Center

SAP's unified inbox—SAP Task Center—seamlessly integrates with different SAP and non-SAP solutions to federate user tasks that can be processed from a single inbox. It eases the day-to-day operations of business users who had to meander through different inboxes, with different authorizations and authentications, having varied user experiences (UXs).

SAP Task Center functionalities bring in these benefits:

- **Working effectively with a single application**
  The greatest benefit lies in the unified UX with a single application and one authorization and authentication technique. Using the SAP Cloud Identity services, the authorization is aligned to manage unified access across different systems. Business users can effectively work on their tasks in a personalized UI, without the hassle of opening different applications, logging in with different user names, and working with varied UXs. This elevates the UX, saves time and resources, and offers task management at a scale where diverse systems are integrated with best-in-class UXs.
- **Efficiently managing task processing from different systems**
  With the SAP Task Center application, business users can manage all their tasks from different systems from one unified application. This web-based application is provided with search, filter, and sort functionalities to easily discover tasks based on different criteria. Business users can view the task in an embedded UI and perform respective actions such as approving the task, validating the information in the task, and so on. To view all the information for a task, there is also a provision to navigate back to the task in the actual system. Notifications, mass approvals, mobile views, and so on are some of the many functionalities that bring added value to using the unified inbox and further helps to improve task processing by business users.
- **Administrating task operations with ease**
  Task administration is often cumbersome, especially when it involves multiple systems. The lifecycle of the human task in SAP Task Center involves federating the task from the source system, assigning or delegating the task to the right user, rendering the task with appropriate data, and completing the task by the respective business user. There can be errors in any stage of the lifecycle. That is when the real power of human task administrators comes into play with their ability to suspend the task; assign another processor to the task; observe the logs for create, update, and other operations; and take the necessary steps to fix the issues. Easy automatic setup of

SAP Task Center with boosters, simple configurations to enable various source systems via destinations, embedded monitoring capability, and integration with SAP Build Work Zone for admin access are key to successful task administration.

Unified UXs, secure connections across systems, single inboxes for user tasks, aligned data models and data semantics for tasks federated from varied systems, embedded analytics, coordinated lifecycle management, and harmonized architecture are other benefits of using SAP Task Center for managing tasks across end-to-end business processes.

16.1.2 SAP Task Center Architecture

Figure 16.2 shows the various components that are involved in task federation. SAP Task Center is dependent on SAP Cloud Identity and SAP Build Work Zone services. SAP Build Work Zone is needed to provide access to the task center web application, and SAP Cloud Identity service is used for authentication for users who are authorized to access the tasks in SAP Task Center. Integration between SAP Task Center and various SAP lines of business happen via the Destination service through APIs.

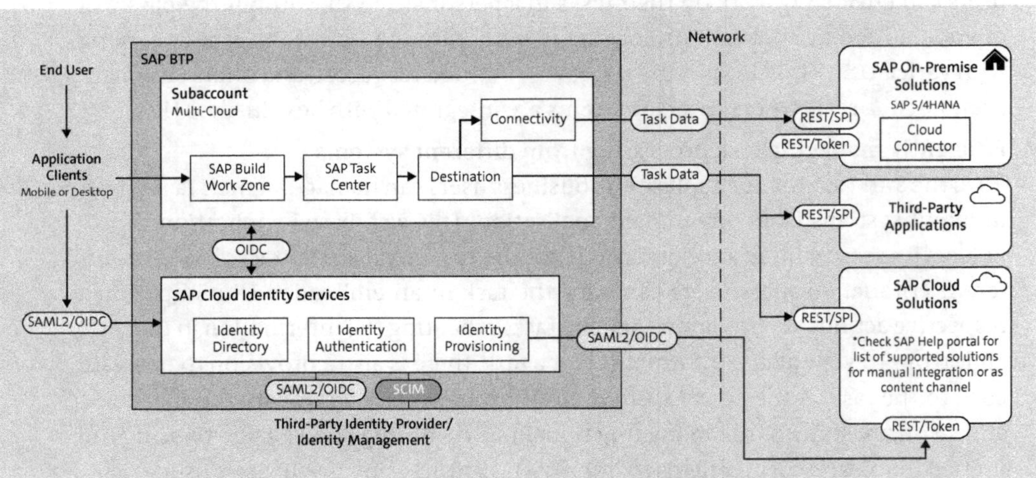

Figure 16.2 Solution Diagram for SAP Task Center Showing How Different Components of SAP BTP Interact for Task Management

As SAP Task Center works on a federation model, the tasks are fetched at a regular time interval from the source systems that are configured with SAP Task Center. These tasks can be newly created tasks, updated tasks, completed tasks, or tasks with errors. They are fetched using the Destination configuration service where a predefined URL and authentication protocol are used to connect to respective source systems, and these destinations are automatically created while setting up the SAP Task Center service in the SAP BTP account.

Note

You can refer to the SAP Task Center help documentation (*https://help.sap.com/docs/task-center*) for more information on the supported scenarios across different systems such as SAP Fieldglass, SAP SuccessFactors, SAP S/4HANA Cloud, SAP Concur, SAP Ariba, and so on.

16.1.3 Installing SAP Task Center

Installing SAP Task Center is a one-click operation by using the **Boosters** functionality in SAP BTP. This booster is available in a global account called **Set up an account for SAP Task Center with SAP Build Work Zone, standard edition**, as shown in Figure 16.3.

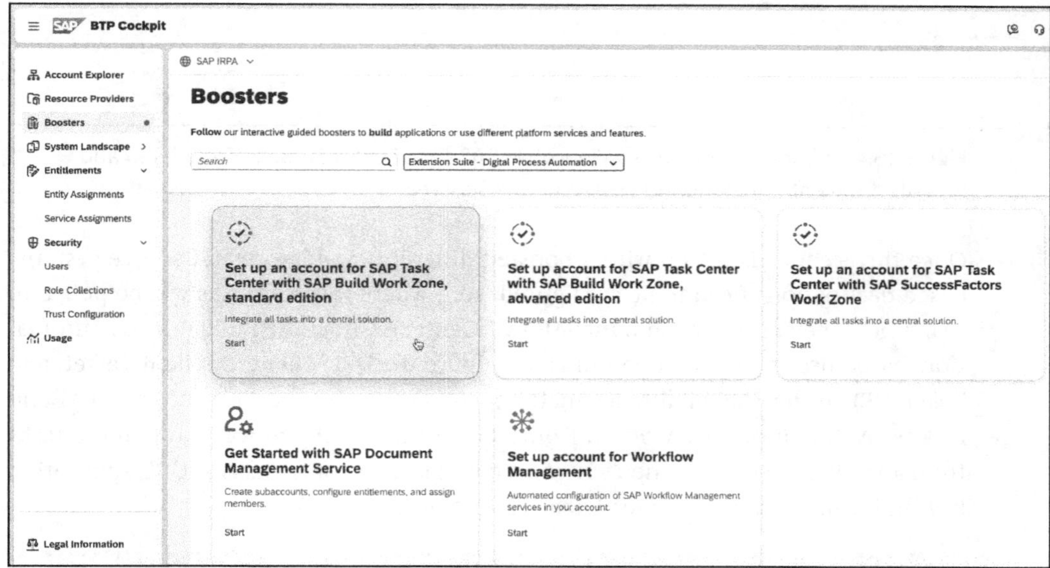

Figure 16.3 SAP Task Center Booster for Automatic Setup of SAP Task Center in the Subaccount

It takes a couple of minutes for the booster to run, which automatically sets up the account by doing the following:

- Adding an SAP Task Center entitlement to the selected subaccount with the appropriate plan
- Creating multiple destinations, one for each source system, such as a destination named **SAPBuildPA** being created for SAP Build Process Automation tasks
- Creating an additional destination called **Task_Center** to enable a web application from SAP Build Work Zone
- Adding a subscription for the SAP Task Center application

- Creating an SAP Task Center service instance and service key to access respective APIs, as shown in Figure 16.4
- Assigning SAP Task Center roles to the users listed as developer and administrator while running the booster

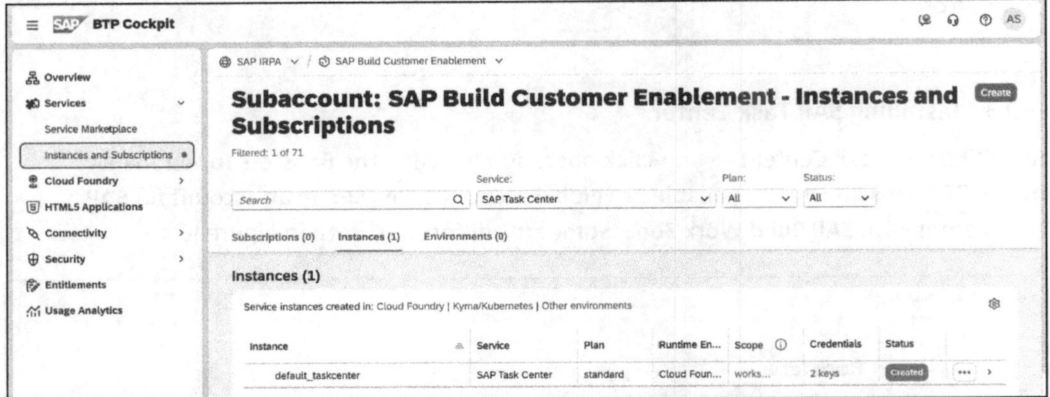

Figure 16.4 Subaccount Instances View after SAP Task Center Setup Is Completed and default_taskcenter Instance Is Created with a Service Key

Once the account is set up with a booster, different mandatory and source system–based destinations need to be configured from where the user tasks will be pulled in real time. For example, if you want SAP Task Center to show SAP Build Process Automation–based user tasks, then you must configure the **URL**, **Client ID**, **Client Secret**, and **Token URL** in the **SAPBuildPA** destination, based on the service instance of SAP Build Process Automation, as shown in Figure 16.5. Similarly, if you want to pull the tasks from an SAP Ariba system, then you must configure the token-service-URL, application key, and realm name in the SAP Ariba destination, for example.

> **Note**
>
> **Identity_Authentication_Connectivity_IDS** and **Task_Center_global_settings** are mandatory destinations that need to be configured. The former connects to the SAP Cloud Identity Provider, which will be used to authenticate users while accessing the SAP Task Center application.

> **Note**
>
> SAP Build Process Automation tasks can be accessed from the SAP Task Center service installed in the same tenant or a different tenant. For the same tenant, you must configure the **SAPBuildPA** destination, and for a remote tenant, you must configure the **SAPBuildPA_rem** destination. With this option, you can install SAP Task Center in one account and access tasks from different accounts.

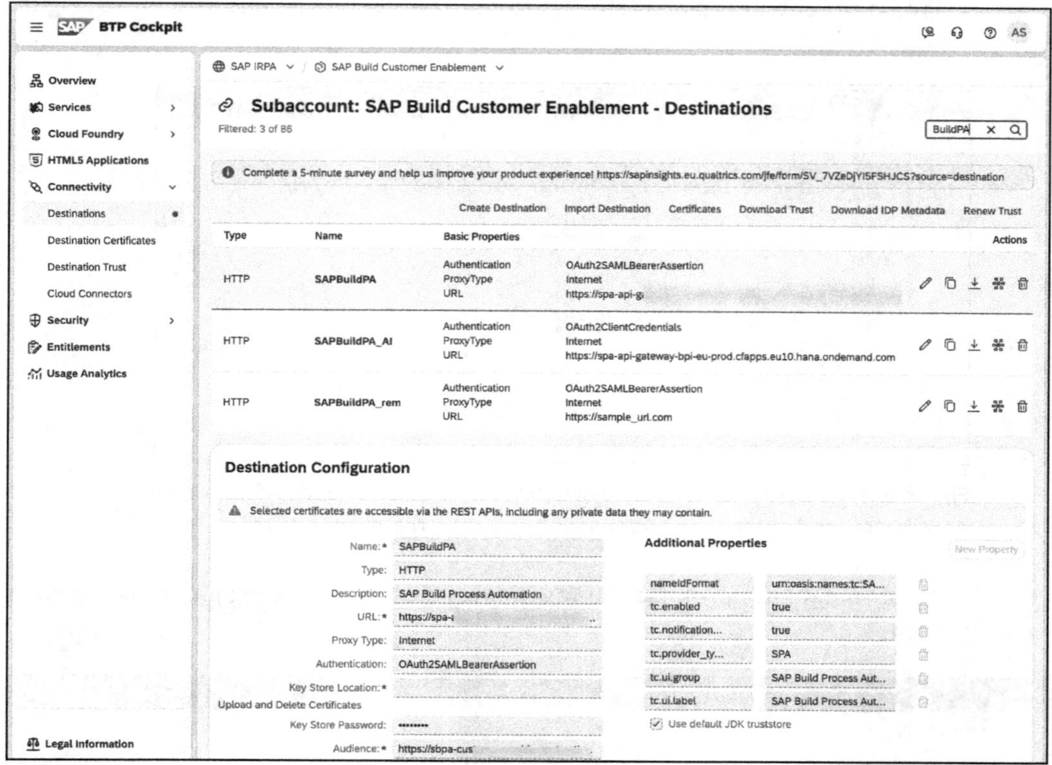

Figure 16.5 SAPBuildPA Destination Configured to Pull Tasks from the SAP Build Process Automation Service

16.1.4 Accessing Tasks

The next step after installing SAP Task Center in its respective subaccount is to enable the access of the SAP Task Center web application from the SAP Build Work Zone, standard edition or the SAP Build Work Zone, advanced edition. To add the **SAP Task Center** tile in the SAP Build Work Zone workspace, follow these steps:

1. In the SAP Build Work Zone content manager, open the **Content Channels** to refresh the HTML applications, as shown in Figure 16.6.

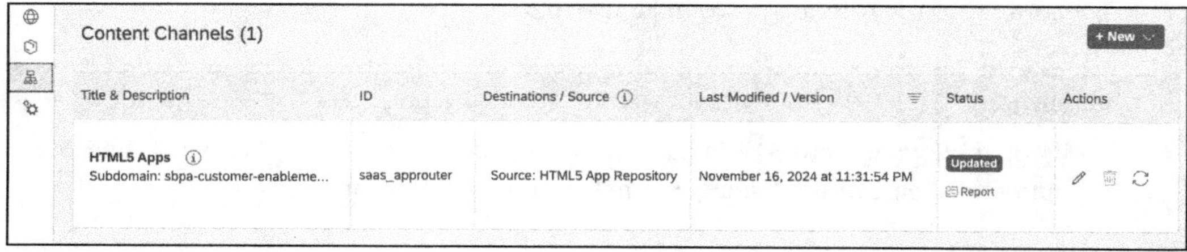

Figure 16.6 Refresh HTML Apps Channel from Content Channels in SAP Build Work Zone

16

2. From the **Content Explorer** tab, click to add **Task Center** content, as shown in Figure 16.7.

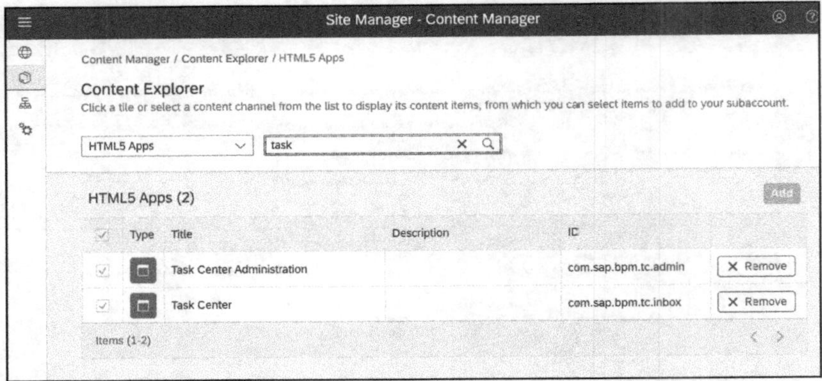

Figure 16.7 Content Explorer View Showing SAP Task Center and SAP Task Center Administration Content to Be Added to the Content Manager

3. Create a new group, or add the **Task Center** tile to any existing group, as shown in Figure 16.8.

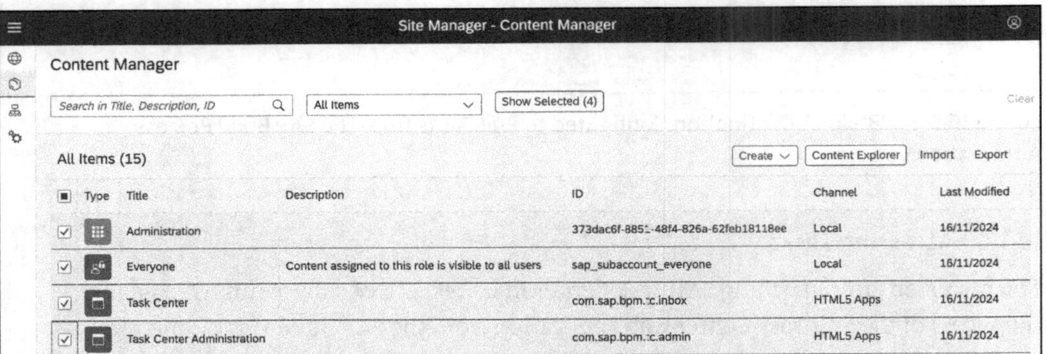

Figure 16.8 Task Center and Task Center Administration Content to Be Added to the SAP Build Work Zone Group and Everyone Role

4. Open the default **Everyone** role, and add **Task Center** and **Task Center Administration** content to the role, as shown in Figure 16.9.

> **Note**
> If you're working with SAP Build Work Zone, advanced edition, then you'll find the option to open **Content Manager** from the **Administrative** option.

With this, the **SAP Task Center** tile should be available in the SAP Build Work Zone workspace. When a business user with the `ProcessAutomationParticipant` role logs in to the workspace and clicks the **SAP Task Center** tile, a web application opens from which the user tasks, created after execution of business processes, can be managed.

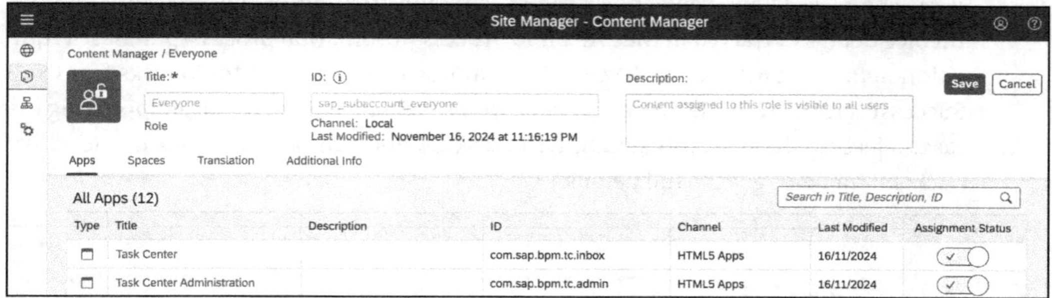

Figure 16.9 Task Center and Task Center Administration Content Assigned to the Everyone Role

Task Overview Page

As the main page of the SAP Task Center application, the task overview page lists all the tasks from different source systems in one single view, as shown in Figure 16.10. Business users can click on respective business area tabs such as **SAP Ariba**, **SAP Concur**, **SAP Fieldglass**, and so on to explicitly filter and view the tasks from that system. Besides filtering, search fields are provided based on respective business areas, which can be used to refine the search of specific tasks. There are generic search amount fields such as **Priority**, **Status**, submitted date (**Created Within**), and **Task Type**, as well as business area–specific fields such as **Currency**, **Amount**, **Supplier**, and **Customer**, which you can access by clicking **Adapt Filters**.

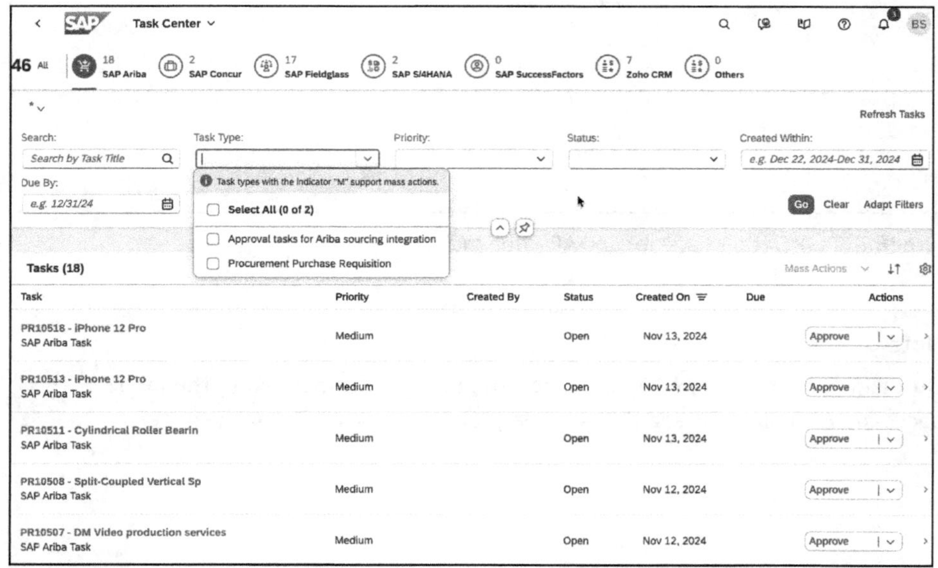

Figure 16.10 Task Overview Page Showing All the Tasks from Different Provider Systems

Task Details Page

Clicking a task from the overview page leads to the task detail page, as shown in Figure 16.11. The task details page shows the specific information of the task, for example, invoice details extracted in the SAP Build Process Automation process, purchase requisition details in SAP S/4HANA waiting for approval, or details of to-do tasks from SAP SuccessFactors. From the task detail page, you can take an action, such as approving the task, rejecting the task, or validating the task, or you can open the task in the actual backend inbox to get the full details.

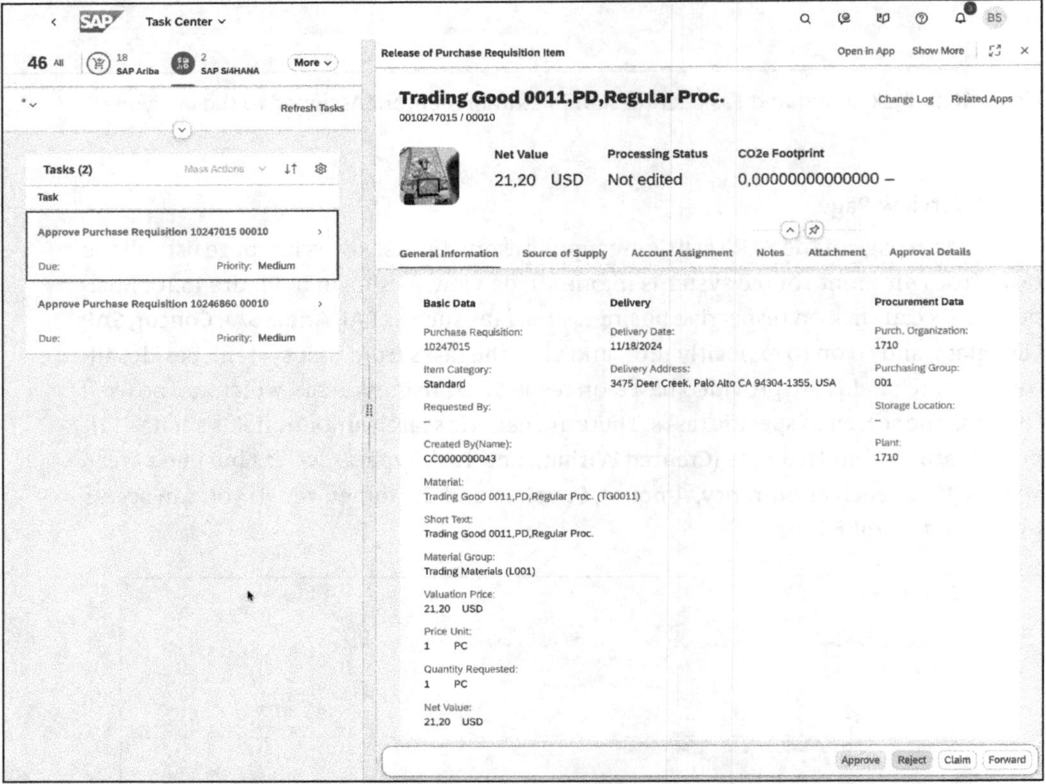

Figure 16.11 Task Details Page for an SAP Ariba task

> **Note**
>
> On taking action from SAP Task Center, the task gets completed in the inbox of the backend system as well, and the business process progresses to the next step.

Task Notifications

The SAP Alert Notification service for SAP BTP is embedded in SAP Build Work Zone for user task notifications. These alerts appear as the tasks are received in SAP Task Center, under a bell icon in the top-right corner of the application (see Figure 16.12).

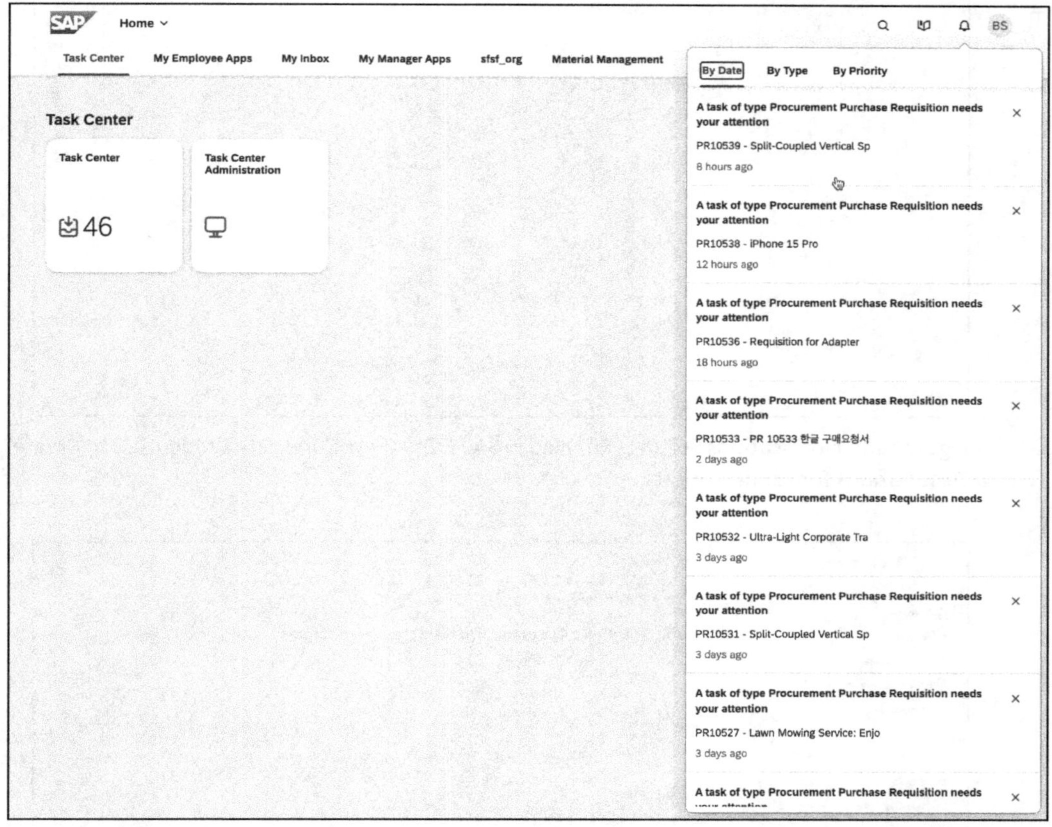

Figure 16.12 Task Notifications on the Task Center Page

To enable the task notifications for SAP Build Process Automation tasks in SAP Build Work Zone, follow these steps:

1. Enable notifications in SAP Build Work Zone settings and enter the values for **Host**, **Client ID**, **Client Secret**, and **Token Endpoint**, as shown in Figure 16.13.
2. Create a destination named **Alert_Notification_Connectivity_ANS** with the **OAuth2-ClientCredentials** protocol and connection properties from the previous step, as shown in Figure 16.14.

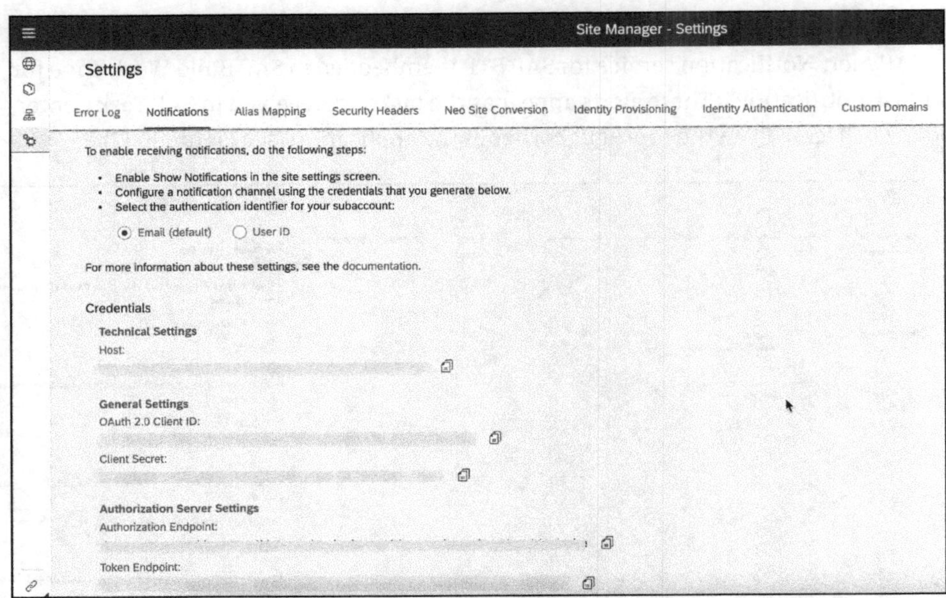

Figure 16.13 Notifications Settings Enabled in SAP Build Work Zone with Credentials to Create Destinations for Sending Alerts

Figure 16.14 ANS Destination Created in the Subaccount Based on the Credentials Obtained from SAP Build Work Zone

3. Set **tc.notifications.enabled** to **true** as an additional property in the SAP Build Process Automation destination when you want the notification to be enabled. This will start showing the notifications for any newly created or updated tasks from SAP Build Process Automation.

Intelligent Recommendations

Embedding AI enables business users to get personalized recommendations while working on the tasks in the form of confidence levels. These tailored suggestions are provided to the business users based on their historical decisions and helps improve decision-making in task management, as shown in Figure 16.15. Such analytical data not only helps the business user make better and more informed decisions but also increases overall productivity.

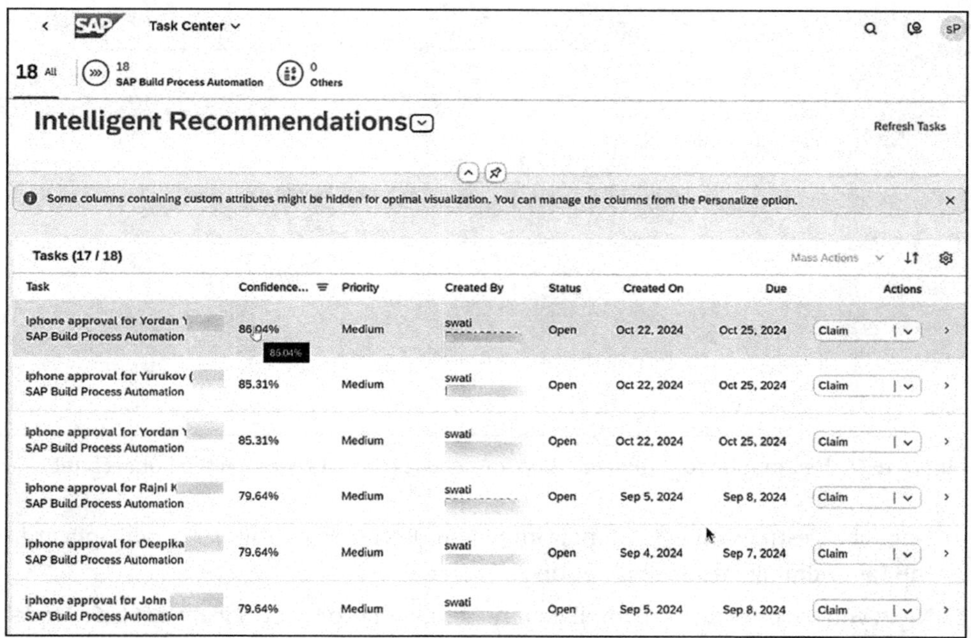

Figure 16.15 Confidence Column Shown in the SAP Task Center for Intelligent Recommendations

To enable intelligent recommendations, follow these steps:

1. Assign the `TaskCenterBusinessAdministrator` role to the user.
2. Ensure the **Task_Center** destination is configured with the **OAuth2ClientCredentials** protocol and credential details such as **Client ID**, **Client Secret**, **URL**, and **Token URL** from the SAP Build Process Automation service instance key.
3. Create a destination named **SAPBuildPA_AI** in the account with the **OAuth2ClientCredentials** protocol and credential details such as the **URL**, **Client ID**, **Client Secret**,

and **Token URL** from the SAP Build Process Automation service instance key. Additional property **HTML5.DynamicDestination** must be set to **true**, as shown in Figure 16.16.

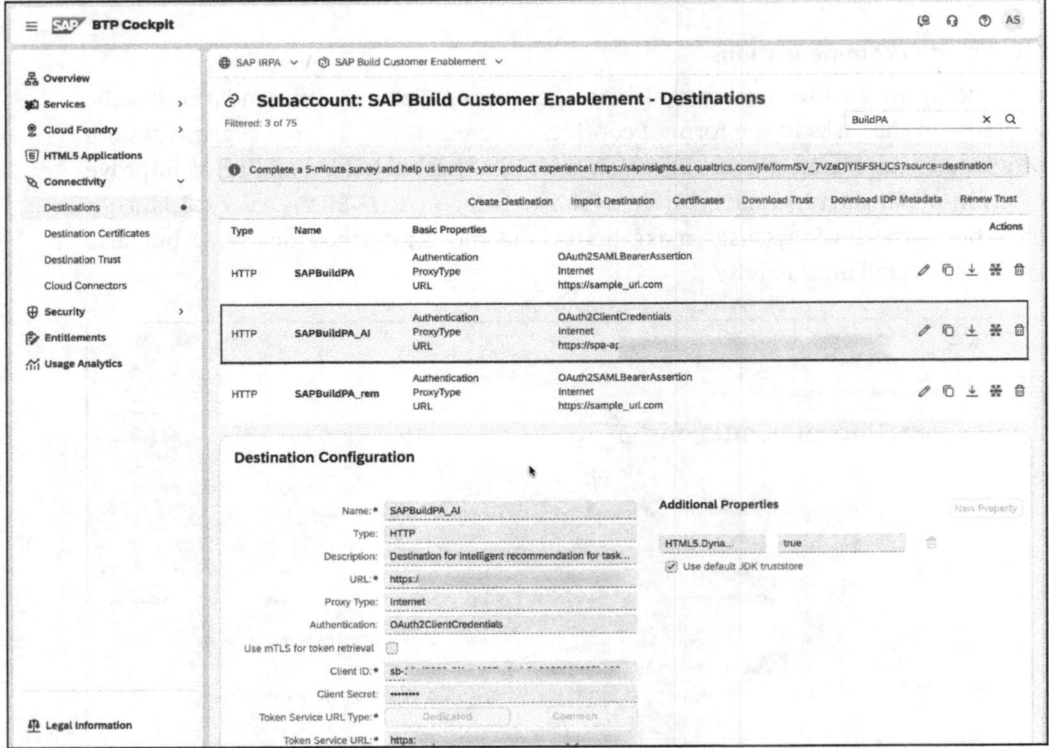

Figure 16.16 Destination for Intelligent Recommendations as Created in the Subaccount

4. Once the destination is set, a new **Intelligent Recommendations** tab will appear in the **Task Administration** application.
5. Next, you have to create an intelligent recommendation model via the **Create Model** option. While creating the new model, define the **Task Provider** from the available options: the **Task Type** and any **Model Name** of your choice. Additionally, configure the task type attributes such as **Numerical Attribute** and **Categorical Attribute**. At least five numerical and categorical attributes must be selected for recommendations. These attributes come from user task attributes defined in the process.
6. Once the recommendation model is created, it will take couple of minutes to be ready. In the background, the available data is trained for the given model, and you'll notice the change in status from **In Training** to **Ready for Activation**.
7. Once ready, click **Activate** to activate the model, as shown in Figure 16.17. The status changes to **Active**.

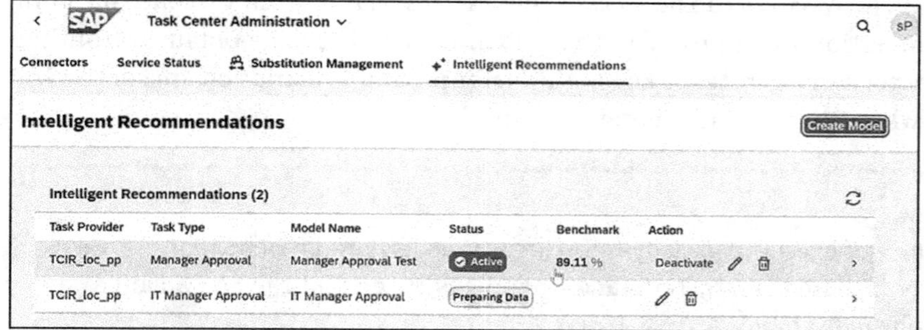

Figure 16.17 Models Created and Activated for Intelligent Recommendations

With this, the intelligent recommendations will be made available to the business users in SAP Task Center while working on the tasks. Using the AI suggestions based on the historical data, business users can make better informed and more efficient decisions.

16.1.5 Using the SAP Task Center Application Programming Interface

Besides working with the SAP Task Center application, you can also use SAP Task Center APIs published in SAP Business Accelerator Hub to work with user tasks and their respective definitions in your solutions.

SAP Task Center APIs are used for monitoring the tasks, working with jobs to pull the tasks from the cache, and exporting all the task data and users as a zip file, as shown in Figure 16.18.

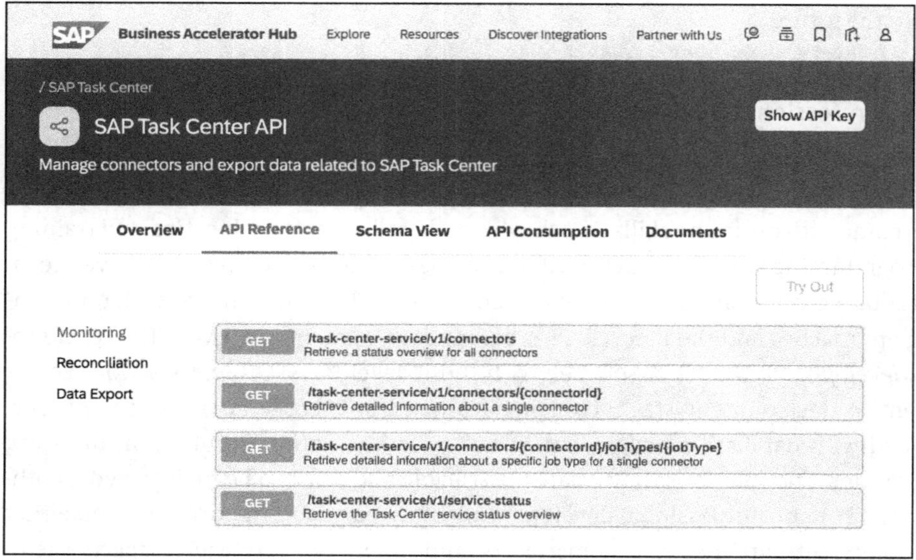

Figure 16.18 SAP Business Accelerator Hub View for SAP Task Center API

These are RESTful APIs and can be accessed by OAuth 2.0 (authorization code and SAML 2.0 Bearer Assertion Flow) protocol. The credential details can be obtained from the SAP Task Center service instance key in the **Instances and Subscription** area of the SAP BTP tenant where the SAP Task Center is setup.

Note

There is limit of execution based on the license models for different source systems. Refer to the updated "Restrictions and Limitations" section of the SAP Build Process Automation guide on the SAP Help Portal.

You've now seen how to involve the business user or end user in working on tasks, besides using the My Inbox application in SAP Build Process Automation. Let's now go to the other angle—the developers' side of the house.

16.2 Enabling Fusion Development for Professional and Citizen Developers

This isn't the first time you've heard about fusion development in this book as we've used this term several times throughout different chapters. So far, we've only scratched the surface of the topic and the term. Now, with the following sections, we'll go one step further, with a kind of characterization of the personas to see how collaboration could or should work here. We'll pick some examples of fusion development, bring out the benefits of this approach without neglecting the potential risks, make how it's supported with SAP Build Process Automation more tangible, and strive for a small outlook to the future.

16.2.1 Professional Developers

By now, you should know that fusion development brings together professional developers and citizen developers. What are their characteristics? Professional developers often come with profound skills that they've gathered in formal education and training for programming languages such as ABAP, Java, JavaScript, or Python. They have a technical IT background, and some of them even love to code. Being familiar with programming approaches and the lifecycle of program development is a given. The term *pro-code development* is often associated with these developers. Often, they work across the entire development stack, covering frontend and backend development, surrounded with database management, security, or integration, resulting in full-stack applications. The solutions they create are scalable and can also be maintained by others in the IT team. Professional developers are often part of a larger IT team and either collaborate with architects, administrators, and others or cover these aspects themselves.

16.2.2 Citizen Developers, Key Users, and Business Process Experts

On the other hand, we see the group of citizen developers, sometimes also called key users or business process experts. They are *the* experts in their business domain but are also interested in technology—not necessarily to the degree to write code, although some of them most likely are able to do so. But in general, you wouldn't expect a technical background here. They are motivated to translate their business requirements in a business application, business process, or workspace on their own. This could happen with support by the IT department and the professional developers within some given guardrails, or it can happen without that support. In the latter case, they make use of low-code or no-code environments. Empowered by their business context and knowledge, they are also close to the customers and know what's needed in tailored solutions to fulfill their business requirements.

Luckily, with SAP Build based on SAP BTP, both personas will find their tools of choice there and even collaboration capabilities, but we'll come to this later.

You might also have heard about the democratization of development, as with low-code or no-code environments, the entry barrier to directly turn business requirements into executable processes, for example, has been lowered. This is absolutely true and opens many new opportunities as well as challenges.

Some tend to draw a strict line between these personas, while others see the line between them blurring. We want to focus here on how these groups come closer together, work together, and drive great business results at the end—with fusion development.

16.2.3 Challenges of Traditional Approaches

In the past, or in more traditional development approaches, there have been several challenges. The business guys knew the customer and the business requirements, and the IT guys knew the technology to develop the needed application or business process. Although sometimes, they didn't really speak the same language, resulting in misunderstandings of what was required or the end results differing from the actual need. Even worse, there have been silos between business and IT teams. The requirements were just thrown over the fence and a few months later, white smoke rose, and the application or business process was implemented.

Often, these kind of IT projects came with high costs and sometimes have been slowed down in the development cycles. This led to missed deadlines, disappointed project participants, and a loss of trust to get the things done in a timely and cost-effective manner.

There has always been time pressure to complete IT projects and cover the business requirements, but these time frames are becoming shorter every year. New business requirements are coming, and changing circumstances and new challenges arise with the expectation to provide results faster. Traditional approaches, with business on one side of the fence and IT on the other side not talking or just barely talking to each other in different languages became more and more difficult to handle. And, the technology has evolved to support new approaches.

16.2.4 Drive Interoperability and Accelerate Business Transformation

The goal of fusion development is to overcome these hurdles and the strict differentiation. In short, bring the people together and stay secure and compliant. With SAP Build, this collaboration is supported. From a tooling perspective, professional developers as well as citizen developers, will find what they need, as shown in Figure 16.9, and even have a joined starting point, the SAP Build lobby.

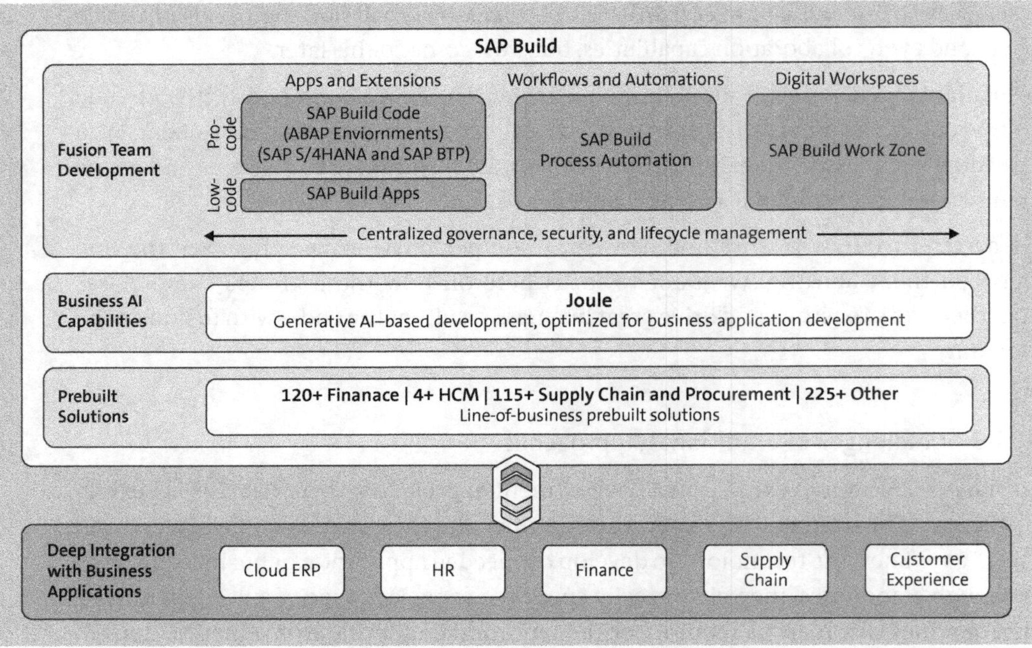

Figure 16.19 SAP Build: Driving Fusion Development

For apps and extensions, they can use SAP Build Code, ABAP environments, or SAP Build Apps. The workflows and automations area (or process extensions) is supported by SAP Build Process Automation, while digital workspaces can be crafted with SAP Build Work Zone.

Supported by business AI capabilities and leveraging prebuilt solutions or content packages, both professional developers and citizen developers will participate in an accelerated developer experience that is rounded out with deep integration into both SAP and non-SAP business applications.

The traditional gap between business and IT is bridged with SAP Build. Of course, it's not only about the right tooling, as the mindset and organizational structures also need to support a fusion development approach. This might be a challenge on its own, although we're quite sure it's worth the effort. It's always better when people come together and overcome boundaries. For this to happen, both the IT people and business people need to be empowered to do so. Technically, they are empowered with pro-code or no-code/low-code tools. The rest must come from the organization.

It's also not only about bridging the gap between IT and business but also between different development stacks, as shown in Figure 16.20. Backend development should go hand in hand with frontend development. In Chapter 15, we talked in detail about on-stack extensions and side-by-side extensions. There are projects out there that require the combination of both and the collaboration of the related developers and experts.

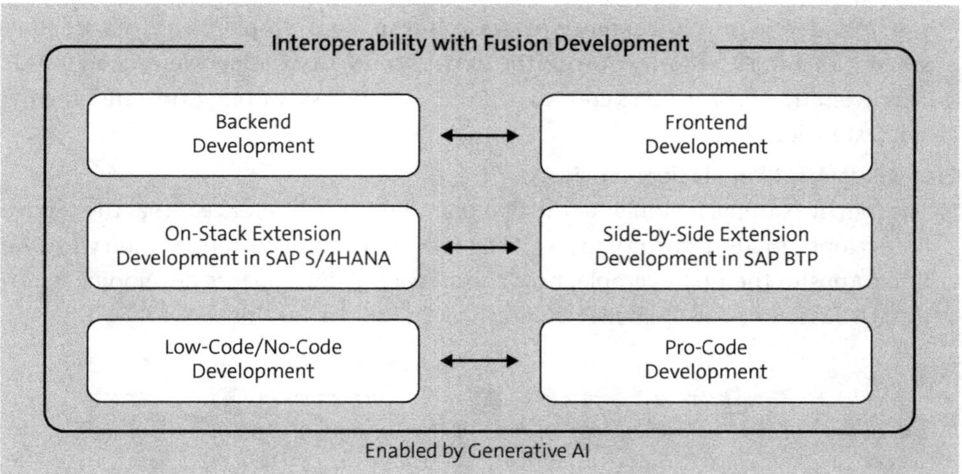

Figure 16.20 Interoperability Lays the Groundwork for Successful Fusion Development

Fusion development could also mean one single person is working on a solution and needs to integrate these various technologies, tools, and development stacks.

Fusion development done right will accelerate time to market for applications and business processes. Let's go through some examples of what this could look like in an SAP environment.

16.2.5 Examples for Fusion Development

The examples give you a first glance at what is possible with a fusion development approach. There are more possibilities, and you'll discover more once you join this journey. But for now, let's consider these four examples, as shown in Figure 16.21:

❶ **Combining low-code and pro-code extensions**
In the first example, it's all about a side-by-side extension approach and focus in cre-

ating an application with SAP Build Apps. This application will be developed by a key user or citizen developer in a low-code/no-code environment. A professional developer creates the service, which will be used by this application following the SAP Cloud Application Programming Model in SAP Build Code.

❷ **Develop ABAP side-by-side extension**
The second example also focuses on a side-by-side extension, but this time, the developer isn't creating a Java-based application, but a service following the ABAP Cloud development model. This service is probably gathering some data. The consumption of the service will be done in a business process designed in SAP Build Process Automation. You saw in Chapter 6 that for these kinds of use cases, you would create an action project in SAP Build Process Automation.

❸ **Combine side-by-side and on-stack extensions**
In Chapter 15, we mentioned that sometimes the combination of both approaches—side-by-side and on-stack extensions—is useful. In this example, the on-stack extension is done in SAP S/4HANA with the extension of an existing event using ABAP. This event triggers a process created in SAP Build Process Automation—the side-by-side extension.

❹ **Develop ABAP on-stack extension**
The fourth example is similar to the first one. The main difference is that the service is developed in ABAP and not in Java. While we've developed an application with SAP Build Apps in the first example, now a professional developer is developing a custom UI based on SAPUI5 and leveraging SAP Build Code for this.

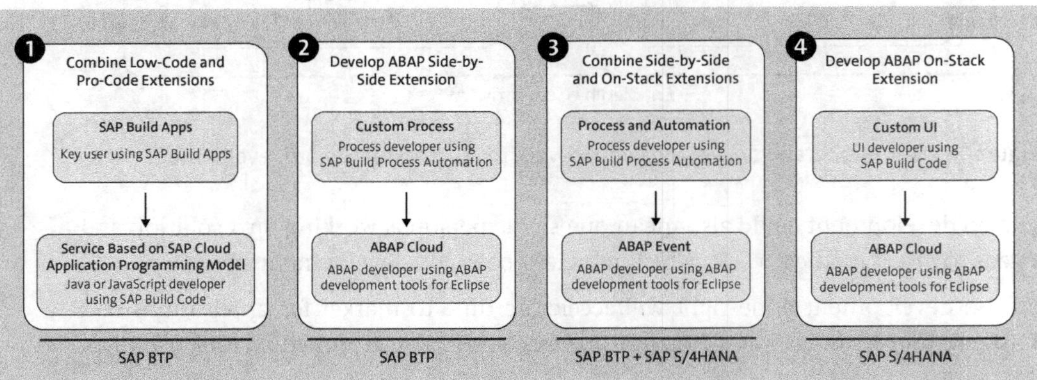

Figure 16.21 Examples of Fusion Development with SAP

In all the examples, there is a strong collaboration and alignment between professional developers and citizen developers or key users. This collaboration is technically enabled via the capabilities of SAP Build, which provides one access point and one place to store and work on projects, offering the flexibility to connect different tools and systems.

16.2.6 Benefits of a Fusion Development Approach

Before gaining the benefits of a fusion development approach, some effort is required. Besides the tooling also the mindset of the people needs to be open, although their concerns need to be taken seriously and addressed as well. There's certainly no need for either side to be wary. For the IT department, it's important to emphasize that this approach isn't about getting their job done by someone else but rather about having allies in the business departments and empowering them to build solutions on their own. The needed guardrails, such as access control to business-critical systems, security, governance, standards, and integration are still being handled by IT. But with well-trained and skilled citizen developers, business process experts, or key users, a new dynamic will come into your organization.

With the fusion development approach, more people within an organization will be able to build solutions. Business experts can take care of their own developments and not slow down the IT department with new requests for business process implementations that the business experts could do themselves.

This approach will help you scale instead of just react to ever-changing business requirements. In fact, you could drive the changes, leverage the full innovation potential of your complete organization, and use a flexible platform, such as SAP Build, to adapt this new working model.

Faster time to value for your application development and business process creation will provide more confidence in your organization and stakeholders, such as customers or suppliers. The development cycles will be accelerated. Another factor not to be underestimated is that you'll be able to create first prototypes much faster because you not only have the tool support but also the experts in IT and business at the same table, talking the same language, and working closely together.

It's fair to say that with the recent and upcoming developments in AI, especially in the area of generative AI, fusion development will also benefit. This will again lower the entry barrier to create business processes, applications, UIs, extensions, and so on. The first attempt was done with the introduction of low-code/no-code development environments. Fusion will also speed up the whole development approach, providing guidance to help improve developer efficiency.

16.3 Summary

In this chapter, we've set our focus on how to scale SAP Build Process Automation and how to involve different personas. Starting with the end user or business user, you've learned how to connect SAP Build Process Automation with SAP Task Center and consume the user tasks there. You're aware now of the benefits of using SAP Task Center after learning about the architecture, what it takes to set it up, how to access the tasks, and the usage of the SAP Task Center API. The second part of this chapter dealt with the fusion development approach, the involved personas, and which benefits you can expect by following this approach. This takes us to the last chapter of this book, where we'll provide a conclusion of the covered topics.

Chapter 17
Conclusion

In this last chapter, we'll go through the topics covered in this book and provide a summary of the highlights.

Congratulations! You made it to the end of this book about SAP Build Process Automation and learned a lot of new stuff. This deserves a round of applause.

Now that you've mastered this comprehensive guide, let's take a final look back. We'll revisit the chapters, pick some of the highlights, and get a better understanding of how they relate to each other. Ready for this last mile of the journey to start becoming a builder with SAP Build Process Automation? Let's start:

- **Chapter 1: Introduction to Enterprise Automation**
  We started the book with a general introduction into enterprise automation. Though not knowing yet what is behind SAP Build Process Automation, we increased the scope by adding SAP Integration Suite to integrate SAP and non-SAP applications, as well as SAP Signavio solutions for process analysis, leveraging prebuilt content and AI capabilities, as shown in Figure 17.1.

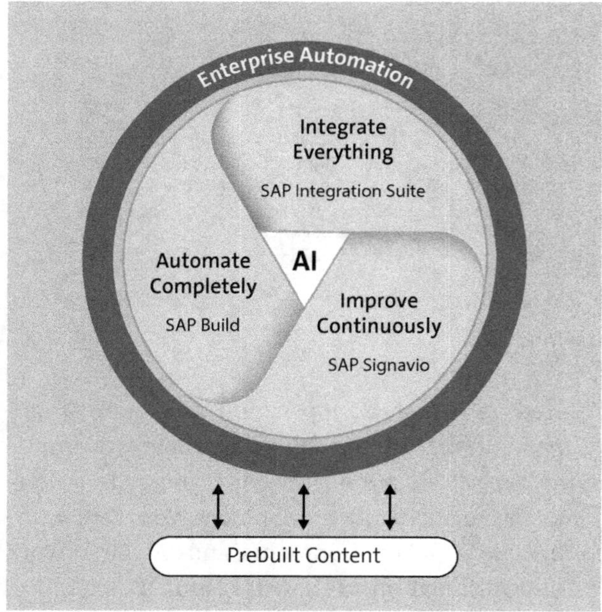

Figure 17.1 Enterprise Automation with SAP

While process automation came with promises, there are also issues and challenges such as fragmented approaches, costly point-to-point integrations, and a zoo of bots where the approach with enterprise automation will help. Guided by concrete use cases, you got your first understanding of how SAP supports this and which features are provided in the products.

- **Chapter 2: SAP Business Technology Platform**
  Chapter 2 gave you the context of where SAP Build Process Automation is in the SAP product offering, that is, SAP Business Technology Platform (SAP BTP). Several services are part of this platform and focus on providing the latest technology supporting the business and laying the technical foundation. You learned what's behind application development, automation, integration, data and analytics, and AI, as shown in Figure 17.2. We touched on the clear core concept for the first time and positioned SAP BTP as the side-by-side platform for innovations.

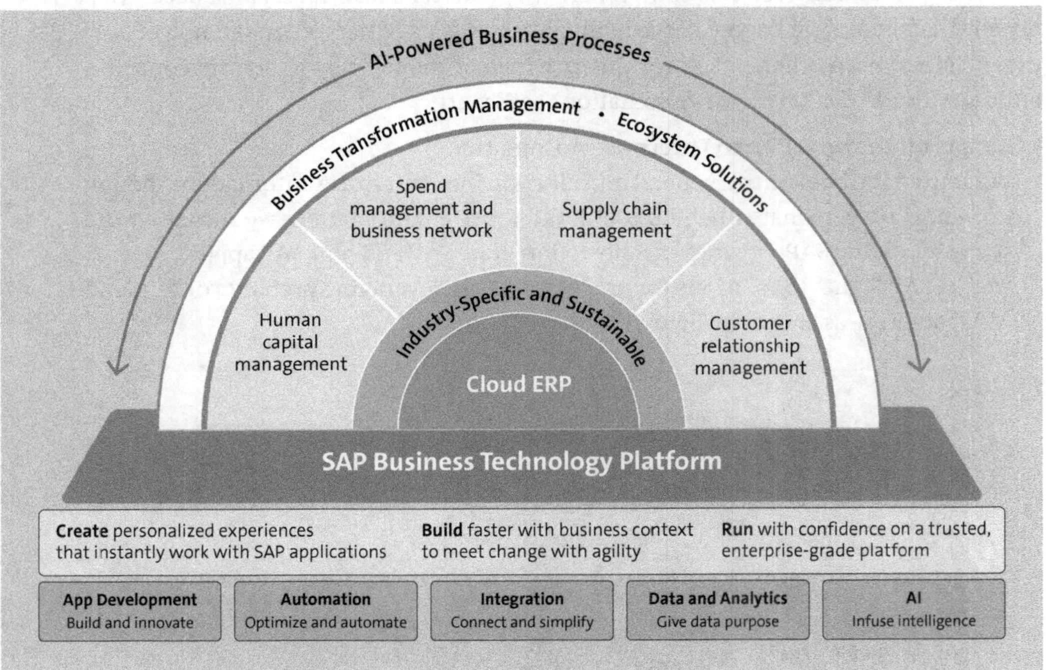

Figure 17.2 SAP BTP Supporting Digital Innovations

But we didn't stop here and focused on two areas, application development and automation, as part of SAP Build, *the* solution for extensions, which helps you implement the differentiating factor in your organization, either in a pro-code or low-code/no-code approach or combined as fusion team development. You were introduced to SAP Build Apps, SAP Build Code, ABAP environments, and SAP Build Work Zone, but not yet to SAP Build Process Automation, as shown in Figure 17.3.

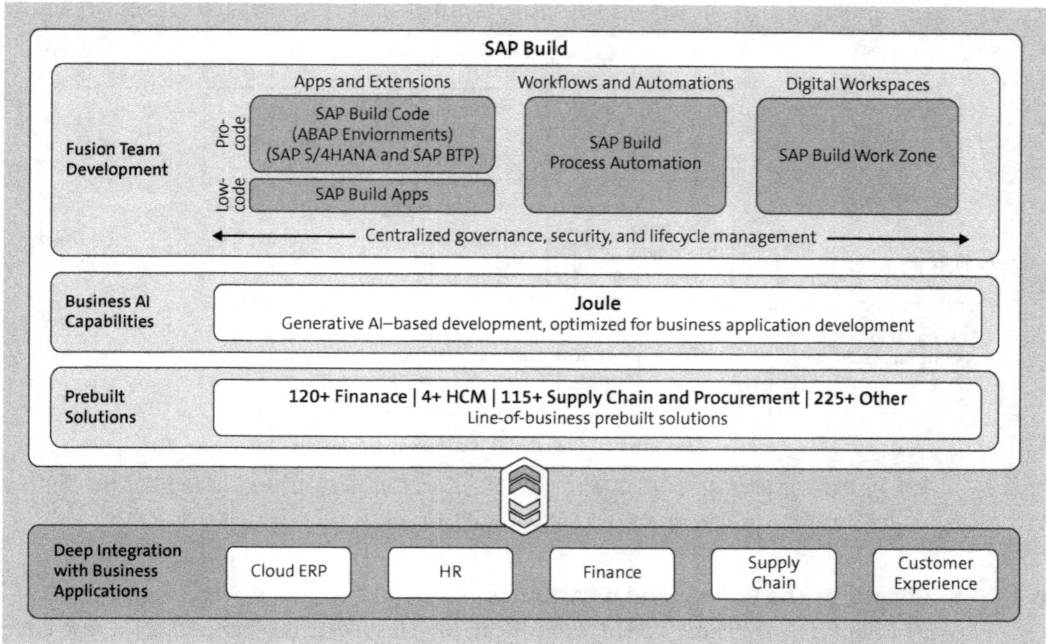

Figure 17.3 SAP Build Supporting Fusion Team Development for Apps, Processes, and Digital Workspace

17

- **Chapter 3: Overview of SAP Build Process Automation**
  Then, the chapter you were all waiting for provided an overview of SAP Build Process Automation. We agreed that business processes are the lifeline of every organization. You learned there must be quite a lot of use cases out there, as it covers the extension of standard business processes, creating net-new, automating tasks, and even supporting the orchestration of activities across organizational and system boundaries. This is possible because SAP Build Process Automation unites features to automate processes and workflows, design forms and multistep approvals, automate tasks, define business rules, use AI for document understanding, provide visibility into the processes, gives you a central inbox to work on tasks, and leverages prebuilt content packages, as shown in Figure 17.4.

  We shared an architectural view, went through the applications on a high level, saw the different artifacts and what they are good for, and left you with a good basic understanding of the tool. But we had only just begun.

- **Chapter 4: Creating a Business Process Project**
  The practitioners among you were cheering for this chapter, as it described how to create a business process project. This also marks the starting point to create processes here. You discovered the different project types: for process, decision, automation, or process visibility.

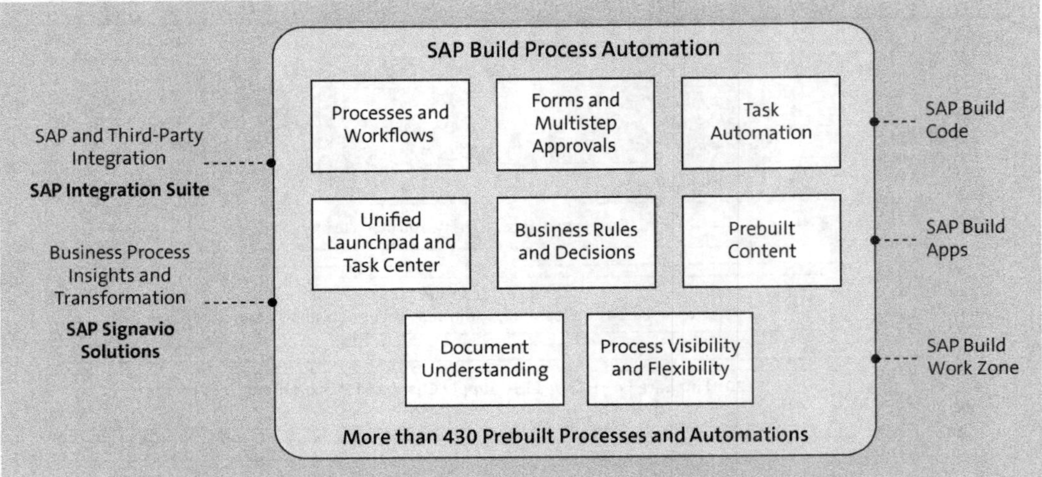

Figure 17.4 Capabilities of SAP Build Process Automation

- **Chapter 5: The Process and Its Artifacts**
  In Chapter 5, we came to the definitions of a process, the benefits of a low-code design tool, and the process editor, as shown in Figure 17.5. Variables and attributes, process triggers, forms, and some more details around actions, automations, decisions, and much more were described. The importance of events, branches, which role mapping plays, the design of forms, and a first glimpse of process variants were also brought to your attention. Additionally, we discussed the mail task, why you would use subprocesses, and the script task. With the script task, we discovered another artifact supporting fusion development, as you can write code here. With this, you were well on your way to being equipped to create a business process for whatever use case in SAP Build Process Automation.
- **Chapter 6: Action Artifacts**
  And finally, we get to the action! The action artifact is used to connect to external systems, SAP or non-SAP, as it has an application programming interface (API) inside. We discussed the different connectivity options, such as Live API with SAP Cloud Application Programming Model, ABAP RESTful application programming model, OData destinations, or SAP Integration Suite, and we explored the usage of API specifications coming from SAP Business Accelerator Hub, uploaded or created from scratch by you. You saw how to create an action project with all the needed configurations. Luckily, there's not too much effort needed to provide reusable actions to be consumed in a process in SAP Build Process Automation.
- **Chapter 7: Automation Artifacts**
  What would SAP Build Process Automation be without the automation artifact? You could already tell from the size of the chapter as the biggest in this book just how important the automation artifact capability is. In short, the automation artifact

allows you to automate repetitive, mundane, rule-based tasks in a bot, freeing up human resources for higher value activities. You've been guided through the automation editor and experienced how to add an activity to the automation flows, as shown in Figure 17.6.

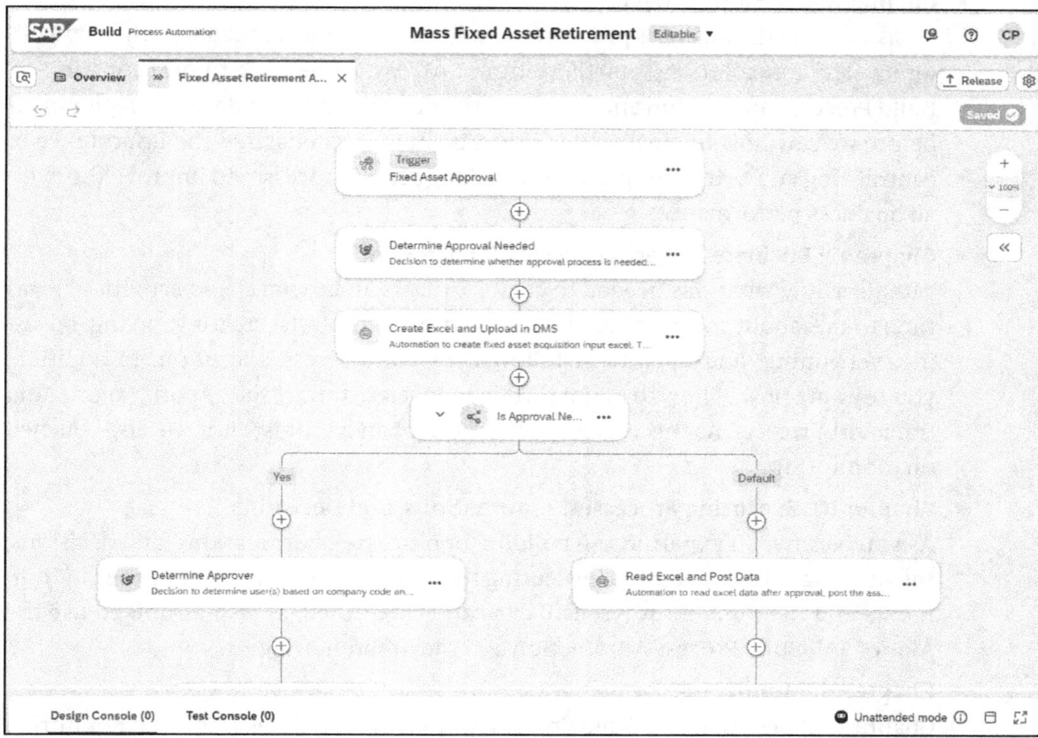

Figure 17.5 The Process Editor

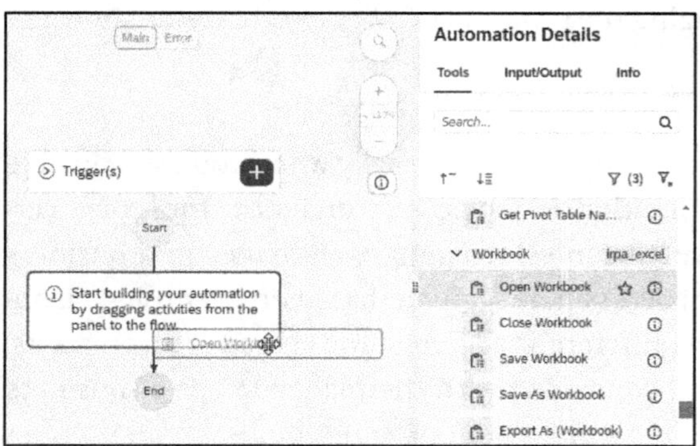

Figure 17.6 Adding an Activity to the Automation Flow

You went from the creation to testing and debugging, learned how to automate screens via capturing, and got recommendations for several applications. The software development kits (SDKs) were introduced. Document information processing and the huge set of controls have been explained in detail. Furthermore, the usage of a desktop agent and why there are environments became much clearer for you.

- **Chapter 8: Decision Artifacts**
  Decisions rule the business process. To underpin this, you learned why there are different decision solutions out there from SAP. We came back to decisions in SAP Build Process Automation and discussed the different types of decision deployment before we saw how to create a decision. Finally, we emphasized the importance of centralizing and automating decision-making across processes to improve the overall business performance.
- **Chapter 9: Business Project Lifecycle Management**
  After learning what was needed to create artifacts and their related activities, it was time to talk about the lifecycle of business process projects: creating, saving, releasing, versioning, and deploying, followed by the process execution. In addition, you're aware now of how to maintain dependencies, import and export projects, and transport projects. As this is close to the work of an administrator, we also touched on monitoring.
- **Chapter 10: Executing Processes, Automations, and Decisions**
  As you became an expert in the building of processes, automations, and decisions, the next logical step followed: executing them. You learned about runtime environments and the different ways artifacts can be executed, as well as how to use the APIs of SAP Build Process Automation and the different triggers.
- **Chapter 11: Monitoring**
  Chapter 11 shared that SAP Build Process Automation provides technical monitoring and business monitoring. As an integral part of business process management, the monitoring capabilities were explained in detail for process owners, administrators, and any other stakeholders, leveraging the process visibility capability, as shown in Figure 17.7.
- **Chapter 12: Governance**
  Providing the right set of boundaries is important, especially when you use a flexible tool such as SAP Build Process Automation. So, it was no wonder the topic of governance resided in Chapter 12. Starting from establishing governance in a business process, we went to the handling of access management based on user roles, elaborated on the environments concept, differentiated public and shared environments, visited transport management, and concluded with the management of business process projects.
- **Chapter 13: Authentication, Authorization, Audit Logging**
  A topic similar to governance waited for you in Chapter 13. It was all about authentication, authorization, and audit logins. The different roles were introduced, along with how to connect securely to cloud and on-premises systems.

- **Chapter 14: Working with Store Content**
  One way to accelerate developer efficiency is to avoid building everything from scratch. With a wide variety of prebuilt content packages across different industries and lines of business for common business scenarios, SAP Build Process Automation takes care of this. You can import, adapt, configure, and deploy these content packages quite easily. If you just want to explore this content, you now know SAP Business Accelerator Hub is the right place to do so, as shown in Figure 17.8.

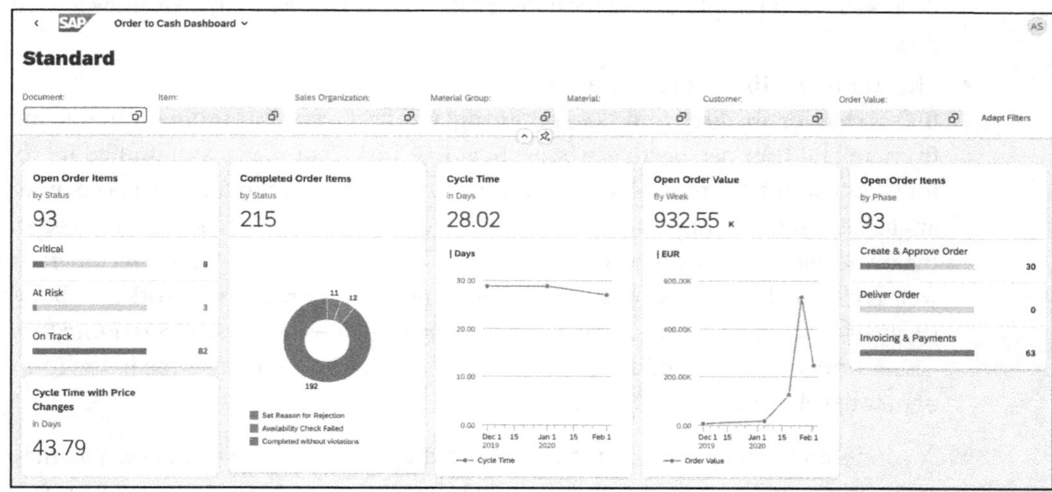

Figure 17.7 Process Visibility Dashboard

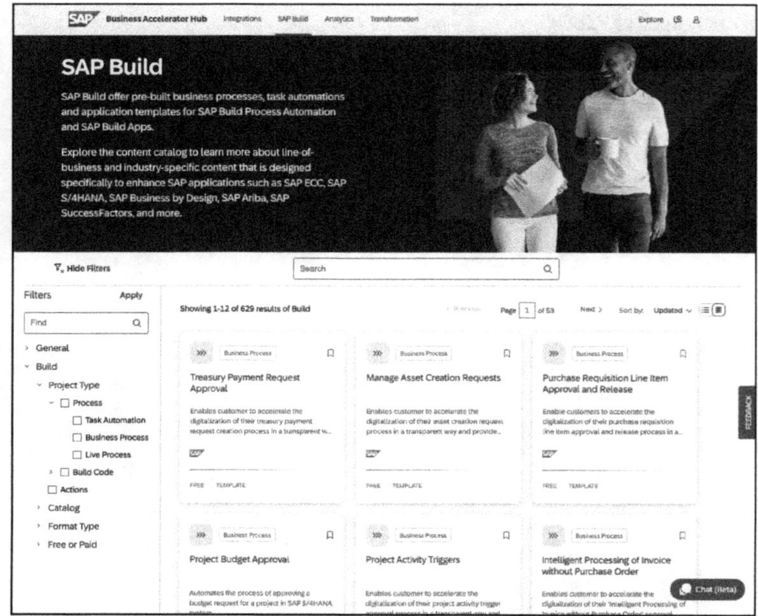

Figure 17.8 Content Packages in SAP Business Accelerator Hub

- **Chapter 15: Extending Standard Processes**
  Chapter 15 took up another important topic in more depth: the extension of standard processes. First, we clarified what is meant by standard business processes and how they are supported by SAP applications. We noticed there are ways to configure and customize these standard processes, but there are cases where this isn't sufficient. Extensions are needed, either on-stack or side by side, with SAP Build as the tool of choice; for processes, SAP Build Process Automation is the tool you need. Process variant templates and the extensibility wizard in SAP S/4HANA show how to do so.
- **Chapter 16: Scaling Process Automation**
  In this chapter, we discovered two other angles to scale process automation. Starting from an end-user perspective, we saw how SAP Task Center acts as a unified inbox for tasks coming from several applications and gathers them in one place. The other angle was quite the opposite, as we've worked out which personas are involved in fusion development. By now, you know it brings together professional developers as well as citizen developers, business process experts, and key users to work collaboratively on a business solution, for example, a new business process in SAP Build Process Automation. SAP Build supports this approach with more to come in the form of new developments.

We're done, and you're done. At the end of the chapter and end of this book, you may be wondering what's next, how to get more information, and how to become an SAP Builder in SAP Build Process Automation. You're invited to make the next steps in your process automation journey and join the great community of SAP Builders in SAP Community at *http://s-prs.co/v592804*.

And, of course, you're now ready to start you own process automation project with SAP Build Process Automation!

The Authors

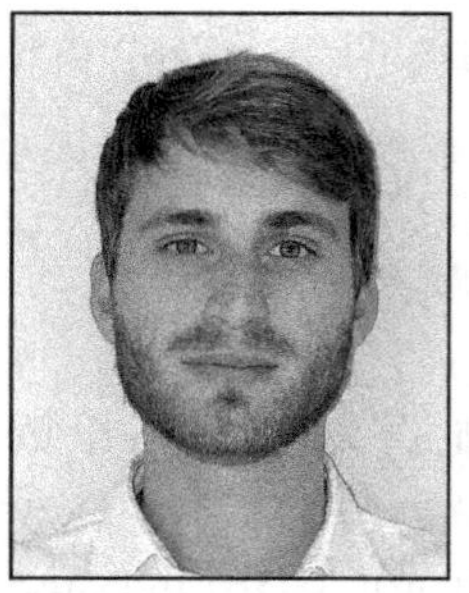

Steven Jacobson is a product manager at SAP who has in-depth knowledge of and expertise with SAP Build Process Automation. Steven started his journey with SAP in 2019 as a consultant helping customers adopt automation and machine learning solutions. He continues to work with customers today while also conducting adoption-focused and product-led initiatives related to SAP Build.

Chaitanya Priya Puvvada is a senior product manager in the human capital management (HCM) domain with SAP SuccessFactors. She has more than 17 years of experience, with her expertise spanning across product management, business analysis, and consulting. With a deep expertise in intelligent automation, Priya has played a pivotal role in driving the adoption of SAP Build Process Automation, empowering professionals through knowledge sharing. She has conducted numerous webinars and workshops, created learning content, and authored several blogs and technical articles. She also serves as an advisor for developers and the architect community for SAP Build Process Automation. As a sought-after speaker and thought leader, she has established herself as a respected voice in the industry, presenting at both national and international conferences like SAP TechEd. Her contributions to the product and innovations in technology have earned her several accolades and prestigious awards from Women in Technology, Women in Robotic Process Automation, and SAP Community.

Seshadri Sreenivas Ramanarayanan is a senior product manager for SAP Build Process Automation at SAP SE, based in Walldorf, Germany. Since joining SAP in 2007, he has specialized in workflows and business rules, shaping the automation solutions. A seasoned speaker at conferences like TechEd, Sapphire, and bpmNext, Seshadri is known for his insights on process automation and digital transformation. Over the years, Seshadri has played a pivotal role in driving SAP's product strategy in the process automation area, helping various enterprises successfully streamline their operations and achieve efficiency through automation. Beyond the world of SAP, Seshadri is an International Cricket Council (ICC) certified coach and an avid sports enthusiast who is passionate about playing cricket and tennis whenever he can.

Stephan Schluchter is a product manager for SAP Build. His focus is on enabling customers, partners, and colleagues on SAP Build and its related services. Stephan has more than 15 years of experience in the business process management domain and is a regular speaker at international conferences like SAP Sapphire, SAP TechEd, ASUG, and DSAG.

Archana Shukla is a senior director of product management at SAP. She has more than 19 years of experience and plays a key role in product strategy, product definition, and customer success of SAP Build Process Automation. She has led multiple strategic projects at SAP and was instrumental in defining and executing content strategy working across different LoBs, partners, and customers. She also heads the product leadership forum of SAP Labs, India, where she is responsible for building programs to nurture future product managers, user experience designers, and product support teams. Archana has conducted numerous workshops and webinars, hosted SAP BTP podcasts, authored blogs and articles and is a well-known speaker at multiple international conferences. She has been recognized for her contributions to coaching young aspirants and leaders across the country. In addition to her work at SAP, she also provides expert technical advisory to developers, consultants, and solution architects and is an eminent member of SAP Community.

Index

C

D

O

P

Q

R